Rare Vascular Plants

of Alberta

Rare Vascular Plants of Alberta

by The Alberta Native Plant Council

Edited by
Linda Kershaw
Joyce Gould
Derek Johnson
Jane Lancaster

Principal Contributors
Alberta Natural Heritage Information Centre
Cheryl Bradley
Bonnie Smith
Anne Weerstra

University of Alberta Press

Natural Resources Canada
Canadian Forest Service

Ressources naturelles Canada
Service canadien des forêts

Published by
The University of Alberta Press and **The Canadian Forest Service**
Ring House 2   Northern Forestry Centre
Edmonton, Alberta T6G 2E1   5320–122 Street
                                      Edmonton, Alberta T6H 3S5

Printed in Canada 5 4 3 2 1
Copyright © The University of Alberta Press 2001

Canadian Cataloguing in Publication Data

Main entry under title:

Rare vascular plants of Alberta

Includes bibliographical references and index.
ISBN 0-88864-319-5

1. Rare plants—Alberta. I. Kershaw, Linda J., 1951-
QK86.C2R37 2001   581.68'097123   C00-910171-3

All rights reserved.
No part of this publication may be produced, stored in a retrieval system, or transmitted in any forms or by any means, electronic, mechanical, photocopying, recording, or otherwise, without the prior permission of the copyright owner.
∞ Printed on acid-free paper.
Colour scanning and prepress by Elite Lithographers Co. Ltd., Edmonton, Alberta.
Printed and bound in Canada by Kromar Printing Ltd., Winnipeg, Manitoba.
Book design by Carol Dragich.
Proofreading by Jill Fallis.

The University of Alberta Press acknowledges the financial support of the Government of Canada through the Book Publishing Industry Development Program for its publishing activities. The Press also gratefully acknowledges the support received for its program from the Canada Council for the Arts.

Contents

Abstract	vii
Preface	ix
Introduction	xiii
Rare Plants in Alberta	xiii
Types of Rare Plants	xv
Lists of Rare Plants	xvi
The Nature Conservancy Element Ranks	xvi
Conservation of Rare Plants	xviii
Rare Plant Surveys	xix
How You Can Help	xxi
Natural Regions of Alberta	xxii
Canadian Shield Natural Region	xxii
Boreal Forest Natural Region	xxv
Rocky Mountain Natural Region	xxix
Foothills Natural Region	xxxii
Parkland Natural Region	xxxiv
Grasslands Natural Region	xxxvii
About the Species Accounts	xxxix
Rare Vascular Plants of Alberta	1
TREES AND SHRUBS	3
MONOCOTS	27
DICOTS	57
GRASS-LIKE PLANTS	257
FERNS AND FERN ALLIES	347
ADDENDUM	373
Appendices	
One: Keys to rare *Botrychium* and *Isoetes* species found in Alberta	389
Two: Rare vascular plants of Alberta by natural region	393
Three: Species in this guide that are not found in the *Flora of Alberta* (second edition)	409
Four: Taxa previously reported as rare for Alberta, but not included in this book	411
Five: Species reported from or expected to occur in Alberta but not yet verified	415
Six: Alberta taxa classified as nationally rare by Argus and Pryer (1990)	417
Seven: Rare native plant report form	423
Illustrated Glossary	425
References	451
Index	467
Illustration Credits	481
Photo Credits	482
Contributors	483
About the Editors	484

Abstract

Kershaw, L.; Gould, J.; Johnson, D.; Lancaster, J. 2001. *Rare vascular plants of Alberta*. Univ. Alberta Press, Edmonton, Alberta and Nat. Resour. Can., Can. For. Serv., North. For. Cent., Edmonton, Alberta.

This book describes about 485 species of rare native vascular plants found in the province of Alberta. Species descriptions are divided into four main sections based on growth form (woody plants, broad-leaved herbs, grass-like plants, and ferns and fern allies). Common, scientific and family names are included for all primary entries. A general description of each plant is followed by descriptions of habitat and maps showing the distribution of most species in both Alberta and North America. Each species description is concluded with notes on synonymy, descriptions of related or similar-looking plants, etymology of the scientific names and, for some species, medicinal or aboriginal uses of the plants. The appendices include keys to some of the more difficult-to-identify genera, listing of the species by natural region, species new to Alberta, species previously considered rare for the province but not included here, and species classified as rare in Canada. An illustrated glossary, a reference list, and an index conclude the book.

Preface

This book has been several years in the making, and many people have contributed to its compilation and production. The Alberta Native Plant Council (ANPC) has been interested in rare plants since its inception in 1987 and has always supported the idea of producing a book about these species.

Alberta Environmental Protection has supported the project from the beginning, through the Recreation and Protected Areas Program and later the Alberta Natural Heritage Information Centre (ANHIC), initially headed by Peter Lee and then by Dan Chambers. Staff at, and contributors to, ANHIC have provided a great deal of assistance to the production of this book. They include Peter Achuff, Lorna Allen, Dana Bush, Patsy Cotterill, Kathi De, Joyce Gould, Graham Griffiths, Coral Grove, Roxy Hastings, Miranda Hoekstra, Julie Hrapko, Duke Hunter, Derek Johnson, Linda Kershaw, Jane Lancaster, Mike Luchanski, Karen Mann, Marge Meijer, Gavin More, Bill Richards, John Rintoul, Christy Sarafinchin, Cindy Verbeek, Dragomir Vujnovic, Cliff Wallis, Kathleen Wilkinson and Joan Williams. Joyce Gould has played a key role in providing information from the ANHIC database for this guide.

The ANPC was able to contribute to the ANHIC database by hiring summer students in 1993 (Kim Krause), 1995 (Ksenija Vujnovic) and 1996 (Heather Mansell) to help with the huge job of entering rare plant information. This work was funded in part by the Summer Career Placement Program of Human Resources Development Canada.

We thank the Canadian Forest Service, the co-publishers of this book for their efforts. In particular, the editorial and production assistance of Brenda Laishley (with assistance from Denise Leroy in editing the reference list) and the manpower and logistical support provided by Derek Johnson and Dan MacIsaac.

Many specimens were examined in different herbaria as information was collected for the database. A special thank-you to the staff who facilitated access to specimens for this essential work: Brij Kohli (University of Alberta Herbarium), Mike Luchanski and Roxy Hastings (Provincial Museum of Alberta Herbarium), C.C. Chinnappa and Bonnie Smith (University of Calgary Herbarium), Derek Johnson (Northern Forestry Centre Herbarium, Edmonton), Walter Willms and Harriet Douwes (University of Lethbridge Herbarium), William Cody and Paul Catling (Agriculture Canada

Athyrium alpestre

Preface

Herbarium, Ottawa), George Argus, Mike Shchepanek and Albert Dugal (National Museum Herbarium, Ottawa) and Warren Wagner (University of Michigan Herbarium). The assistance of staff in the collection of information from the herbaria at various national and provincial parks is also greatly appreciated.

Many volunteers came forward to collect information for different groups of plants. This was coordinated by Linda Kershaw, and contributors included Peter Achuff, Cheryl Bradley, Debra Brown, Dana Bush, Adrien Corbiere, Patsy Cotterill, Dave Ealey, Gina Fryer, Joyce Gould, Graham Griffiths, Julie Hrapko, Derek Johnson, Linda Kershaw, Jane Lancaster, Dan MacIsaac, Debra Nicholson, Bill Richards, Art Schwarz, Bonnie Smith, Joan Williams and Kathleen Wilkinson.

Several ANPC members helped to produce the Guidelines for Rare Plant Surveys. This project was coordinated by Jane Lancaster, with contributions from Lorna Allen, Joyce Gould, Anne Weerstra, Kathleen Wilkinson and Joan Williams.

When the bulk of the data had been collected, work began on writing the species accounts for the book. This was coordinated by Linda Kershaw, with Cheryl Bradley, Joyce Gould, Derek Johnson, Linda Kershaw, Cindy Verbeek and Anne Weerstra writing text for different species. Peter Achuff, Joyce Gould, Derek Johnson, Linda Kershaw and Cliff Wallis wrote the text for the introduction, Linda Kershaw wrote and illustrated the glossary, Coral Grove compiled the appendices (these were updated and revised by Derek Johnson and Marie Paton), and Elisabeth Beaubien compiled the Alberta phenological information. Linda Kershaw edited the text, and it was then reviewed by Adolf Ceska, Bill Crins, Vern Harms, Derek Johnson and John Packer. Bruce Ford and John Hudson also reviewed the sedges.

As the ANHIC database became more complete, it was possible to use its information to generate maps electronically, and the Alberta distribution maps in the book were generously provided by ANHIC. A special thank-you to Joyce Gould, Duke Hunter, John Rintoul and Dragomir Vujnovic for their work. Another group, working in the Biota of North America Program (BONAP) at the University of North Carolina, studies plant distributions across the continent, and they kindly provided us with North American maps for our rare species. Our thanks to

Tanacetum bipinnatum

Preface

John Kartesz and Amy Farstad for their help. The production of current maps is a complicated process, requiring intensive research and the accumulation of large amounts of information. Without the support of ANHIC and BONAP, we could not have included these important elements in the book.

The collection of photographs and illustrations continued throughout the project. Jane Lancaster coordinated the collection of photographs and illustrations, with some help from Linda Kershaw and Ruth Johnson. Original illustrations by John Maywood and Joan Williams were provided by ANHIC and commissioned by the ANPC. Previously published illustrations in the book are used, with kind permission, from the University of Washington Press, the New York Botanical Garden, the Royal British Columbia Museum, Stanford University Press, the National Museum of Canada, William Wagner and Bill Merilees. Illustration sources are listed on pp. 481–82. Photographers who donated slides for use in the book include Peter Achuff, Lorna Allen, Ralph Bird, Terry Clayton, Bill Crins, Teresa Dolman, Joseph Duft, Joyce Gould, Dierdrie and Graham Griffiths, Bonnie Heidel, Julie Hrapko, Derek Johnson, John Joy, Linda Kershaw, Tulli Kerststetter, Fred Korbut, Jane Lancaster, Archie Landals, Peter Lee, Peter Lesica, William Merilees, Bob Moseley, Sandra Myers, Jim Pojar, John Rintoul, Steve Shelly, Kathy Tannas, Jim Vanderhorst, Warren Wagner Jr., Cliff Wallis, Cleve Wershler and Steve Wirt. ANHIC, the Devonian Botanic Garden, Lone Pine Publishing, the Montana Natural Heritage Program, the Recreation and Protected Areas Division of Alberta Environmental Protection, and Waterton Lakes National Park also contributed photographs. Photo sources are listed on page 482.

Fund-raising and financial management have been necessary throughout the life of the project. Our thanks to ANPC treasurers Dave Downing, Joyce Gould and Dan MacIsaac; intrepid secretary Lorna Allen; and the people who helped with applications for funding, Dana Bush, Derek Johnson, Linda Kershaw, Gavin More and Margaret Zielinski. The Canadian Forest Service provided substantial financial support toward the completion of this book. Several other groups have provided financial support over the years for the rare plants project. The Federation of Alberta Naturalists provided a generous grant in 1993, which has supported most aspects of the project since its

Erigeron trifidus

Preface

Carex pseudocyperus

inception. We also received grants and donations from the Alberta Sports, Recreation, Parks and Wildlife Foundation to place copies of the book in public and high school libraries, Canada Trust Friends of the Environment, Canadian 88 Energy Corporation, Gulf Canada, Quadra Environmental Services, Axys Environmental Consulting, the Red Deer River Naturalists and several private supporters.

The search for a publisher also required the energies of dedicated workers. Thanks to Elisabeth Beaubien, Joyce Gould, Derek Johnson, Linda Kershaw, Dan MacIsaac and Carla Zelmer for their efforts in this. Thank you to our co-publishers, the University of Alberta Press and the Canadian Forest Service. Glenn Rollans, Carol Dragich and Leslie Vermeer at the University of Alberta Press, and Brenda Laishley at the Canadian Forest Service transformed our manuscript, photos, drawings and maps into this beautiful book.

Most of the work for this guide has been done by dedicated volunteers, and more than 100 individuals and organizations have been involved in the rare plants project since its beginning. We hope that this book will help Albertans to learn more about our rare vascular plants, and that it will play a part in conserving the rich diversity of species and habitats in this province.

—Linda Kershaw, Joyce Gould,
Derek Johnson and Jane Lancaster

Introduction

The word 'rare' sparks the imagination. To many people, it suggests mystery, adventure and great value. If something is rare, it is precious—perhaps even priceless. Rare things may be difficult to find, but are cherished when discovered.

Many rare plants and animals can be discovered in Alberta's wild and not-so-wild places. Some are big and beautiful, but more often the discovery and appreciation of rare wild things depend on your ability to go slowly and look closely.

This book is for anyone interested in learning more about rare plants in Alberta. Teachers, students and natural history enthusiasts can use it to learn more about the wide variety of rare plants in Alberta. Land-use planners, foresters, environmental consultants and researchers can determine which rare species to expect in different habitats or regions, and how to identify these plants during fieldwork.

Montia linearis

Rare Plants in Alberta

Approximately 30 percent (about 485 species) of Alberta's native vascular plants are classified as rare in the province. Similarly, recent work on bryophytes has shown that approximately 25 percent of the moss flora is rare (Vitt and Belland 1996). Of the rare vascular species, approximately 268 (55 percent) are restricted to an area of less than 3 percent of the province, and less than 10 percent of the province supports approximately 153 (31 percent) of Alberta's rare plants. The Waterton–Crowsnest area in southwestern Alberta has by far the highest concentration of rare vascular plants in the province. Other areas with large numbers of rare species include the northern Rocky Mountains (Jasper National Park, Cardinal Divide and north) and parts of extreme southeastern Alberta.

Many factors determine the distribution of rare species. Some rare Alberta species may have developed or persisted as isolated populations. The flora of Alberta is young. Much of this province was covered by a massive ice sheet 15,000 to 16,000 years ago. Most species that survived glaciation moved south in advance of the ice sheet, and when the ice sheet finally retreated, they moved north again. However, some species were able to persist in small areas (refugia) that were not covered by ice. Refugia have been identified along the Front Range of the Rocky Mountains and in the Cypress Hills. It is difficult to determine which species survived in these areas and how important these populations were to re-colonization, but genetic research may eventually provide some of this information.

Adenocaulon bicolor

Introduction

Carex pseudocyperus

As new areas are explored, new species are discovered. In fact, 31 species of vascular plants that are new to Alberta have been discovered since the publication of the second edition of the *Flora of Alberta* (Moss 1983). These species are listed in Appendix Three. Many recently discovered species—and many rare species in general—are small, ephemeral plants that either grow inconspicuously for most of the year or grow in rarely visited habitats (high on mountains or in the middle of ponds, for example). As more of these sites are studied, more plants are located, and we may learn that some species are not as rare as we now think they are. Other species are new to the flora because of recent taxonomic revisions. These species are also listed in Appendix Three. For example, western false-asphodel (*Triantha occidentalis* ssp. *brevistyla* and ssp. *montana*, p. 44) has recently been reclassified as a species separate from sticky false asphodel (*Triantha glutinosa*) (Packer 1993), and Siberian polypody (*Polypodium sibiricum*) has been split from rock polypody (*Polypodium virginianum*, p. 370) (Haufler et al. 1991).

Plants may be rare because of one or more factors in a myriad of variables that affect geographic range, habitat specificity and population size (Rabinowitz 1981). Changes in both biotic (living) and abiotic (non-living) factors affect plant populations. Biotic changes might involve the introduction of aggressive non-native species that out-compete local native plants. Abiotic changes may include alteration of drainage patterns or fluctuations in climate. Both direct and indirect effects can contribute to species rarity. For example, changes in nearby habitats may have no obvious, direct impact on a rare species, but if they result in the loss of the insects that pollinate the plant's flowers, they can have a devastating, though less obvious, effect.

Many human activities can contribute to species' rarity. Agricultural and urban expansion, roads, mining, oil and gas development, long-term fire suppression, global warming and stabilization of sand dunes and surface water levels are all examples of human activities that could cause some plants to become rare. Plant populations may also decline as a result of over-collecting. In some cultures, a plant may be considered valuable, and therefore a target for collection, simply because it is rare. A recent study in the United States found that up to ten percent of the rare species in that country were rare because of collecting (Schemske et al. 1994). Fortunately, over-collecting does not yet appear

Introduction

to be a serious problem in Alberta, but it does introduce some important questions about the use and sharing of rare species information.

Types of Rare Plants

Most of Alberta's rare vascular plants are 'peripherals'—that is, species at the edge of their geographic range. The North American distribution maps featured with each species account show which species are widespread outside of Alberta. Increasing evidence suggests that peripheral populations are often genetically different from those at the centre of a species' range. This genetic diversity can be important in enabling a species to adapt to change. Soapweed (*Yucca glauca*) and western blue flag (*Iris missouriensis*) are examples of peripheral species.

Spergularia salina

Species that are restricted to a particular geographic area are called 'endemics.' Alberta and its immediate region have only a few local endemics. For example, dwarf alpine poppy (*Papaver pygmaeum*) is known only from northern Montana, southeastern British Columbia and southwestern Alberta. Some very dynamic habitats, such as the Athabasca sand dunes, support rapidly evolving taxa. For example, sand-dune chickweed (*Stellaria arenicola*) evolved recently from the common species long-stalked chickweed (*Stellaria longipes*) (Purdy et al. 1994).

Some rare Alberta species are 'disjuncts'—that is, populations separated from the main range of their species by 500 km or more. Wood anemone (*Anemone quinquefolia*) is a good example of an Alberta disjunct. Here, it is known only from near Nordegg, and the nearest population to this is in east-central Saskatchewan.

A few of Alberta's rare vascular plants have widespread distributions within North America but are uncommon wherever they are found. Bog adder's-mouth (*Malaxis paludosa*) is an example of this type of rare plant.

Papaver radicatum

To monitor and manage rare species successfully, we must understand both the biology of the species and the reasons for its rarity. For example, genetically variable (plastic) plants, such as those found in some peripheral populations, may be able to adapt to new environments, whereas plants with less variability (e.g., some disjunct or endemic populations) may be unable to survive seemingly minor changes in their specific habitats.

Rare *Vascular* Plants of Alberta

Introduction

Draba macounii

Lists of Rare Plants

In 1978, Argus and White produced the first publication on rare Alberta plants: *The Rare Vascular Plants of Alberta*. This listed 350 species (approximately 20 percent of the known vascular flora) as rare, and defined a rare plant as 'one with a small population size within the province or territory.' In 1984, Packer and Bradley published *A Checklist of the Rare Vascular Plants in Alberta*, which included 360 species. They considered a species rare if it had been collected in five or fewer localities in the province. Argus and Pryer (1990) used these lists, and similar publications from other jurisdictions, to assess rarity on a national basis in *Rare Vascular Plants in Canada*. This nation-wide list classified 125 Alberta species (about 7 percent of the province's vascular flora) as nationally rare.

Lists of rare species are constantly being revised and updated. Additional collections may show that a plant is too common or widespread to be classified as rare. Taxonomic revisions may result in the 'lumping' of two or more species, which together are too widespread to be considered rare. Field surveys and re-identification of plant collections may show that a plant was misidentified and that a species was thus falsely reported.

There are many differences between the list of rare species included in this book and lists from other publications. Appendix Four (Taxa previously reported as rare for Alberta but not included in this book) and Appendix Five (Species reported from or expected to occur in Alberta but not yet verified) address some of these discrepancies.

In 1996, Alberta joined the Nature Conservancy network of Conservation Data Centres (known as Natural Heritage Programs in the United States). Each centre or program assesses species rarity using number of occurrences and population size as the main criteria. Status is assessed using the following system of classification.

The Nature Conservancy Element Ranks

Species (and sometimes other taxa such as varieties and subspecies) are ranked using the following numbers and letters. Each number or letter is preceded by either a G for global rank or S for subnational (provincial or state) rank.

Introduction

1 — 5 or fewer occurrences and with low population size

2 — 6–20 occurrences or with low population size

3 — 21–100 occurrences

4 — apparently secure, >100 occurrences

5 — abundant and demonstrably secure, >100 occurrences

F — falsely reported

H — known historically, may be rediscovered

P — potentially present, expected in the province or state but not yet discovered

Q — questionable taxonomic rank

R — reported but without documentation for either accepting or rejecting the report

U — status uncertain, more information needed

X — apparently extinct or extirpated, not expected to be rediscovered

? — no information available, or the number of occurrences estimated

Sagina nivalis

Because each conservation data centre uses the same criteria to determine rarity, results can be pooled to assess status at subnational, national and global levels. Many of Alberta's rare vascular plants are considered globally abundant and are common in adjacent provinces and states. For example, soapweed (*Yucca glauca*) is ranked S1 in Alberta, because it is known from fewer than six locations, and because these are the only Canadian sites, it is ranked N1 in Canada. However, soapweed is common from Montana south, and therefore it has a global ranking of G5. The populations in this province are important components of the flora and overall biodiversity of both Alberta and Canada, and are worthy of conservation.

This guide includes taxa that have a ranking of S1 to S3, species that are reported from or otherwise expected in the province, and species that are considered extirpated from the province. The numbers of taxa in each group are as follows: S1, 211 species; S1?, 2 species; S1S2, 25 species; S2, 186 species; S2?, 1 species; S2S3. 13 species; S3, 26 species; SH, 4 species; SRF, 4 species; SU, 14 species; SX, 1 species;

Nymphaea leibergii

Introduction

Diphasiastrum sitchense

S?, 1 species. Appendix Two lists all of the rare species included in this book, the natural regions in which they are found and their provincial ranking.

Conservation of Rare Plants

Biological variation (biodiversity) can be viewed at many different levels, from broad landscapes to specific habitats, local communities and even the genetic variation within a single species. Variability within a species is usually assessed on the basis of differences in form, but unseen changes in genetic make-up from one plant to the next are equally important. It is essential to consider diversity at all levels to conserve Alberta's natural biodiversity.

Management activities have traditionally focussed on individual species, but other levels of biodiversity (e.g., landscape, community and habitat) are increasingly being taken into consideration. No species lives in isolation. Most rare plants reflect the presence of rare habitats, where other organisms may be equally or even more threatened. The distribution of small plants and animals in Alberta is largely unknown. We cannot monitor, let alone understand the specific requirements of, every species. However, the recognition of rare species may help to identify threatened habitats that should be given special attention. Conservation of habitat through good land-stewardship is the most effective and efficient way to ensure the long-term protection of species.

Legislation and policy at both federal and provincial levels can provide protection for plants. At the provincial level, Alberta's endangered species are now included in an amendment to the Wildlife Act. The minister responsible for the act may designate a plant to an endangered species list. If designated, a plant would enjoy protection under the endangered classification of this act. The Endangered Species Conservation Committee and Scientific Subcommittee have been established to evaluate the status of species that may be at risk in Alberta, to list those that are threatened or endangered, and to advise the Alberta minister of environmental protection on appropriate legal designation and protection in the Wildlife Act. The committees will also facilitate the planning and implementation of recovery programs for species at risk. Other legislation that may be used to protect plants include the Historical Resources Act, Public Lands Act, Provincial Parks Act, National Parks Act and Wilderness Areas, Ecological Reserves and Natural

Introduction

Areas Act. There is no federal endangered species legislation at this time.

Several plants that are rare in Alberta have been listed as vulnerable, threatened or endangered by the Committee on the Status of Endangered Wildlife in Canada (COSEWIC). COSEWIC has representatives from the federal and provincial governments and from selected non-government organizations. This committee determines the national status of plants and animals in Canada on the basis of scientific data summarized in a status report. Such information is invaluable for the management of sites and species, but the listing of a species by COSEWIC has no tie to legislation.

Brasenia schreberi

Rare Plant Surveys

Rare plant surveys document the occurrence of rare species in a given area, both through literature and through field surveys. The first step in conducting a rare plant survey is to gather existing rare plant information for the area. Lists of rare species that are known to occur in specific regions are available from the Alberta Natural Heritage Information Centre (ANHIC). Other sources of information—such as local naturalists, distribution maps and published reports—should also be reviewed to determine rare species potentially in the area. Flowering and fruiting dates and ecological information pertaining to these species should be collected wherever possible to assist in determining appropriate timing for the survey and potentially significant habitats that may require more intensive study; herbarium labels, published literature and local naturalists are valuable sources for this type of information. If significant habitats in the study area are known, these should be located on aerial photographs and topographic maps. Interpretation of photos and maps may also identify other potentially significant habitats.

Rare plant field surveys should be conducted in a full range of habitats at several times of year (spanning the growing season). Ideally, sites should be surveyed over a number of growing seasons to take biological and climatic fluctuations into consideration. For example, some species do not produce stems or leaves, let alone flowers, in dry years, so they are difficult to find and identify during droughts. Other factors, such as grazing and insect predation, can reduce vigour and visibility in some years but not in others. Lack of discovery does not necessarily mean that a species is absent.

Romanzoffia sitchensis

Rare *Vascular* Plants of Alberta

Introduction

Cypripedium acaule

Both 'random' and 'directed' surveys (including intensive hands-and-knees ground surveys in potential significant habitats) can be done, but most field time should be devoted to directed surveys. Random surveys are generally used to determine the rare-plant potential of major habitat types. They typically involve inspecting all species within a randomly selected, 1-m wide, 250- to 500-m straight-line transect across a habitat type. The spacing of these transects depends on the density of the vegetation cover, the visibility through it and the size of the plants in it. Directed surveys are tailored to the habitat being surveyed. They use a meandering pattern to sample specific microhabitats in habitats with high rare-plant potential. In some wetlands, this may involve seeking out small hummocks and areas of standing water or exposed mud, while in others it may focus on the interface between the wetland and the adjacent woodland. Hands-and-knees ground surveys are used along random and directed survey routes in high-potential microhabitats. They involve close examination of ground cover and parting the vegetation to inspect all plants, including small, less conspicuous species.

When we document rare species, the location, habitat, plant community, aspect, slope, relative abundance, soil type, texture, drainage and date should be recorded as precisely as possible. Any additional information on phenology, vigour, size or age classes, and factors affecting the plants such as moisture conditions, competition, insect pests, current land use, grazing pressure and other threats should also be recorded if possible. A map showing the detailed location and extent of the population is extremely valuable for relocating the population. A photograph of the plant and its distinguishing characteristics is an essential part of this documentation. Collection of rare plants is discouraged but may be necessary in order to confirm the identification of plants in some of the more difficult groups. A specimen should be taken only when more than twenty individual plants can be seen at the site *and* the proper permit has been obtained (where applicable). Collections should always be filed in a recognized herbarium (e.g., University of Alberta Herbarium, University of Calgary Herbarium or Canadian Forest Service Herbarium) so that they are available for study by others. In some cases, it may be necessary to bring a taxonomic expert into the field to confirm an identification.

ANHIC collects information related to the location, phenology, population size and biology of rare and

Introduction

uncommon species. It also tracks changes in species status (e.g., new sites) and taxonomy (e.g., name changes as a result of reclassification). New sightings should be always reported to ANHIC. A Rare Native Plant Report Form (Appendix Seven) should be completed and submitted whenever a rare plant is sighted. This information will then be added to the central data bank and used to update species status and distribution. The latest rare species list (known as the tracking list) is available from ANHIC, care of Alberta Environmental Protection (Second Floor, 9820–106 Street, Edmonton, Alberta T5K 2C6, 780-427-5209).

Two publications from the California Native Plant Society (Nelson 1984, 1987) discuss how to carry out rare plant surveys, and the Alberta Native Plant Council (ANPC) has also produced a set of guidelines (ANPC 1997). These publications provide more detailed descriptions of rare plant surveys.

Ranunculus occidentalis

How You Can Help

There are many ways that we can contribute to the conservation of rare plants in Alberta. A valuable first step is to learn more about these unusual species and their habitats, and to share this information with others. This increases our appreciation of Alberta's wild plants and encourages the land-stewardship practices that accommodate the conservation of rare species. We can often accomplish more as a group than we could on our own, and it can be fun to share ideas and experiences with others. Groups such as the ANPC and the Federation of Alberta Naturalists play important roles in the conservation of biodiversity in Alberta. Their activities often increase public awareness of threatened species and habitats. They also share an appreciation of the beauty and importance of natural systems through talks, papers, field trips and books.

ABOUT THE ANPC

The ANPC is an active group of volunteers striving to promote knowledge of Alberta's native plants and to conserve these plants and their habitats. Members include people from all walks of life: from naturalists, gardeners and students to teachers, ecologists and botanists. There are four main action committees. The Education and Information Committee organizes field trips, courses and speakers' programs. The Rare Species Committee gathers information on rare Alberta species and, most recently, has compiled

Arabis lemmonii

Introduction

Pedicularis flammea

this book. The Reclamation and Horticulture Committee sets out guidelines for collecting and planting native species, provides information on reclamation and produces lists of suppliers of native seed and nursery stock. The Conservation Action Committee increases public awareness and approaches industry and government agencies regarding pressing environmental issues, provides stewardship for protected areas, and monitors habitats affected by development. If you would like more information about our group, write to us.

Alberta Native Plant Council
Box 52099, Garneau Postal Outlet
Edmonton, Alberta T6G 2T5

Natural Regions of Alberta

Alberta is a large, diverse province. It encompasses 662,948 km^2, from the Northwest Territories in the north (60°N) to the United States in the south (49°N), and from the heights of the Continental Divide in the Rocky Mountains in the west to the rolling prairies and boreal forest in the east. This broad area encompasses six natural regions and twenty subregions, each with a distinctive combination of physical features (climate, geology, soils, hydrology) and biological features (plant and animal species, vegetation communities). A vast patchwork of habitats supports approximately 1,800 species of vascular plants, 600 species of mosses and 650 species of lichens, along with untold numbers of fungi.

The distinguishing features of each natural region and subregion are described below, along with notes about some of the rare plants. Many rare species occur in more than one natural region. Appendix Two provides a complete list of rare species and their natural regions.

Canadian Shield Natural Region

The Canadian Shield extends into the far northeastern corner of Alberta and contains two quite different subregions. The Kazan Upland Subregion, north of Lake Athabasca, is characterized by exposed, glaciated Canadian Shield bedrock. The Athabasca Plain Subregion includes part of the north shore of Lake Athabasca and an area south of Lake Athabasca where shallow glacial outwash deposits cover Canadian Shield bedrock.

Introduction

Natural Subregions

BOREAL FOREST
Central Mixedwood
Dry Mixedwood
Wetland Mixedwood
Sub-Arctic
Peace River Lowlands
Boreal Highlands

ROCKY MOUNTAIN
Alpine
Subalpine
Montane

FOOTHILLS
Upper Foothills
Lower Foothills

CANADIAN SHIELD
Athabasca Plain
Kazan Upland

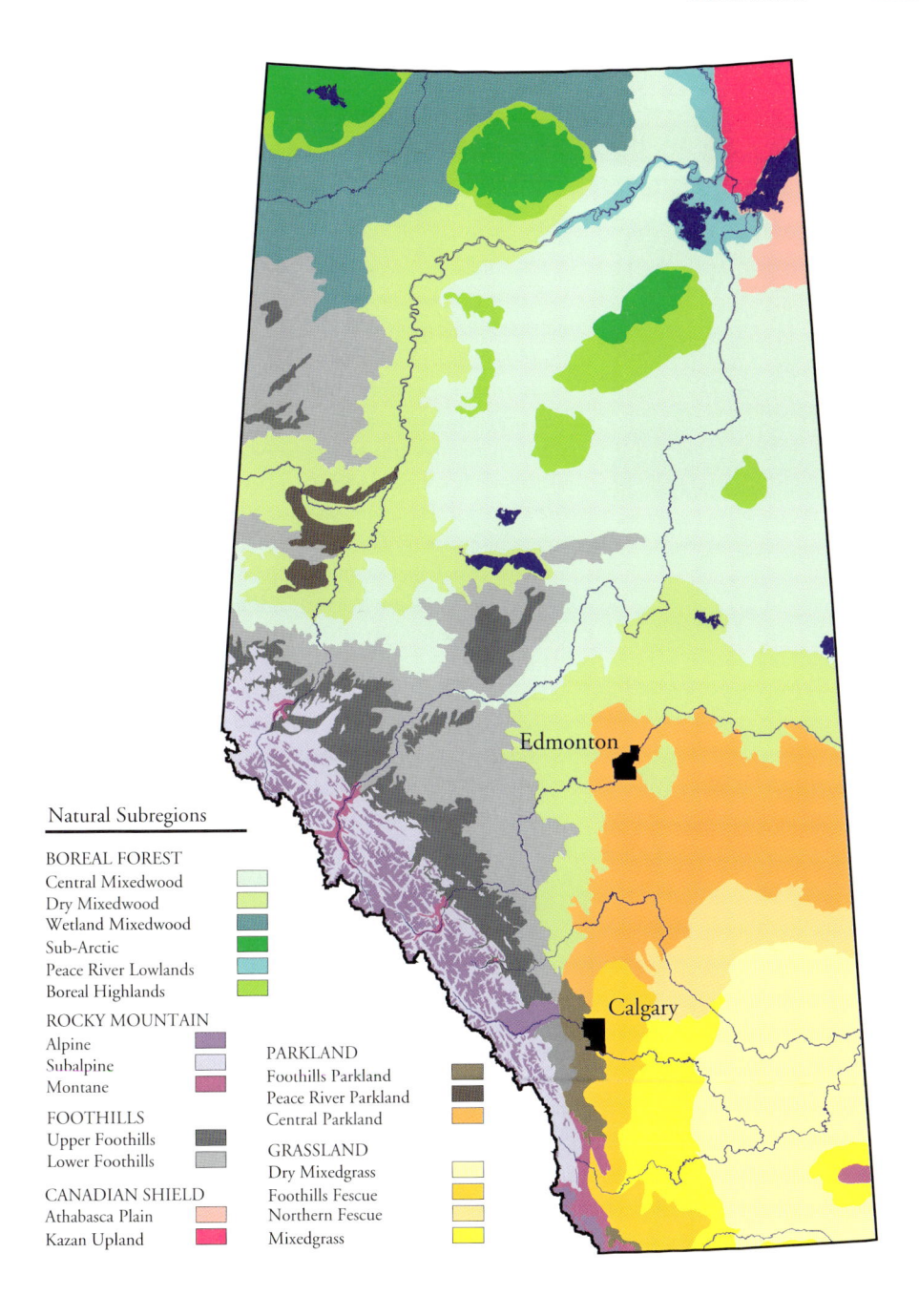

PARKLAND
Foothills Parkland
Peace River Parkland
Central Parkland

GRASSLAND
Dry Mixedgrass
Foothills Fescue
Northern Fescue
Mixedgrass

Rare *Vascular* Plants of Alberta

Introduction

Kazan Upland Subregion at Wylie Lake

KAZAN UPLAND SUBREGION

Extensive outcrops of acidic bedrock and scattered patches of glacial deposits determine the vegetation pattern in this area. Much of the rolling terrain has either exposed bedrock or too little weathered material over the bedrock to qualify as soil. Organic soils and permafrost occur in low, wet areas.

The upland vegetation is a mosaic of barren rock, open jack pine (*Pinus banksiana*) forests on sand plains and rocky hills, and black spruce (*Picea mariana*) stands in wet peatlands (muskegs). Jack pine forests are most widespread, especially in the northeast, and often contain aspen (*Populus tremuloides*). The understorey on sandy sites is quite simple, and lichens are often as important as vascular plants. The forests of rocky sites typically contain jack pine, black spruce and white birch (*Betula papyrifera*), with saskatoon (*Amelanchier alnifolia*), common bearberry (*Arctostaphylos uva-ursi*), three-toothed saxifrage (*Saxifraga tricuspidata*), pasture sagewort (*Artemisia frigida*) and reindeer lichen (*Cladina mitis*) in the understorey. On deeper soils, forests of jack pine, white birch and black spruce have more diverse understoreys, with green alder (*Alnus viridis* ssp. *crispa*), bog cranberry (*Vaccinium vitis-idaea*), common bearberry, common blueberry (*Vaccinium myrtilloides*), lesser rattlesnake plantain (*Goodyera repens*), crowberry (*Empetrum nigrum*), bunchberry (*Cornus canadensis*), common Labrador tea (*Ledum groenlandicum*), reindeer lichens (*Cladina mitis, C. rangiferina* and *C. stellaris*) and snow lichen (*Cetraria nivalis*).

Low areas support peatlands. These are mainly acidic, mineral-poor bogs dominated by black spruce, common Labrador tea, reindeer lichens and peat mosses (*Sphagnum* spp.). Fens (mineral-rich peatlands) are rare because of the acidic bedrock and soils.

ATHABASCA PLAIN SUBREGION

This broad, flat subregion is covered predominantly by water- and wind-deposited materials in the eastern part and by glacial deposits in the west. Most deposits are derived from Athabasca sandstone, and there are extensive sandy beaches along Lake Athabasca. Distinctive landscape features include large areas of kame deposits and active dunes. The kames, at over 60 m in height, are among the largest in the world, and the active dune system is the largest in Alberta.

Introduction

Extensive jack pine forests cover the sandy uplands, sometimes co-dominant with white spruce (*Picea glauca*). Typical understorey species include common bearberry and reindeer lichens on dry sites, and common blueberry and feathermosses (stair-step moss [*Hylocomium splendens*] and Schreber's moss [*Pleurozium schreberi*]) on moister sites. Open, 'park-like' white spruce forests occur along the shore of Lake Athabasca.

Peatlands range from dry bogs, dominated by black spruce, common Labrador tea and reindeer lichens, to wetter peatlands with black spruce, tamarack (*Larix laricina*), common Labrador tea and peat mosses. Shrubby peatlands typically contain common Labrador tea, northern laurel (*Kalmia polifolia*), water sedge (*Carex aquatilis*), reed grasses (*Calamagrostis* spp.) and peat mosses.

Dunes in the Athabasca Plain

Rare Plants

About 30 rare species (6 percent of Alberta's rare vascular plants) occur in the Canadian Shield Natural Region; of these, 7 are restricted to this region in Alberta. Several grow on sand dunes along Lake Athabasca and farther south in the Athabasca Plain Subregion. These include American dune grass (*Leymus mollis*), Indian tansy (*Tanacetum bipinnatum* ssp. *huronense*) and sand-dune chickweed (*Stellaria arenicola*). Other notable species from this region include horned bladderwort (*Utricularia cornuta*), Greenland wood-rush (*Luzula groenlandica*) and stemless lady's-slipper (*Cypripedium acaule*).

Boreal Forest Natural Region

The Boreal Forest constitutes the largest natural region in Alberta. It covers broad lowland plains and discontinuous hill systems where the bedrock is deeply buried by glacial deposits or, along major rivers, by fluvial deposits. This vast, diverse region is divided into six subregions—Dry Mixedwood, Central Mixedwood, Wetland Mixedwood, Boreal Highlands, Peace River Lowlands, and Subarctic—distinguished by differences in vegetation, geology and landform, but many of the changes from one subregion to the next are gradual and subtle.

Introduction

Dry Mixedwood habitat at Heart River

Wetland Mixedwood habitat at McClelland Lake

DRY MIXEDWOOD, CENTRAL MIXEDWOOD, WETLAND MIXEDWOOD AND BOREAL HIGHLANDS SUBREGIONS

These four subregions have similar vegetation. Aspen (*Populus tremuloides*) is widespread, in both pure and mixed stands, and balsam poplar (*Populus balsamifera*) often grows with aspen in moister areas. With time, white spruce (*Picea glauca*) or balsam fir (*Abies balsamea*) gradually replace the aspen and balsam poplar, but frequent fires seldom permit this progression. Pure deciduous stands are common in the southern part of the Dry Mixedwood Subregion, and mixed stands of aspen and white spruce are widespread farther north.

Mixedwood forests contain a mosaic of deciduous and coniferous stands, with species typical of each. The diverse understoreys of upland aspen forests commonly include low-bush cranberry (*Viburnum edule*), beaked hazelnut (*Corylus cornuta*), prickly rose (*Rosa acicularis*), red-osier dogwood (*Cornus sericea*), bluejoint (*Calamagrostis canadensis*), wild sarsaparilla (*Aralia nudicaulis*), dewberry (*Rubus pubescens*), cream-coloured vetchling (*Lathyrus ochroleucus*), common pink wintergreen (*Pyrola asarifolia*) and twinflower (*Linnaea borealis*). Both balsam poplar and white birch (*Betula papyrifera*) also grow in these forests. Spruce and spruce-fir forests are not common, and they generally have a less diverse understorey with large patches or carpets of feathermosses including stair-step moss (*Hylocomium splendens*), Schreber's moss (*Pleurozium schreberi*) and knight's plume moss (*Ptilium crista-castrensis*).

Dry, sandy sites usually support open jack pine (*Pinus banksiana*) forests with a prominent carpet of lichens. Common understorey species include common bearberry (*Arctostaphylos uva-ursi*), common blueberry (*Vaccinium myrtilloides*), bog cranberry (*Vaccinium vitis-idaea*) and prickly rose.

Peatlands occur throughout the six subregions, but they are most extensive in the Wetland Mixedwood Subregion. Peatland complexes typically contain both mineral-poor, acidic bogs—dominated by black spruce (*Picea mariana*), common Labrador tea (*Ledum groenlandicum*) and peat mosses (*Sphagnum* spp.)—and mineral-rich fens—featuring tamarack (*Larix laricina*), dwarf birches (*Betula* spp.), sedges (*Carex* spp.), tufted moss (*Aulacomnium palustre*), golden moss (*Tomenthypnum nitens*) and brown mosses (*Drepanocladus* spp.).

Climate and vegetation vary slightly from one subregion to the next. The Dry Mixedwood Subregion has a drier, warmer climate than the other Boreal Forest subregions. Its vegetation is transitional between the Central Parkland (Parkland Natural Region) and Central Mixedwood subregions, with pure aspen forests being more common and peatlands (muskegs) being less extensive. The Central Mixedwood Subregion has a slightly cooler, moister climate than the Dry Mixedwood Subregion. Mixedwood and white spruce forest and peatlands are more common than in the Dry Mixedwood Subregion, and white birch is widespread, forming nearly pure stands in some areas. The Wetland Mixedwood Subregion is characterized by a greater proportion of wetlands, including peatlands, willow-sedge complexes (on mineral soil) and upland black spruce forests. Permafrost occurs in many peatlands in this relatively flat, moist, cool region. The Boreal Highlands Subregion includes rolling uplands and steep slopes on hill masses and plateaus in the Central Mixedwood and Wetland Mixedwood subregions. It has a relatively cool, moist climate (especially during the summer), and permafrost is common. Balsam poplar, white spruce and black spruce are more common than in the mixedwood subregions, with coniferous forests covering much of the landscape.

Boreal Highlands habitat in the Cameron Hills

Peace River Lowlands Subregion

This subregion encompasses terraces and deltas along the lower Peace, Birch and Athabasca rivers, including the Peace-Athabasca Delta, which is one of the largest freshwater deltas in the world. Fluvial processes and landforms are most important in this area.

White spruce forests with large trees (16–23 m tall) are characteristic of moist to wet river terraces, but little remains of these very productive forests, because of heavy logging. Drier upland sites typically support jack pine forests with understoreys of green alder (*Alnus viridis* ssp. *crispa*), bog cranberry, reindeer lichen (*Cladina* spp.) and feather-mosses. Mixedwood forests of aspen, balsam poplar and white spruce occupy mesic upland sites. Non-forested communities include a complex mosaic of aquatic, shoreline, meadow, shrub and marsh vegetation, maintained in large part by periodic erosion and deposition of river sediments, especially in the Peace-Athabasca Delta area.

Peace River Lowlands habitat

Introduction

The Caribou Mountains in the Subarctic Subregion

SUBARCTIC SUBREGION

The Subarctic Subregion occurs on the Birch Mountains, Caribou Mountains and Cameron Hills, all of which are erosional remnants rising above the surrounding plain as flat-topped hills. Peat deposits and permafrost are widespread, with occasional deposits of glacial till. The climate is cold, and the 45-day frost-free period is about half that of the mixedwood subregions. Open black spruce forest dominates the landscape, with an understorey of common Labrador tea, northern Labrador tea (*Ledum palustre*), cloudberry (*Rubus chamaemorus*), bog cranberry, peat mosses and reindeer lichens. Fires have resulted in large areas of heath shrub and lichen vegetation with scattered, young black spruce. Upland sites with mineral soil typically support black spruce, white spruce–aspen and white spruce–white birch forests. Stands of black spruce with lodgepole pine (*Pinus contorta*) (jack pine in the Birch Mountains) also occur in small areas with warmer, drier sites.

RARE PLANTS

More than 110 rare species (about 25 percent of Alberta's rare vascular plants) occur in the Boreal Forest Natural Region, and about 20 of these are found in this natural region only. Several species in the Subarctic Subregion have their main range to the north; these include Russian ground-cone (*Boschniakia rossica*), bog bilberry (*Vaccinium uliginosum*), polar grass (*Arctagrostis arundinacea*) and purple rattle (*Pedicularis sudetica*). Another group of rare boreal species is associated with saline or alkaline soils, such as those of the 'Salt Plains' in northern Wood Buffalo National Park; these include sea-side plantain (*Plantago maritima*) and mouse-ear cress (*Arabidopsis salsuginea*). These habitats also contain disjunct populations of species that are common in southern Alberta, such as green needle grass (*Stipa viridula*), needle-and-thread (*Stipa comata*), wild bergamot (*Monarda fistulosa*), purple reed grass (*Calamagrostis purpurascens*) and prairie sagewort (*Artemisia ludoviciana*). Other notable rare plants in this region include white adder's-mouth (*Malaxis monophylla* var. *brachypoda*), bog adder's-mouth (*Malaxis paludosa*), pitcher-plant (*Sarracenia purpurea*), northern slender ladies'-tresses (*Spiranthes lacera*) and water lobelia (*Lobelia dortmanna*).

Introduction

Rocky Mountain Natural Region

The Rocky Mountain Natural Region is part of a major geological uplift that trends along the western part of Alberta forming the Continental Divide. It is distinguished from the Foothills Natural Region by its geology. The Rocky Mountain Natural Region has upthrust and folded carbonate and quartzitic bedrock, whereas the Foothills Natural Region has mostly deformed sandstone and shale. This is the most rugged natural region in Alberta, and it ranges from about 10 km wide near Waterton Lakes National Park to more than 100 km wide in its central portion. Elevations rise from 1,000 to 1,500 m in the east to 3,700 m along the Continental Divide in the west. The three subregions—Montane, Subalpine and Alpine—reflect environmental differences associated with changes in altitude.

Montane habitat on the Kootenay Plains

MONTANE SUBREGION

South of the Porcupine Hills, this subregion occurs mostly on ridges that extend out from the foothills, but to the north, the Montane Subregion is found mostly along major river valleys, with small, disjunct areas at Ya-Ha-Tinda on the Red Deer River and in the Cypress Hills. Elevations range from 1,000 to 1,350 m in Jasper National Park to more than 1,600 m along the east slopes south of Calgary. Chinooks keep this region intermittently snow-free in the winter. The landscape is covered by a mosaic of open forests and grasslands.

Characteristic montane trees include Douglas-fir (*Pseudotsuga menziesii*), limber pine (*Pinus flexilis*) and white spruce (*Picea glauca*). Typical understorey plants in Douglas-fir forests include pine reed grass (*Calamagrostis rubescens*), hairy wild rye (*Leymus innovatus*), low northern sedge (*Carex concinnoides*), common bearberry (*Arctostaphylos uva-ursi*), junipers (*Juniperus* spp.) and snowberry (*Symphoricarpos albus*). Distinctive species in the Waterton Lakes National Park area include creeping mahonia (*Berberis repens*), mallow-leaved ninebark (*Physocarpus malvaceus*), mountain maple (*Acer glabrum*), purple clematis (*Clematis occidentalis*) and bluebunch wheat grass (*Elymus spicatus*).

Limber pine forms open stands on rock outcrops and eroding slopes. Common understorey species include common bearberry, junipers, bluebunch wheat grass, bluebunch fescue (*Festuca idahoensis*), northern bedstraw (*Galium boreale*), field mouse-ear chickweed (*Cerastium arvense*), crested beardtongue (*Penstemon eriantherus*) and scorpion-

Introduction

Typical Subalpine habitat

weed (*Phacelia* spp.). Grasslands are typically dominated by bluebunch wheat grass, fescue grasses (*Festuca* spp.) and oat grasses (*Danthonia* spp.) with a variety of forbs. Lodgepole pine (*Pinus contorta*) forests cover many upland areas and are similar to those of the adjacent Subalpine Subregion. White spruce forests occur on mesic sites, especially along stream terraces, and aspen (*Populus tremuloides*) forests typically occupy fluvial fans and terraces.

Subalpine Subregion

This cool, moist subregion lies between the Montane and Alpine subregions in the south (at 1,600 to 2,300 m) and the Upper Foothills (Foothills Natural Region) and Alpine subregions in the north (at 1,350 to 2,000 m). Below-freezing temperatures occur in all months, the frost-free period is usually less than 30 days, and winter precipitation is higher than in any other part of Alberta.

The Subalpine Subregion consists of two portions: the Lower Subalpine, characterized by closed forests of lodgepole pine, Engelmann spruce (*Picea engelmannii*) and subalpine fir (*Abies bifolia*), and the Upper Subalpine, characterized by intermixed open and closed forests. Engelmann spruce–subalpine fir forests typically occur on higher, moister sites. Pure Engelmann spruce is characteristic of high-elevation stands. At lower elevations, spruce trees (often white spruce–Engelmann spruce hybrids) are more common, and burned areas are covered by extensive lodgepole pine forests.

Forests in the Waterton Lakes National Park area support many species that do not occur farther north. These include bear grass (*Xerophyllum tenax*), thimbleberry (*Rubus parviflorus*), mountain wood-rush (*Luzula piperi*), yellow angelica (*Angelica dawsonii*) and mountain-lover (*Pachystima myrsinites*).

Upper Subalpine open forests are transitional to the treeless Alpine Subregion above. Dominant trees include Engelmann spruce, subalpine fir, whitebark pine (*Pinus albicaulis*) and, south of Bow Pass, subalpine larch (*Larix lyallii*). Some forests contain dwarf shrubs that also grow in the Alpine Subregion, including red heather (*Phyllodoce empetriformis*), yellow heather (*Phyllodoce glanduliflora*), white mountain-heather (*Cassiope mertensiana*), grouseberry (*Vaccinium scoparium*) and rock willow (*Salix vestita*).

High-elevation grasslands are mostly found on steep, south- and west-facing slopes in the Front Ranges. Dominant species include hairy wild rye (*Leymus innovatus*), June grass (*Koeleria macrantha*) and common bearberry.

Introduction

ALPINE SUBREGION

The Alpine Subregion includes all areas above treeline, including vegetated slopes, rock fields, snow fields and glaciers; there are extensive areas of unvegetated bedrock. This is the coldest subregion in Alberta, with essentially no frost-free period. Alpine vegetation typically forms a fine-scaled mosaic, in which minor variations in microclimate can produce marked changes in species composition. Significant factors include aspect, wind exposure, time of snow melt, soil moisture and snow depth.

Deep, late-melting snow beds are occupied by black alpine sedge (*Carex nigricans*) communities. Moderate snow-bed communities typically support dwarf-shrub heath tundra, which is dominated by heathers (*Phyllodoce* spp.), mountain-heathers (*Cassiope* spp.) and grouseberry. Common species in shallow snow areas on ridge tops and other exposed sites include white mountain avens (*Dryas octopetala*), snow willow (*Salix reticulata*), moss campion (*Silene acaulis*) and bog-sedge (*Kobresia myosuroides*). Diverse, colourful herb meadows blanket moist sites below melting snowbanks and along streams. Communities at the highest elevations consist mainly of lichens on rocks and shallow soil.

Some floristic differences are apparent south of Crowsnest Pass. Mountain-heathers are absent, heathers are less common than farther north, and bear grass meadows occupy some low-elevation sites.

Typical Alpine habitat at Pipestone Pass

RARE PLANTS

More than 300 rare species (about 65 percent of Alberta's rare vascular plants) occur in the Rocky Mountain Natural Region; of these, about 140 are restricted in Alberta to this natural region. More than 25 species grow in the Alpine Subregion and are connected by a series of high-altitude habitats to the Arctic tundra to the north. These 'arctic-alpine' plants include alpine harebell (*Campanula uniflora*), alpine bitter cress (*Cardamine bellidifolia*), seven sedge species (*Carex glacialis, C. haydeniana, C. incurviformis* var. *incurviformis, C. lachenalii, C. misandra, C. petricosa* and *C. podocarpa*), alpine mouse-ear chickweed (*Cerastium beeringianum*), alpine gentian (*Gentiana glauca*), alpine alumroot (*Heuchera glabra*), alpine sweetgrass (*Anthoxanthum monticola*), koenigia (*Koenigia islandica*), arctic lousewort (*Pedicularis langsdorfii*), woolly lousewort (*Pedicularis lanata*), hairy cinquefoil (*Potentilla villosa*) and snow buttercup (*Ranunculus nivalis*).

Introduction

Lower Foothills habitat at Running Lake

The mountains of southwestern Alberta, especially in the Waterton Lakes National Park and Castle River areas, contain many species that are not found farther north in Alberta. These include Jones' columbine (*Aquilegia jonesii*), big sagebrush (*Artemisia tridentata*), large-flowered brickellia (*Brickellia grandiflora*), mountain hollyhock (*Iliamna rivularis*), bitter-root (*Lewisia rediviva*), dwarf bitter-root (*Lewisia pygmaea*), biscuit-root (*Lomatium cous*), Brewer's monkeyflower (*Mimulus breweri*), small yellow monkey-flower (*Mimulus floribundus*), dwarf alpine poppy (*Papaver pygmaeum*), Lewis' mock-orange (*Philadelphus lewisii*), mallow-leaved ninebark (*Physocarpus malvaceus*), pink meadowsweet (*Spiraea splendens*), western yew (*Taxus brevifolia*) and western wakerobin (*Trillium ovatum*). This area is also a continental 'hot spot' for rare species of grape fern (*Botrychium* spp.), with eight species currently known from this area, including the Waterton moonwort (*Botrychium* x *watertonense*), which is endemic to Waterton Lakes National Park. Front-range fleabane (*Erigeron lackschewitzii*) and mountain dwarf primrose (*Douglasia montana*) are found in Canada only in Waterton Lakes National Park.

Another group of rare mountain species extends into western Alberta from moister areas in BC. These include oval-leaved blueberry (*Vaccinium ovalifolium*), spreading stonecrop (*Sedum divergens*), western red cedar (*Thuja plicata*), western hemlock (*Tsuga heterophylla*) and western larch (*Larix occidentalis*).

Foothills Natural Region

The Foothills Natural Region is transitional between the Rocky Mountains and Boreal Forest natural regions. It consists of two subregions: Lower Foothills and Upper Foothills.

LOWER FOOTHILLS SUBREGION

The Lower Foothills Subregion includes rolling hills created by deformed bedrock along the edge of the Rocky Mountains and flat-topped erosional remnants, such as the Swan Hills. Although it is cooler in summer than the adjacent, lower Boreal Forest subregions, it is warmer in winter because it is less affected by cold arctic air masses. Mixed forests of white spruce (*Picea glauca*), black spruce (*Picea mariana*), lodgepole pine (*Pinus contorta*), balsam fir (*Abies balsamea*), aspen (*Populus tremuloides*), white birch (*Betula papyrifera*) and balsam poplar (*Populus balsamifera*) dominate, with lodgepole pine stands indicating the lower boundary (adjacent to the Boreal Forest) and the nearly pure coniferous forests (without aspen, balsam poplar or birch) of the Upper Foothills indicating the upper boundary. Hybrids of lodgepole pine, a mountain species, and jack pine (*Pinus banksiana*), a boreal species, occur along the eastern boundary and at lower elevations.

Lodgepole pine forests dominate in the upland, especially after fire, with common understorey species including Canada buffaloberry (*Shepherdia canadensis*), white meadowsweet (*Spiraea betulifolia*), junipers (*Juniperus* spp.), common bearberry (*Arctostaphylos uva-ursi*) and common blueberry (*Vaccinium myrtilloides*). On more mesic sites, white spruce and aspen are more frequent and understoreys are more diverse, including prickly rose (*Rosa acicularis*), common Labrador tea (*Ledum groenlandicum*), bunchberry (*Cornus canadensis*), twinflower (*Linnaea borealis*), common fireweed (*Epilobium angustifolium*), bog cranberry (*Vaccinium vitis-idaea*) and feathermosses (stair-step moss [*Hylocomium splendens*], Schreber's moss [*Pleurozium schreberi*] and knight's plume moss [*Ptilium crista-castrensis*]).

Black spruce forests occupy wet organic soils (muskegs) and some moist upland sites, with typical understorey species including common Labrador tea, dwarf birch (*Betula* spp.), bracted honeysuckle (*Lonicera involucrata*), horsetails (*Equisetum* spp.), bishop's-cap (*Mitella nuda*), twinflower, peat mosses (*Sphagnum* spp.), tufted moss (*Aulacomnium palustre*) and golden moss (*Tomenthypnum nitens*). Fens, both patterned and unpatterned, are common, with dwarf birch, common Labrador tea, willow (*Salix* spp.), sedges (*Carex* spp.), buck-bean (*Menyanthes trifoliata*), rough hair grass (*Agrostis scabra*), tufted moss, golden moss, peat mosses, and scattered black spruce and tamarack (*Larix laricina*).

Upper Foothills habitat at Switzer Provincial Park

Upper Foothills Subregion

The subregion generally lies between the Lower Foothills and Subalpine (Rocky Mountain Natural Region) sub-regions on strongly rolling hills along the eastern edge of the Rocky Mountains and in the Swan Hills and Clear Hills. It has the highest summer precipitation in Alberta. Its winters are colder than those of the Lower Foothills Subregion, but are similarly less affected by cold Arctic air masses.

Upland forests of the Upper Foothills Subregion are dominated by white spruce, black spruce, lodgepole pine and, occasionally, subalpine fir (*Abies bifolia*). Hybrids between white spruce and Engelmann spruce (*Picea glauca–*

Introduction

Central Parkland habitat at Rumsey

P. engelmannii) and between subalpine fir and balsam fir (*Abies bifolia–A. balsamea*) are sometimes found. The understorey resembles that of a lodgepole pine forest, with a well-developed carpet of feathermosses (stair-step moss, Schreber's moss and knight's plume moss) in older stands.

Black spruce dominates wet sites with organic or mineral soils. Typical understorey species include common Labrador tea, dwarf birch, bracted honeysuckle, horsetails, bishop's cap, twinflower, peat mosses, tufted moss and golden moss.

Rare Plants

Roughly 80 rare species (about 18 percent of Alberta's rare vascular plants) occur in the Foothills Natural Region. Only four of these (wood anemone [*Anemone quinquefolia*], northern oak fern [*Gymnocarpium jessoense*], western white lettuce [*Prenanthes alata*] and small twisted-stalk [*Streptopus streptopoides*]) are restricted in Alberta to this natural region, reflecting the transitional nature of the Foothills Natural Region. Many of its rare species also occur in the Rocky Mountain and Boreal Forest natural regions. Some rare species of interest include goldthread (*Coptis trifolia*), northern beech fern (*Phegopteris connectilis*) and rose mandarin (*Streptopus roseus*).

Parkland Natural Region

The Parkland Natural Region (with the exception of the Peace River Parkland Subregion) forms a broad transition between the grasslands of the plains and the coniferous forests of the Boreal Forest and Rocky Mountain natural regions. This region is found only in the prairie provinces of Canada, except for a few small tongues that extend into the northern United States. Its three subregions—Central Parkland, Foothills Parkland and Peace River Parkland—are distinguished by location, soil and vegetation.

Central Parkland Subregion

This subregion gradually changes from grassland with groves of aspen (*Populus tremuloides*) in the south, to aspen parkland, to closed aspen forest in the north. True aspen parkland, with continuous aspen forest broken by grassland openings, is now rare as a result of agricultural clearing. The two major forest types—aspen and balsam poplar (*Populus balsamifera*) stands—have dense, species-rich understoreys. Snowberry (*Symphoricarpos albus*), saskatoon (*Amelanchier alnifolia*), beaked hazelnut (*Corylus cornuta*), choke cherry (*Prunus virginiana*), bunchberry (*Cornus canadensis*), wild lily-of-the-valley (*Maianthemum canadense*) and purple oat grass (*Schizachne purpurascens*) are typical of aspen stands, whereas red-osier dogwood (*Cornus sericea*), pussy willow (*Salix discolor*), northern gooseberry (*Ribes oxyacanthoides*), green alder (*Alnus viridis* ssp. *crispa*), bracted honeysuckle (*Lonicera involucrata*), tall lungwort (*Mertensia paniculata*), palmate-leaved coltsfoot (*Petasites palmatus*), bishop's-cap (*Mitella nuda*) and red baneberry (*Actaea rubra*) are characteristic of the moister balsam poplar stands.

The grassland vegetation of the 'parks' is essentially the same as that of the Northern Fescue Subregion (Grasslands Natural Region). Plains rough fescue (*Festuca hallii*) dominates most sites, with western porcupine grass (*Stipa curtiseta*) important on south-facing slopes and saline soils, along with June grass (*Koeleria macrantha*) and western wheat grass (*Elymus smithii*).

Shrub communities of buckbrush (*Symphoricarpos occidentalis*), wild roses (*Rosa* spp.), choke cherry, pin cherry (*Prunus pensylvanica*), saskatoon and silverberry (*Elaeagnus commutata*) are more extensive in northern sections, often extending out from the forest in belts.

Foothills Parkland habitat at the Palmer Ranch

Foothills Parkland Subregion

This narrow (1–5 km wide) transitional band between the Grasslands and Rocky Mountain natural regions lies along the eastern edge of the southern foothills. It is very diverse, with rougher topography than the Central Parkland Subregion and rapid changes in elevation and climate over short distances. Its northern boundary (near Calgary) is the northern limit for many distinctive southwestern species, including lupines (*Lupinus* spp.), oat grasses (*Danthonia* spp.) and bluebunch fescue (*Festuca idahoensis*).

The grasslands of this subregion are similar to those of the Foothills Fescue Subregion (Grasslands Natural Region): fescue–oat grass communities with many forb and grass species. The aspen forests resemble those of the Central Parkland Subregion, but several Central Parkland species (e.g., beaked hazelnut, high-bush cranberry [*Viburnum opulus* ssp. *trilobum*], wild sarsaparilla [*Aralia nudicaulis*], bunchberry and wild lily-of-the-valley) are absent, and some plants (e.g., yellow angelica [*Angelica dawsonii*]) are restricted to southwestern Alberta. Glacier lily (*Erythronium grandiflorum*) blankets much of the forest floor in south-western sectors in early to mid May.

Introduction

Peace River Parkland habitat at Silver Valley

Peace River Parkland Subregion

This northern subregion has relatively short, cool summers, long, cold winters, high precipitation, less wind and lower evaporation than other Parkland subregions. The till deposits of the uplands are covered by forests similar to those of the surrounding mixedwood subregions of the Boreal Forest, dominated by aspen and white spruce (*Picea glauca*), with some balsam poplar (especially on wetter sites). Grassland soils are mostly saline (forest soils are not), and this is an important factor in maintaining these grasslands, with fire, and possibly climate, playing a secondary role. The upland grasslands are dominated by sedges (*Carex* spp.), timber oat grass (*Danthonia intermedia*), western porcupine grass, slender wheat grass (*Elymus trachycaulus*), inland bluegrass (*Poa interior*), three-flowered avens (*Geum triflorum*) and low goldenrod (*Solidago missouriensis*). On steep, south-facing slopes, western porcupine grass, sedges, and pasture sagewort (*Artemisia frigida*) dominate. Other common species include Columbia needle grass (*Stipa columbiana*), June grass, green needle grass (*Stipa viridula*), bastard toadflax (*Comandra umbellata*) and mountain goldenrod (*Solidago spathulata*). Several species that usually grow farther south or west are found here. These include brittle prickly-pear (*Opuntia fragilis*), Richardson needle grass (*Stipa richardsonii*), Columbia needle grass, Drummond's thistle (*Cirsium drummondii*) and low goldenrod.

Rare Plants

More than 100 rare species (about 20 percent of Alberta's rare vascular plants) occur in the Parkland Natural Region; most of these are also found in the Boreal Forest, Grasslands or Rocky Mountain natural regions, but 10 are restricted in Alberta to this transitional region. These include few-flowered aster (*Aster pauciflorus*), field grape fern (*Botrychium campestre*), sand millet (*Panicum wilcoxianum*), marsh gentian (*Gentiana fremontii*) and few-flowered salt-meadow grass (*Torreyochloa pallida*). Several rare plants are found only in southwestern Alberta, in the Foothills Parkland Subregion and, in some cases, also in the adjacent mountains. These include Geyer's wild onion (*Allium geyeri*), alpine foxtail (*Alopecurus alpinus*), blue camas (*Camassia quamash*), intermediate hawk's-beard (*Crepis intermedia*), mountain lady's-slipper (*Cypripedium montanum*), small-flowered rockstar (*Lithophragma parviflorum*), linear-leaved montia (*Montia linearis*) and small baby-blue-eyes (*Nemophila breviflora*).

Introduction

Grasslands Natural Region

This southern natural region is flat to gently rolling with a few hill systems. Most of the bedrock is covered with thick glacial till, with many areas of fine-textured materials associated with glacial lake basins. Deep river valleys, carved into bedrock, contain badlands with many coulees and ravines. The four subregions—Dry Mixedgrass, Mixedgrass, Northern Fescue and Foothills Fescue—are distinguished by their climate, soil and vegetation.

Dry Mixedgrass Subregion

This is the most extensive grassland subregion. It has the warmest, driest climate in Alberta, with the lowest summer precipitation and little snow cover. The characteristic soils are Brown Chernozems.

Dry Mixedgrass habitat at Milk River

Mixedgrass regions have both short and mid-height grasses. Mid-height grasses (needle grasses [*Stipa* spp.]), western wheat grass [*Elymus smithii*] and June grass [*Koeleria macrantha*]) and a short grass (blue grama [*Bouteloua gracilis*]) are most common. Northern wheat grass (*Elymus dasystachyus*) and western porcupine grass (*Stipa curtiseta*) are characteristic of moister sites. Communities of needle-and-thread (*Stipa comata*) and blue grama are widespread, but western wheat grass and northern wheat grass are also important on hummocky moraines. Sand dunes are usually dominated by needle-and-thread, sand grass (*Calamovilfa longifolia*), June grass and low shrubs such as silver sagebrush (*Artemisia cana*), silverberry (*Elaeagnus commutata*), buckbrush (*Symphoricarpos occidentalis*) and prickly rose (*Rosa acicularis*).

Mixedgrass Subregion

This subregion has slightly moister, cooler summers than the Dry Mixedgrass Subregion, and its characteristic soils are Dark Brown and Brown Chernozems. Vegetation is similar to that of the Dry Mixedgrass Subregion, but communities generally produce about 25 percent more biomass and have a greater abundance of species that favour cooler and moister sites (e.g., western porcupine grass and northern wheat grass). Fine-textured soils in glacial lake basins are usually covered by communities of northern wheat grass and June grass, and blue grama is more common on drier, exposed sites. Extensive narrow-leaf cottonwood (*Populus angustifolia*) forests grow on terraces of the Oldman, Belly, Waterton and St. Mary's rivers.

Mixedgrass habitat at the Turin Dunes

Introduction

Northern Fescue habitat at the Hand Hills

NORTHERN FESCUE SUBREGION

This narrow belt of grassland is characterized by gently rolling terrain with Dark Brown and Black Chernozems. The climate is transitional between the Mixedgrass and Central Parkland (Parkland Natural Region) subregions.

Plains rough fescue (*Festuca hallii*) dominates, but June grass, western porcupine grass, northern wheat grass and Hooker's oat grass (*Helictotrichon hookeri*) are also important. Common forbs include prairie crocus (*Anemone patens*), prairie sagewort (*Artemisia ludoviciana*), field mouse-ear chickweed (*Cerastium arvense*), wild blue flax (*Linum lewisii*), smooth fleabane (*Erigeron glabellus*), northern bedstraw (*Galium boreale*), harebell (*Campanula rotundifolia*) and three-flowered avens (*Geum triflorum*). In sand-dune areas, rough fescue grasslands have scattered silverberry shrubs or thickets of rose and buckbrush.

FOOTHILLS FESCUE SUBREGION

This is the highest (up to 1,400 m), most western grassland subregion, forming a narrow band between the Mixedgrass and Foothills Parkland (Parkland Natural Region) subregions, with disjunct sites in the Cypress Hills and on the Milk River Ridge. Chinooks are frequent, so winters are relatively mild, but there is relatively high snowfall in late winter and early spring. The soils are predominantly Dark Brown and Black Chernozems.

Foothills rough fescue (*Festuca campestris*), bluebunch fescue (*Festuca idahoensis*), Parry's oat grass (*Danthonia parryi*) and timber oat grass (*Danthonia intermedia*) dominate, but June grass, northern wheat grass, western porcupine grass, Columbia needle grass (*Stipa columbiana*), early bluegrass (*Poa cusickii*) and Hooker's oat grass are also common. Forbs are more common and more diverse here than in the Northern Fescue Subregion and typically include sticky purple geranium (*Geranium viscosissimum*), prairie crocus, woolly gromwell (*Lithospermum ruderale*), golden bean (*Thermopsis rhombifolia*), prairie sagewort, low larkspur (*Delphinium bicolor*), heart-leaved buttercup (*Ranunculus cardiophyllus*), shooting stars (*Dodecatheon* spp.) and western wild parsley (*Lomatium triternatum*). Balsamroot (*Balsamorhiza sagittata*) is characteristic of the region near the foothills but is absent from the Cypress Hills. Many species grow in the Foothills Fescue Subregion but not in the Northern Fescue Subregion, including Parry's oat grass, bluebunch fescue, Columbia needle grass,

Introduction

sticky purple geranium, silky perennial lupine (*Lupinus sericeus*), kittentails (*Besseya wyomingensis*), western wild parsley, American thorough-wax (*Bupleurum americanum*), squawroot (*Perideridia gairdneri*), woolly gromwell and Williams' conimitella (*Conimitella williamsii*).

Rare Plants

Almost 125 rare species (about 27 percent of Alberta's rare vascular plant species) occur in the Grasslands Natural Region; about 55 of them are restricted in Alberta to this region. Most rare plants in the Grasslands Natural Region are at the northern or western edge of their range, extending into Alberta from widespread populations in Saskatchewan and the United States.

About half of the rare species in this subregion grow in grasslands. Many others occupy more specialized habitats. Water hyssop (*Bacopa rotundifolia*), smooth boisduvalia (*Boisduvalia glabella*), chaffweed (*Anagallis minima*), downingia (*Downingia laeta*), hairy pepperwort (*Marsilea vestita*) and woollyheads (*Psilocarphus elatior*) grow in pools and on the edges of sloughs. Shadscale (*Atriplex canescens*), spatula-leaved heliotrope (*Heliotropium curassavicum*), scratch grass (*Muhlenbergia asperifolia*), Moquin's sea-blite (*Suaeda moquinii*) and poison suckleya (*Suckleya suckleyana*) grow on alkaline or saline soils. Sand verbena (*Tripterocalyx micranthus*), smooth narrow-leaved goosefoot (*Chenopodium subglabrum*), bur ragweed (*Ambrosia acanthicarpa*) and western spiderwort (*Tradescantia occidentalis*) are typical of sandy habitats, whereas Watson's goosefoot (*Chenopodium watsonii*), small cryptanthe (*Cryptantha minima*), nodding umbrella-plant (*Eriogonum cernuum*) and greenthread (*Thelesperma subnudum*) are found in badlands and on other eroding slopes.

Foothills Fescue habitat just east of the Porcupine Hills

About the Species Accounts

This guide includes all of the vascular plant species that are currently recognized as rare in Alberta. The species accounts are divided into four major sections on the basis of growth form: woody plants (trees and shrubs), herbs (most non-woody plants), grass-like plants (grasses, sedges, rushes), and ferns and fern allies (horsetails, ferns). The coloured bars at the top of each page identify these sections. Within each section, plants are organized by family, with similar families grouped together, and within each family similar genera and species appear together. Each species account is organized as follows.

Hydrophyllum capitatum

Introduction

Ranunculus uncinatus

Plantago maritima

NAMES

Common and scientific names (with the scientific authority) appear at the top of the account; the family name is displayed beside the colour bar. Most common names are taken from the master list published by Alberta Environmental Protection in 1993, whereas scientific names generally follow the *Flora of Alberta* (Moss 1983) except where there have been recent taxonomic revisions. *Flora of North America* (Flora of North America Editorial Committee 1993, 1997, in press) and *A Synonomized Checklist of the Vascular Flora of the United States, Canada and Greenland* (Kartesz 1994) are the main sources of information for such revisions. This guide includes rare vascular plant species only. Rare subspecies, varieties, forms and hybrids are not covered. However, a variety or subspecies name may be noted with the species name, if it is the only one found in Alberta.

DESCRIPTIONS

Many books and papers provided information on different species, but five floras were the main sources used in composing the plant descriptions. These were *Flora of Alberta* (Moss 1983), *Budd's Flora of the Canadian Prairie Provinces* (Looman and Best 1979), *Vascular Plants of the Pacific Northwest* (Hitchcock et al. 1955–69), *Vascular Plants of Continental Northwest Territories, Canada* (Porsild and Cody 1980) and *Flora of the Yukon Territory* (Cody 1996).

Whenever possible, plants are described in everyday language, with few technical terms (e.g., 'caespitose' and 'glabrous' have been replaced with 'tufted' and 'hairless'). In some cases, technical terms are included in brackets after a replacement word or phrase; e.g., 'awn' has been replaced with 'bristle (awn)' and 'panicle' with 'branched cluster (panicle).' Terms that are not easily replaced are retained in the descriptions but are also included in the illustrated glossary (see pp. 425–50).

Diagnostic features vary from group to group and from species to species. For example, leaf size and shape may be important in one group, whereas seed characteristics may be diagnostic in another. These key characteristics are highlighted with bold text. Some plants can change dramatically with changes in their environment. For example, a species may have loose clusters of large, spreading leaves in moist, shady sites, and compact clusters of much smaller leaves when growing on exposed ridges. The size ranges included in the descriptions should help to identify the degree of variability that can be expected. Sizes observed for Alberta plants are presented in normal text, but if larger or smaller sizes have been recorded from areas outside the province, these are presented in braces before or after the Alberta range. For example, if leaves on Alberta plants are 5–10 cm long, but the same species' leaves are 2–8 cm in Alaska and 5–15 cm in Washington, leaf length will appear as 5–10 {2–15} cm.

The first descriptive section, 'Plants,' describes the plant as a whole: its longevity (annual, biennial, perennial), size, colour, branching, hairiness and rooting structure. The next section, 'Leaves,' describes the leaves: their arrangement (alternate, opposite, whorled), shape, size, texture, colour, edges and stalks. The third section, 'Flowers,' encompasses the greatest variety of structures and varies the most from one group of plants to another. In most woody plants and herbs, it focusses on individual flowers, but in the aster family, 'Flowerheads' (rather than single florets) are described; in grass-like plants, 'Flower Clusters' are described; and in the ferns and fern allies, 'Spore Clusters' are described. These descriptions include the colour, shape, size and hairiness of flowers and flower parts, and the size, shape and arrangement of flower clusters.

Viola pallens

The 'Flowers' section ends with the flowering period, which indicates when to look for flowers and, by extension, when you are likely to find fruits. Alberta information is provided whenever possible, but most of these data pertain to other regions, such as the Pacific Northwest and the Rocky Mountains. Flowering dates based solely on published information for regions outside Alberta are presented in braces. For example, if plants have been observed blooming only in July in Alberta but are reported to bloom from May to July in the United States and from July to August in Alaska, the flowering period appears as July {May–August}. However, it is important to remember that this information is only a general guide. Phenology can vary from year to year and from site to site. Spring may arrive early one year and late the next. In years with ideal conditions, plants may flower profusely, but during droughts or unusually short summers, the same plants may not bloom at all. Similarly, plants might bloom in June on sheltered, south-facing hillsides at low elevations and in August on exposed alpine slopes.

The fourth and final descriptive section, 'Fruits' (or 'Cones'), describes the type, colour, shape, size and texture of the fruits and seeds. A description of their arrangement may follow, but only if this is notably different from that of the flower clusters.

Introduction

Erigeron radicatus

Aristida purpurea

Each species account is accompanied by a line drawing to illustrate the habit (the appearance of the whole plant) and small details that may help with identification. Whenever possible, a colour photograph is also included. If available, the order of the illustrative material is as follows: photograph(s) of primary species, Alberta and North American distribution maps of primary species, line drawing(s) of primary species, photograph(s) of secondary species, line drawing(s) of secondary species, Alberta distribution map of secondary species. If no photographs were available, line drawings are substituted in their place.

Habitat

A description of the general habitat for the species follows the plant description. Alberta information is presented first, but if the species has been found in different habitats outside the province, this information is included at the end of the section as habitats 'elsewhere.' The habitat notes may help you to determine whether you are looking at the plant described. For example, if the habitat is described as 'dry, rocky, exposed ridges' and you are in a moist, shady forest, chances are that you are looking at a different species. However, once again, the notes are only a general guide.

Distribution

Two maps showing species distributions are presented: one for Alberta and one for North America. The records from Alberta are plotted using two symbols indicating the time period during which the collection was made, as follows: ○ (open circles) prior to 1950 and • (closed circles) 1950 to the present. If more than one collection exists for a given locality, only the most recent collection is mapped. Not all species have an Alberta distribution map; in some cases, the information supplied with the collection was not specific enough to be able to locate a dot on a map. The Alberta Natural Heritage Information Centre provided the Alberta maps in 1998–99. They are meant to help you to compare the location of your find with other records from the province. Perhaps you will discover that you have found a new location for a plant, or perhaps the Alberta distribution will suggest that you have misidentified your specimen. John Kartesz at the University of North Carolina provided the North American distribution maps (unless otherwise noted). If the species has been found in a state, province or territory but is not considered rare there, that political

Introduction

■ Extirpated
■ Rare
■ Not rare
▨ Status uncertain

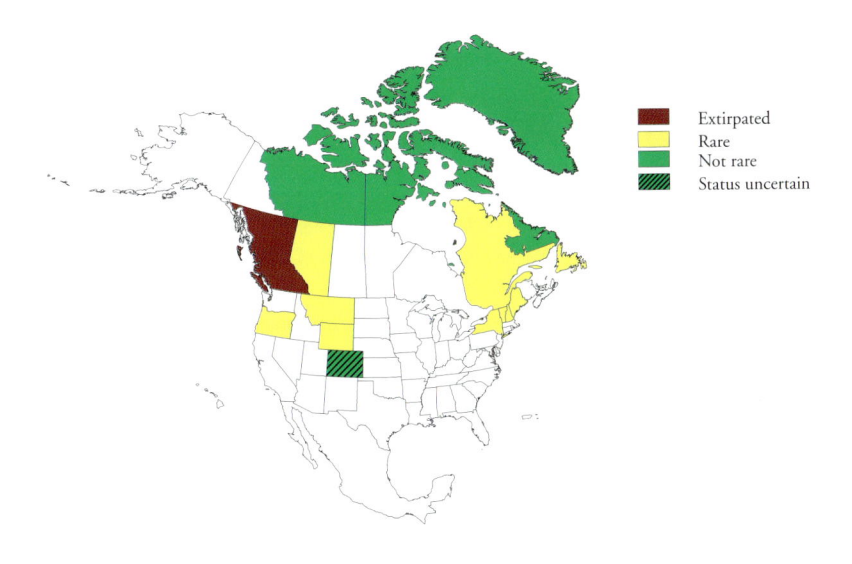

jurisdiction is shaded green. If the species has been found in a state province or territory and is considered rare there, that political jurisdiction is shaded yellow. If the species has been extirpated (no longer occurs) in that state, province or territory, that political jurisdiction is shaded red. Diagonal lines indicate that the status shown for the species in that particular political jurisdiction is uncertain. The North American maps are very helpful in identifying widespread species that are rare in Canada only because their ranges extend just across the Alberta border from the north, south, east or west.

NOTES

This section encompasses a wide range of topics, including synonyms, other rare species, similar or related common species, ethnobotanical information, natural history notes and etymology (name derivation). The type and amount of information available varies greatly with the species.

The notes often begin with other names for the species (synonyms) that have appeared recently in published papers and floras. Plant names often change as more is discovered about the relationships between different groups. Two species may be lumped into one, two varieties may be recognized as separate species, or two species may be reclassified in separate genera. Many groups have been revised recently, several in association with work on the *Flora of North America*, and these revisions have been

Introduction

Leymus mollis

incorporated into the text. For example, when the name in the *Flora of Alberta* is different from that in the *Flora of North America*, the *Flora of North America* name appears at the top of the account and the *Flora of Alberta* name appears in the notes section.

The 'Notes' section sometimes includes descriptions of similar or related plants that are also rare in Alberta. These descriptions are not as detailed as those presented for the primary species. Instead, they focus on key characteristics that can be used to distinguish the species from their close relatives. Each description is followed by a general outline of the species' habitat and flowering period, and is accompanied by illustrations and distribution maps. The 'Notes' section may also include information about common plants that could be confused with the rare species, highlighting the characteristics that distinguish the two.

In some cases, information about historical uses (aboriginal and/or European), edibility, toxicity, natural history and plant lore is also presented, but the type and amount of information available for rare species is usually limited. Information about historical uses of plants comes from areas outside Alberta, where the species is common. It is presented to give the reader a better sense of our rich natural heritage and is *not* meant to encourage the gathering of rare plants in Alberta.

Rare Vascular Plants

# of Alberta

Trees & Shrubs

CEDAR FAMILY   **Trees and Shrubs**

WESTERN RED CEDAR

Thuja plicata Donn *ex* D. Don
CEDAR FAMILY (CUPRESSACEAE)

Plants: Coniferous trees to 40 {75} m tall; **trunks straight, vertically lined, grey, buttressed at base**, 1–3 {5} m in diameter; **branches spreading to hanging** but upturned at tips; crown cone-shaped when young, irregular with age; bark thin, cinnamon to greyish, fissured, easily peeled off in long, fibrous strips.

Leaves: Opposite, evergreen, **scale-like**, 3 {1–6} mm long, pointed, overlapping, **shiny yellowish green**, aromatic; arranged in 4 rows with upper–lower pair **flattened** and side pair **folded, in flat, fan-like sprays**.

Cones: Male and female cones on same tree in clusters at branch tips; pollen (male) cones reddish, almost round, 2 {1–3} mm long; seed (female) cones narrowly egg-shaped, 8–12 {14} mm long, with **6–8 dry, brown scales**; seeds reddish brown, with 2 membranous wings; {April–May}.

Habitat: Rich, moist to wet (often saturated) foothill and montane slopes.

Notes: Aboriginal peoples twisted, wove and plaited the long, soft strings of inner bark to make a wide range of items, including baskets, blankets, clothing, ropes and mats. The trunks were used to make dugout canoes and rafts, and frames for birch-bark canoes. Fine roots were split, peeled and used to make water-tight baskets for cooking. Larger roots were formed into coils which were alternately bound together in pairs using the split roots. Bark was woven into baskets where air could circulate around stored berries, or whole sheets were formed into containers. The light, easily worked wood splits easily and resists decay. It was used by many tribes to make cradle boards, bowls, roofing, siding and a variety of implements. Today, it is widely used for siding, roofing, panelling, doors, patio furniture, chests and caskets. Western red cedar has been heavily logged for its lumber, and it is disappearing from parts of its range where trees are not replanted. • Western red cedar is very shade tolerant, but it has a low resistance to drought and frost. It indicates relatively wet regions in the Rockies. In BC, this tree can reach over 1,000 years of age. • The specific epithet *plicata* means 'folded,' in reference to the flattened, overlapping leaves.

Trees and Shrubs   PINE FAMILY

WESTERN LARCH

Larix occidentalis Nutt.
PINE FAMILY (Pinaceae)

Plants: Slender coniferous trees, usually 25–50 m tall; trunks straight, usually 1–1.3 {2} m in diameter; **branches short, well spaced with sparse leaves**; twigs orange-brown, hairy at first, but usually hairless by the end of the first year; **bark cinnamon-brown**, becoming 10–15 cm thick, deeply furrowed with **large, flaky scales**; crown short and cone-shaped.

Leaves: Tufts of 15–30 soft, **deciduous needles on stubby twigs**, 2.5–5 cm long, pale yellow-green in spring and summer, bright yellow in fall, shed for winter.

Cones: Male and female cones on same tree; pollen (male) cones yellow, about 1 cm long; seed (female) cones brown, egg-shaped, **2–3 cm long**, with **3-pointed bracts** projecting beyond the scales, borne on curved stalks; cone **scales wider than they are long**, smooth-edged, **woolly below when young**; seeds reddish brown, 3 mm long, with a 6-mm wing; {May–June}, mature in a single season.

Habitat: Moist to dry, often gravelly or sandy sites in upper foothill and montane zones; elsewhere, often with ponderosa pine and Douglas-fir.

Notes: Some tribes hollowed out cavities in western larch trunks and allowed the sweet syrup to accumulate. This sugary liquid could be concentrated by evaporation to make it even sweeter, but some warned that eating too much 'cleans you out.' The sweet inner bark was eaten in spring, and lumps of pitch were chewed like gum year-round. The sap and gum contain galatin, a natural sugar with a flavour like slightly bitter honey. Dried larch gum was also ground and used as baking powder. • The wood was used for making bowls, and rotted logs were collected for smoking buckskins. Today, it is often used in heavy construction, and it also makes good firewood. Arabino galatin, a water-soluble gum, is extracted from the bark for use in paint, ink and medicines. • Mature western larch trees have thick bark and few lower limbs, and as a result they are able to survive low-intensity fires, sometimes living 700 to 900 years. Larch seeds germinate readily and young trees grow rapidly on fire-blackened soils, but fire suppression and selective cutting have reduced populations in many areas.

PINE FAMILY

Trees and Shrubs

WESTERN WHITE PINE, SILVER PINE

Pinus monticola Dougl. *ex* D. Don.
PINE FAMILY (Pinaceae)

Plants: Evergreen trees usually less than 30 m tall, but sometimes up to 50 m; trunks straight, 1 m or more in diameter; young bark silvery-grey, thin, smooth; mature bark greyish, furrowed, with small, rectangular, flaky scales, cinnamon-brown under the outer scales; **branches symmetrical, arranged in circles** (whorled); relatively wind-firm, with spreading, shallow roots and a few deep roots; crowns symmetrical, cone-shaped, short-branched.

Leaves: Bundles of 5 very slender, flexible evergreen needles, {3} **5–10 cm long**, blue-green with whitish lines, blunt or slightly pointed (not sharp-pointed), **finely toothed** along the edges.

Cones: Male and female cones on same tree; pollen (male) cones yellow, scarcely 1 cm long, clustered; **seed (female) cones yellow-brown** (green to yellowish or purplish when young), **cylindrical**, **slightly curved, 10–25 cm long**, 6–9 cm thick, pitchy, **hanging on stalks** at upper branch tips; cone scales thin, spreading, deep red to chocolate-brown at the base, blunt at the yellowish brown tips, lacking prickles; seeds reddish brown, 7–10 mm long, with a large, narrow wing 2–3 times as long as the seed, widest above the middle and pointed at the tip; {May–June}, mature in the fall of the second year.

Habitat: Open rocky slopes in the mountains; elsewhere, on moist, north-facing mountain slopes, in mixed forest or pure stands.

Notes: Other 5-needle pines in Alberta have smooth-edged needles about 12.5 cm long and egg-shaped, short-stalked cones with thick scales. • This species was named by David Douglas in 1825, during his exploration of the west coast of North America. • The bark was boiled to make medicinal teas for treating stomach disorders and purifying the blood. • Western white pine trees are very susceptible to white pine blister rust, a fungus that was introduced from Europe. • Western white pine requires more than 500 mm of precipitation annually and a precipitation/evaporation ratio of greater than 1. Low relative humidity and rainfall are doubtless limiting factors in Alberta. • The wood is easily worked and is good for carving and construction.

Rare *Vascular* Plants of Alberta

Trees and Shrubs

PINE FAMILY

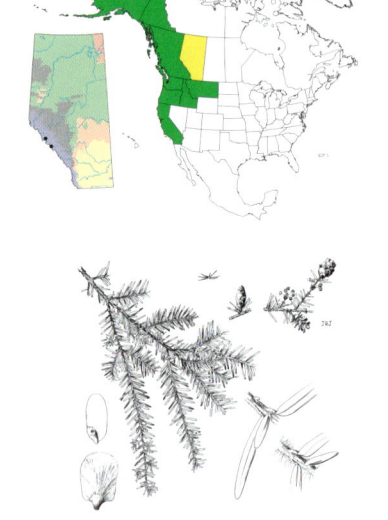

WESTERN HEMLOCK

Tsuga heterophylla (Raf.) Sarg.
PINE FAMILY (Pinaceae)

Plants: Coniferous evergreen trees, usually 30–50 m tall, sometimes taller; buds 2.5–3.5 mm long; trunks straight, up to 1 {2} m in diameter; branches slender, down-swept, appearing feathery; crown open, narrowly cone-shaped, with a **flexible, nodding tip**; bark yellowish brown and finely hairy when young, becoming dark greyish brown with age and developing deep furrows and flat ridges; **inner bark dark orange with purple streaks**.

Leaves: Flat evergreen **needles, 5–20 {30} mm long**, of **unequal lengths**, blunt-tipped, shiny yellowish green above, whitish beneath with 2 broad bands of stomata, borne on small stubs, **held at right angles to branches in 2 opposite rows**.

Cones: Male and female cones on same tree; pollen (male) cones 3–4 mm long, yellowish; seed (female) cones oblong to **egg-shaped, 15–25 {10–30} mm long, hanging**, purplish green when young, light brown when mature; scales thin, wavy-edged; mature in a single year and shed intact.

Habitat: Moist, shady montane forests; elsewhere, at lower elevations in moist woodlands mixed with hardwoods.

Notes: Western hemlock is very shade tolerant. • The boughs were used for bedding and as a disinfectant and deodorizer. The wood is hard and strong, with an even grain and colour. It is widely used to make doors, windows, staircases, mouldings, cupboards and floors. It also provides construction lumber, pilings, poles, railway ties, pulp, and alpha cellulose for making paper, cellophane and rayon.
• This attractive, feathery tree is a popular ornamental, and several cultivars have been developed. • Western hemlock is also known as Pacific hemlock and west coast hemlock. The generic name *Tsuga* is a Japanese name that has been adopted as a scientific 'Latin' name. The specific epithet *heterophylla*, from the Latin *hetero* (variable) and *phyllum* (leaf), refers to the variably sized needles.

YEW FAMILY Trees and Shrubs

WESTERN YEW

Taxus brevifolia Nutt.
YEW FAMILY (TAXACEAE)

Plants: Evergreen shrubs or small trees, 10–15 m tall; trunks often twisted; **bark thin, scaly, red-brown or purplish**, sloughing off to expose rose-coloured underbark; branches spreading or drooping; **young twigs** dark green or olive-green, **hairless**.

Leaves: Flat, needle-like, 1–2 cm long, 1–2 mm wide, shiny, dark green on upper surface, yellowish green beneath, **tipped with a tiny spine**; arranged spirally along the branches but **twisted into 2 opposite rows**, giving the branches a flattened appearance.

Flowers: Male (staminate) and female (pistillate), from leaf axils, usually on separate trees; male cones inconspicuous, tiny (2 mm wide), tipped by a yellow lobed cluster of stamen-like structures on a short stalk; female cones with single ovules in green, cup-like structures on short, scale-covered stalks; {April–June}.

Fruits: Red, fleshy, cupped 'berries' 4–5 mm broad, with single hard seeds; seeds brown or dark blue, egg-shaped, 5–6 mm long.

Habitat: Moist woods, under the canopy of larger conifers.

Notes: The seeds, bark and leaves are **highly poisonous**, containing taxine, a potent alkaloid. • Yew bark is the source of the drug taxol, used to treat ovarian and breast cancers. • The flexible, strong wood was used by native people to make bows, canoe paddles, clubs, combs, shovels, harpoons, sewing needles, fish hooks, knives, spoons and snowshoe frames. The berries were sometimes eaten in small quantities as a contraceptive. • At high elevations, western yew is reduced to a low, spreading shrub. It does not form pure stands in forests. • The berries are eaten by a wide variety of birds, and the needles are browsed by moose in winter. • *Taxus* is from the Greek name for yews, which was perhaps derived from *taxon* (a bow), a common use for yew wood in earlier times. *Brevifolia* is Latin for 'short-leaved.'

Trees and Shrubs

WILLOW FAMILY

NARROW-LEAVED COTTONWOOD

Populus angustifolia James
WILLOW FAMILY (Salicaceae)

Plants: Slender **deciduous** trees, up to 12 {20} m tall; branches directed upward; twigs orange-brown; **buds sticky, aromatic**, hairless, covered by **several scales**; crown nearly cone-shaped; bark whitish green to yellowish green, smooth when young, slightly furrowed with low ridges on lower trunk when mature.

Leaves: Alternate, narrowly to broadly lance-shaped, **3–5 times as long as they are wide**, 5–8 {4–11} cm long, **gradually tapered to a pointed tip**, wedge-shaped to rounded at base, finely round-toothed; **stalks** slightly flattened or channelled on the upper side and rounded beneath, **short**, up to ⅓ as long as the blade.

Flowers: Tiny, in the axils of conspicuously fringed scales that are soon dropped, forming long, slender, **loosely hanging clusters** (catkins); male and female catkins on separate trees; male (staminate) catkins 2–3 cm long, with about 12–18 stamens per flower; **female** (pistillate) **catkins 4–10 cm long**; {April} May.

Fruits: Narrowly egg-shaped, wrinkled capsules, hairless, 5–6 mm long, on 1–2 mm stalks, in **hanging catkins**, splitting into 2 parts when mature; seeds minute, with a tuft of soft, white hairs at tip, often dispersed in **large fluffy masses**.

Habitat: River valleys and streambanks.

Notes: Typical narrow-leaved cottonwood is seldom found in Alberta, because it hybridizes readily with black cottonwood (*Populus balsamifera* L. ssp. *trichocarpa* (T. & G. *ex* Hook.) Brayshaw), balsam poplar (*Populus balsamifera* L. ssp. *balsamifera*) and plains cottonwood (*Populus deltoides* Marsh), producing hybrids that resemble both parents and that are difficult to identify. Its bright white branches and twigs distinguish it from other trees during winter. • At first glance, this small, slender tree looks more like a large willow (*Salix* spp.) than a cottonwood, but willows have only a single scale covering each bud, their male flowers have few (often 2) anthers, and their female catkins tend to be shorter, rarely hanging. • Cottonwood trees produce millions of tiny, fragile seeds with fluffy tufts of hairs that carry them on the wind.

Rare *Vascular* Plants of Alberta

WILLOW FAMILY

Trees and Shrubs

Raup's willow

Salix raupii Argus
WILLOW FAMILY (SALICACEAE)

Plants: Erect shrubs, 1–2 m tall; **branches hairless**, glossy, chestnut-brown when young, greyish brown with flaky bark by the third year.

Leaves: Alternate; **blades widest at or above the middle, 3.2–5.8 cm long, 1.2–1.9 cm wide**, blunt to pointed at tip, rounded to pointed at base, **lacking teeth or with inconspicuous, rounded teeth**, **hairless** (sometimes with hairs at the edges when young), **bluish grey with a thin, waxy coat on the lower surface; stalks yellowish**, 5–9 mm long, hairless; stipules narrow, 1–3 mm long.

Flowers: Tiny, in the axils of **pale lemon-yellow to bicoloured** (yellow with pink or brownish tips), 1.6–2.5 mm long, **scale-like bracts**, forming dense, elongating clusters (catkins), with male and female flowers on separate plants; male (staminate) catkins 2.5–3.5 cm long, on leafy branchlets 8–12 mm long; female (pistillate) catkins 2–3.5 cm long, on leafy branchlets 5–18 mm long; **styles 0.6–0.8 mm long**, stigmas 0.4 mm long; **in spring, with the leaves**.

Fruits: Greenish brown capsules, **hairless** or with 2 longitudinal bands of short, fine hairs, **4.4–8 mm long, on short, hairy stalks**.

Habitat: Lush mountain meadows and willow fens.

Notes: This species is critically imperilled due to extreme rarity in all parts of its range. • Raup's willow is very similar to extreme variants of Barclay's willow (*Salix barclayi* Anderss.) and Farr's willow (*Salix farriae* Ball). Barclay's willow is distinguished by its soft-hairy branches, distinctly toothed leaves, long stipules (up to 2 cm long) and hairy floral bracts. The young leaves of Farr's willow often have white and reddish hairs on the midvein; the capsules tend to be smaller (4–5 mm) and the bracts darker (brown to tawny) than those of Raup's willow. Raup's willow may also be confused with bog willow (*Salix pedicellaris* Pursh), but bog willow is easily distinguished by its firmer, narrower leaves, which are bluish grey with a thin, waxy coating on both surfaces, and by its shorter styles and stigmas (0.1–0.2 mm).

Rare *Vascular* Plants of Alberta

Trees and Shrubs

WILLOW FAMILY

WOOLLY WILLOW, LIME WILLOW

Salix lanata L. ssp. *calcicola* (Fern. & Wieg.) Hult.
WILLOW FAMILY (SALICACEAE)

Plants: Erect, low shrubs, up to 50 cm tall; branches stout, gnarled, reddish brown, **covered with long, soft hairs when young**, slightly hairy when mature.

Leaves: Alternate; **blades broadly elliptic to egg-shaped**, widest at or below the middle, 2–5 cm long, 1.5–3.5 cm wide, abruptly pointed at the tip, usually rounded at the base, smooth or with fine, gland-tipped teeth along the edges, sparsely long-hairy to almost hairless, bluish grey with a thin, waxy coat on the lower surface and with a broad, conspicuous midvein; stalks short; **stipules prominent**, rounded or sometimes long-tapering, dotted **with tiny glands** on edges and upper surface, **persisting** after the growing season; **overwintering buds large** and rounded, with long, soft hairs.

Flowers: Tiny, on very short stalks in the axils of dark, scale-like bracts with long, soft hairs, forming dense, elongating clusters (catkins), with male and female flowers on separate plants; **female** (pistillate) **catkins stout, 5–6 cm long, stalkless**; styles up to 2.5 mm long; in spring, before the leaves.

Fruits: Hairless, reddish brown capsules, 7–9 mm long.

Habitat: Calcareous riverbanks, floodplains and meadows.

Notes: This species has also been called *Salix calcicola* Fern. & Wieg. • The specific epithet *lanata*, from the Latin *lana* (wool), refers to the woolly leaves; *calcicola* refers to the calcareous habitat.

Rare *Vascular* Plants of Alberta

Alaska willow, felt-leaved willow

Salix alaxensis (Anderss.) Coville ssp. *alaxensis*
WILLOW FAMILY (SALICACEAE)

Plants: Gnarled to erect, deciduous shrubs or small trees, 1–3.7 {9} m tall; branches reddish brown, sometimes with persistent greyish hairs, conspicuously white-woolly or grey-woolly when young.

Leaves: Alternate, **elliptic to egg-shaped**, widest above middle, 4–10 cm long, 2–2.5 cm wide, pointed at tip, narrowly wedge-shaped at base, dull green and sparsely hairy above, **densely white-woolly beneath**; edges rolled under, toothless or glandular and wavy; stalks 3–15 mm long, yellowish, sometimes enlarged around the bud, with 2 persistent, linear bracts (stipules) at the base.

Flowers: Tiny, in the axils of dark brown to black, silky-hairy scales, forming **dense, stalkless clusters** (catkins), with male and female on separate plants; male (staminate) catkins about 4 cm long, with 2 anthers per flower; female (pistillate) catkins 6–10 {12} cm long, with green, hairy pistils; {May} June, before the leaves.

Fruits: Sparsely hairy capsules, 4–5 mm long, with stalks less than 1 mm long; seeds minute, tipped with a tuft of silky hairs that allows them to be carried by the wind.

Habitat: Alpine slopes; elsewhere, on gravel bars and river terraces, on glacial moraines and in young forests.

Notes: Changeable willow (*Salix commutata* Bebb) is another rare Alberta willow. It is smaller (0.2–2 m tall) than Alaska willow, but its twigs are often densely white-woolly. Changeable willow is most easily identified by its leaves, which have rounded teeth along their edges, green lower surfaces (not coated with a whitish bloom, as is the case with most willows), and silky to woolly hairs on both sides (especially when young). It forms thickets in subalpine areas and produces nearly hairless, tawny, 3.5–7 cm long female catkins in July {August–September}, with or following leaf growth. • Hoary willow (*Salix candida* Fluegge *ex* Willd.) is a much more common, more widespread species that also has white, woolly hairs on its lower leaf surfaces. Hoary willow is distinguished by its narrower (0.5–1.5 cm wide) leaves and its leafy-stalked catkins which appear with the leaves.

S. COMMUTATA

Trees and Shrubs

WILLOW FAMILY

SITKA WILLOW

Salix sitchensis Sanson *ex* Bong.
WILLOW FAMILY (SALICACEAE)

Plants: Shrubs or small trees, 1–6 m tall, sometimes lying on the ground in exposed areas but otherwise erect; branches **brittle**, smooth, dark reddish brown and **densely velvety-hairy when young**, becoming grey, furrowed and scaly with age.

Leaves: Alternate; blades elliptic to narrowly egg-shaped, generally widest toward the tip, 4–9 {18} cm long, 1.5–3.5 {6} cm wide, short-pointed or rounded at the tip, tapered to a narrow base, toothed; upper surface green, with greyish, flat-lying hairs; **lower surface pale and satiny, with silky hairs** and evident veins; stalks yellowish, velvety, 5–15 {20} mm long; stipules variable, small and soon shed or well developed, leaf-like and persistent.

Flowers: Tiny, in the axils of brown, dark-tipped, scale-like bracts with long, soft hairs; forming dense, elongating clusters (catkins), with male and female flowers on separate plants; male (staminate) catkins 2.5–5 cm long, 1–1.5 cm thick on 2–10 {20} mm long leafy stalks, **single stamen** per floret; **female** (pistillate) **catkins 3–9 cm long** when mature, on **short, leafy stalks**; styles 0.4–0.8 {0.3–1.2} mm long, stigmas 0.1–0.3 mm long; May, usually as the leaves appear (rarely before).

Fruits: Tawny, **silky-hairy capsules**, {1} 3–5.5 mm long, on short (0.5–1.4 mm long) stalks.

Habitat: Alluvial soil on the Athabasca River flood plain.

Notes: The common, widespread species Scouler's willow (*Salix scouleriana* Barr. *ex* Hook.) has a hairy form that resembles Sitka willow, but its styles are shorter (0.2–0.6 mm), its stigmas are longer (0.2–1 mm), its leaves have a bluish grey, thin, waxy coating on their lower surface, and its twigs are not brittle. • The satiny sheen on the lower surface of the leaves is diagnostic of Sitka willow in Alaska. • The wood has been used by native people in drying fish, because its smoke does not have a bad odour. The pounded bark has been applied to heal wounds. • *Sitchensis* means 'of Sitka,' Alaska, near which it was collected by Karl Heinrich Mertens in 1827.

WILLOW FAMILY   **Trees and Shrubs**

STOLONIFEROUS WILLOW

Salix stolonifera Coville
WILLOW FAMILY (SALICACEAE)

Plants: Dwarf shrubs; branches rooting at joints (nodes), trailing on the ground or spreading underground, reddish brown, hairless and glossy, sometimes with a waxy, bluish grey coating.

Leaves: Alternate; **blades broad**, widest at or above the middle, **1–2.5 cm long, 0.5–2 cm wide**, rounded at the tip, pointed to rounded at the base, usually without teeth, **hairless or sparsely hairy**, green on the upper surface, bluish grey with a thin, **waxy coating on the lower surface**; stalks slender; stipules tiny.

Flowers: Tiny, in the axils of dark brown, sparsely hairy, scale-like bracts, forming dense, elongating clusters (catkins), with male and female flowers on separate plants; female (pistillate) **catkins 1.5–4 cm long on long** (1–6 cm), **leafy branchlets**; styles about 0.8–1.4 mm long; in spring, when the leaves appear.

Fruits: Smooth, essentially hairless capsules, usually **reddish purple** (sometimes greenish or whitish) **when young**, becoming brown with age, 5–7 mm long, short-stalked.

Habitat: Wet hummocks at alpine elevations.

Notes: This species hybridizes with arctic willow (*Salix arctica* Pallas) and Barclay's willow (*Salix barclayi* Anderss.) in BC. Arctic willow is distinguished by its silky-woolly pistils, and Barclay's willow is usually a larger (1–3 m tall), erect shrub with longer (3–7 cm long) leaves and catkins (4–7 cm long) and obvious stipules. • Snow willow (*Salix reticulata* L.) is another dwarf willow, but it has shorter pistils (4–5 mm long) and styles (0.2–0.3 mm long), and its leaves have distinctive netted veins. • Smooth willow (*Salix glauca* L.) is occasionally of a dwarf stature, but its pistils have long, soft or woolly hairs, its bracts have short, wavy hairs, and its branchlets are hairy. • *Stolonifera* is Latin for 'bearing stolons,' referring to the horizontal, rooting stems (stolons) of this species.

Rare *Vascular* Plants of Alberta

SHADSCALE, SALTBUSH

Atriplex canescens (Pursh) Nutt. var. *aptera* (A. Nels.) C.L. Hitchc.
GOOSEFOOT FAMILY (CHENOPODIACEAE)

Plants: Low, somewhat rounded **shrubs**, 10–40 cm tall; branches numerous, spreading, white-scaly when young, eventually smooth and shedding outer layers of bark.

Leaves: Alternate, or the lower ones opposite; **blades** variable, elliptic to spatula-shaped or linear, 1–5 cm long, **greyish** with a covering of fine, **bran-like particles, veins in a netted pattern visible under magnification** (kranz-type venation), toothless; stalks short or absent.

Flowers: Small, lacking petals, with male and female flowers on separate plants; male (staminate) flowers tawny (sometimes reddish-tinged), with a {3} 5-parted calyx densely covered in fine scales, numerous, borne in many narrow, elongated, dense (sometimes interrupted) clusters (spikes) that usually form leafy, branched clusters (panicles); female (pistillate) flowers lacking a calyx, but with a fused **pair of bracts completely enclosing the ovary**, 4–6 {8} mm long in flower (up to 20 {25} mm long in fruit), **prominently 4-winged**, the wings broader than the body, wavy-edged to irregularly toothed, few to many flowers in less congested spikes and panicles; {June–August}.

Fruits: Single, vertical seeds, 1.5–2.5 mm long, enclosed by enlarged, winged bracts.

Habitat: Saline flats, sometimes on sandy soil.

Notes: This variety has been considered a species, *Atriplex aptera* A. Nels., by some taxonomists. • Shadscale is a valuable browse species for wildlife and livestock in arid regions; however, it is **mildly poisonous**. It can tolerate high concentrations of salt, and because of this it is being studied for revegetation of disturbed sites in dry areas. Shadscale is known to form mycorrhizal relationships with soil fungi.

HYDRANGEA FAMILY

Trees and Shrubs

LEWIS' MOCK-ORANGE

Philadelphus lewisii Pursh
HYDRANGEA FAMILY (Hydrangeaceae)

Plants: Erect, deciduous shrubs, 1–3 m tall; **branches opposite**, loose, spreading, with brown, flaky bark; twigs reddish and hairless.

Leaves: Opposite, broadly lance-shaped, 2–6 {7} cm long, with **3–5 main veins** from the base, **slightly rough** to touch, hairless to somewhat hairy (especially on lower side of veins), **fringed** with short, curved hairs, short-stalked.

Flowers: White, about 3 cm across, cross-shaped, with both male and female parts (perfect); sepals fused to ovary for most of their length; 4 or 5 broad, rounded or notched, spreading petals surround a cluster of 25–40 bright yellow stamens and 4 styles; 3–11 flowers in **showy, fragrant clusters** (racemes) at branch tips; {May} July–August.

Fruits: Woody, oval capsules, 6–10 mm long, splitting into 4 parts (valves), with many tiny, rod-shaped seeds.

Habitat: Well-drained sites in montane forests; elsewhere, on rocky hillsides with sagebrush or ponderosa pine.

Notes: Lewis' mock-orange is the state flower of Idaho. • The stiff, hard wood has been used to make combs, knitting needles, rims for birch-bark baskets and cradle hoods. Dried, powdered leaves or charcoal from burned branches was mixed with pitch or bear grease to make salves for treating sores and swellings. Branches were boiled to make a medicinal tea for treating sore chests and for washing or soaking eczema and bleeding hemorrhoids. • Mock-orange leaves were mashed and used as soap. • Deer and elk browse on the branches of Lewis' mock-orange, but it is not a preferred food. • The specific epithet *lewisii* honours Captain Meriwether Lewis, of the Lewis and Clark Expedition, who first collected this plant in Idaho.

Trees and Shrubs

CURRANT FAMILY

MOUNTAIN CURRANT, TRAILING BLACK CURRANT

Ribes laxiflorum Pursh
CURRANT FAMILY (GROSSULARIACEAE)

Plants: Sprawling shrubs up to 1 m tall; branches usually **spreading** or running along the ground, grey or brown, without prickles (unarmed).

Leaves: Alternate, maple-leaf-shaped, with a **strong odour** when crushed; blades 6–8 cm long and 7–10 {2.5–12} cm wide, heart-shaped at base and tipped with **5–7 toothed, triangular lobes**, hairless on the upper surface, light green and slightly hairy with small, yellow glands near base on the lower surface; stalks 5–7.5 cm long.

Flowers: Red to purple, saucer-shaped, about 6–8 mm across, with both male and female parts (5 stamens; 2 styles); 5 triangular, 3–4 mm long petals; 5 egg-shaped to rounded, {2.5} 3–4 mm long, **hairy sepals, glandular-hairy on the ovary and stalk**; {6} 10–20 flowers in erect elongated clusters (racemes) 10–14 cm long; {May} June, as the leaves expand.

Fruits: Purplish black berries with a whitish, waxy covering (bloom), **short-stalked** (less than 0.5 mm long), **glandular-hairy**, {8} 12–15 mm in diameter, unpleasant-smelling when crushed; {late July–early August}.

Habitat: Moist subalpine woods; elsewhere, also in disturbed areas such as open meadows and logged areas.

Notes: A much more common species, skunk currant (*Ribes glandulosum* Grauer), resembles mountain currant and also grows in moist woods and openings. It is distinguished by its red fruit, which is covered with long-stalked glands (1.2 mm long), and by its shorter (2–2.5 mm), essentially hairless sepals. • The generic name *Ribes* comes from the Arabic or Persian *ribas* (acid-tasting), in reference to the acidic fruit of many species. *Laxiflorum* means 'with loose flowers,' in reference to the open flower clusters.

Rare *Vascular* Plants of Alberta

GINSENG FAMILY

Trees and Shrubs

DEVIL'S CLUB

Oplopanax horridus (Sm.) T. & G. *ex* Miq. var. *horridus*
GINSENG FAMILY (ARALIACEAE)

Plants: Strong-smelling, spiny, coarse deciduous shrubs, 1–3 m tall; stems erect to sprawling, thick, **armed with many long spines**, crooked, **seldom branched**, often entangled.

Leaves: Alternate, **broadly maple-leaf-shaped with 5–7 toothed lobes**, 10–30 {40} **cm wide**, prickly on the ribs beneath, **spreading on long, bristly stalks near stem tips**.

Flowers: Greenish white, 5–6 mm long, with both male and female parts (5 stamens; 2 styles), with 5 petals and 5 sepals, nearly stalkless, in dense heads (umbels) forming 10–30 cm long, branched, pyramid-shaped clusters (panicles) at stem tips; {May} June–August.

Fruits: Bright red, slightly flattened, berry-like drupes, 4–5 {6} mm long, 2-seeded, in **showy clusters**.

Habitat: Moist to wet, shady sites.

Notes: The berries are **not edible**, but the young, fleshy stems have been eaten. • Some people have an **allergic reaction** to the spines of these plants. The spines break off easily and can cause infections when they remain embedded in the skin. • The wood has a distinctive, sweet smell. The aromatic roots were mixed with tobacco and smoked to relieve headaches. • The stems have been used medicinally. Dried pieces, scraped free of prickles, were steeped in boiling water to make medicinal teas for treating flu and other illnesses, relieving stomach troubles, improving appetite and aiding weight gain. Stem sections were boiled to make a tonic and blood purifier with laxative effects. Root tea was used to treat diabetes. Dried stems were burned and the ashes were mixed with grease to make a salve for healing swellings and running sores. • These large, attractive plants are sometimes grown as ornamentals in gardens, but they grow slowly, and gardeners must be careful to avoid their dangerous spines. • The generic name *Oplopanax* was derived from the Greek *hoplon* (weapon) and *panax* (ginseng), a closely related, large-leaved plant.

Rare *Vascular* Plants of Alberta

ROSE FAMILY

MALLOW-LEAVED NINEBARK

Physocarpus malvaceus (Greene) Kuntze
ROSE FAMILY (ROSACEAE)

Plants: Deciduous shrubs, {0.5} 1–2 m tall, with **star-shaped hairs**; branches arching, with **brown bark shredding** in strips.

Leaves: Alternate, **somewhat maple-leaf-shaped**, 3–5 lobed, about **2–6 cm long** and wide, edged with round teeth, **strongly veined, shiny and dark green above**, paler beneath, hairless or with branched hairs.

Flowers: White, saucer-shaped, about 1 cm across, with both male and female parts (many stamens; 2 styles), with 5 rounded petals and 5 persistent sepals, several, in rounded clusters (corymbs) with star-shaped and woolly hairs, borne at branch tips; {June} July–August.

Fruits: Reddish, **egg-shaped pairs of flattened, keeled, pod-like fruits** (follicles), with erect styles, joined on lower half, **fuzzy**, about 5 mm long, usually containing 2 seeds per pod.

Habitat: Dry, open or lightly wooded montane slopes and ravines; also in canyon bottoms and thickets south of Alberta.

Notes: The Okanagan used mallow-leaved ninebark as a good-luck charm to protect their hunting equipment. • Mallow-leaved ninebark makes an attractive garden addition. It is best grown from cuttings. Seeds should be sown in the autumn, but they are slow to become established. • The common name 'ninebark' was given to these shrubs because they were believed to have 9 layers of shredding bark. The generic name *Physocarpus* was taken from the Greek *physa* (bladder) and *carpos* (fruit), because of the inflated pods of most species.

PINK MEADOWSWEET

Spiraea splendens Baumann *ex* K. Koch var. *splendens*
ROSE FAMILY (ROSACEAE)

Plants: Slender shrubs, 0.5–1.5 m high with ascending, densely clustered, reddish brown branches.

Leaves: Alternate, short-stalked, simple; blades **egg-shaped or elliptic**, 1.5–4 cm long, **rounded at both ends, toothed toward the tip**, bright green above, paler beneath, essentially hairless.

Flowers: Pink or rose-coloured, sweet-scented, with both male and female parts (15–many stamens; 5 styles); **5 petals 1.5–2 mm long**; 5 sepals, egg-shaped, blunt-tipped; many flowers in **dense, flat-topped clusters 2–5 cm across**; {June–August}.

Fruits: Pod-like fruits that split open down 1 side (follicles), firm, shiny, **2.5–3 mm long, in erect groups of 4 or 5**, each usually containing 4 narrow, tapered seeds.

Habitat: Moist woods, wet meadows and boggy areas near timberline.

Notes: This species has also been called *Spiraea densiflora* Nutt. *ex* T. & G. • White meadowsweet (*Spiraea betulifolia* Pallas) is a much more common species. It also has flat-topped flower clusters, but its flowers are white and its leaves are larger (2–8 cm long) with larger teeth (2 mm vs. 1 mm or less). • The generic name *Spiraea* is from the Greek *speiraira*, a plant used in wreaths and garlands. Some garden varieties of *Spiraea* are called 'bridal-wreath.'

Trees and Shrubs

HEATH FAMILY

LAPLAND ROSE-BAY

Rhododendron lapponicum (L.) Wahl.
HEATH FAMILY (ERICACEAE)

Plants: Small shrubs, matted and 5–25 cm tall or bushy and up to 80 cm tall (in protected sites); twigs densely covered with tiny, brown, bran-like scales.

Leaves: Alternate, elliptic to oblong, widest above the middle, blunt-tipped, 6–18 {5–20} mm long, **persisting for more than a single growing season**, thick and **leathery**, **brownish beneath with dense, rusty-brown scales**, greenish above with scattered scales, short-stalked; buds scaly and conspicuous.

Flowers: Deep rose-purple, very fragrant, broadly cupped, 1–2 cm across, 8–12 mm long, with 5 spreading lobes, slightly bilaterally symmetric (zygomorphic), with both male and female parts (5–10 stamens; 1 style); calyx 0.5–1.5 mm long, with 5 tiny, triangular lobes, fringed with hairs; 5–10 stamens, projecting from the mouth of the flower on hairless filaments; stalks 6–10 mm long, scurfy; 3–6 flowers **in flat-topped clusters** (umbel-like) at the branch tips; June.

Fruits: Capsules, 4–8 mm long, scurfy, splitting lengthwise into 5 parts; seeds tiny and edged with small wings; June {July–August}.

Habitat: Moist alpine slopes and upper subalpine sites near timberline; elsewhere, on dry, rocky tundra and heathlands and occasionally in woodlands.

Notes: The only other species of this genus that is native to Alberta, white-flowered rhododendron (*Rhododendron albiflorum* Hook.), is easily distinguished by its thin, pale-green, deciduous leaves and its white flowers, which are borne in small (1–3-flowered) clusters from leaf axils along the branches. • Some sources report that the leaves and flowers of Lapland rose-bay have been used to make tea, but all rhododendrons are **poisonous**, as is the honey made from their nectar, so these plants should be treated as **toxic**. • Many beautiful rhododendrons are cultivated around the world in flower gardens. • The generic name *Rhododendron* was derived from the Greek *rhodon* (rose) and *dendron* (tree). The specific epithet *lapponicum* means 'of Lapland.'

HEATH FAMILY   **Trees and Shrubs**

OVAL-LEAVED BLUEBERRY

Vaccinium ovalifolium J.E. Smith
HEATH FAMILY (ERICACEAE)

Plants: Spreading, diffusely **branched, deciduous shrubs**, 20–60 cm tall (sometimes to 100 cm); **twigs angled**, hairless, yellowish green, often reddish, greyish with age.

Leaves: Alternate; blades elliptic to oblong-oval, blunt-tipped, 2–4 {5} cm long, **hairless, thin, greyish green** with a thin, waxy coating on the lower surface, **essentially toothless** (sometimes slightly wavy or very finely toothed below the middle); stalks 1–2 mm long.

Flowers: Pinkish, urn-shaped, with 5 petal lobes, about 7–9 mm long, usually **longer than they are wide** and broadest below the middle, with both male and female parts (10 stamens; 1 style); stamens with short, hairless filaments (shorter than the anthers) and large anthers tipped with a pair of slender tubes and bearing 2 spreading bristles (awns); styles usually included within the flower; **borne singly** from leaf axils on 1–5 mm long stalks; {May–July}, before the leaves are half expanded.

Fruits: Hanging, **blackish blue berries with a whitish, waxy covering** (bloom), 6–9 mm wide; **stalks strongly bent downward, not enlarged immediately below the berry**, 1–5 mm long; {July–September}.

Habitat: Moist subalpine thickets and woods; elsewhere, on peaty soils and in coastal forests.

Notes: This species has also been called tall huckleberry, oval-leaved huckleberry and blue bilberry. • Low bilberry (*Vaccinium myrtillus* L.) and tall bilberry (*Vaccinium membranaceum* Dougl. *ex* Hook.) are 2 common species that are also nearly hairless and that have angled twigs and scattered bluish black berries. They are easily distinguished from oval-leaved blueberry by their finely toothed, pointed leaves. • Another rare Alberta species, bog bilberry (*Vaccinium uliginosum* L.), has rounded (not clearly angled) reddish or brownish twigs and firm, dull green, toothless leaves with rounded tips and wedge-shaped bases. Its flowers are commonly tipped with 4 (rather than 5) petal lobes and contain 8 (rather than 10) stamens. Bog bilberry grows in bogs and on alpine slopes, and is reported to bloom from May to July. • The fruit of both species is edible and delicious.

V. ULIGINOSUM

Rare *Vascular* Plants of Alberta

Trees and Shrubs

HEATH FAMILY

ALPINE AZALEA

Loiseleuria procumbens (L.) Desv.
HEATH FAMILY (ERICACEAE)

Plants: Dwarf evergreen shrubs about 5–6 cm tall; stems much branched, tufted, spreading on the ground (prostrate), up to 30 cm long, mat-forming.

Leaves: Opposite, evergreen, narrowly elliptic to oblong, **2–8 mm long**, leathery, hairless on the upper surface, short-hairy beneath, rolled under along the edges, borne on short stalks that partly clasp the stem.

Flowers: Pink or sometimes whitish, 3–4 {5} **mm long**, with both male and female parts (5 stamens; 1 style), with **5 spreading petals** that are fused together at the base; 5 sepals, red, hairless, 1.5–2.5 mm long; **5 stamens**; 2–6 flowers in small, flat-topped clusters (corymbs) at the branch tips; June–July {August}.

Fruits: Egg-shaped capsules, 3–5 mm long, containing many seeds in 2 or 3 compartments; July–August.

Habitat: Moist alpine meadows and rocky subalpine slopes; elsewhere, in dry, stony heathlands.

Notes: The generic name *Loiseleuria* commemorates Jean Louis Auguste Loiseleur-Deslongchamps, a French botanist who lived 1774–1849. The specific epithet *procumbens* (to fall forward or bend over) refers to the prostrate growth form of this small shrub.

ASTER FAMILY **Trees and Shrubs**

BIG SAGEBRUSH

Artemisia tridentata Nutt.
ASTER FAMILY (ASTERACEAE [COMPOSITAE])

Plants: **Greyish**, wintergreen shrubs, **aromatic**, much branched, erect, {0.5} 1–2 m tall; bark greyish and shredding on older branches, densely silvery-hairy on young twigs.

Leaves: Alternate, **narrowly wedge-shaped, 3-toothed at tip** and tapered to base, 1–2 cm long, densely silvery-hairy.

Flowerheads: Small yellow or brownish heads, 2–3 mm across, of 5–8 disc florets, with both male and female parts (perfect); involucres 3–5 mm high; numerous, in **long, narrow, branched clusters** (panicles) 1.5–7 cm wide; August–September {July–October}.

Fruits: Sparsely hairy, seed-like fruits (achenes), not tipped with a tuft of hairs (pappus).

Habitat: Dry, open sites from the plains to near timberline, locally abundant, often covering many hectares.

Notes: Although they can be quite bitter, the seeds of several sagebrushes were used for food, and those of big sagebrush were considered the best. They were eaten raw or dried, often ground into meal and cooked. • Sagebrush leaves were boiled to make a tea for treating colds and pneumonia and aiding in childbirth. It was said to act like Epsom salts and to cure indigestion and flatulence, and the Navajo drank it before long hikes and athletic contests to cleanse the body. In the sweat bath, moistened branches are sometimes thrown onto hot rocks, and the resulting vapours are grimly inhaled until the participants are nearly prostrate. • The leaves were burned as a fumigant or smudge, and leafy branches were bound together to make brooms. The shredding bark was woven into mats, bags and cloaks. The wood was burned when no other fuel was available. • Livestock do not like to eat big sagebrush, and overgrazing of grasslands can result in dramatic increases in the abundance of this shrub.

Monocots

BUR-REED FAMILY

Monocots

GLOBE BUR-REED

Sparganium glomeratum (Beurling *ex* Laest.) L.M. Neuman
BUR-REED FAMILY (SPARGANIACEAE)

Plants: Aquatic perennial herbs; stems slender, erect or floating, up to 2.5 m long; from fibrous roots and creeping buried stems (rhizomes).

Leaves: In 2 vertical rows on the stems, **ribbon-like, 3–10 mm wide, mostly floating** on the surface of the water, stalkless, slightly expanded at the base and sheathing the stem.

Flowers: Tiny, male or female (unisexual); female flowers with 3–6 sepals (no petals) and 1 ovary; numerous flowers in dense, round male or female heads, **crowded, with 1 {2} male head(s) above 3–5 female heads**, held above the water on erect stalks; July {August}.

Fruits: Dry, dull brown, nut-like 'seeds' (achenes), **ellipsoid, 4–5 mm long, tipped with a straight beak about 1–1.5 mm long**, forming dense, 1–1.5 cm, **bur-like heads**; mid to late summer.

Habitat: Cool lakes, ponds and slow streams; often in water 1–2 m deep; possibly introduced.

Notes: Another rare Alberta species, floating-leaved bur-reed (*Sparganium fluctuans* (Engelm. *ex* Morong) B.L. Robins.), grows in ponds and flowers in July and August. It is similar to globe bur-reed, but its cross-veins form a distinct pattern on the lower surface of the leaves, it has branched flower clusters with 2 or more male heads per branch (rather than solitary), and its achenes are tipped with stout, flat, strongly curved (not straight) beaks. • Many shorebirds and waterfowl, including common snipes, rails, black ducks, wood ducks, mallards and coots, eat these nut-like fruits. Muskrats feed on the leaves.

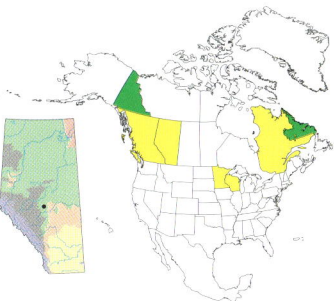

S. FLUCTUANS

Rare *Vascular* Plants of Alberta

NORTHERN BUR-REED

Sparganium hyperboreum Beurling *ex* Laest.
BUR-REED FAMILY (SPARGANIACEAE)

Plants: Slender **aquatic perennial** herbs; stems weak, submerged or floating, 10–30 cm long; from fibrous roots and creeping buried stems (rhizomes).

Leaves: In 2 vertical rows on the stems, ribbon-like, 10–30 {50} cm long and **1–5 mm wide, opaque, often yellowish**, mostly floating on the surface of the water, stalkless, sheathing the stem.

Flowers: Tiny, male or female (unisexual); female flowers with 3–6 scale-like sepals (no petals) and 1 ovary; numerous flowers in dense, round male or female heads, arranged in **small, unbranched clusters** with a **single, stalkless male head** at the tip (touching the uppermost female head) and **2–4 female heads below**, held above the water on erect stalks; July–August.

Fruits: Dry, **nut-like, ribbed** 'seeds' (achenes), **spindle-shaped to club-shaped, 4–5 mm long**, tipped with a **short point** up to 0.5 mm long, forming dense, 10 {5–13} mm wide, bur-like heads.

Habitat: Alpine lakes, in shallow water up to 50 cm deep; elsewhere, widespread at lower elevations across boreal North America and Eurasia.

Notes: The thick roots of some bur-reeds have been collected, cooked and eaten, but they are usually too small and scattered to warrant gathering. It is generally unwise to gather the roots of aquatic plants for food, because wetlands can be very sensitive to disturbance. In addition, poisonous plants commonly share these environments, and their roots could be confused with those of edible species.

SLENDER NAIAD, SLENDER WATER-NYMPH

Najas flexilis (Willd.) Rostk. & Schmidt
WATER-NYMPH FAMILY (NAJADACEAE)

Plants: Delicate, pale green, aquatic annual herbs; stems **submerged**, about 1 mm thick (or less), 30–60 {10–150} cm long, with **many alternate branches** giving the plants a **tufted** appearance.

Leaves: Crowded at stem tips in sub-opposite pairs, 1–3 cm long, 0.5–1 mm wide but with **enlarged, clasping bases** and slender, tapered tips, edged with tiny, sharp teeth.

Flowers: Tiny, male or female (unisexual), **stalkless**; male flower with 1 stamen in a membranous sheath; female flowers with a single, exposed pistil (not enclosed); flowers **solitary** (sometimes with 1 male and 1 female together) in the axils of the lower pair of leaves; {April–June} July–August.

Fruits: Membranous-walled 'seeds' (achenes), **3–4 mm long**, tipped with a slender, 1–2 mm long style, containing pale brown, shiny, 3 mm long seeds.

Habitat: Ponds and streams; elsewhere, in clear, shallow to deep, fresh or brackish water.

Notes: Slender naiad is usually an annual plant, reproducing by seed each year. In some cases, however, the base of the plant may survive over winter and produce new plants from shoots the following spring. • This is a true aquatic plant. Its tiny flowers are pollinated underwater. • When abundant, slender naiad is an important food for waterfowl, which eat all parts of the plants. • The generic name *Najas* is taken from the Greek *naias* (a water nymph). The specific epithet *flexilis* means 'pliant.'

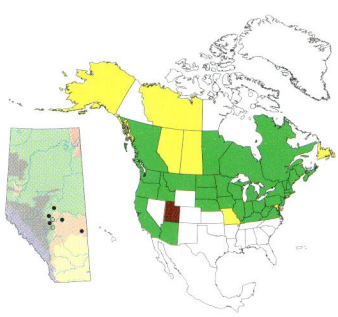

Monocots PONDWEED FAMILY

FLOATING-LEAVED PONDWEED

Potamogeton natans L.
PONDWEED FAMILY (POTAMOGETONACEAE)

Plants: Aquatic perennial herbs; **stems submerged, 1.5 m long**, 1–2 mm thick, with few branches; from extensive, slender underground stems (rhizomes), often with overwintering tubers.

Leaves: Alternate, leathery, of 2 kinds; floating leaves flat, egg-shaped to elliptic, 4–9 {10} cm long, rounded to heart-shaped at the base, 13–37-nerved, with stalks longer than blades; **submerged leaves linear**, 10–20 cm long, 0.8–2 mm wide, obscurely 3–5-nerved; **stipules free from rest of leaf**, 4–10 cm long, stiff, **fibrous**.

Flowers: Greenish, inconspicuous, with 4 minute petals (tepals); **many**, in 10–12 whorls, forming dense **spikes {1} 2–5 cm long and about 1 cm thick, held above water** on long (4–14 cm) stalks; June–July {August–September}.

Fruits: Greenish, shiny, short-beaked 'seeds' (achenes) 3–5 mm long, slightly fleshy at first, dry and hard when mature.

Habitat: Still or slow-moving, shallow water.

Notes: The thickened, buried rhizomes of these plants are a source of starch for both humans and animals. • The generic name *Potamogeton* was derived from the Greek *potamos* (a river) and *geiton* (a neighbour), in reference to the habitats of many species. The specific epithet *natans* means 'floating,' a reference to the leathery floating leaves of this species.

BLUNT-LEAVED PONDWEED

Potamogeton obtusifolius Mert. & W.D.J. Koch
PONDWEED FAMILY (POTAMOGETONACEAE)

Plants: Green or often reddish, **submerged aquatic perennial** herbs; stems slightly zigzagged, 50–100 cm long, slender, with pairs of conspicuous glands at each joint, much branched; from slender underground stems (rhizomes); producing winter buds.

Leaves: Alternate, translucent, all broadly **linear**, 3–9 {10} cm long, **2–3 {1–4} mm wide**, usually **3-nerved** (sometimes 5-nerved), with 2–4 rows of white air channels (lacunae) bordering the midvein; **stipules 1–2 cm long, open** (not sheathing), translucent, eventually shredding into fibres from the tips.

Flowers: Greenish, **inconspicuous**, with 4 minute petals (tepals); **many**, in dense whorls, forming egg-shaped to cylindrical **spikes 8–17 {20} mm long, held above water** on short (1–4 cm long), stout stalks; {July–September}.

Fruits: Rounded or slightly flattened, dry 'seeds' (achenes) **2.5–4 mm long**, often tipped with a short (up to 0.7 mm) beak.

Habitat: Shallow lakes and ponds, often growing in organic sediment.

Notes: Another rare species, leafy pondweed (*Potamogeton foliosus* Raf.), is very similar to blunt-leaved pondweed, but its leaves are slightly narrower (usually less than 1.5 mm wide), its stems have no glands at their joints, and its fruits are smaller (2–3 mm long), with a sharp, wavy ridge along the back. Leafy pondweed grows in shallow, standing water, and it is reported to bloom from July to September. • These aquatic species often overwinter by producing dense, leafy winter buds at the tips of their branches. These are dropped in the autumn and sink to the bottom of the water, where they lie in the mud through the winter. In spring, the buds sprout and grow into new plants.

P. FOLIOSUS

Monocots

PONDWEED FAMILY

LINEAR-LEAVED PONDWEED

Potamogeton strictifolius A. Bennett
PONDWEED FAMILY (POTAMOGETONACEAE)

Plants: Olive-green to brownish, **submerged aquatic perennial** herbs; stems leafy, 50–100 cm long, very slender, slightly flattened, rather sparingly branched toward the top, sometimes with paired glands at the joints (nodes); from slender underground stems (rhizomes); producing slender winter buds on non-flowering branch tips.

Leaves: Alternate, crowded, **stiff** with firm, down-rolled edges, all **linear**, 2–6 cm long, **0.5–2.5 mm wide**, tapered to a slender point, **3** {5}**-nerved** but the side nerves faint; **stipules 1–2 cm long**, often overlapping and thus covering the stem, **white, strongly ribbed**, soon shredding into fibres.

Flowers: **Greenish, inconspicuous**, with 4 minute petals (tepals); **many**, in 3 or 4 well-spaced whorls or pairs, forming slender **spikes 1–1.5 cm long**, **held above water** on very slender stalks 1–9 cm long; {July–September}.

Fruits: Dry 'seeds' (achenes) **2–3 mm long**, sometimes with a very low ridge (keel) along the back.

Habitat: Shallow lakes and ponds.

Notes: Linear-leaved pondweed is distinguished from other slender-leaved pondweeds by its stiff, 3-nerved leaves and its whitish, strongly fibrous stipules.

Rare *Vascular* Plants of Alberta

PONDWEED FAMILY   **Monocots**

WHITE-STEMMED PONDWEED

Potamogeton praelongus Wulfen
PONDWEED FAMILY (POTAMOGETONACEAE)

Plants: Large, **submerged aquatic perennial** herbs; stems whitish to olive-green, 2–3 {6} m long and 2–3 mm thick, zigzagged, freely branched near the top; from stout (2–6 mm thick), white underground stems (rhizomes) with rusty spots; producing winter buds.

Leaves: Alternate, shiny, all **oblong to lance-shaped**, tapering gradually to round, somewhat hooded tips, stalkless, rounded and slightly clasping at the base, 10–30 {7–35} cm long, {1} **2–3 cm wide**, with **3–5 main nerves** and several fainter veins; **stipules whitish, 2–8 {10} cm long, conspicuous and persistent**, free of the leaf blades and **open** (not sheathing).

Flowers: Greenish, **inconspicuous**, with 4 minute petals (tepals); **many**, in 6–12 dense whorls, forming **spikes 3–4 {2.5–5} cm long, held above water** on stout, 10–30 {50} cm long stalks; {May–June} July–August.

Fruits: Dry 'seeds' (achenes) **4–5 {7} mm long**, egg-shaped, widest above middle, tipped with a thick, short (0.5–1 mm long) beak, prominently ridged down the back.

Habitat: Deep, clear water in lakes and ponds; in water up to 6 m deep.

Notes: In the autumn, these plants produce winter buds beneath conspicuous white sheaths, in the form of short, densely leafy branches. These buds begin to grow at an amazing rate the following spring. In southern BC, some plants can reach up to 5 m in length by the first week of May. • Pondweeds are very important to the ecology of many lakes and ponds. Their seeds are a favourite and important food of many wild birds, and most parts of the plants are eaten by waterfowl, shore birds, marsh birds, muskrats, beaver and deer. Most species provide food, shelter and shade for fish and small aquatic organisms. They are especially important in providing habitat for many insects that, in turn, are eaten by fish.

Rare *Vascular* Plants of Alberta

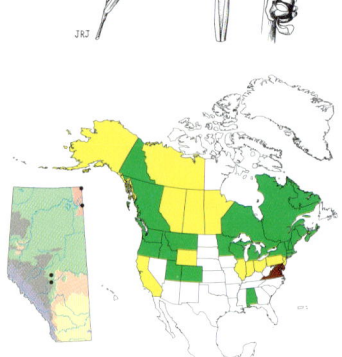

Robbins' pondweed

Potamogeton robbinsii Oakes
PONDWEED FAMILY (POTAMOGETONACEAE)

Plants: Leafy, submerged aquatic perennial herbs; stems stout, round, 0.5–3 m long, freely branched, creeping on mud and rooting at lower joints; from underground stems (rhizomes); producing winter buds.

Leaves: Alternate, **crowded, in 2 opposite vertical rows**, widely **spreading**, dark brownish green, all **linear to lance-shaped**, pointed at the tip, stalkless and somewhat clasping at the base, 2–8 {12} cm long, **3–5 {8} mm wide**, edged with **fine, sharp, white teeth, with 20–35 fine nerves** (or more); **stipules** pale, with their bases **joined to the leaf blades in a {5} 10–15 mm long sheath**, shredding into white fibres, which often cover the stem between the leaves.

Flowers: Greenish, **inconspicuous**, with 4 minute petals (tepals); **many**, densely whorled, with about 4 whorls per spike and **several 7–20 mm long spikes** per branched cluster, **held above water** on flat, stiff, 3–5 {7} cm long stalks; {August–September}.

Fruits: Blackish 'seeds' (achenes) **3–4 mm long**, egg-shaped, widest above middle, tipped with an inconspicuous beak, sharply ridged down the back.

Habitat: Shallow to deep (1–3 m), quiet water in lakes and ponds, usually growing on organic material or muck.

Notes: Robbins' pondweed rarely produces mature seed. Instead, it reproduces vegetatively by growing from stem (stolon) tips and joints and by producing coarse, leafy winter buds. • Pondweed seeds can be carried for thousands of miles by migrating waterfowl. These birds not only carry the seeds for great distances and deposit them in favourable habitats, but they also improve the seeds' ability to germinate. In fact, germination of the seeds of at least one species proved possible only after the seeds had passed through a bird's digestive tract or been exposed to similar conditions.

DITCH-GRASS FAMILY   **Monocots**

WIDGEON-GRASS

Ruppia cirrhosa (Petagna) Grande
DITCH-GRASS FAMILY (RUPPIACEAE)

Plants: Submerged aquatic perennial **herbs; stems slender, forked**, up to 80 cm long; from elongated underground stems (rhizomes) and runners (stolons).

Leaves: Alternate, spreading, in 2 vertical rows (ranks), thread-like, 5–20 cm long, scarcely 0.5 mm wide; basal appendages (stipules) 5–15 mm long, **sheathing the stem** and with the tips sometimes free for 1–2 mm, thin and translucent along the edges.

Flowers: Tiny, **lacking obvious sepals and petals**, consisting of 2 stamens and 4 ovaries; borne in small, stalked, **spike-like** structures with 2 flowers at the tip, from the sheaths of the uppermost leaves, the thick stalk (spadix) elongating to carry the flowers to the surface of the water; July {August}.

Fruits: Egg-shaped or pear-shaped, somewhat fleshy, **drupe-like achenes**, 3–4 mm long, borne **on slender stalks in flat-topped** (umbel-like) **clusters at the tips of slender, 3–50 cm long, coiled stalks** (elongated spadixes) from upper leaf axils.

Habitat: Saline and alkaline lakes, ponds and ditches; elsewhere, in brackish or salt water along the coast, rarely in fresh water.

Notes: This species has also been called *Ruppia maritima* L. and *R. occidentalis* S. Wats. • Some taxonomists include widgeon-grass in the pondweed family (Potamogetonaceae). Although the unique clusters of flowers and fruits on their long, coiled stalks are unmistakable, non-flowering plants can scarcely be distinguished from some of the narrow-leaved pondweeds, especially sago pondweed (*Potamogeton pectinatus* L.), which is widespread in Alberta. • Widgeon-grass flowers are usually pollinated at the water surface by floating pollen, but some straight-stalked flowers are pollinated underwater. This is a highly variable species, with many forms. • The generic name *Ruppia* commemorates Heinrich Bernhard Ruppius (1688–1719), a German botanist who wrote *Flora Jenensis*. The specific epithet *cirrhosa* means 'tendrilled,' referring to the long, coiled flower stalks.

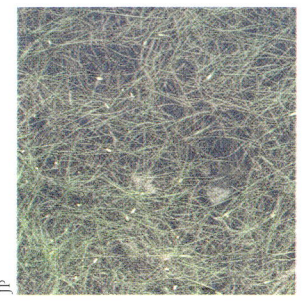

Rare *Vascular* Plants of Alberta

Monocots — ARROW-GRASS FAMILY

FLOWERING-QUILLWORT
Lilaea scilloides (Poir.) Hauman
ARROW-GRASS FAMILY (JUNCAGINACEAE)

Plants: Small, densely tufted, **semi-aquatic annual herbs**; stems very short, essentially absent; from fibrous roots.

Leaves: Basal, rush-like, cylindrical (terete), linear to awl-shaped, {3} 8–35 cm long, 1–4 mm thick, soft, with dry, translucent, sheathing bases.

Flowers: Tiny, lacking petals and sepals, each with 1 stalkless stamen or 1 carpel (or both); flowers in 2 types of clusters; **solitary female flowers enclosed in the leaf bases**, each with a 3-sided carpel tipped by a thread-like style up to 10 cm long and ending in a head-like stigma; **small, 0.5–4 cm long spikes on slender, leafless stalks** 5–20 {3–25} cm long, each spike with female flowers at the base and male flowers at the tip; male flowers in the axils of 2–3 mm long bracts but these are soon dropped; female flowers tipped with short, stout styles; July {June–August}.

Fruits: Erect, dry, single-seeded, {3} 4–6 mm long, ribbed, those in the stalked spikes **flattened and 2-winged**, those of the lower flowers in the leaf sheaths, **3-sided and tipped with 3 small horns**.

Habitat: Slough edges and mud flats; elsewhere, in sloughs and dry creek beds, shallow water and tidal flats.

Notes: This species has also been called *Lilaea subulata* Humb. & Bonpl. and has been placed in the flowering-quillwort family (Lilaeaceae). • Flowering-quillwort usually grows in shallow water, where the stigmas of the lower female flowers and the spikes of mixed flowers float on the surface or are held above the water. They are also often stranded in mud at the edges of receding ponds. The non-flowering plants might be mistaken for quillworts (*Isoetes* spp., pp. 352–53) at first glance, but they are truly stalkless, without the bulbous stocks found at the bases of quillwort plants. • The genus *Lilaea* was named in honour of Alire Raffeneau Delile (1778–1850), a French authority on the flora of north Africa and Asia Minor. The specific epithet *scilloides* refers to the similarity between the leaves of flowering-quillwort and those of a squill (*Scilla* sp.), a member of the lily family (Liliaceae).

WATER-PLANTAIN FAMILY | **Monocots**

BROAD-LEAVED ARROWHEAD, SWAMP POTATO, WAPATO
Sagittaria latifolia Willd.
WATER-PLANTAIN FAMILY (ALISMATACEAE)

Plants: Emergent perennial herbs; stems 30–60 {20–80} cm tall; from slender underground stems (rhizomes) that often end in **tubers**.

Leaves: Of two types; emergent leaves **arrowhead-shaped** with the **basal lobes equal to or shorter than the upper part of the blade**, up to 25 cm long, on long, angled stalks; submerged leaves (when present) thin and ribbon-like, about 4–10 mm wide.

Flowers: White, 2–4 cm wide, each with **3 petals**, 3 small, green sepals, male or female (unisexual), from the axil of a blunt-tipped, **egg-shaped bract 5–10 {15} mm long**; flowers arranged in **2–8 whorls of 3**, usually with male flowers above female flowers on the same plant (sometimes with male or female plants), in 20–50 {90} cm long clusters; August {July–September}.

Fruits: Dry, **egg-shaped** 'seeds' (achenes), 2.3–3.5 {5} mm long, with 0.5–1.5 mm long **beaks projecting horizontally** from their tips, flattened, broadly winged, **in round, dense heads** that are **usually over 1.5 cm in diameter**; mid to late summer.

Habitat: Ponds, lakes and ditches.

Notes: This species has also been called *Sagittaria sagittifolia* L. var. *latifolia* Muhl. • Arum-leaved arrowhead (*Sagittaria cuneata* Sheld.) is a widespread species that resembles broad-leaved arrowhead, but it is generally a smaller plant (20–40 cm tall) with smaller (less than 1.5 cm) heads of smaller (2–2.5 mm long) achenes bearing small (less than 0.5 mm), erect (not horizontal) beaks. The bracts in its flower clusters are lance-shaped and slender-pointed. • The leaf shape of broad-leaved arrowhead varies with changes in water level and other environmental factors, but it is less 'plastic' than that of arum-leaved arrowhead. • The seeds and tubers (especially the tubers) are valuable food for waterfowl and marsh birds. Muskrats and porcupines eat the leaves and tubers. • Many native peoples used the large, chestnut-sized tubers for food. Swamp potatoes were roasted and eaten, or were cooked and dried for later use.

Rare *Vascular* Plants of Alberta

Monocots WATERWEED FAMILY

TWO-LEAVED WATERWEED, LONG-SHEATHED WATERWEED
Elodea bifoliata St. John
WATERWEED FAMILY (Hydrocharitaceae)

Plants: **Aquatic** perennial herbs, **submerged**, with only the flowers reaching above the water; stems slender, sparingly branched in 2s; with roots anchored in mud or free floating.

Leaves: Opposite and well spaced on the lower stem, occasionally more crowded on the upper stem, linear, blunt-tipped (rarely pointed), 1–2 {4} mm wide, edged with fine, sharp teeth; **larger leaves** {17} **20–26 mm long**.

Flowers: Inconspicuous, either male or female on each plant, each surrounded by an enlarged bract (spathe) and consisting of **3 sepals**, **3 petals** and the reproductive organs **at the tip of a slender, stalk-like tube** (hypanthium) **that holds the flower at or above the water surface**; male (staminate) **flowers** with a 2–15 mm long spathe, a 4–30 cm long hypanthium tube, 3 broad, 3.5–5 mm long and 2–2.5 mm wide sepals, 3 slender, white, 5 mm long and 0.5 mm wide petals, 3 sterile stamens (staminodia) and 6 fertile stamens with 3.1–4.5 mm long anthers; **female (pistillate) flowers with a 3–7 cm long spathe**, a 15–40 cm long (or longer) hypanthium tube, 3 white, spoon-shaped, 4 mm long and 1.3 mm wide petals, 3 sepals about 3 mm long and 1.3 mm wide, and a single, inferior, 3–4 mm long ovary tipped with 3 stigmas 1–2 mm long; solitary flowers in leaf axils; July–August {September}.

Fruits: Capsules, about 10 mm long, containing 6 seeds, each 5–6 mm long.

Habitat: Sloughs, ponds and lakes, in quiet or running water.

Notes: This species has also been called *Elodea longivaginata* St. John, and it was called *Elodea canadensis* Michx. in the first edition of the *Flora of Alberta*. • Canada waterweed (*Elodea canadensis* Michx.) has not yet been found in Alberta, but it grows in BC, Montana and eastern Saskatchewan, so it may be discovered here eventually. Canada waterweed is distinguished by its shorter (rarely over 15 mm long) leaves, which are mostly crowded and borne in whorls of 3 (rather than in pairs). It grows in similar habitats and is reported to flower from July to September.

DUCKWEED FAMILY **Monocots**

WATERMEAL, COLUMBIAN DUCKMEAL

Wolffia columbiana Karsten
DUCKWEED FAMILY (LEMNACEAE)

Plants: Tiny aquatic plants, consisting of a **round to egg-shaped body** (thallus) about **1 mm long, floating low in the water** with the centre of the upper side exposed above the surface of the water, greenish, without white dots on the rounded upper surface, **lacking roots**; pores (stomata) few (usually fewer than 10); can be found from June to early October.

Leaves: None.

Flowers: Tiny, inconspicuous, either male or female, with 1 of each sex on the upper surface of the plant; male flowers with 1 stamen; female flowers with 1 ovary; **none** yet found in Alberta, apparently reproducing only by budding here.

Fruits: Single, tiny, thin-walled 'seeds' (utricles); not yet found in Alberta.

Habitat: Beaver ponds in hummocky moraines; elsewhere, in moderately to extremely nutrient-rich ponds.

Notes: This species has also been called *Wolffia arhiza* Wimm. • Watermeal is readily carried on the feet and feathers of waterfowl and doubtless reached Alberta in this way. However, these plants are so tiny and inconspicuous that they are easily overlooked. They are often discovered by phycologists studying water samples under a microscope. • Watermeal overwinters at the bottom of ponds as young bulblets (turions). In northern parts of its range, it reproduces largely or entirely by budding (asexual reproduction, without flowers), so some Alberta populations could be clones of single plants. It can form virtual monocultures or can grow intermixed with similar small floating plants, such as northern duckmeal (*Wolffia borealis* (Engelm. *ex* Hegelm.) Landolt), common duckweed (*Lemna turionifera* Landolt, also called *Lemna minor* L.) and larger duckweed (*Spirodela polyrhiza* (L.) Schleiden). • The stabilization of water levels as a result of beaver activity is very beneficial to watermeal. Stable water levels and sufficient depth of water to ensure that the pond does not freeze to the bottom in winter are essential for survival, because these plants cannot survive drying or prolonged freezing. • Like northern duckmeal, watermeal provides superior feeding for dabbling ducks in the autumn.

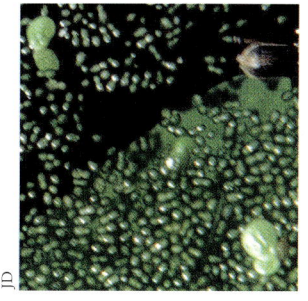

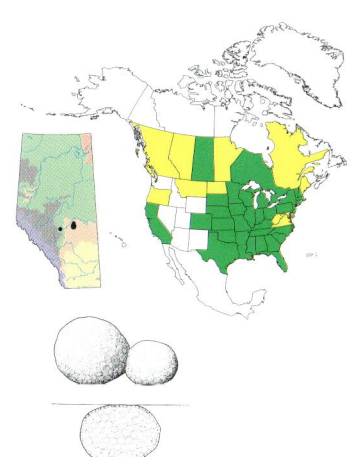

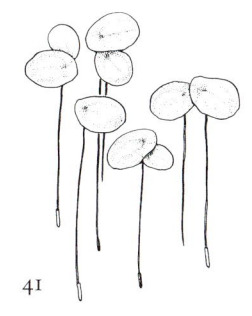

LEMNA TURIONIFERA

Rare *Vascular* Plants of Alberta

Monocots

SPIDERWORT FAMILY

WESTERN SPIDERWORT

Tradescantia occidentalis (Britt.) Smyth var. *occidentalis*
SPIDERWORT FAMILY (COMMELINACEAE)

Plants: Blue-green perennial herbs; stem erect, smooth, almost succulent, greyish with a waxy coat (glaucous), 10–50 {60} cm tall, branched, with 2–5 joints (nodes); roots thickened.

Leaves: Alternate, **linear to narrowly lance-shaped, parallel-veined**, often folded lengthwise or with edges rolled inward, 10–30 cm long and 4–12 mm wide, **with loose, inflated sheaths**; sheaths 2–4 times the width of the blade, hairless, prominently parallel-veined.

Flowers: **Blue**, sometimes purplish or rose-coloured, with **3 petals and 3 sepals; petals egg-shaped, 7–15 mm long**; sepals green with purplish edges, glandular-hairy, elliptic, sharp-pointed, 6–12 mm long; **6 stamens**, with bearded stalks (filaments); stigmas single, head-like (capitate), at the tip of a very thin style; few to many flowers on 1–2 cm long, glandular-hairy stalks in **spreading, flat-topped clusters** (cymes) at branch tips or from upper joints (nodes) **above 2 leaf-like, 5–15 cm long, strongly sheathing bracts**; June {July}.

Fruits: Oblong capsules with 3 cavities, each containing 1 or 2 seeds, splitting lengthwise along 3 lines; seeds ridged and pitted, grey; late July–August.

Habitat: Sand dunes, sandy plains and dry grassland.

Notes: Often only a single flower in each cluster opens each day and lasts only a few hours. • The generic name *Tradescantia* commemorates John Tradescant Sr., an English horticulturist who lived ca. 1580–1638. The specific epithet *occidentalis* is Latin for 'western.' The common name 'spiderwort' refers to sticky threads that can be pulled from the broken stem tips; these were likened to the strands of a spider's web. The suffix 'wort' comes from the Old English for 'plant.'

LILY FAMILY · **Monocots**

Geyer's wild onion

Allium geyeri S. Wats.
LILY FAMILY (LILIACEAE)

Plants: Slender perennial herbs with an onion-like aroma; single stems, {10} 20–50 cm tall; from egg-shaped (or more elongate) bulbs, usually clustered, with whitish inner layers, outer layers persisting as fibrous, **rather coarse meshed netting** (reticula) **enclosing 1 or more bulbs**.

Leaves: Alternate on lower stem, usually **3 or more, narrow, channelled**, concave-convex in cross section, 1–5 mm broad, with bases sheathing the stem.

Flowers: Pink, rarely white, with **6 tepals** (3 petals and 3 sepals), **often replaced with small, pointed, bulb-like structures** (bulbils); tepals 6–8 {4–10} mm long, egg-shaped to lance-shaped, with **erect**, blunt to long-tapered tips, often obscurely toothed along the edges and with tiny bumps (papillae) on the midrib, becoming thickened along the midrib and permanently covering the capsule; stigma head-like (capitate), sometimes with 3 obscure lobes; **10–25** (sometimes more) **flowers in erect, flat-topped clusters** (umbels) with 2–3 bracts at the base; bracts egg-shaped to lance-shaped with slender-pointed tips, mostly 1-nerved; stalks (pedicels) often less than twice as long as the tepals; {May} June–July.

Fruits: Egg-shaped capsules tipped with 6 inconspicuous, rounded knobs, on rigid and stiffly spreading stalks; seeds black, shiny, **honeycombed on the surface, with a tiny blister** (pustule) **in each hollow**.

Habitat: Wet meadows and streambanks.

Notes: In Alberta, there are 2 varieties of this species, var. *geyeri*, which has only normal flowers, and var. *tenerum* M.E. Jones (also called *Allium rubrum* Osterh. and *Allium rydbergii* Macbr.), which has most flowers replaced with slender-pointed, egg-shaped bulbils. • This species could be confused with other wild onions. Nodding onion (*Allium cernuum* Roth) has nodding flowers and lacks fibrous bulb coats. Prairie onion (*Allium textile* Nels. & Macbr.) has netted outer bulb layers, but its plants usually have only 2 leaves per stem and its tepals are usually white and spreading. The honeycombed seeds also lack tiny blisters. • All onions are edible, and many species (e.g., scallions, garlic, leeks and chives) are cultivated.

VAR. *GEYERI*   VAR. *TENERUM*

Rare *Vascular* Plants of Alberta

Monocots LILY FAMILY

WESTERN FALSE-ASPHODEL

Triantha occidentalis (S. Wats.) Gates
LILY FAMILY (LILIACEAE)

Plants: Slender perennial herbs, **sticky with more or less cylindrical, glandular hairs; stems leafless** or with 1–3 leaves near the base, glandular or glandular-hairy, especially near the top, erect, 10–80 cm tall; from short underground stems (rhizomes).

Leaves: Basal, **linear**, up to 50 cm long and 8 mm wide.

Flowers: **White or greenish**, **3–7 mm long**, with **6 tepals** (3 petals and 3 slightly shorter and wider sepals), 6 stamens and 3 styles; 3–45 flowers, usually in 3s in the axils of **deeply 3-lobed bractlets**, forming dense, round or egg-shaped **spikes** or sometimes open or interrupted clusters 1–8 cm long; {June–August}.

Fruits: Egg-shaped to broadly ellipsoid **capsules**, 4–9 mm long, clearly longer than the persistent tepals at the base, **papery, 3-cavitied**; **seeds** about 1 mm long, covered **with an inflated, white, net-like envelope** and tipped with a tail-like appendage at one or both ends (rarely lacking).

Habitat: Wet, calcareous sites.

Notes: Some taxonomists have called this species *Tofieldia occidentalis* S. Wats., while others have included it in *Tofieldia glutinosa* (Michx.) Pers., where it comprised 3 subspecies, ssp. *montana* Hitchc., ssp. *brevistyla* Hitchc. and ssp. *occidentalis* Hitchc. Two subspecies have been found in Alberta. *Triantha occidentalis* ssp. *brevistyla* (C.L. Hitchc.) Packer has seeds with a strongly inflated netted envelope about 1–2 times as long as it is wide, and its stems are covered with a mixture of stubby glands (½–2 times longer than wide) and stout, cylindrical hairs (2–4 times longer than wide). *Triantha occidentalis* ssp. *montana* (C.L. Hitchc.) Packer has less swollen seed envelopes (3–4 times longer than wide), and its stems have coarse, cylindrical hairs (4–6 times longer than wide) and no stubby glands. • Western false-asphodel could be confused with the more common species, sticky false-asphodel (*Triantha glutinosa* (Michx.) Baker). Sticky false-asphodel is distinguished by the predominance of conical or dome-shaped glands (dubbed 'haycocks') on its stems and by the scarcely lobed bractlets beneath its flowers. Its seeds do not have a loose, white, netted envelope.

LILY FAMILY   **Monocots**

BLUE CAMAS

Camassia quamash (Pursh) Greene var. *quamash*
LILY FAMILY (LILIACEAE)

Plants: Perennial herbs; stems {20} 30–60 cm tall, leafless; from **egg-shaped bulbs** 1–3 cm thick, covered with brown or black scales.

Leaves: Basal, grass-like, linear, 6–15 mm wide, 20–40 cm long.

Flowers: Pale to deep blue or violet, star-shaped, about 4–5 cm across, with **6 linear, 3–9-nerved petals** (tepals); stamens 6, with anthers attached at the middle to thread-like filaments; stigmas 3-lobed, on thread-like styles; many flowers in showy clusters 5–30 cm long; May–July.

Fruits: Ellipsoid, 3-sided capsules, {12} 15–18 mm long, containing shiny black seeds.

Habitat: Moist to wet meadows.

Notes: Camas was a prized root crop in many areas, and tribes fought for the right to collect in certain meadows. Its role was likened to that of cereal plants in Europe. Settlers and explorers used it less, because large quantities caused vomiting and diarrhea in the uninitiated and, as Father Nicholas Point observed, 'very disagreeable effects for those who do not like strong odours or the sound that accompanies them.' • Girls competed in the annual camas harvest to show their worth as future wives. One young woman was reported to have collected and prepared 60 42-litre sacks of camas roots. Men were excluded from the harvest because they would bring 'bad luck,' and if a woman burned any bulbs, some of her relatives would soon die. The roots were roasted, pounded, made into cakes, eaten raw or stone-boiled, but most were cooked and dried. The bulbs were baked in pits with hot stones for several days, and when they were done, they were dark brown, with a glue-like consistency and a sweet taste, like that of molasses. They were then mashed together and made into cakes which were sun-dried for storage. During the cooking process, insulin (an indigestible sugar) breaks down to fructose; cooked, dried bulbs are 43 percent fructose by weight. Camas was the principal sweetening agent of many tribes before the introduction of sugar. Cooked, dried bulbs kept indefinitely. David Thompson reported eating 36-year-old bulbs that still had good flavour.

Rare *Vascular* Plants of Alberta

Monocots LILY FAMILY

ROSE MANDARIN

Streptopus roseus Michx.
LILY FAMILY (LILIACEAE)

Plants: Perennial herbs; stems simple or occasionally branched, 15–50 cm tall, **with a sparse fringe of hairs at each joint** (node) opposite the leaf base; from extensive underground stems (rhizomes).

Leaves: Alternate; blades **egg-shaped**, pointed at the tip, 3–10 cm long, 1–5 cm wide, stalkless, **smooth-edged or fringed with several-celled hairs**.

Flowers: Rose-coloured with white tips, or white to greenish yellow and streaked or spotted **with reddish purple**, bell-shaped, 6–10 mm long, with 6 petal-like segments (tepals) curving outward at their tips; stalks 5–20 mm long, curved but not bent, sparsely coarse-hairy; flowers **single** (sometimes 2) **in leaf axils**; June–July.

Fruits: Round, **red berries**, 5–6 mm across, several-seeded.

Habitat: Moist, coniferous woods and streambanks.

Notes: Small twisted-stalk (*Streptopus streptopoides* (Ledeb.) Frye & Rigg) is another species of moist, coniferous woods that is rare in Alberta. It differs from rose mandarin in having saucer-shaped flowers, with petal-like segments that curve outward from their bases (rather than from their tips), and its leaves are shorter (3–5 cm long) with single-celled, translucent teeth crowded along the edges. The flowers of small twisted-stalk are rose to reddish brown with yellowish green tips, and they are reported to appear in June or July. • The common twisted-stalk in Alberta, clasping-leaved twisted-stalk (*Streptopus amplexifolius* (L.) DC.), has larger plants (50–100 cm tall) with greenish white flowers on kinked or bent stalks, and its leaf joints (nodes) are hairless. • Native people believed that twisted-stalk berries were good food for grizzly bears but not for humans.
• *Streptopus* is derived from the Greek *streptos* (twisted) and *pous* (foot), referring to the bent or twisted flower stalks. The specific epithet *roseus* is Latin for 'rose-coloured,' a reference to the flowers.

S. STREPTOPOIDES

Rare *Vascular* Plants of Alberta

LILY FAMILY   **Monocots**

WESTERN WAKEROBIN

Trillium ovatum Pursh
LILY FAMILY (LILIACEAE)

Plants: Low, hairless perennial herbs; stems stout, erect, {10} 20–40 cm tall; from stout, short, fleshy rootstocks (rhizomes).

Leaves: In a whorl of 3 (sometimes 4 or 5) **at the stem tip**, stalkless, 6–12 {5–15} cm long, 4–8 cm wide, **broadly egg-shaped**, tapered to a short, slender point.

Flowers: White (pink to purplish with age), **about 6–9 cm across**, with **3 broad, white petals alternating with 3 small, green sepals**; single, erect, on a long (4–5 cm) stalk from the centre of the leaf whorl; May–June {March–July}.

Fruits: Yellowish green, berry-like capsules, oval, slightly winged; seeds numerous, shed in a sticky mass.

Habitat: Moist to wet, shady mountain woods and thickets.

Notes: Some tribes used these thick rhizomes to help mothers during childbirth—hence another common name, 'birthroot.' The leaves have been eaten as cooked greens.
• Trillium seeds each have a small, oil-rich body that attracts ants. The ants carry the seeds to their nests, where they eat the oil-rich part or feed it to their young, and discard the rest. This is a very effective way to disperse seeds, and North American forests contain many 'ant plants' (plants whose seeds contain oil bodies), including wild ginger (*Asarum* spp.), bleeding hearts (*Dicentra* spp.) and many violets (*Viola* spp.), as well as trilliums. • The generic name *Trillium*, from the Latin *tri* (three), refers to the 3 leaves, 3 petals, 3 sepals and 3 stigmas of these plants. The flowers bloom in the spring, just as the robins return or 'wake up' from their winter absence—hence the name 'wakerobin.'

Rare *Vascular* Plants of Alberta

Monocots AGAVE FAMILY

SOAPWEED

Yucca glauca Nutt. *ex* Fraser
AGAVE FAMILY (Agavaceae)

Plants: Coarse, evergreen perennial herbs; flowering stems erect, 50–100 {150} cm tall; from 1 or more short, woody root crowns.

Leaves: Basal, densely tufted, stiff, linear, stiffly sharp-pointed, 20–40 {60} cm long; edges rolled inward, whitish, with a few frayed fibres.

Flowers: Cream-coloured to greenish white, showy, bell-shaped, 3–5 cm long and wide, with 6 leathery, egg-shaped petals (tepals); many **nodding** flowers **in long clusters** (racemes); {May–July}.

Fruits: Hardened, oblong capsules, 5–7 cm long, with many thin, flat, black seeds.

Habitat: Dry, open grassland slopes and coulees.

Notes: Soapweed has sometimes been included in the lily family (Liliaceae). • The flower petals were eaten raw in salads and the young seed pods were roasted in ashes and eaten. Ripe fruits were split in half, their seeds and fibre were scraped out, and the remaining pulp was then baked and eaten. • The fibrous leaves were split and used as all-purpose ties, but they were not twisted or plaited to make cord or rope. • The Navajo pounded the roots with rocks to remove bark and soften them, then vigorously stirred the softened mass of fibres in warm water to whip up suds. This was used as soap to wash wool, clothing, hair and the body. It was also said to reduce dandruff and baldness. • Soapweed and a small, white, night-flying moth (the yucca moth, *Pronuba*) depend on one another for survival. The flowers open fully only at night, when they are visited by a female yucca moth. She takes a ball of pollen and flies to another flower, where she eats through the ovary wall and deposits an egg inside. She then climbs to the stigma, deposits the pollen from the first flower and moves to the anthers to collect a second ball of pollen before flying to the next plant. By fertilizing the flowers, she assures the development of seeds, which will provide both food for her young when the eggs hatch and new plants for future generations of moths. Ripe yucca pods almost always have a tiny hole, where the grub ate its way out.

IRIS FAMILY   **Monocots**

WESTERN BLUE FLAG

Iris missouriensis Nutt.
IRIS FAMILY (IRIDACEAE)

Plants: Perennial herbs; stems **clumped, about 20–50 cm tall**; from **thick, spreading underground stems (rhizomes)**.

Leaves: Mainly basal, linear, **sword-shaped**, 20–40 {50} cm long, 5–10 mm wide.

Flowers: Pale to deep blue, sometimes pale with purple lines, about **6–7 cm across**, with **3 backward-curved, purple-lined sepals, 3 erect, narrower and paler petals and 3 flattened, petal-like style-branches**; 2–4 flowers on stout, leafless stalks up to 6 cm long; {May} June–July.

Fruits: Erect capsules, 3–5 cm long.

Habitat: Open, moist to wet (at least in spring) meadows and streambanks.

Notes: Western blue flag roots are **poisonous**, and they should never be taken internally. People gathering edible rootstocks in wetlands, such as those of cattails (*Typha* spp.) or sweetflag (*Acorus americanus* (Raf.) Raf., previously called *A. calamus* L.), must be careful not to confuse them with blue flag rhizomes. Iris rootstocks are odourless and unpleasant-tasting, whereas those of cattails are odourless and bland, and those of sweetflag are pleasantly aromatic.
• Historically, small amounts of blue flag roots were taken to induce vomiting and cleanse the system. Blue flag was an official drug in the US *National Formulary* until 1947 for the treatment of syphilis, as an alternative to the use of bismuth, arsenic, mercury and so on, but its effectiveness was questionable. The fresh roots were also said to be effective in treating *Staphylococcus* sores, when applied as a poultice.

Rare *Vascular* Plants of Alberta

Monocots IRIS FAMILY

PALE BLUE-EYED GRASS

Sisyrinchium septentrionale Bicknell
IRIS FAMILY (IRIDACEAE)

Plants: Slender, tufted perennial herbs; stems 2-edged, flattened, 10–20 cm tall, 1–2 mm wide; from short underground stems (rhizomes) and fibrous roots.

Leaves: Alternate, near stem base, attached edgewise to the stem (like iris leaves) in 2 vertical rows, linear, grass-like, 1–2 mm wide.

Flowers: Pale violet-blue to almost white, delicate, about 8 mm long, with 6 spreading tepals; **tepals blunt or tapered to slender points**; flowers in small clusters from the axils of 2 erect bracts, with **flower stalks usually shorter than the inner bract**; outer bract 2.5–4 cm long; {April} May–July.

Fruits: Rounded capsules, 3–4 mm high, containing many black seeds.

Habitat: Moist meadows and grassy streambanks.

Notes: This species was called *Sisyrinchium sarmentosum* Suksd. *ex* Green in the first edition of the *Flora of Alberta*, but that name now applies to a distinct species. • Common blue-eyed grass (*Sisyrinchium montanum* Greene) is widespread in Alberta. It is very similar to pale blue-eyed grass, but its leaves are generally wider (up to 4 mm wide), its flower stalks are longer (longer than the inner bract), and its tepals are either notched or abruptly bristle-pointed. Some taxonomists have included both of these species in *Sisyrinchium angustifolium* Mill., but that name now applies to an eastern species. • The generic name *Sisyrinchium* was given to an iris-like plant by the Greek philosopher Theophrastus (ca. 372–c. 287 BC) and was later adopted by Linnaeus. The specific epithet *septentrionale* is Latin for 'northern' or 'of northern regions.'

Rare *Vascular* Plants of Alberta

ORCHID FAMILY **Monocots**

STEMLESS LADY'S-SLIPPER

Cypripedium acaule Ait.
ORCHID FAMILY (ORCHIDACEAE)

Plants: Perennial herbs, 10–30 {60} cm tall; stems **single, erect, leafless**, hairy; from short underground stems (rhizomes) and coarse, thick roots.

Leaves: Basal, 2, nearly opposite, narrowly **oblong or egg-shaped**, 10–28 cm long, 5–15 cm wide, stalkless, strongly **pleated**, sparsely hairy, bright green.

Flowers: Deep to pale pink, showy, with **yellow-green to purplish brown, 3–5 cm long, lance-shaped sepals** (3) **and side petals** (2) above a hanging, inflated, **3–6 cm long, pouch-like lip** that is usually **pink** (occasionally white) with purple veins and **deeply grooved** down the middle; **solitary**; late June–July.

Fruits: Erect capsules, widest at the middle, 3 cm long, 1 cm wide, strongly ribbed, containing thousands of tiny seeds.

Habitat: Wetlands, woods and sand dunes; elsewhere, in habitats with sterile, acidic soil and light shade, including sand ridges, jack pine woods and sphagnum bogs.

Notes: Stemless lady's-slipper is also known as pink lady's-slipper and moccasin-flower. It is the only lady's-slipper in Alberta that has only basal leaves (none on the flowering stalk) and a pink, deeply grooved lip petal. • The hairy plants **can cause a rash** in sensitive individuals. • Stemless lady's-slipper has been used by native peoples as a sedative. • Pollination is carried out primarily by bees, which enter the lip through a slit at the upper end and exit through an opening at the base of the pouch. These beautiful flower pouches can also conceal danger: they often harbour predators such as crab spiders. • Colonies of stemless lady's-slipper produce few seed capsules. This is due in part to the mechanism of pollination. Bees are initially attracted to the flowers by smell and colour, but they soon discover that the flowers contain no nectar or available pollen and avoid these flowers. Each capsule contains tens of thousands of seeds, and this high seed set helps to offset the small number of mature capsules. • Picking and trampling can limit population size, especially if the flower that is killed is one of the few to set seed. • Stemless lady's-slipper is the provincial floral emblem of Prince Edward Island.

Monocots

ORCHID FAMILY

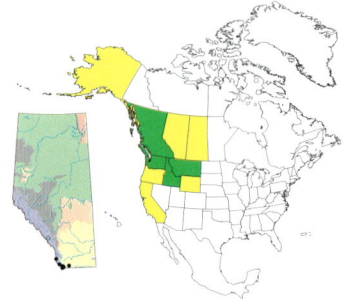

MOUNTAIN LADY'S-SLIPPER

Cypripedium montanum Dougl. *ex* Lindl.
ORCHID FAMILY (Orchidaceae)

Plants: Glandular-hairy perennial herbs; **stems leafy**, 20–50 {70} cm tall; from short underground stems (rhizomes) and coarse, thick roots.

Leaves: Alternate, 4–6, egg-shaped to broadly lance-shaped, 5–16 cm long, 3–8 cm wide, with many parallel veins, stalkless, sheathing at the base.

Flowers: White and purple or purplish green, showy, sweet-scented, about 8–10 cm across, with **3 greenish brown to brownish purple**, lance-shaped **sepals** (3–6 cm long, 1–1.5 cm wide), **2 narrowly lance-shaped, twisted, brownish purplish petals** (4–7 cm long, 3–6 mm wide) spreading to the sides, and a **broad, white, pouched lower lip petal**, tinged and veined with pink or purple near the base and spotted with purple on the inside (2–3 cm long, 1.3–1.7 mm deep and wide), also with an **egg-shaped, purple-dotted, yellow** lobe (staminode) at the mouth of the pouch; flowers usually **in pairs** but sometimes single or 3 per plant; June–August.

Fruits: Oblong capsules, 2–3 cm long, 1 cm wide, erect or ascending, containing many tiny seeds.

Habitat: Moist open places in the mountains at elevations below 1,700 m.

Notes: Mountain lady's-slipper most closely resembles yellow lady's-slipper (*Cypripedium calceolus* L.), but yellow lady's-slipper has single, bright-yellow flowers with a triangular staminode at the mouth. The only other lady's-slipper in Alberta with a white pouch is sparrow's-egg lady's-slipper (*Cypripedium passerinum* Richards.), and that species has much smaller flowers with short (1–1.5 cm long), egg-shaped sepals. • Orchids, and especially members of the genus *Cypripedium*, are ancient erotic and sexual symbols. The common name 'orchid' and the scientific names *Orchis* and Orchidaceae all stem from the Greek *orchis* (testicle). The generic name *Cypripedium* is derived from the Latin *Cypris* (Venus, the goddess of love and beauty) and *podion* (small foot). This refers to the delicate, slipper-like pouches of these beautiful flowers. The specific epithet *montanum* means 'of the mountains.'

ORCHID FAMILY · **Monocots**

BROAD-LIPPED TWAYBLADE

Listera convallarioides (Sw.) Nutt.
ORCHID FAMILY (Orchidaceae)

Plants: Delicate perennial herbs 8–20 {35} cm tall; stems slender, glandular-hairy on upper parts; from fibrous roots.

Leaves: Opposite (or nearly so), **1 pair near the middle of the stem, broadly egg-shaped** to almost round, blunt or abruptly pointed, 3–5 {2–8} cm long, hairless.

Flowers: Yellowish green, with 3 lance-shaped to linear, **4–5 mm long sepals** and 2 slightly smaller petals, all **bent backward above a broad lower lip petal**; lip wedge-shaped, 8–10 {13} **mm long, fringed** with fine hairs, **notched at the tip**, narrowed abruptly to a **slender basal section** (claw) and usually with 2 small teeth near the base, **projecting** outward almost **horizontally**; stalks slender, 4–8 {10} mm long; 5–25 flowers in slender, elongated clusters (racemes); {June} July–September.

Fruits: Many-seeded, oval capsules, usually slightly glandular.

Habitat: Boggy woods and meadows; elsewhere, in deep woods, on streambanks and by lakes.

Notes: This species has also been called *Ophrys convallarioides* (Sw.) Wight. • Broad-lipped twayblade might be confused with another rare species, western twayblade (*Listera caurina* Piper, also called *Ophrys caurina* (Piper) Rydb.), but the flowers of western twayblade have a smaller (4–6 mm long) lower lip with 2 prominent teeth at their base and without the fringing hairs and slender claws found in broad-lipped twayblade. Its ovaries and capsules are also hairless. Western twayblade grows in moist woods and flowers from June to July {August}. • Charles Darwin studied the fascinating pollination mechanisms of twayblades. These tiny flowers shoot their pollen out in a sticky drop of fluid that glues the pollinia to visiting insects. • The name 'twayblade' originated from the archaic word *tway* (two), in reference to the 2 leaf blades on these delicate plants. *Listera* honours Dr. Martin Lister, an English naturalist who lived 1638–1711.

L. CAURINA

Rare *Vascular* Plants of Alberta

Monocots

ORCHID FAMILY

BOG ADDER'S-MOUTH

Malaxis paludosa (L.) Sw.
ORCHID FAMILY (Orchidaceae)

Plants: Slender perennial herbs; **stems** erect, 5–15 {4–20} cm tall; from a swollen 'bulb' (corm).

Leaves: Alternate, near the stem base, **2–5**, broadly lance-shaped, blunt-tipped, 5–12 mm long, 3–10 mm wide, stalkless, sheathing the stem, **often producing small bulbils in their axils**.

Flowers: Yellowish green or pale green, inconspicuous, **1–1.5 mm long** and less than half as wide, **rotated 180°** and therefore with the lip petal pointing upward; **sepals broadly lance-shaped, 2 pointing upward and 1 pointed downward (like a lip)**; petals about as long as the sepals, with **2 slender side petals and a broader, green-striped lip petal pointing upward** (erect); 15–30 flowers in very slender, loose, elongated clusters (racemes); {June–August}.

Fruits: Erect capsules, 4 mm long, 2 mm wide, splitting into 3 parts to release many tiny seeds.

Habitat: Mossy ground (usually on peat moss [*Sphagnum* spp.]) in bogs and fens.

Notes: This species has also been called *Ophrys paludosa* L. and *Hammarbya paludosa* (L.) Kuntze. • The genus *Malaxis* can be distinguished from other orchids by its swollen stem bases. The 2 members of this genus that occur in Alberta both have leaves near the stem base. This feature can be used to distinguish them from other tiny orchids such as the twayblades (*Listera* spp.). • White adder's-mouth (*Malaxis monophylla* (L.) Sw. var. *brachypoda* (A. Gray) Morris & Ames, also known as *Malaxis brachypoda* (A. Gray) Fern.) is also rare in Alberta. It has a single leaf (rarely 2), and the lip petal has 2 small lobes (auricles) at its base and points downwards. White adder's-mouth typically grows in slightly drier, less acidic habitats such as damp woods, thickets and the drier parts of bogs and fens. It is reported to flower from mid June to August. • Bog adder's-mouth is thought to be one of the rarest orchids in North America, but because of its small size and inconspicuous colour, it may often be overlooked. • Both adder's-mouths are pollinated by small insects.

M. MONOPHYLLA var. BRACHYPODA

Rare *Vascular* Plants of Alberta

ORCHID FAMILY — **Monocots**

SLENDER BOG ORCHID

Platanthera stricta Lindl.
ORCHID FAMILY (ORCHIDACEAE)

Plants: Slender, hairless perennial herbs; stems erect, 20–50 {110} cm tall; from clusters of fleshy, swollen roots.

Leaves: Alternate, several, oblong to lance-shaped, often widest above the middle, 4–12 {3–18} cm long, slightly less than 3 times as long as they are wide, stalkless, clasping or sheathing the stem; the lowermost 2 or 3 leaves reduced to bladeless sheaths.

Flowers: Green, sometimes tinged with purple or brown, odourless or with a faint fragrance, about 1 cm long, somewhat 2-lipped, with 3 sepals, 3 petals and a hollow basal appendage (spur); sepals and petals 3–7 mm long, with the egg-shaped upper sepal and 2 upper petals curved together in a hood, and the 2 narrower side sepals flanking the slender lip petal; **spur pouch-shaped**, often purplish, ½–⅔ **as long as the lip petal** and often hidden beneath it; many stalkless flowers **in loose, narrow, elongating clusters** (spike-like racemes) ⅓–½ as long as the stem; late June–August.

Fruits: Many-seeded capsules about 1 cm long.

Habitat: Wet meadows and forests.

Notes: This species has also been called *Habenaria saccata* Greene. • The bog orchids are recognized by their small, slender-lipped flowers, each with a hollow spur projecting back from the base, and by their numerous leaves (although some species have only 1–3 basal leaves). Slender bog orchid might be confused with northern green bog orchid (*Platanthera hyperborea* (L.) Lindl., also called *Habenaria hyperborea* (L.) R. Br.), but northern green bog orchid has slender, cylindrical spurs that are 5–8 mm long (as long as the lip petal). Its flower clusters also tend to be much more densely flowered.

Monocots

ORCHID FAMILY

NORTHERN SLENDER LADIES'-TRESSES

Spiranthes lacera (Raf.) Raf. var. *lacera*
ORCHID FAMILY (ORCHIDACEAE)

Plants: Slender perennial herbs; stems erect, 20–80 cm tall; from clusters of 2–several fleshy, tuberous roots.

Leaves: In a **basal rosette** of {2} 3–5, light green, shiny, **oval to elliptic**, 1–4 {5.5} cm long, 1–2 cm wide, short-stalked, usually wilted or withered by flowering time.

Flowers: Crystalline white, fragrant, {2.5} 3.5–5.5 mm long, somewhat 2-lipped; sepals 4–5 mm long, with the egg-shaped upper sepal joined to the 2 side petals, and the 2 narrower side sepals flanking the lower lip; petals 4–5 {6} mm long, the 2 upper petals linear and blunt-tipped, and the lower **lip** petal wider (oblong, 2–3 mm wide), fringed at the tip, and marked **with a** distinctive **green spot** at the centre; 13–25 {9–35} flowers in a single, vertical, spiralling row, forming a **slender spike**, often with the lowermost flowers separated; mid July {August}.

Fruits: Egg-shaped capsules, 3–5 mm long and 2 mm wide.

Habitat: Dry, rocky, open woods and grassy areas in or near jack pine–lichen forest, often with common blueberry (*Vaccinium myrtilloides* Michx.).

Notes: This species has erroneously been called *Spiranthes gracilis* (Bigel.) Beck. • Hooded ladies'-tresses (*Spiranthes romanzoffiana* Cham.) is the only other ladies'-tresses in Alberta. It has narrower, linear leaves, its lip petal does not have a green spot, and its flowers form compact spikes with several vertically spiralling rows of flowers. Where they occur together, these species produce hybrids known as *Spiranthes* x *simpsonii*. • Northern slender ladies'-tresses grows for 3–5 years before it begins flowering. It can spread rapidly once it invades an area, and colonies are known to persist for several decades. It was first reported for Alberta in the early 1990s, where it is known only from the northeastern corner of the province. • Northern slender ladies'-tresses is pollinated by several species of small bees. • The generic name *Spiranthes* was taken from the Greek *speira* (coiled) and *anthos* (flower), in reference to the spiralling rows of flowers. The specific epithet *lacera* means 'lashed or torn,' a reference to the ragged edge of the lip petal. The common name 'ladies'-tresses' probably stemmed from the resemblance of the twisted flower spike to a braid of hair.

Dicots

BUCKWHEAT FAMILY **Dicots**

NODDING UMBRELLA-PLANT

Eriogonum cernuum Nutt.
BUCKWHEAT FAMILY (POLYGONACEAE)

Plants: Slender annual herbs, 10–40 cm tall; stems freely branched, usually divided in 3s (trichotomous) near the base and in 2s (dichotomous) on the upper parts, **slender**, hairless or somewhat woolly near the base; from slender taproots.

Leaves: Basal, spreading; blades **round to oval**, 1–2 cm wide, densely **white-woolly beneath**, usually greenish and slightly woolly on the upper surface; stalks shorter to longer than the blades, **lacking stipules**.

Flowers: **White to pink**, cone-shaped, **about 2 mm across** and 2 mm long, with **3 broad, wavy-edged lobes** (sepals) notched at the tip, **alternating with 3 narrower lobes** (petals); 3 styles and 3 stigmas; **9 stamens**, ¾ as long as the sepals and petals; borne on short stalks in **several-flowered, flat-topped** (umbel-like) **heads from hairless, 5-lobed cones** of fused bracts (involucres) about 1.5–2 mm long, these heads borne singly **on slender branches about 5–15 mm long** that are often **bent sharply downward** in open, freely branched clusters (cyme-like); {June} July–September.

Fruits: Dry, **3-sided**, seed-like fruits (achenes), enclosed in the calyx.

Habitat: In coarse-grained sand in badlands, on dry valley rims and slopes, and on active (though usually partially stabilized) sand dunes; elsewhere, on sandy desert hills.

Notes: The total population of nodding umbrella-plant in Alberta is estimated at fewer than 10,000 plants. A few small populations are found in Dinosaur and Writing-on-Stone provincial parks, but none of the larger populations are yet protected formally. With dune stabilization, other species will eventually crowd out some populations. • Umbrella-plants (*Eriogonum* spp.) are an important source of nectar for bees, and the honey produced from these flowers is said to be of excellent quality. • The specific epithet *cernuum* means 'nodding,' in reference to the nodding flower clusters.

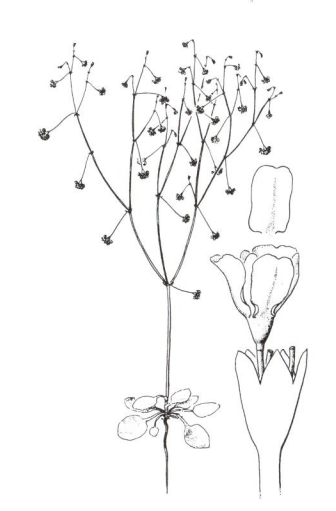

Rare *Vascular* Plants of Alberta

Dicots

BUCKWHEAT FAMILY

SILVER-PLANT

Eriogonum ovalifolium Nutt. var. *ovalifolium*
BUCKWHEAT FAMILY (POLYGONACEAE)

Plants: Tufted, white-woolly and silky-hairy perennial herbs; stems **5–15 cm tall**, leafless; often **forming mats** from densely branched, woody root crowns.

Leaves: Basal, many, **egg-shaped to round**, widest at or above the middle, **about 1 cm long**, densely white-woolly, stalked.

Flowers: Creamy to pale yellow, pink-tinged with age, hairless, 4 mm long, with 3 oval to round outer tepals and 3 spatula-shaped inner tepals; on small, slender stalks in hairy, cup-like involucres which are **stalkless and form a dense head-like cluster** (umbel) above a whorl of very small involucral bracts; June–July {May–August}.

Fruits: 3-sided, seed-like fruits (achenes).

Habitat: Dry, open, often rocky sites.

Notes: Another species that may occur in Alberta, few-flowered umbrella-plant (*Eriogonum pauciflorum* Pursh, also called *Eriogonum multiceps* Nees), is a low (usually less than 20 cm tall), tufted, perennial plant with small clusters of whitish or pale-pink flowers at the tips of slender stems. However, it has slender, lance-shaped leaves and the outer surface of its flowers is covered with long hairs. Few-flowered umbrella-plant grows on eroded banks and rocky ridges in prairie badlands. It flowers in July {August}. • Silver-plants can do well in sunny rock gardens, but they are often difficult to grow. They are best propagated from seed. • Prospectors and miners believed that this species indicated the presence of silver ore, so they called it 'silver plant.' • The generic name *Eriogonum* was derived from the Greek *erio* (woolly) and *gony* (knee or joint), in reference to the woolly stems and jointed flower stalks of many species.

E. PAUCIFLORUM

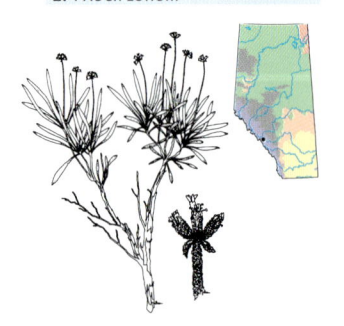

Rare *Vascular* Plants of Alberta

KOENIGIA

Koenigia islandica L.
BUCKWHEAT FAMILY (POLYGONACEAE)

Plants: Tiny, hairless annual herbs; stems slender, simple or branched, 1–4 {15} **cm tall, often reddish**; from slender taproots.

Leaves: Alternate or opposite, somewhat **fleshy**, broadly elliptic to lance-shaped, widest above the middle, **2–5 {9} mm long**, 1.5–3 mm wide, with translucent, **sheathing stipules** (ochreae).

Flowers: Green, sometimes whitish or reddish, inconspicuous, **1–1.4 mm long**, with 3 sepals and no petals; few to several in small, flat-topped clusters (umbels) immediately above 2–4 leaf-like bracts; July–August.

Fruits: Hairless, **3-sided**, seed-like fruits (achenes), enclosed by the sepals.

Habitat: Moist silt or mud in alpine areas; elsewhere, by snow beds and on wet moss.

Notes: Koenigia is found all around the world; in western North America, its range extends from Alaska to Colorado, but its tiny plants are easily overlooked. It is classified as rare in the continental Northwest Territories, British Columbia, Alberta, Manitoba, Ontario and Colorado. Most annual species penetrate only short distances into the Arctic from the south. However, koenigia is found far to the north, reaching Devon Island in the Canadian Arctic Archipelago, both coasts of Greenland, and Peary Land. It seems unlikely that it ever grows as a biennial. All viable seeds sprout promptly in their first spring, even though a partial delay of germination beyond the first year would be an effective safeguard against disastrous summers. This tiny plant is usually confined to wet sites in areas with irregular topography, where evaporation keeps plant temperatures below the lethal level in summer, and where variations in relief both ensure dynamic warming of the air and allow stratification of the air close to the surface.

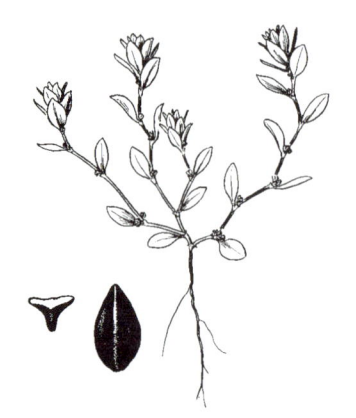

Dicots

BUCKWHEAT FAMILY

LEAST KNOTWEED

Polygonum minimum S. Wats.
BUCKWHEAT FAMILY (POLYGONACEAE)

Plants: Dwarf annual herbs, 3–15 {25} cm tall; stems **wiry**, less than 1 mm thick, leafy, usually **branched** at the base and **ascending**, sometimes unbranched and erect, often zigzagged, round or slightly angled, with rough lines of bran-like particles; from shallow taproots.

Leaves: Alternate, usually crowded and only slightly smaller on the upper stem and concealing the flowers; **blades jointed** at the base, **oblong-elliptic to egg-shaped**, blunt-tipped or slightly pointed, 5–15 {20} **mm long**, some usually at least ½ as wide as long; stalks short or absent; sheathing stipules (ochreae) usually 2-lobed, becoming **torn** along the edge.

Flowers: Greenish with white or pinkish (not yellow) **edges**, cone-shaped, usually **less than 2.5 mm long**, cut about ⅔ of the way to the base into **5 broad lobes**; 3 stigmas, nearly stalkless; **in groups of 2–3** {1–4} **on erect, 1–2 mm long stalks from leaf axils**, from the tip of the stem almost to the base; July–August {September}.

Fruits: Dry, shiny, dark brown or nearly black seed-like fruits (achenes) **with 3 concave sides**, 2–2.5 mm long, only slightly longer than the petal lobes, **erect** or nearly so.

Habitat: Dry ground in alpine to subalpine zones, rarely below 1,100 m elevation; elsewhere, on sandy soil and rock outcrops.

Notes: Least knotweed plants may flower when they are only 2–3 cm tall. Leaf production continues gradually throughout the lifespan of the plant and declines only slightly during fruit development. After a leaf has expanded, a flower develops in its axil. In this way, sexual reproduction continues throughout the life of the plant, and flowers and fruits at all stages of development can be found on a single plant. Drought delays reproduction and slows seed production, but least knotweed has greater drought tolerance than most other annuals in these habitats. • The generic name *Polygonum* was taken from the Greek *poly* (many) and *gony* (knee or joint), in reference to the many swollen joints of some species.

WHITE-MARGINED KNOTWEED

Polygonum polygaloides Meissn. ssp. *confertiflorum*
(Nutt. *ex* Piper) Hickman
BUCKWHEAT FAMILY (Polygonaceae)

Plants: Small, **hairless annual** herbs, 3–15 {2–25} cm tall; stems **erect or ascending**, unbranched or branched in 2s, usually sharply angled; from slender taproots.

Leaves: Alternate, linear to **linear–lance-shaped**, **1–3 {0.5–4} cm long** and about 1–2 mm wide, with obscure side veins, jointed at the base above 4–7 {2–8} mm long, **short-sheathing stipules** that are **deeply cut into slender lobes**.

Flowers: Greenish with white or pinkish edges, broadly **cone-shaped, 2–2.5 {3.5} mm long** and wide, **cut over ⅔ of the way to the base** into **5 oblong to lance-shaped lobes**, the outer 3 clearly broader and longer than the inner 2; usually **8 (sometimes 3) fertile stamens**; 3 styles, about 0.3 mm long; numerous, almost stalkless, erect or spreading, in small clusters of {1} 2–4 **in the axils of crowded, linear, leaf-like bracts** that are less than 3 times as long as the flowers, forming **dense spikes** about 1 cm long and ⅓–⅔ as wide **at stem tips**; June {May–August}.

Fruits: Dry, **3-sided**, seed-like fruits (achenes), 1.5–2 mm long, greenish yellow to brown, shining and almost smooth, to (commonly) **dull blackish and prominently ribbed** (often on the same plant).

Habitat: Moist meadows and flats; elsewhere, in damp silty or gravelly areas, in seasonally wet sites (e.g., vernal pools).

Notes: This species has also been called *Polygonum watsonii* Small, *P. imbricatum* Nutt., *P. kelloggii* Greene var. *confertiflorum* (Nutt. *ex* Piper) Dorn and *P. confertiflorum* Nutt. *ex* Piper. • This is a complex species, which includes several taxa that were once recognized as separate species. *Polygonum polygaloides* and *P. watsonii* were distinguished from *P. confertiflorum* and *P. kelloggii* primarily on the basis of the number of fertile stamens (8 in the former 2 species and 3 in the latter 2) and by their shorter bracts, which lack white edges.

Dicots BUCKWHEAT FAMILY

ALPINE SHEEP SORREL, MOUNTAIN SHEEP SORREL

Rumex paucifolius Nutt. *ex* Wats.
BUCKWHEAT FAMILY (POLYGONACEAE)

Plants: Hairless perennial herbs; stems loosely tufted, 1–few, slender, 20–60 {15–70} cm tall, unbranched below the flower cluster; from short underground stems (rhizomes) on thick taproots.

Leaves: Several, basal and {1} 2–3 alternate on the stem, **lance-shaped or elliptic** (never arrowhead-shaped), **tapered to a blunt tip and a wedge-shaped base**, 4–10 {13} cm long and about 2.5 cm wide; stalks long on basal leaves, but gradually shorter and eventually absent on upper leaves, **sheathing the stem with fused, membranous stipules** at their bases.

Flowers: Greenish, usually **red-tinged**, with **male and female flowers on separate plants** (dioecious), cupped, about 2–3 mm long and wide, with **3 spreading to erect, lance-shaped outer lobes** (never bent sharply backward) and **3 rounded inner lobes; 3 stigmas**; stalks very slender, clearly jointed near the base; numerous, in branched clusters (panicles) with many vertical branches, the whole cluster often as long as the stem; {June–July} August.

Fruits: Smooth, 3-sided, seed-like fruits (achenes), about 1.5 mm long, with 3 sharp angles, enclosed by **3 reddish, veiny, round to heart-shaped flaps** (valves) {2.5} **3–4 mm long**, nodding on slender stalks at least as long as the valves; valves veined, rarely with a tiny (never large and conspicuous) bump (callosity, tubercle) near the base.

Habitat: Moist meadows from montane to alpine elevations.

Notes: Many dock and sorrel species (*Rumex* spp.) are eaten as vegetables, but most are high in oxalic acid, which can be toxic in large amounts. Alpine sheep sorrel is said to be one of the more acid-tasting species. • Although alpine sheep sorrel is widely distributed, its populations are usually very local and sporadic. It is easily distinguished from other dioecious docks by its tapered leaf bases and upward-curved sepals at the base of its flowers and fruits. Other species have pointed lobes at the base of their leaf blades and their 3 small outer sepals are bent sharply backward. • These plants are pollinated by wind, so they produce large amounts of pollen. Dock plants are notorious among allergy sufferers as a source of irritation. For example, green sorrel (*Rumex acetosa* L.) can produce about 4 million pollen grains per plant.

GOOSEFOOT FAMILY **Dicots**

WEDGESCALE, SALTBUSH

Atriplex truncata (Torr.) A. Gray
GOOSEFOOT FAMILY (CHENOPODIACEAE)

Plants: **Annual** herbs; stems erect, 15–40 {100} cm tall, often with many **ascending branches**; from slender taproots.

Leaves: Alternate, or the lower opposite; blades broadly oval to triangular, with wedge-shaped to heart-shaped bases, 10–25 {40} mm long, nerveless, **covered with fine, grey granules** (sometimes few on the upper surface), **veins in a netted pattern visible under magnification** (kranz-type venation), **toothless**, though sometimes wavy-edged; stalks short or absent.

Flowers: Small, **inconspicuous**, without petals, either male or female but with flowers of both sexes on the same plant; male (staminate) flowers with a 5-parted calyx; female (pistillate) flowers lacking a calyx, but with a pair of **triangular bracts 2–2.5 {3.5} mm long** with broad tips and tapered bases, usually **tipped with 3 shallow teeth**, usually **smooth** (sometimes with 1 or 2 obscure bumps (tubercles) on the sides); in compact, head-like clusters in leaf axils and at the branch tips; {June–July} August.

Fruits: Dry seeds, about 1.5 mm long, enclosed by the united, enlarged bracts.

Habitat: Alkaline flats and disturbed areas.

Notes: This species is very similar to Powell's saltbush (*Atriplex powellii* S. Wats.), also considered rare. Powell's saltbush has oblong flowers and fruit bracts (widest above the middle) with a flattened lobe at the tip and small bumps on the sides. Its leaves are prominently 3-nerved. Powell's saltbush also resembles silver saltbush (*Atriplex argentea* Nutt.), a common species, but the latter has teeth all along the edges of the flower or fruit bracts and much larger, toothed leaves. Powell's saltbush is a species of alkaline flats and badlands that flowers from {July} August to September. • Also rare in Alberta is four-wing saltbush (*Atriplex canescens* (Pursh) Nutt.). Four-wing saltbush is a shrub, 10–40 {200} cm tall, with spreading, woody branches. This widely distributed, highly variable complex grows on saline flats and flowers {June to September}. It is distinguished from Alberta's other shrubby *Atriplex*, salt sage (*Atriplex nuttallii* S. Wats. in the *Flora of Alberta*), by the 4 conspicuous, longitudinal, wavy-edged to irregularly toothed wings on the bracts enclosing its seeds.

A. POWELLII

Rare *Vascular* Plants of Alberta

Dicots

GOOSEFOOT FAMILY

NARROW-LEAVED GOOSEFOOT

Chenopodium leptophyllum (Nutt. *ex* Moq.) Nutt. *ex* S. Wats.
GOOSEFOOT FAMILY (CHENOPODIACEAE)

Plants: Greyish green to whitish annual herbs **covered with fine granules** (farinose); **stems erect** or nearly so, up to 40 {80} cm tall; from slender taproots.

Leaves: Alternate; blades linear to lance-shaped, up to 30 mm long and **less than 15 mm wide, 1-veined from the base**, not toothed, sometimes with 1 or 2 lobes above the base, fleshy; stalks short.

Flowers: Small, **lacking petals**, with both male and female parts; **calyx split** almost to the base **into 5 lobes with a sharp ridge down the back** of each, remaining attached to the mature fruit; numerous, in loose to dense head-like clusters, widely spaced on a single stem or on several branches; {May} June–August.

Fruits: Thin-walled, single-seeded nutlets, **covered by the calyx lobes when mature, mostly horizontal**, lens-shaped, 0.9–1.1 mm long and 0.9–1.1 mm wide, black, with a slightly wrinkled seed coat.

Habitat: Open, slightly disturbed, sandy areas; elsewhere, in sandy blowouts under deciduous vegetation and on shale cliffs.

Notes: Another rare Alberta species, dark-green goosefoot (*Chenopodium atrovirens* Rydb., also called *Chenopodium fremontii* S. Wats. var. *atrovirens* (Rydb.) Fosberg), also has moderately to densely mealy plants with slender leaves (less than 15 mm wide and 1–5 times longer than wide) and horizontal, lens-shaped seeds (rarely vertical). However, its calyx lobes are much smaller, and they do not cover the mature seeds. Dark-green goosefoot grows in open, disturbed areas, usually at higher elevations, and is reported to flower from June to September. It could be confused with a more common species, meadow goosefoot (*Chenopodium pratericola* Rydb.), but meadow goosefoot is more erect, with flowerheads in elongated clusters. Its leaves have 1 or 2 lobes or teeth above the base and its calyx closely covers the mature fruit. Meadow goosefoot grows along the edges of sloughs and in open, disturbed, moist alkaline areas in the parklands and prairies.• The seeds of narrow-leaved goosefoot were used for food, mixed with cornmeal and salt. Raw or cooked plants were also eaten.

C. ATROVIRENS

GOOSEFOOT FAMILY

Dicots

SMOOTH NARROW-LEAVED GOOSEFOOT

Chenopodium subglabrum (S. Wats.) A. Nels.
GOOSEFOOT FAMILY (CHENOPODIACEAE)

Plants: Erect or semi-erect annual herbs; stems light green, 10–50 {60} cm tall, with **many ascending branches**, smooth or sparsely mealy (farinose) on the upper parts; from slender taproots.

Leaves: Alternate; **blades** fleshy, linear, 1–5 cm long, **1–4 mm wide**, **1-veined**, greenish, only **lightly mealy**; stalks slender.

Flowers: Small, lacking petals, with both male and female parts in the same flower; **flowers in dense, head-like clusters, widely spaced on 1 stem or on several branches**; calyx lobes egg-shaped, sharp-pointed, broadest above the middle, with a sharp ridge down the back, remaining attached to the fruit; July {May–August}.

Fruits: Thin-walled, single-seeded nutlets, **covered by the calyx lobes when mature** but eventually falling free, **mostly horizontal**, lens-shaped, **1.3–1.6 mm long** and 1.2–1.4 mm wide, black, shiny.

Habitat: Sand dunes.

Notes: Another rare Alberta species, dried goosefoot (*Chenopodium dessicatum* A. Nels.), is similar to smooth narrow-leaved goosefoot, but dried goosefoot is branched from the base, which gives its plants a spreading appearance. Dried goosefoot grows on undisturbed saline soils and flowers in {May–July} August. • Both of these species are very similar to narrow-leaved goosefoot (p. 66), and some taxonomists included all 3 in the same species. Smooth narrow-leaved goosefoot has 1-veined, green, slightly mealy leaves and larger (1.5 mm) seeds that eventually fall free; narrow-leaved goosefoot has 1-veined, greyish, densely mealy leaves and smaller (1 mm) seeds that remain firmly attached to the outer wall of the fruit; dried goosefoot has 3-veined, greyish, densely mealy leaves and smaller (1 mm) seeds that eventually fall free. • The generic name comes from the Greek *chen* (goose) and *podos* (foot) because of the shape of the leaves of some species.

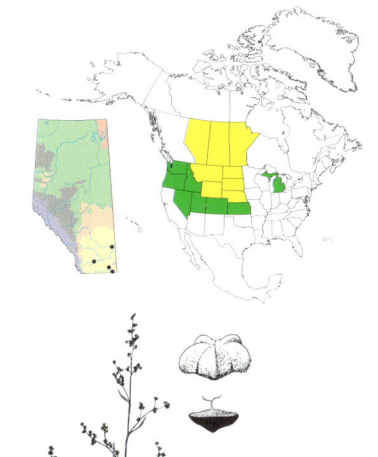

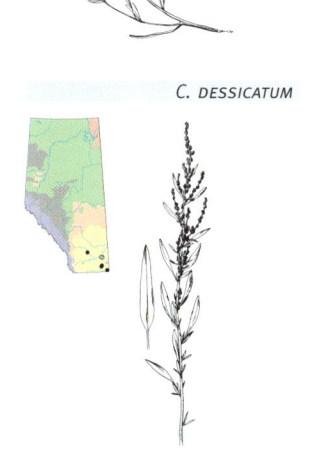

C. DESSICATUM

Rare *Vascular* Plants of Alberta

Dicots

GOOSEFOOT FAMILY

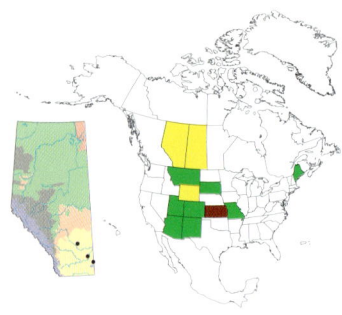

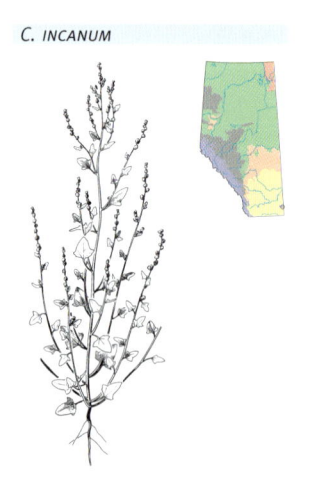

WATSON'S GOOSEFOOT

Chenopodium watsonii A. Nels.
GOOSEFOOT FAMILY (CHENOPODIACEAE)

Plants: Foul-smelling annual herbs; **stems angular**, erect or ascending, **2–15 cm tall**, freely branched, covered with fine granules (farinose); from slender taproots.

Leaves: Alternate, **broadly diamond-shaped to egg-shaped**, blunt-tipped, **wedge-shaped at the base**, 15–35 mm long, **more than 15 mm wide**, thick, **densely mealy** (farinose) **on both surfaces**, toothless.

Flowers: Small, lacking petals, with both male and female parts; 5 calyx lobes, egg-shaped or oblong, densely farinose; numerous, in large heads (glomerules), closely to widely spaced along leafy upper branches, forming branched, spike-like clusters (paniculate spikes); {June–September}.

Fruits: Thin-walled, single-seeded nutlets, **covered by the calyx lobes when mature, mostly horizontal**, lens-shaped, 1–1.3 mm long, 0.9–1.3 mm wide, with a **coarsely wrinkled seed coat**.

Habitat: Open areas, mainly in badlands.

Notes: Another species reported for Alberta, hoary goosefoot (*Chenopodium incanum* (S. Wats.) Heller, also called *Chenopodium fremontii* S. Wats. var. *incanum* S. Wats.), resembles Watson's goosefoot in having leaves 10–30 mm long that are diamond-shaped to round and densely covered with granules (though sometimes smooth above). It also has very similar flower clusters, though sometimes these lack leaves. However, hoary goosefoot is a larger plant (to 50 cm tall) and it lacks angular stems; it is not ill-smelling, and its seeds fall free of the calyx rather than remaining attached to it. It grows on dry plains, hillsides and open, sandy ground, and is reported to flower from June to September.

C. INCANUM

Moquin's Sea-blite

Suaeda moquinii (Torr.) Greene
GOOSEFOOT FAMILY (CHENOPODIACEAE)

Plants: Perennial herbs; stems erect to ascending, 20–70 cm tall, **hairless or short hairy**, branched; from a woody base and stout taproot.

Leaves: Alternate, crowded; blades fleshy, very narrow, flat to almost cylindrical, 1–2 cm long, **narrow at the base**.

Flowers: Small, without petals; **5 sepals, fused at the base, fleshy, of approximately equal lengths, rounded on the back**; 1–3 in the axils of upper leaves, stalkless, each usually with 2 small, translucent bracts; {June} July.

Fruits: Black nutlets, **1.5–2 mm wide**, smooth or slightly net-veined, usually horizontal, enclosed by the sepals.

Habitat: Moist saline or alkaline soil.

Notes: This species has also been called *Suaeda intermedia* S. Wats. • Western sea-blite (*Suaeda calceoliformis* (Hook.) Moq., also called *Suaeda depressa* (Pursh) S. Wats.), an annual species, is much more common in Alberta. Its sepals are unequal in length, with hood-like tips and a lengthwise ridge on the back (like the keel of a boat). Its leaves are 1–4 cm long and they have enlarged (rather than wedge-shaped) bases. The plants are usually green, but they can become almost black at maturity.

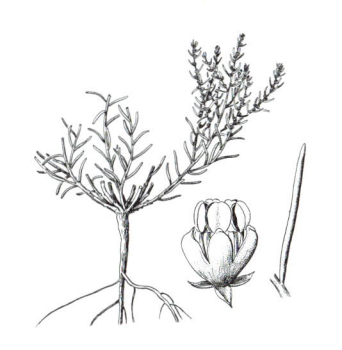

Dicots

GOOSEFOOT FAMILY

POISON SUCKLEYA

Suckleya suckleyana (Torr.) Rydb.
GOOSEFOOT FAMILY (CHENOPODIACEAE)

Plants: Low, somewhat **fleshy annual herbs**; stems reddish, hairless or with a few small scales, {10} 20–40 cm long, **spreading near the ground, with ascending branches**; from taproots.

Leaves: Alternate, essentially hairless; **blades round to broadly diamond-sided, tapered to a wedge-shaped base**, 1–3 cm wide, wavy-toothed along the upper edge; stalks equal to or longer than the blades.

Flowers: Small, **inconspicuous**, lacking petals, **male or female** (unisexual), but with both sexes on the same plant (monoecious); male flowers with a 3- or 4-lobed calyx and 3–4 stamens, in leaf axils toward stem tips; female flowers lacking sepals but **with 2 small, stiff bracts** that enlarge in fruit, in leaf axils **below the male flowers**; June {July–August}.

Fruits: Small, thin-walled and single-seeded (utricles), 5–6 mm long, abruptly pointed at tip, enclosed by a **flattened pair of 5–6 mm wide bracts with toothed wings and notched tips**.

Habitat: Moist, saline lake shores; elsewhere, along streams.

Notes: These plants contain hydrocyanic acid, which makes them **toxic to livestock**—hence the name 'poison suckleya.' They are reported to have caused the poisoning and death of cattle.

Rare *Vascular* Plants of Alberta

Californian Amaranth

Amaranthus californicus (Moq.) S. Wats.
AMARANTH FAMILY (AMARANTHACEAE)

Plants: Low, **mat-forming annual** herbs, green to reddish purple; stems freely **branched**, up to 30 cm long, prostrate; from weak, slender taproots.

Leaves: Alternate; **blades** elliptic to egg-shaped, usually widest above middle, **5–15 mm long**, with wavy, somewhat thickened edges; stalks very slender, 5–15 mm long.

Flowers: Male or female (unisexual) but with both sexes on the same plant (monoecious), **small** and inconspicuous, with **2 lance-shaped, 1–1.5 mm long bracts**, no petals, and **0 or 1** (female flowers) **or 3** (male flowers) **pointed sepals**; in small clusters in leaf axils; {July–October}.

Fruits: Ovoid, single-seeded capsules with 2 or 3 beaks, opening by a cap at the tip (circumscissile); **seeds round, lens-shaped, about 0.7 mm wide**.

Habitat: Moist, often disturbed ground on lake shores and roadsides; elsewhere, on moist alkaline flats.

Notes: This species may have been introduced to Alberta. • Many species of *Amaranthus* around the world have been used as pot herbs, but others are reported to be **toxic to livestock** and have caused the poisoning and deaths of pigs and cattle. • The generic name *Amaranthus* was taken from the Greek *amarantos* (unfading), in reference to the long-lasting flowers, which keep their appearance after they have been picked and dried. Some poets used amaranth as a symbol of immortality, and in ancient Greece it was used to decorate tombs and images of the gods.

Dicots

FOUR O'CLOCK FAMILY

SAND VERBENA

Tripterocalyx micranthus (Torr.) Hook.
FOUR O'CLOCK FAMILY (NYCTAGINACEAE)

Plants: Annual herbs, lying on the ground; **stems brittle, trailing but** with **tips pointing upward**, up to 60 cm long, almost hairless or with short glandular hairs.

Leaves: Opposite pairs; blades elliptic to egg-shaped, widest at or below the middle, 2–6 cm long, 1–3 cm wide; stalks usually slightly shorter than the blades.

Flowers: Greenish white, **tubular**, tipped with 5 wide-spreading lobes, 10–12 mm long, 3–5 mm across at tip, lacking petals but with **5 glandular-hairy, petal-like sepals**; in **flat-topped clusters** (umbels) with a whorl of small (5–10 mm long), lance-shaped bracts at the base, clusters on stalks shorter than the leaves and growing from leaf axils; {May–July}.

Fruits: Dry, seed-like fruits (achenes) enclosed in **2 or 3 wide, papery, conspicuously veined wings**, 20 {15–28} **mm long**.

Habitat: Sandy ground, usually on hard-packed level ground, but also on active, south-, west- and east-facing slopes of dunes and dune ridge tops.

Notes: This species has also been called *Abronia micrantha* Torr. • The encroachment of vegetation on active sand-dune areas is reducing the suitable habitat of this species. • Sand verbena superficially resembles wild begonia (*Rumex venosus* Pursh), but wild begonia is a perennial plant with running, elongated underground stems (rhizomes) and stout (not brittle), erect stems; its flowers are borne in leafy, branched clusters (panicles) at the stem tips. • Young sand verbena plants sometimes resemble members of the goosefoot family (Chenopodiaceae) in shape, colour and mealiness (of the underside of the leaves). • Several *Abronia* species have been grown in borders, rockeries and baskets. They make colourful displays in gardens and as cutflowers. Bougainvillea, a cultivated showy vine, is also in the four o'clock family. • The wings on these fruits aid dispersal by the gusty winds of sand-dune fields. • The four o'clock family is so named because the flowers tend to open in the late afternoon. The Greek word *nyx* means 'night,' a reference to the late opening of the flowers. *Micranthus* means 'small-flowered.'

BITTER-ROOT

Lewisia rediviva Pursh
PURSLANE FAMILY (Portulacaceae)

Plants: Low perennial herbs; stems leafless, 1–3 cm tall, with a whorl of **5–7 slender, leaf-like bracts**; from deep, **fleshy taproots**.

Leaves: Basal, **1–5 cm long**, club-shaped, **fleshy** and **nearly cylindrical**.

Flowers: Deep to light rose-pink (sometimes whitish) with a yellow or orange centre, **4–6 cm across**, with **12–18** lance-shaped **petals** and **6–9** pinkish, egg-shaped **sepals**, on long, slender stalks jointed just above the whorl of bracts; solitary; {April–June} July–August.

Fruits: Egg-shaped capsules with 6–20 dark, shiny seeds.

Habitat: Dry, open, montane sites; elsewhere, on desert flats.

Notes: Dwarf bitter-root (*Lewisia pygmaea* (A. Gray) B.L. Robins. ssp. *pygmaea*) is another rare Alberta species. It resembles bitter-root, but it has longer (5–15 cm), flatter leaves and smaller flowers (about 1.5–2 cm across) with 6–8 petals and only 2 sepals. Dwarf bitter-root grows on dry, rocky, alpine slopes and is reported to flower from late May to August. • These roots were one of the most important crops of the Flathead and Kutenai. Europeans found them too bitter, but native peoples considered them a treat. They were collected early in the spring, because their brownish black covering was most easily removed before the plants flowered. Roots were peeled and washed, then boiled until soft or dried for 1–2 days for storage. The inner core or 'heart' was removed to reduce bitterness, and storage for 1–2 years was also said to help. The cooked roots were eaten alone, mixed with sweeter foods such as berries and camas roots, or added to stews, soups and gravies as a thickener. • It took a woman 3–4 days to fill a 23-kg bag with these small roots, but this was enough to sustain a person through winter.

L. PYGMAEA

Dicots

PURSLANE FAMILY

SMALL-LEAVED MONTIA, SMALL-LEAVED SPRING BEAUTY
Montia parvifolia (Moc. *ex* DC.) Greene
PURSLANE FAMILY (PORTULACACEAE)

Plants: Succulent **perennial** herbs, **often producing bulblets** in leaf axils; stems several, very slender, 5–20 {25} cm long, erect or spreading at the base; from **short, thick root crowns** and **slender underground stems** (rhizomes), often forming patches from spreading, runner-like (stoloniferous) stems.

Leaves: Mainly basal, but also alternate on the stems, several, thick and fleshy; blades 1.5–3 {6} cm long, varying from **spoon-shaped** and 2–3 mm wide **to almost round** and 2 cm wide; stalks usually as long as the blades or longer; **stem leaves** gradually **reduced** to bracts on the upper stem.

Flowers: Rose-coloured or white with pink veins, rather showy, funnel-shaped to broadly cupped, about 1.5–2 cm across, with **5 spreading, 6–10 {15} mm long petals** and **2 unequal sepals**, 1.5–3 mm long; **6–10 stamens**; nearly erect in the axils of small bracts, usually 3–8, in elongating clusters (racemes); July {May–August}.

Fruits: Capsules, almost entirely covered by the persistent calyxes, splitting into 3 segments (valves) to release 2 {1–3} black, shiny, 1–1.5 mm long seeds covered with tiny bumps.

Habitat: Moist montane to alpine slopes and ledges, preferring sites that are moist in early summer and dry later; elsewhere, on lake shores, seepage areas and mossy streambanks.

Notes: This species has also been called *Claytonia parvifolia* Moc. • Linear-leaved montia (*Montia linearis* (Dougl. *ex* Hook.) Greene), also known as slender-leaved spring beauty, is also rare in Alberta. It is an annual species with fibrous roots and its stems are branched near the base. Its leaves are thin, linear and alternate on the stem, and it has white, nodding flowers. Linear-leaved montia flowers from May to July on moist to dry, open sites on sandy plains and hills at lower elevations. It also grows in disturbed habitats and in open woodlands. This is the most common, widespread species of the genus. • When a capsule opens, its 3 sections curl inward and downward. This presses the seeds firmly against one another and eventually throws them outward, up to 2 m from the plant.

M. LINEARIS

PINK FAMILY   **Dicots**

LOW SANDWORT

Arenaria longipedunculata Hult.
PINK FAMILY (CARYOPHYLLACEAE)

Plants: Low, tufted or loosely matted perennial herbs; stems short, **2–4 cm high**; from thread-like, underground, light-coloured runners (stolons).

Leaves: Opposite in **1 or 2 pairs**, overlapping, lance-shaped to egg-shaped (**about twice as long as they are wide**), blunt or somewhat pointed, slightly fringed near the base, **2–6 mm long.**

Flowers: White, inconspicuous, about 3–5 mm long, with 5 petals and 5 sepals; **petals rounded or shallowly notched** at the tip, equal to the sepals; sepals egg-shaped, blunt or slightly pointed, hairless or with short glandular hairs, indistinctly 3-nerved; **10 stamens**; **3 styles**; **solitary**, on erect, 1–2 cm long, finely hairy stalks with 1 pair of pointed, egg-shaped bracts below the middle.

Fruits: Egg-shaped capsules, slightly longer than the sepals, **splitting open from the tip into 6 tooth-like segments**; seeds several, brown, round, shiny, slightly wrinkled, 0.7–0.8 mm long.

Habitat: Moist gravelly areas at higher elevations; elsewhere, in wet, often mossy, places along rivers and streams and in tundra, on moist calcareous or serpentine gravel and in moist rock crevices.

Notes: This species has also been included in *Arenaria humifusa* Wahl. and called *Arenaria cylindricarpa* Fern. • In sheltered and shaded situations, low sandwort may grow taller and produce flowers and capsules that are raised well above the plant mat. This form has been called *Arenaria pedunculata* Hult. • *Arenaria* species are very similar to *Minuartia* species. The two genera are distinguished mainly by their capsules, which split into 3 parts in *Minuartia* (1 segment per style) and into 6 parts in *Arenaria* (2 segments per style) • The generic name *Arenaria* is from the Latin *arena* (sand) because of the sandy habitats of some species. The specific epithet *longipedunculata* refers to the long flowering stalks (peduncles).

Rare *Vascular* Plants of Alberta   75

PINK FAMILY

PURPLE ALPINE SANDWORT

Minuartia elegans (Cham. & Schlecht.) Schischkin
PINK FAMILY (CARYOPHYLLACEAE)

Plants: Loosely tufted, often purplish, **hairless** perennial herbs **about 2–4 cm tall**; stems numerous, from a branched root crown, forming tufts up to 30 cm in diameter.

Leaves: Opposite and in compact clusters (fascicles) in the axils of the main leaves, linear, blunt-tipped, **3–10 mm long**, **1-nerved** or appearing **nerveless**.

Flowers: White, saucer-shaped, about 5–6 mm across, with **5 petals** and **5 sepals**; petals oblong to egg-shaped, widest above the middle, blunt or shallowly notched at the tip, no longer than the calyx, rarely absent; **sepals separate, purple, 3–3.5 {2–4} mm long, broadly lance-shaped**, pointed at the tip, slightly ridged down the back (keeled), weakly **3-nerved**; 10 stamens; 3 styles; solitary, on **1–4 cm long stalks** at the stem tips; July–August.

Fruits: Ovoid or oblong **capsules, 2–4 mm long, splitting open from the tip into 3 tooth-like segments** to release tiny, minutely bumpy or spiny, kidney-shaped seeds.

Habitat: Moist, gravelly alpine slopes; elsewhere, on dry, turfy, gravelly or sandy calcareous barrens.

Notes: This species has also been called *Arenaria rossii* (R. Br. *ex* Richards.) Graebn. ssp. *elegans* (Cham. & Schlecht.) Maguire and *Minuartia rossii* R. Br. var. *elegans* (Cham. & Schlecht.) Hult. • This species is a member of the *Minuartia rossii* complex, which includes *Minuartia rossii*, *M. austromontana* and *M. elegans*. These 3 species have distinct geographical distributions. Ross' sandwort (*Minuartia rossii* (R. Br.) Graebn.) is a high-arctic to alpine species that has not yet been found in Alberta. Green alpine sandwort (*Minuartia austromontana* Wolf & Packer) is the most common species in Alberta, where it grows in the Rockies. It is distinguished from purple alpine sandwort by its green, linear to lance-shaped sepals and its usual lack of petals. Its leaves are awl-shaped and fleshy, and its plants are densely tufted. • The generic name *Minuartia* commemorates Juan Minuart (1693–1768) of Barcelona. The specific epithet *elegans* means 'elegant.'

PINK FAMILY   Dicots

KNOTTED PEARLWORT

Sagina nodosa (L.) Fenzl ssp. *borealis* Crow
PINK FAMILY (CARYOPHYLLACEAE)

Plants: Slender, **tufted** perennial herbs; **stems** ascending to loosely spreading to lying on the ground, **5–25 cm long**, **thread-like**, hairless or hairy at the joints; from branched root crowns.

Leaves: Opposite on the stem and basal (but not in rosettes), linear, 1.5–3 cm long on the lower stem, 1–2 mm **long and forming dense clusters** (fascicles) in leaf axils **on the upper stem**, giving the plant a knotted appearance.

Flowers: White, showy, saucer-shaped, about 6–10 mm across, with **5 petals** and **5 sepals** (sometimes 4 petals and 4 sepals); **petals** separate, **rounded or shallowly notched** at the tip, 3–5 mm long; **sepals separate**, often purple-tipped, elliptic, 2–3 mm long, hairless or with gland-tipped hairs at the base; **8 or 10 stamens**; **4 or 5 styles**; single, at branch tips; July {August–September}.

Fruits: Capsules, 3–4 mm long, with sepals pressed to the outer surface, **splitting into 4 or 5 thick segments** (valves) to release **many** dark brown to black, smooth to distinctly pebbled, egg-shaped to kidney-shaped **seeds** 0.5 mm long.

Habitat: Moist rocky, gravelly, sandy or peaty places, especially on shores; elsewhere, on damp sand in dune slacks, on dry sand on the lee side of fixed dunes and on salty ledges along the coast.

Notes: Snow pearlwort (*Sagina nivalis* (Lindbl.) Fries, also known as *Sagina intermedia* Fenzl) is a similar plant reported to occur in Alberta. It is distinguished by its smaller (less than 5 mm wide) flowers with fewer (mostly 4), shorter (about as long as the sepals) petals. It forms low (1–5 cm tall), compact cushions without leaf clusters in the upper leaf axils. Snow pearlwort grows in moist, gravelly alpine areas. • Spreading pearlwort (*Sagina decumbens* (Ell.) T. & G.) is an annual species without basal leaf clusters. It has small, 5-petalled flowers with purple-edged sepals, and its capsules split only halfway to the base. Spreading pearlwort is reported from Alberta and grows on sandy, alpine sites. • Double-flowered forms of knotted pearlwort are sometimes used as a filler in rock terraces and walls, and as a ground cover, but the plants do not withstand traffic as well as grass, and they are apt to invade nearby areas and become a troublesome weed.

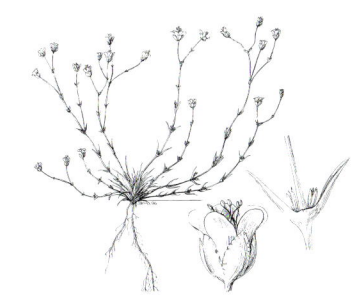

S. DECUMBENS

S. NIVALIS

Rare *Vascular* Plants of Alberta

Dicots   PINK FAMILY

ALPINE BLADDER CATCHFLY

Silene involucrata (Cham. & Schlecht.)
Bocquet ssp. *involucrata*
PINK FAMILY (Caryophyllaceae)

Plants: Perennial herbs, often purplish; stems single, erect, 5–35 cm tall, covered with long, **purple-banded, gland-tipped hairs** in the upper parts; from branched crowns of taproots.

Leaves: Mainly basal, lance-shaped, widest toward the tip, stalkless, 1.5–5.5 cm long, 2–8 mm wide, fringed with hairs; stem leaves in 1–3 pairs, narrower.

Flowers: Erect, **single** (sometimes 2 or 3); **5 petals, white to pinkish, broadly 2-lobed** at the tip and **projecting** {2} 3–7 mm **past the sepals**, narrowed abruptly to a slender base; 5 sepals, hairy, **united to form a 1–2 cm long bell or urn with 10 purple nerves**; 5 (sometimes 4) styles; July.

Fruits: Erect capsules, opening at the tip by 4 or 5 (sometimes 8–10) teeth; seeds kidney-shaped to round, edged with a narrow, flattened section (wing), 1–1.5 mm long.

Habitat: Gravelly and turfy alpine slopes; elsewhere, dry sites in meadows and on rock outcrops, river terraces and floodplains in arctic tundra.

Notes: This species has also been called *Silene furcata* Raf., *Lychnis furcata* (Raf.) Fern., *Lychnis affinis* J. Vahl and *Melandrium affine* J. Vahl. Its Alberta population is isolated (disjunct) from the widespread northern populations.
• Nodding cockle (*Silene uralensis* (Rupr.) Bocq. ssp. *attenuata* (Farr) McNeill, also called *Lychnis apetala* L. and *Melandrium apetalum* (L.) Fenzl.) is a more widespread, small species of rocky alpine slopes. However, its flowers nod, and their purplish petals are usually shorter than (and therefore contained within) its urn-shaped calyxes. • Alpine bladder catchfly often grows near animal dens. • The derivation of *Silene* is uncertain. It is either from the Greek *sialon* (saliva), referring to the sticky bands on the stem, or from *seilenos*, a type of woodland deity or satyr.

S. URALENSIS

PINK FAMILY | **Dicots**

SALT-MARSH SAND SPURRY

Spergularia salina J. & K. Presl var. *salina*
PINK FAMILY (CARYOPHYLLACEAE)

Plants: Small **annual herbs**, more or less glandular-hairy; **stems 5–20 cm long**, much branched, widely spreading or lying on ground; from slender taproots.

Leaves: Opposite; blades **somewhat fleshy**, slender to thread-like, **cylindrical**, 1–2.5 cm long, 0.6–1.5 mm wide, sharply pointed at the tip, bearing **triangular, membranous stipules** up to 4 mm long at the base.

Flowers: White or pinkish, with 5 small petals (shorter than the sepals) and 5 sepals 2.5–5 mm long; **usually 3 styles**; numerous, in loose, spreading clusters (cymes); May–August.

Fruits: Egg-shaped capsules, **3.5–6 mm long**, usually splitting to the base into **3 parts**, containing smooth, **brown or reddish seeds** 0.5–0.9 mm long.

Habitat: Brackish or saline mud and sands.

Notes: This species has also been called *Spergularia marina* (L.) Griseb. var. *leiosperma* (Kindb.) Ghrke. • Two-stamened sand spurry (*Spergularia diandra* (Guss.) Boiss.) has been introduced to Alberta from Europe. It has narrower leaves (0.5–1 mm wide), round capsules and black, 0.4–0.5 mm long seeds. • Salt-marsh sand spurry appears to be native to Alberta, but there are both native and non-native populations elsewhere. In southern Ontario, introduced populations are spreading along highways in response to the accumulation of salt used for melting ice in winter. • The generic name is derived from *Spergula*, a similar genus. *Spergula* was derived from the Latin *spargo* (to scatter), in reference to the wide scattering of the seeds.

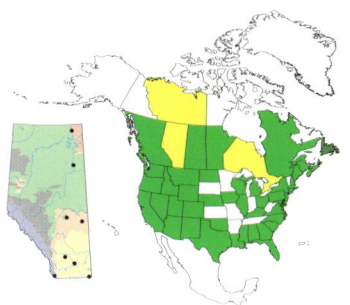

Rare *Vascular* Plants of Alberta

Dicots PINK FAMILY

AMERICAN CHICKWEED

Stellaria americana (Porter *ex* B.L. Robins.) Standl.
PINK FAMILY (CARYOPHYLLACEAE)

Plants: Delicate perennial herbs; stems slender, **sticky-hairy**, leafy, 10–20 cm long, lying on the ground; forming **loose mats** from elongated underground stems (rhizomes).

Leaves: Opposite, **sticky**, stalkless; blades **egg-shaped**, 1–3 cm long, usually blunt-tipped.

Flowers: White, with **5 deeply notched petals** 1½–2 times as long as the sepals; 5 sepals, blunt-tipped, 3–4 mm long, with glands; styles commonly 3; few, in branching, leafy clusters (cymes); {July} August.

Fruits: Elliptic capsules, shorter than the sepals, splitting lengthwise into 6 sections, several- to many-seeded.

Habitat: Rocky slopes, usually at high elevations.

Notes: Shining chickweed (*Stellaria nitens* Nutt.) is a delicate annual species with thread-like stems up to 20 cm tall. Its leaves are mainly on the lower half of the stem, and its flowers are borne in large, open clusters with membranous (not leaf-like) bracts. Shining chickweed has been reported to occur in Alberta. It grows on gravelly sites or grassy slopes in prairies and along streams, and is reported to flower from April to June. • Chickens and other poultry are said to like the seeds of these plants, hence the name 'chickweed.' The generic name *Stellaria* was derived from the Latin *stella* (star), in reference to the shape of the flower. *Americana* means 'of America.'

S. NITENS

PINK FAMILY | **Dicots**

MEADOW CHICKWEED
Stellaria obtusa Engelm.
PINK FAMILY (CARYOPHYLLACEAE)

Plants: Delicate, **hairless perennial** herbs; stems numerous, branched, lying on the ground, 5–15 cm long; forming **loose mats**.

Leaves: Opposite, stalkless; blades **egg-shaped, 4–8 mm long**, pointed at the tip, **fringed** with hairs along lower edges.

Flowers: Petals usually absent, 5, white and tiny when present; 4 or 5 **sepals**, sometimes fused at the base, 2–3 mm long, **blunt-tipped**; styles commonly 3; **single, in leaf axils**; {June–July}.

Fruits: Broadly **egg-shaped capsules**, blunt-tipped, longer than the sepals, splitting lengthwise into 6 sections, containing several to many wrinkled seeds.

Habitat: Damp meadows, on lake shores and stream sides.

Notes: Wavy-leaved chickweed (*Stellaria crispa* Cham. & Schlecht.) is a similar species that is also rare in Alberta. It too has single flowers in the leaf axils with either no or tiny petals, but its sepals are longer (3–4 mm long) and pointed and they have dry, translucent edges. Its leaves are usually hairless and often longer than those of meadow chickweed (up to 3 cm long), and they tend to have wavy edges—hence the specific epithet *crispa*, which is Latin for 'curled.' Wavy-leaved chickweed flowers in June and July in moist woods and clearings. • The blunt-tipped sepals of meadow chickweed give rise to its specific epithet *obtusa*, which means 'blunt' in Latin.

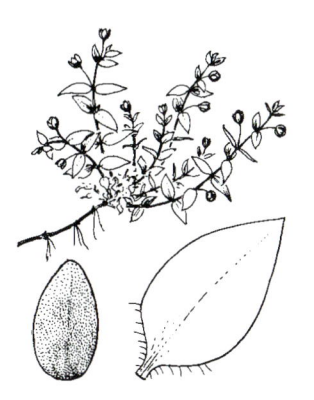

S. CRISPA

Dicots   PINK FAMILY

UMBELLATE CHICKWEED

Stellaria umbellata Turcz. *ex* Kar. & Kir.
PINK FAMILY (CARYOPHYLLACEAE)

Plants: Delicate perennial herbs; stems slender, erect to ascending, 10–30 cm tall; from creeping underground stems (rhizomes).

Leaves: Opposite, stalkless; blades **narrowly lance-shaped to elliptic**, 5–20 mm long, hairless or fringed with hairs on the lower edges.

Flowers: Green to whitish, lacking petals or with 5 tiny, white petals; sepals usually 5, 2–3 mm long, with dry, translucent edges; styles commonly 3; numerous, in upper leaf axils above **dry, translucent bracts**, and in open, spreading clusters (cymes) with branches eventually bent backward; {July–August}.

Fruits: Egg-shaped capsules, 4–5 mm long, splitting lengthwise into 6 teeth, containing several to many slightly wrinkled seeds.

Habitat: Moist woods, meadows and banks from subalpine to alpine elevations.

Notes: Another rare Alberta species, sand-dune chickweed (*Stellaria arenicola* Raup), also has whitish, translucent bracts in its flower clusters, but unlike umbellate chickweed, it has showy flowers on erect or ascending stalks and its petals are usually longer than their sepals. Each petal is deeply cut into 2 slender lobes, so at first glance, these flowers seem to have 10 linear petals. Sand-dune chickweed is very similar to the much more common species long-stalked chickweed (*Stellaria longipes* Goldie), but sand-dune chickweed has straw-coloured capsules that open by 6 teeth at the tip which are bent sharply back. Long-stalked chickweed usually has purplish black capsules with erect teeth. Sand-dune chickweed grows on the shifting sand of sand dunes, and flowers in July and August. • The specific epithet *umbellata*, from the Latin *umbella* (parasol), refers to the loose, umbrella-shaped flower clusters.

S. ARENICOLA

WATER-LILY FAMILY

Dicots

PYGMY WATER-LILY

Nymphaea leibergii Morong
WATER-LILY FAMILY (NYMPHAEACEAE)

Plants: Aquatic perennial herbs; leaf and flower stalks arising from the tip of unbranched, cylindrical, erect, **thick, fleshy underground stems** (rhizomes).

Leaves: Usually **floating, broadly heart-shaped**, with a deep, **V-shaped hollow** at the base between 2 pointed lobes, **5–12 {3–19} cm long, 3–7 {2–15} cm wide**, with 7–13 main veins, green on the upper surface, often deep purplish beneath; stalks hairless, long, slender.

Flowers: White, without fragrance, broadly cupped, **3–4 {7.5} cm across**, with **8–15 white, purple-veined petals** in whorls of 4 and **4 greenish sepals**, all about **2–3 cm long**; stigmatic disc with upcurved, tapered or slightly boat-shaped **appendages 0.6–1.5 mm long around the edges**, yellow or purplish-tipped; 20–40 stamens, yellow, sometimes purplish-tipped; **solitary, floating**, attached to buried stems by a long stalk; August {June–September}.

Fruits: Round, leathery, berry-like capsules, about 1 cm wide, containing egg-shaped, 2–3 mm long seeds, each tipped with a small appendage that becomes sticky when wet.

Habitat: Quiet streams, ponds and lakes, usually in deep water; elsewhere, in water 25–200 cm deep.

Notes: This species has also been called *Nymphaea tetragona* Georgi ssp. *leibergii* (Morong) Porsild. • Small white water-lily (*Nymphaea tetragona* Georgi) is similar to pygmy water-lily, and these 2 rare taxa have often been included in the same species. Small white water-lily has stigmatic discs edged with larger (usually over 3 mm long) appendages, and its sepals have prominent lines of attachment that leave a tetragonal pattern on the receptacle. Small white water-lily grows in ponds, lakes and quiet streams, and is reported to flower throughout the summer.
• Water-shield (*Brasenia schreberi* J.F. Gmelin), a member of the water-shield family (Cabombaceae), occurs in northeastern Alberta. It is easily recognized by its floating, elliptic, 5–10 cm long leaves, with long, slimy stalks attached at the centre of the blade and covered with a thick gelatinous sheath. Its small, inconspicuous flowers are reported to appear in late July to September. The leathery, 1- or 2-seeded pods do not split open. Water-shield grows in shallow, calm or slow-moving water.

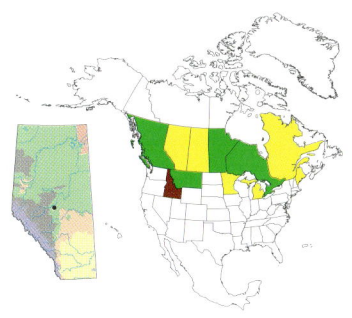

N. TETRAGONA

BRASENIA SCHREBERI

Rare *Vascular* Plants of Alberta

BUTTERCUP FAMILY

WOOD ANEMONE

Anemone quinquefolia L. var. *bifolia* Farw.
BUTTERCUP FAMILY (RANUNCULACEAE)

Plants: Delicate perennial herbs, 10–20 {5–30} cm tall; stems slightly hairy; growing from crisp, slender (1–4 mm thick), white, elongated underground stems (rhizomes).

Leaves: Basal leaves single or absent, long-stalked, divided into 3–5 leaflets, coarsely and unevenly toothed or deeply and irregularly cut, with the terminal leaflet diamond-shaped to lance-shaped and 1–4.5 cm long; **stem leaves** (involucral bracts) **usually 3, in a circle** (whorl) below the flower, stalked, divided into 3–5 leaflets; leaflets lance-shaped to egg-shaped and widest above middle, tapered to a wedge-shaped base, deeply and irregularly cut and lobed.

Flowers: White to pinkish or purplish-tinged, with 5 {4–6} **petal-like sepals** 1–2 {0.6–2.5} cm long; no petals; 30–60 stamens; conspicuous, **single**, on slender stalks with long, soft, spreading hairs; {April–June} July.

Fruits: Dry, seed-like fruits (achenes), 3–4 {2.5–4.5} mm long, spindle-shaped (fusiform) with a stiff, curved to straight, 0.5–2 mm long tip, densely covered with short, stiff hairs, numerous, in dense, round, head-like clusters on 1–6 cm long stalks.

Habitat: Moist woods in the main foothills; elsewhere, in thickets, on open hillsides, in aspen and mixed woods, in clearings, and on sandy stream sides and riverbanks.

Notes: This variety also encompasses *Anemone nemorosa* L. var. *bifolia* (Farw.) B. Boivin and *A. quinquefolia* L. var. *interior* Fern. • All anemones contain acrid, poisonous compounds that **may irritate the skin**. They **should never be eaten**. • The word *Anemone* is commonly assumed to be derived from the Greek word *anemos*, which means 'wind,' probably in reference to the exposed, windy places where many of these plants grow. Many species are early bloomers, and it was once believed that their flowers opened at the command of the first mild breezes of spring.

GOLDTHREAD

Coptis trifolia (L.) Salisb.
BUTTERCUP FAMILY (RANUNCULACEAE)

Plants: Small perennial herbs; flowering stems slender, {4} 5–15 cm tall; from thread-like, **bright yellow or orange underground stems** (rhizomes).

Leaves: Basal, evergreen, shiny, on slender 1.5–12 cm long stalks, divided into **3 leaflets**; leaflets short-stalked or stalkless, egg-shaped, 7–20 mm long, wedge-shaped at base, broadly 3-lobed and sharply toothed at the tip.

Flowers: White with a yellowish base, **10–16 mm across**, with 5–7 elliptic to spoon-shaped sepals and 5–7 inconspicuous petals; **petals reduced** to small, fleshy, club-shaped structures (staminodia) tipped with nectar; single, erect; July {May–August}.

Fruits: Pod-like, **splitting open along 1 side** (follicles), **7–12 mm long** (including a prominent, 2–4 mm long beak), **on long stalks** (equal to or longer than the follicle), in a spreading **whorled cluster of** {3} **4–7 at the stem tip**; seeds 1–1.5 mm across, shiny, black.

Habitat: Damp, mossy woods; elsewhere, in muskeg, willow scrub and tundra.

Notes: Goldthread has been used in preparations to combat alcoholism, because it is said to relieve the craving for alcohol. The roots are very astringent, and they were either chewed or boiled to make medicinal teas for treating sore throats, canker sores and other mouth irritations—hence another common name, 'canker-root.' It was also used as an eyewash and as a topical anesthetic on the gums of teething children. Goldthread was listed in the *US Pharmacopoeia* from 1820 to 1882, and at the turn of the century 500 g of the root fetched about a dollar. Today, these attractive, shade-tolerant plants are sometimes grown as ground cover in wooded areas. • The generic name *Coptis* was taken from the Greek *kopto* (to cut), in reference to the deeply cut leaves. The common name 'goldthread' refers to the long, slender, yellow underground stems.

Dicots — BUTTERCUP FAMILY

EARLY BUTTERCUP

Ranunculus glaberrimus Hook.
BUTTERCUP FAMILY (RANUNCULACEAE)

Plants: Rather fleshy, **hairless** perennial herbs with acrid juice, usually less than 10 cm tall; stems usually several, **erect to lying on the ground, 4–18 {20} cm long**; from clusters of thick, fleshy roots.

Leaves: Mainly basal; basal leaves nearly round to narrowly egg-shaped, often widest above the middle, rounded at the base or tapered to long, slender stalks, smooth, wavy or toothed along the edges, sometimes 3-lobed; stem leaves similar to the basal leaves but short-stalked.

Flowers: Yellow (sometimes white with age), **saucer-shaped, 10–25 mm across**, with **5 {6–8} separate petals and 5 sepals**; petals 6–15 mm long, broad and conspicuous, each with a nectar-producing pit at the base; sepals 5–8 mm long, usually tinged with purple, with or without hairs, soon shed; 40–80 stamens; 1–6 on long stalks (up to 10 cm long), in open clusters (corymbs); {March} May–June.

Fruits: Finely hairy 'seeds' (achenes), 1.5–2 mm long, widest toward the short (0.5–0.8 mm), straight beak; 75–100 {30–150} borne on round receptacles in large (1–2 cm wide), round heads.

Habitat: Grassland and meadows in the prairies; elsewhere, in ponderosa pine woodland and sagebrush desert.

Notes: Western buttercup (*Ranunculus occidentalis* Nutt. var. *brevistylis* Greene) is also rare in Alberta. It is a larger plant (stems 20–70 cm long) with hairy, deeply cut, 3–5-lobed and toothed basal leaves. Its showy yellow flowers are about 1.5–2 cm wide, and its flattened achenes are edged with a distinct, narrow wing and tipped with a long, nearly straight beak (hooked at the tip only). Western buttercup grows in moist alpine and subalpine meadows, and is reported to flower from April to June.
• The generic name *Ranunculus* was derived from the Greek *rana* (a little frog), in reference to the aquatic habit of many species.

R. OCCIDENTALIS

BUTTERCUP FAMILY **Dicots**

ALPINE BUTTERCUP

Ranunculus gelidus Karel. & Kiril. ssp. *grayi* (Britt.) Hult.
BUTTERCUP FAMILY (RANUNCULACEAE)

Plants: Small, tufted, grey-green perennial herbs, 5–12 {2–22} cm tall, essentially **hairless; stems** usually several, simple or forked in 2s, with a cluster of small leaves at or somewhat below the middle, **often arched** when in fruit; forming small, firmly rooted tussocks from short, branched crowns on slender, 0.5–1 mm thick roots.

Leaves: Basal, simple; **blades heart-shaped** to kidney-shaped, 5–15 {18} mm long and 8–20 {30} **mm wide, twice deeply cut in 3s** into blunt, egg-shaped to lance-shaped lobes widest above the middle; stalks curved, white, 1–6 {8} cm long; stem leaves stalkless or short-stalked.

Flowers: Pale yellow, often purplish-tinged, with 5 sepals and 5 petals; sepals 2.5–4.5 {5} mm long, soon bent backward and then shed; petals 3.5–5 {3–6} mm long; stalks 3–7 cm long; 1–3 {5}; {June} July–September.

Fruits: Plump, hairless, lens-shaped, seed-like fruits (achenes), **2–2.5 mm long**, with a slender, curved or hooked, 0.5 {0.4–0.8} mm beak; borne in **cylindrical heads, 5–9** {4–13} **mm long** and 4–6 mm wide.

Habitat: Dry, rocky alpine slopes; elsewhere, on moist, gravelly alpine slopes.

Notes: This species includes *Ranunculus karelinii* Czern., *R. grayi* Britt. and *R. verecundus* B.L. Robins. *Ranunculus verecundus* was previously separated on the basis of its larger size (10–20 cm tall) and smaller achenes (1–1.5 mm long), but these characteristics appear to vary over the range of the species. • Alpine buttercup is very similar to mountain buttercup (*Ranunculus eschscholtzii* Schlecht.), but the basal leaves of mountain buttercup are usually folded, the petals are often much larger, and the achenes have a slender, straight beak. • Alpine buttercup could be confused with dwarf buttercup (*Ranunculus pygmaeus* Wahl.), but dwarf buttercup has erect stems when in fruit and its achenes are only about 1 mm long. • Another rare Alberta buttercup, hairy buttercup (*Ranunculus uncinatus* D. Don), is similar to alpine buttercup, but it is a larger (20–60 cm tall) plant of moist, shady woodlands at lower elevations, and its achenes have prominent, 1–2 mm long, hooked beaks. Hairy buttercup is reported to bloom from April to July.

R. UNCINATUS

Rare *Vascular* Plants of Alberta

Dicots

BUTTERCUP FAMILY

SNOW BUTTERCUP

Ranunculus nivalis L.
BUTTERCUP FAMILY (RANUNCULACEAE)

Plants: Tufted perennial herbs with acrid juice, **hairless or with sparse brown hairs**; stems slender, unbranched, **erect, 3–15 {30} cm tall** (elongating in fruit), few to several; from fibrous roots.

Leaves: Basal, with {1} 2 or 3 on stem; basal leaves kidney-shaped, 1–2 cm long and usually wider, squared or notched at the base, **3-lobed**, with the side lobes often shallowly cut, borne on 1–5 cm long stalks; **stem leaves similar** but often larger and more deeply lobed and with undivided side lobes, squared to broadly wedge-shaped at the base, usually stalkless.

Flowers: Deep yellow, saucer-shaped, 15–25 mm across, with **5 separate petals** and **5 sepals**; petals 9–12 mm long, broad and conspicuous, each with a nectar-producing dot or pit at the base; sepals 5–9 mm long, **densely brownish black-hairy on the back**; stamens usually numerous; solitary; June–July.

Fruits: Plump seed-like fruits (achenes), 1.5–2 mm long, **without winged edges**, tipped with a straight, firm, slender beak 1–1.5 mm long, borne on essentially hairless receptacles, in 5–15 mm long, **egg-shaped to cylindrical heads**.

Habitat: Moist **alpine** slopes, often near melting snow beds and by streams.

Notes: Snow buttercup might be confused with the common species mountain buttercup (*Ranunculus eschscholtzii* Schlecht.), but mountain buttercup flowers have pale-yellow hairs on their sepals. • The specific epithet *nivalis* is Latin for 'snowy,' in reference to the snow-bed communities where this species is often found.

Rare *Vascular* Plants of Alberta

BUTTERCUP FAMILY **Dicots**

Jones' columbine

Aquilegia jonesii Parry
BUTTERCUP FAMILY (Ranunculaceae)

Plants: Dwarf perennial herbs, finely hairy and more or less glandular (even on flowers), 5–12 {3.5–20} cm tall; from short, branched underground stems (rhizomes).

Leaves: Basal, crowded, finely hairy, greyish with a thin, waxy coating (glaucous) on both surfaces; **blades rarely over 1 cm long**; 1 or 2 times divided in 3s; leaflets crowded, deeply cut into 3 or 4 oblong to rounded lobes; stalks slender, 3–8 {1–12} cm long.

Flowers: Blue or purplish, showy, facing upward (erect), with 5 sepals and 5 petals; **sepals deep blue or purple, petal-like**, oblong–lance-shaped to broadly elliptic, **15–22 {25} mm long; petals often whitish**, expanded at the tip into a **rounded blade about 6–8 {13} mm long**, each prolonged backward at the base as a **stout, straight, blue tube** (spur) 8–10 {5–15} mm long; **single**, on erect stalks about 2–5 cm long; July {June–August}.

Fruits: Erect groups of **5 pod-like fruits that split open along inner side** (follicles), 15–20 {12–25} mm long, tipped with an 8–12 mm long beak, hairless, greyish with a thin, waxy coat; seeds black.

Habitat: Alpine scree slopes; elsewhere, in subalpine and alpine areas on limestone talus slopes and in rock crevices.

Notes: The nectar-producing spurs on these flowers attract long-tongued pollinators, such as moths, capable of reaching the fragrant nectar stored at the spur tips. • The generic name *Aquilegia* was derived from the Latin *aquila* (eagle) because the spurs on the petals were thought to resemble the talons of an eagle. The common name 'columbine' comes from the Latin *columbina* (dove-like) because the spreading sepals and arched spurs were thought to resemble a circle of 5 doves drinking from a central dish. This resemblance is more obvious in species with nodding flowers and curved spurs, where the spurs point upward and inward.

Rare *Vascular* Plants of Alberta

BUTTERCUP FAMILY

SITKA COLUMBINE

Aquilegia formosa Fisch. *ex* DC. var. *formosa*
BUTTERCUP FAMILY (RANUNCULACEAE)

Plants: Stout perennial herbs, {15} 30–100 cm tall; stems freely branched, hairless to copiously hairy or glandular-hairy; from short, branched underground stems (rhizomes).

Leaves: Mainly basal, long-stalked; blades **thin, usually tinged bluish** with a thin, waxy coating (glaucous), **twice divided in 3s** (biternate); **leaflets deeply lobed** or coarsely blunt-toothed, 15–50 {68} **mm long**.

Flowers: Red and yellow, usually nodding, showy, with 5 petals and 5 sepals; **petals** expanded at the tip into **a yellowish, 3–5 {6} mm long, egg-shaped blade** (widest above the middle), each prolonged back from the base in a **straight, 13–21 mm long, reddish tube** (spur) **with a bulbous, nectar-producing tip; sepals pale to deep red, petal-like**, spreading or bent back (reflexed), 14–26 mm long, longer than the blades of the petals; **stamens numerous, projecting** well past the petals; few to several, **nodding** to hanging on slender stalks in elongating clusters (racemes); June–July {May–August}.

Fruits: Erect groups of 5 pod-like fruits that open on the inner side (follicles), shaggy with soft hairs or glandular-hairy, **15–25 {29} mm long**, spreading at the tip, with firm, slender beaks 9–12 mm long; seeds black.

Habitat: Open woods and rocky slopes from lowlands to timberline; elsewhere, in moist sites on partly shaded roadsides and in woods, subalpine meadows and thickets.

Notes: This species has also been called *Aquilegia canadensis* L. var. *formosa* (Fisch.) S. Wats. and *Aquilegia columbiana* Rydb. Other common names include western columbine and crimson columbine. • Sitka columbine is similar to, and often hybridizes with, yellow columbine (*Aquilegia flavescens* S. Wats.), but yellow columbine is a smaller (20–75 cm tall) plant, with wholly yellow flowers, and it usually grows at higher elevations (in alpine areas), although the ranges of these 2 species do overlap to some extent. • In Europe, where columbines were used medicinally, children sometimes died from overdoses of the seeds. In North America, some native peoples rubbed columbine seeds in their hair to control lice. • Sitka columbine is easily grown and maintains itself from seed. Its beautiful flowers attract hummingbirds and butterflies.

DWARF ALPINE POPPY

Papaver pygmaeum Rydb.
POPPY FAMILY (PAPAVERACEAE)

Plants: Loosely **tufted perennial** herbs with milky sap; flowering stems leafless, 5–6 {3–12} cm tall, with spreading yellowish (sometimes purple-mottled) hairs; from taproots.

Leaves: Basal, many, blue-green on both surfaces, 2–5 cm long, hairless on upper surface but often with stiff, flat-lying hairs on the lower surface and on the stalks; blades broadly egg-shaped, much shorter than their stalks, 1–1.5 cm long, deeply cut into **pinnate lobes**; stalks to ⅔ of the leaf length.

Flowers: Yellow, or orange with a yellow spot at the centre (usually drying pinkish), **cupped, 1–2 cm across**, with **4 or more large, thin petals**; 4 or 5 stigmas in a convex disc; buds oval, nodding, with 2 dark-hairy sepals; solitary; July–August.

Fruits: Erect, **egg-shaped to cone-shaped capsules**, 2–2½ times longer than wide, widest at the tip, 1–1.5 cm long, tipped with a disc-like stigma with 4 or 5 rays, conspicuously stiff-hairy with yellow, bulbous-based bristles.

Habitat: Alpine scree slopes and ridges.

Notes: This species has also been called *Papaver alpinum* L., *Papaver nudicaule* L. ssp. *radicatum* (Rottb.) Fedde var. *pseudocorydalifolium* and *Papaver radicatum* Rottb. var. *pygmaeum* (Rydb.) S.L. Welsh. • Poppy flowers do not produce nectar, and most species are self-pollinating, but they are visited by many insects in search of pollen. The expanded, star-shaped stigma provides a good landing platform for visiting insects. The parabolic shape of the flowers focusses radiation on the stigma, increasing temperatures there by an average of 5.9°C, and thus increasing the rate at which pollen grains develop once they have landed. In calm, sunny weather, these flowers constantly face the sun. On exposed sites, flowers may open while they are still touching the basal rosette of leaves, keeping them within the warmest layer of air, near the ground. As the seeds develop, the stalks elongate. Eventually, the seeds are dispersed by the 'censer mechanism,' in which the long, elastic stalks swing to and fro in the wind, shaking out thousands of tiny seeds through small holes at the top of the capsule (like a salt-shaker).

Dicots NETTLE FAMILY

AMERICAN PELLITORY

Parietaria pensylvanica Muhl. *ex* Willd.
NETTLE FAMILY (URTICACEAE)

Plants: Slender, finely hairy annual herbs, sprawling to erect; stems branched or unbranched, 5–50 {4–60} cm long, sparsely to densely hairy; from slender taproots.

Leaves: Alternate, spreading; blades **narrowly egg-shaped or elliptic**, 1–8 {9} **cm long**, 3-veined, toothless; **stalks slender**, 1/6–1/3 as long as the blades; lower leaves soon withered.

Flowers: Green to brownish, inconspicuous, 1.5–2 mm long, with **4 calyx lobes** and **no petals, either male or female**; 4 stamens; 1 ovary, stigma and style; nestled in a tuft of several linear, 2–4 mm long bracts on a short stalk, forming small clusters of intermixed male and female flowers in leaf axils on the upper half of the stem; {May–July}.

Fruits: Shiny, light reddish brown seed-like fruits (achenes), 0.9–1.2 mm long.

Habitat: Shady, gravelly sites, often in disturbed areas; elsewhere, in heavy woods and on shaded banks, on talus slopes and ledges, and around hot springs.

Notes: American pellitory is rather different from its close relatives the nettles (*Urtica* spp.), which have stinging hairs and opposite leaves. • Extracts of the European species pellitory-of-the-wall (*Parietaria officinalis* L.) were highly valued as a remedy for gravel, dropsy, stone of the bladder and other urinary complaints. • The generic name *Parietaria* was derived from the Latin *paries* (a wall) because the European plants of this genus were commonly found growing from crannies in old walls.

CAPER FAMILY

Dicots

CLAMMYWEED

Polanisia dodecandra (L.) DC. ssp. *trachysperma* (T. & G.) Iltis
CAPER FAMILY (Capparidaceae)

Plants: Clammy, foul-smelling, sticky-hairy annual herbs; stems 10–80 cm tall, usually freely branched; from weak taproots.

Leaves: Alternate, divided into **3 narrowly to broadly lance-shaped leaflets** 1–4 cm long and often widest above the middle; stalks 1–4 cm long.

Flowers: **Yellowish white**, about 1 cm across, with **4 showy petals, 4 small sepals** and 5–16 conspicuous stamens; petals {7} 8–12 mm long, tipped with broad, notched lobes abruptly narrowed to a slender, purplish, stalk-like base (claw); sepals purplish-tinged, lance-shaped, 3–4 mm long, soon shed; **stamens purplish**, unequal in length, **projecting to 1 side** of the flower on thread-like filaments 10–20 mm long; ovaries superior, solitary, with styles 4–6 mm long; few to several, on slender, ascending, 10–22 mm long stalks in elongated clusters (racemes); {June} July.

Fruits: Erect, **swollen, veiny, pod-like capsules, with gland-tipped hairs**, cylindrical or slightly flattened, narrowly oblong to elliptic, {2.5} 3–5 cm long, stalkless or nearly so but narrowed to a stalk-like base 1–3 mm long, splitting in 2 to release many slightly roughened seeds.

Habitat: Gravelly or sandy soil, often on disturbed or eroding sites; elsewhere, in sagebrush scrub, in gravel pits, along railways and on sandy stream beds, shores, plains and foothills, often in desert 'washes.'

Notes: This species has also been called *Polanisia trachysperma* T. & G. • At one Alberta site, clammyweed is found growing on a sparsely vegetated sand-dune blowout. • Probably fewer than 10,000 plants grow in Alberta. It is possible that more clammyweed plants in Alberta are growing on man-made disturbances than in natural habitats. Encroachment of vegetation onto active blowouts could eliminate some populations, but at present most Alberta populations appear stable. Over the long term, other plants eventually crowd out clammyweed. • The subspecies *dodecandra* is also reported from Alberta. It grows in the Cypress Hills in Saskatchewan.

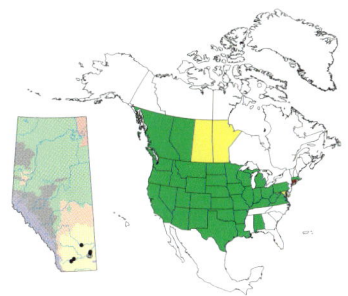

MUSTARD FAMILY

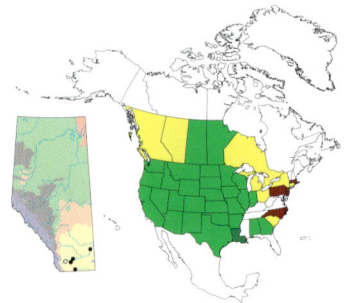

CREEPING WHITLOW-GRASS

Draba reptans (Lam.) Fern.
MUSTARD FAMILY (BRASSICACEAE [CRUCIFERAE])

Plants: Delicate **annual** herbs; stems thread-like, erect, {2} 5–20 cm tall, simple or branched at the base, with bristly hairs near the base, hairless above; from branched root crowns.

Leaves: Mostly in basal rosettes, spatula-shaped to broadly egg-shaped and widest above the middle, 1–3 cm long, 2–10 mm wide, **not toothed**, fringed with simple hairs, bristly with coarse, simple and **long-stalked branched hairs** on both surfaces, the **upper surface with once-forked hairs** and the lower surface with repeatedly branched (dendritic) or forked hairs; **1–3 stem leaves**, crowded near the base.

Flowers: White, about 5 mm across, with **4 sepals and 4 petals**; petals yellow, with a shallow notch at the tip, 3–4 {2–5} mm long; sepals 1.5–2.5 mm long with simple hairs; **6 stamens**; **style 0–0.15 mm long**; branch flowers distinctly smaller, often lacking petals; 3–12 flowers on hairless, spreading to ascending **stalks 2–7 mm long**, in **elongating, rather flat-topped clusters** (subumbellate racemes); April {March–May}.

Fruits: Slightly curved, **linear to narrowly oblong pods** (siliques), 5–20 {22} **mm long**, 1–2 mm wide, **about 6 times as long as they are wide**, hairless or with stiff, unbranched hairs, splitting in 2 to release 15–80 small (0.5–0.8 mm long) seeds.

Habitat: Sparsely vegetated, exposed, sandy or gravelly sites, especially in naturally disturbed areas, in grassland at low elevations; elsewhere, in dry forests and on dry beach strands and limestone outcrops or pavements.

Notes: Creeping whitlow-grass is inconspicuous and easily overlooked. Cultivation of sandy areas poses a potential threat, and one population on the Milk River is near a proposed dam site. Stabilization of dunes by other vegetation could also eliminate some sites. It is unclear whether dunes in native grassland were kept active by periodic drought, fire, grazing or a combination of these factors. • The generic name *Draba* was taken from the Greek *drabe* (acrid or biting), in reference to the bitter sap of many mustard species. Dioscorides first used this name for some unknown member of the mustard family, and it was later applied to this genus. The specific epithet *reptans* means 'creeping.'

Austrian Whitlow-Grass

Draba fladnizensis Wulfen
MUSTARD FAMILY (BRASSICACEAE [CRUCIFERAE])

Plants: Tufted perennial herbs; **stems erect, 1–9 {15} cm tall**, unbranched, usually **hairless**, sometimes with mostly simple hairs near the base, rarely hairy throughout; from simple or branched root crowns.

Leaves: Mainly in basal tufts, lance-shaped, widest above the middle, 3–10 {25} mm long, 1–2 mm wide, toothless, with a prominent midrib, **hairless or sparsely covered with long, unbranched hairs** on both surfaces and fringed with straight hairs; **stem leaves usually absent**, but 1 or 2, small, toothless or minutely toothed leaves are sometimes present.

Flowers: White, about 4 mm across, with **4 petals, 4 sepals** and **6 fertile stamens** (2+4); petals 2–3 mm long; sepals 1–1.5 mm long, with simple hairs; **styles 0.1–0.2 mm long**, evident to nearly obsolete; 3–12, on straight, erect or ascending stalks in elongating clusters (racemes); July {August}.

Fruits: Oblong–egg-shaped to lance-shaped pods (siliques), 3–6 {9} mm long, 1.5–2 mm wide, **mostly hairless** (sometimes with a few simple or forked hairs), containing many small, 0.7–0.9 mm long seeds.

Habitat: Alpine gravel and scree slopes; elsewhere, on turfy tundra and in cold ravines and subalpine fir forests.

Notes: Dense-leaved whitlow-grass (*Draba densifolia* Nutt.) is a similar species that is also rare in Alberta. It is distinguished by its yellow flowers and densely hairy stems and siliques. Its pollen is not fertile. Dense-leaved whitlow-grass grows on alpine ridges and scree slopes and flowers in {June–July} August. The small plants often expand through offsets, and some clumps cover about 1 m². The Alberta populations are apparently very local but stable. • The specific epithet *densifolia* means 'densely leaved.'

D. DENSIFOLIA

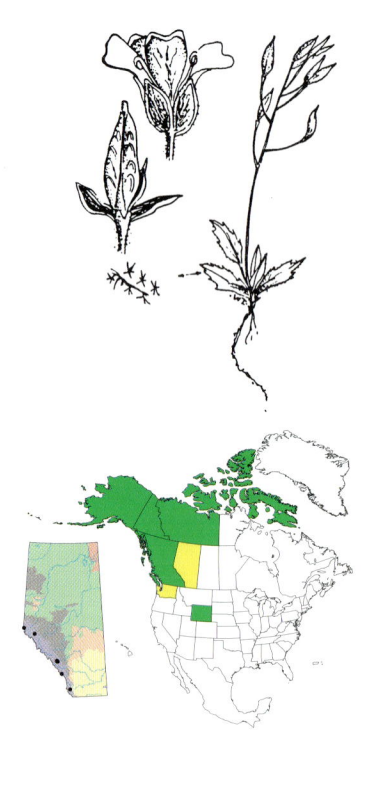

Dicots MUSTARD FAMILY

LONG-STALKED WHITLOW-GRASS
Draba longipes Raup
MUSTARD FAMILY (BRASSICACEAE [CRUCIFERAE])

Plants: Low, **loosely tufted perennial** herbs; **stems slender**, erect or somewhat spreading, 10–15 {25} cm tall, branched, with sparse simple, forked or star-shaped hairs (sometimes hairless); from **branched root crowns**.

Leaves: **Mainly in loose basal tufts**, lance-shaped, widest above the middle, toothless or finely toothed, **1.5–3 cm long**, 1–5 mm wide, with **sparse, 4-branched hairs on both surfaces** (sometimes hairless on the upper surface) and simple or branched hairs along the edges; **stem leaves 1–3 {0–5}**, about as wide as the basal leaves.

Flowers: White to creamy-white or yellowish, about 5 mm across, with **4 petals, 4 sepals** and 6 (4+2) stamens; petals 4–5 mm long; sepals 2.5–3 mm long, hairless or with a few simple hairs; **styles 0.6–0.9 {0.5–1} mm long**; {2} 3–15, on slender, spreading stalks 5–15 mm long, in rather flat-topped (but soon elongating) clusters (racemes); {June} July–August.

Fruits: Narrowly egg-shaped to linear–lance-shaped pods (siliques), 4–13 {3–15} mm long, 2–3 mm wide, **hairless**, containing many seeds about 1 mm long.

Habitat: Moist banks and ledges in alpine areas, often by snow beds; elsewhere, in ravines, near streams and springs, and on gravelly beaches and open grassy slopes.

Notes: A yellow-flowered species, Kananaskis whitlow-grass (*Draba kananaskis* Mulligan), endemic to Alberta and Alaska, has recently been included by some taxonomists in long-stalked whitlow-grass. It was distinguished from long-stalked whitlow-grass by its yellow (rather than white) flowers and its short-stalked (rather than stalkless) leaf hairs. Kananaskis whitlow-grass grows on dry alpine slopes at elevations around 2,200 m and flowers in July. It is normally a self-fertilized species. • Macoun's whitlow-grass (*Draba macounii* O.E. Schulz) is another rare Alberta species. It is recognized by its yellow flowers on short (1–8 cm), leafless stems; its short-stalked, hairless, elliptic to egg-shaped pods (siliques); and its densely tufted basal leaves with predominantly cross-shaped hairs. Macoun's whitlow-grass is found on alpine slopes, where it flowers in July.

D. MACOUNII

D. KANANASKIS

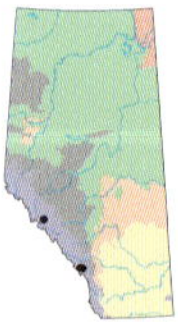

MUSTARD FAMILY   **Dicots**

WINDY WHITLOW-GRASS
Draba ventosa A. Gray
MUSTARD FAMILY (BRASSICACEAE [CRUCIFERAE])

Plants: Small, **tufted perennial** herbs, **silvery-green with a tangle of branched hairs**; stems erect, 0.5–4 {6} cm tall, **leafless**, branched, greyish with many fine, simple hairs and forked hairs; from loosely branched root crowns.

Leaves: **Basal, numerous**, linear to egg-shaped and widest above the middle, tapered to a wedge-shaped base, **4–14 mm long**, 1.5–2 {2.4} mm wide, with a prominent midrib, densely covered with **long-stalked, star-shaped hairs on both surfaces**, also fringed with simple hairs; withered leaves persisting for several years (marcescent).

Flowers: Yellow, about 5 mm across, with **4 petals, 4 sepals** and 6 (2+4) stamens; **petals** 2–4 {5} mm long; **sepals** 2–3 mm long, hairy; **styles 0.5–1 mm long**; 3–10 {20}, in short, condensed clusters that elongate in fruit (racemes); July–August.

Fruits: Flattened, **broadly oval to egg-shaped pods** (siliques) rounded at the tip, 4–5 {2.5–5.5} **mm wide, 3–8 mm long,** on spreading stalks of about the same length, **densely covered with soft, long-stalked, repeatedly branched** (dendritic) **hairs**, containing 10–16 seeds 1.5–2 mm long.

Habitat: Alpine ridges and scree slopes; elsewhere, in alpine tundra and at the base of cliffs.

Notes: These plants are agamospermous, that is to say, they produce seeds asexually (without fertilization). The resulting offspring are genetically identical to their parent. • Whitlow-grasses require sunny locations on open, gritty soil. Soils rich in lime are usually preferred. They are usually propagated by dividing plants, but they can also be grown from seed in the fall or spring.

Rare *Vascular* Plants of Alberta

SMOOTH WHITLOW-GRASS

Draba glabella Pursh
MUSTARD FAMILY (BRASSICACEAE [CRUCIFERAE])

Plants: Tufted or loosely matted perennial herbs, 10–30 {40} **cm tall**; stems simple or freely branched, erect, hairy with few to many **soft, star-shaped hairs** (sometimes hairless on the upper stem); from simple or extensively branched, woody root crowns.

Leaves: Mostly in dense basal clusters, lance-shaped, widest above middle and tapered to slender stalks, 5–50 mm long, 3–18 mm wide, toothless or with a few small, scattered teeth, sparingly to densely covered with **short-stalked, star-shaped hairs with** {7} **9 or more rays**; stem leaves 2–7 {1–10}, lance-shaped, 5–40 mm long, 2–18 mm wide, **usually toothed**, with simple and branched hairs above and below.

Flowers: White or cream-coloured, with **4 sepals and 4 petals**; sepals 2–3 mm long, with simple and 2-forked, flat-lying hairs; petals {3.5} 4–6 mm long; **styles 0.2–0.5 mm long**; 5–20 {30}, on ascending stalks in elongating clusters (racemes); {May–July}.

Fruits: Narrowly lance-shaped pods (siliques), 7–14 {6–15} **mm long**, 2–4 mm wide, **often twisted**, hairless or hairy; **lowest stalks** 3–12 mm long, **shorter than or equal to the pod**; seeds 0.7–1.3 mm long.

Habitat: Moist banks and ledges at alpine elevations; elsewhere, on rocky, sandy or gravelly shores, on rocky, grassy tundra, and on slaty and calcareous cliffs.

Notes: Smooth whitlow-grass could be confused with a more common species, northern whitlow-grass (*Draba borealis* DC.), but northern whitlow-grass has mostly cross-shaped hairs on its leaves and its styles are longer (up to 1 mm long). • Smooth whitlow-grass is strongly nitrophilous (nitrogen-loving). It is often found growing in areas enriched by animal dung. • The specific epithet *glabella* is from the Latin *glaber* (smooth or hairless), perhaps in reference to the hairless pods.

NORTHERN BLADDERPOD

Lesquerella arctica (Wormskj. *ex* Hornem.) S. Wats. var. *purshii* Wats.
MUSTARD FAMILY (BRASSICACEAE [CRUCIFERAE])

Plants: Low **perennial** herbs, **silvery with dense stalkless or short-stalked, star-shaped hairs**; stems single to several, erect to spreading or lying on the ground, 5–25 {30} cm long; from woody crowns on stout taproots.

Leaves: In prominent basal rosettes, 2–6 {1–15} cm long, egg-shaped to lance-shaped, widest above the middle, **not toothed**, tapered to short stalks; stem leaves narrower, 5–15 {30} mm long, essentially stalkless, **not clasping**.

Flowers: Yellow, about 1 cm across, with **4 petals, 4 sepals** and 6 stamens (2+4); **petals 4–7 mm long**, egg-shaped, widest above the middle; **sepals** oblong; **styles generally 1–2.5 mm long**; few to several, in loose, elongated clusters (racemes); June–July {May–August}.

Fruits: Rounded, usually **inflated pods** (siliques), 4–6 {9} **mm long**, hairless or sparsely covered with tiny, stalkless, star-shaped hairs and containing 5–7 {4–8} plump seeds per cavity (locule), erect, on **straight or slightly curved**, 5–20 {40} mm long **stalks**.

Habitat: Dry, sandy or calcareous slopes and ridges, often in alpine areas; elsewhere, on rocky, gravelly or clay soils on river flats.

Notes: The generic name *Lesquerella* commemorates Leo Lesquereux, an American bryologist who lived 1805–89. The specific epithet *arctica* means 'of the northern or polar regions.'

BLUNT-LEAVED YELLOW CRESS

Rorippa curvipes E.L. Greene
MUSTARD FAMILY (BRASSICACEAE [CRUCIFERAE])

Plants: Low, somewhat tufted **annual** or biennial herbs, usually less than 10 cm tall; stems slender, hairless, 10–50 cm long, erect to spreading on the ground, usually branched (beginning near the base), single or several; from taproots.

Leaves: Basal and alternate on the stem, with **scattered, simple hairs** on the upper surface, hairless beneath; blades 4–13 cm long, 0.5–3 cm wide, regularly pinnately divided and angularly toothed; stalks short or absent; **stem leaves** somewhat smaller, oblong to spatula-shaped, sometimes slightly clasping the stem.

Flowers: Yellow, about 2–3 mm across, with **4 petals and 4 sepals**; petals oblong to spatula-shaped, about 1 mm long, 0.2–1 mm wide, **shorter than the sepals**; sepals 1.5 mm long, spreading; stamens usually 6; styles 0.2–1 {1.3} mm long; several to many, erect to nodding, on **short stalks** (2–5.1 {1.6–8} mm long), in elongating clusters (racemes) at stem tips or from leaf axils; {May–September}.

Fruits: Hairless, straight or curved, **egg-shaped to short-cylindrical pods** (siliques) **constricted near the middle**, 2–5 {1.4–8.7} mm long, splitting in 2 to release 10–80 brown, heart-shaped seeds over 0.5 mm long; borne on short stalks half as long as the pods.

Habitat: Moist ground, including mud flats, shores, roadsides, stream beds and wet meadows.

Notes: Two varieties have been found in Alberta: var. *truncata* (Jepson) Rollins (which has also been called *Rorippa truncata* (Jeps.) Stuckey) and var. *curvipes*. Variety *curvipes* has egg-shaped to pear-shaped pods that taper to a pointed or rather blunt tip and that are usually less than twice as long as they are wide at maturity. Variety *truncata* has short, cylindrical pods that have a squared tip and are more than twice as long as they are wide when mature. • Slender yellow cress (*Rorippa tenerrima* E. Greene) is a similar species that is also rare in Alberta. It is distinguished from blunt-leaved yellow cress by its siliques, which are rough with tiny, nipple-shaped bumps (papillae) and which taper to a cone-shaped tip without narrowing at the centre. It flowers in September {May–August} in moist, usually sandy, often alkaline sites.

MUSTARD FAMILY   **Dicots**

SPREADING YELLOW CRESS

Rorippa sinuata (Nutt. *ex* T. & G.) A.S. Hitchc.
MUSTARD FAMILY (BRASSICACEAE [CRUCIFERAE])

Plants: Perennial herbs, essentially hairless, but the stems, lower leaf surfaces and pods often sparsely to densely **dotted with small, hemispheric, air-filled sacs** (modified hairs); stems 10–40 {50} cm long, ascending or spreading on the ground, branched from the base; from slender, creeping underground stems (rhizomes).

Leaves: Basal and alternate on the stem, lance-shaped, widest above the middle, 2–7 {1.6–8.5} cm long, **deeply pinnately cut** into **blunt lobes, with or without short teeth**; upper leaves stalkless, often **clasping** the stem with small basal lobes (auricles), only **slightly smaller** than those below.

Flowers: Light yellow, about **4 mm across**, with **4 sepals and 4 petals**; sepals inconspicuous, much shorter than the petals; petals oblong to narrowly spatula-shaped, 4 {2.5–6} mm long, styles stout, about 1–2 mm long, tapered gradually from base to tip; pointing upward or curved downward on slender stalks 5–9 mm long; several to many, in branched, elongating clusters (racemes); June {April–July}.

Fruits: Linear-oblong to lance-shaped, cylindrical pods (siliques), {5} **7–15 mm long**, {1.5} 2 mm wide, straight or slightly arched, often somewhat wrinkled, hairless, narrowed abruptly to a slender **beak 1–3 mm long**; stalks 3.5–15 mm long, slightly curved to S-shaped; seeds in 2 vertical rows, angular, heart-shaped, about 40–80 per pod.

Habitat: Shores, stream flats, ditches and roadsides; elsewhere, in woods, vacant fields, and wet lowlands, often in alkaline areas.

Notes: Spreading yellow cress could be confused with the introduced species creeping yellow cress (*Rorippa sylvestris* (L.) Besser), but the leaves of creeping yellow cress have pointed lobes that are clearly toothed, and each pod is tipped with a short (up to 1 mm long) beak.

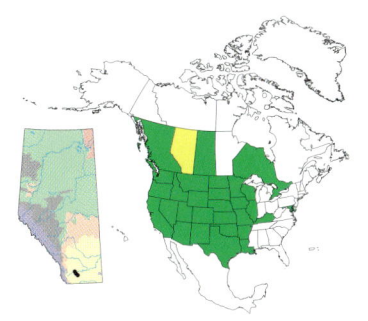

Dicots

MUSTARD FAMILY

AMERICAN WINTER CRESS

Barbarea orthoceras Ledeb.
MUSTARD FAMILY (BRASSICACEAE [CRUCIFERAE])

Plants: Hairless to sparsely hairy biennial herbs (sometimes perennial); stems erect, 20–50 {10–60} cm tall, **often purplish near the base**, angled, usually single, hairless and freely branched; from woody crowns on taproots.

Leaves: In a **basal rosette** (first year), up to 12 cm long, pinnately divided, with a large terminal lobe and 2–8 small side lobes, long-stalked; **stem leaves** (second year) **alternate**, similar to basal leaves but smaller upward on the stem and with clasping bases.

Flowers: Pale yellow, with **4 sepals and 4 petals**; petals spatula-shaped, **2.5–5 mm long**; several to many, in elongating, often branched clusters (racemes); {March–May} June–early August.

Fruits: Narrow, slightly flattened or somewhat **4-angled pods** (siliques), straight or slightly curved, **2–4 {1.5–5} cm long**, 1.5–2.5 mm wide, tipped **with a short** (0.3–1 {2} mm), **beak-like style**, pointing upward on 2–3 mm long, **thick** (up to 1 mm wide) **club-shaped stalks**; seeds flat, brownish, minutely pitted, in a single row.

Habitat: Streambanks, wet meadows and moist woods; elsewhere, on sand bars and rocky cliffs.

Notes: In Alaska and northeast Asia, where this species is common, young plants and buds are eaten raw or cooked. Raw plants have a radish-like flavour and are said to be high in vitamin C. • The genus *Barbarea* was named for St. Barbara, a Christian martyr of the 4th century AD who refused to renounce her belief in God. The protection of St. Barbara was usually invoked against lightning and fire, so she became the patron saint of military architects, artillery men and miners (people who work with gunpowder).

MOUSE-EAR CRESS

Arabidopsis salsuginea (Pallas) N. Busch
MUSTARD FAMILY (BRASSICACEAE [CRUCIFERAE])

Plants: Small **annual** (or biennial) herbs, **hairless, blue-green** with a thin, waxy coating (glaucous); stems erect, 5–15 {20} **cm tall**, usually **branched** above and near the base; from slender taproots.

Leaves: Alternate (sometimes also with a sparse basal rosette of stalked, egg-shaped leaves), oblong to lance-shaped, blunt-tipped, 3–20 mm long, 1–4 {10} mm wide; **without teeth**, stalkless, **clasping the stem with small basal lobes** (auricles).

Flowers: White, about 3 mm across, with **4 petals, 4 sepals** and 6 (4+2) stamens; petals spatula-shaped to egg-shaped, widest above the middle, 2–3 {3.5} mm long; **sepals yellowish to pinkish with translucent edges**, 1.5–2 mm long, **soon shed**; styles 0.1–0.2 mm long; several to many, on straight, slender, spreading stalks 1–9 mm long, in dense, elongating clusters (racemes) without bracts; {late April–June}.

Fruits: Linear, almost cylindrical pods (siliques), 7–16 {20} mm long, 0.8–0.9 mm wide, straight or slightly curved, nearly erect, containing a single row of plump, oblong seeds about 0.6 mm long.

Habitat: Flat, moist saline ground by springs and lakes; elsewhere, in open, sandy alkaline soils in dry lakes and in salt plains and meadows.

Notes: This species has also been called *Thellungiella salsuginea* (Pallas) O.E. Schulz. • This species has a broad range but is seldom collected. One reason for this spotty distribution might be its dependence on saline habitats, which are often isolated from one another in arid regions. • The specific epithet *salsuginea* means 'of salt marshes.'

Dicots MUSTARD FAMILY

LEMMON'S ROCK CRESS

Arabis lemmonii S. Wats.
MUSTARD FAMILY (BRASSICACEAE [CRUCIFERAE])

Plants: Slender, tufted perennial herbs, **greyish with tiny repeatedly branched** (dendritic) **hairs** (at least on the lower parts); stems few to many, unbranched, 5–20 {40} cm tall; from branched root crowns.

Leaves: Basal leaves broadly spatula-shaped, blunt-tipped, 1–2 cm long, sometimes edged with a few teeth, **felted with tiny hairs**; **stem leaves** oblong to lance-shaped, 4–10 {15} mm long, stalkless, **with basal lobes** (auricles) **slightly clasping** the stem.

Flowers: Pink to purple, with **4 sepals and 4 petals**; sepals 2–3 {3.5} mm long, blunt-tipped, often pink or purple-tinged; petals 4–6 mm long, spatula-shaped; styles very short; few to several, on 2–5 mm long stalks in elongating clusters (racemes); July–August.

Fruits: Hairless, straight or slightly curved, **linear pods** (siliques), 2–5 cm long, **2–3.5 mm wide, evidently flattened**, pointing upward to nodding (**usually horizontal**), containing a single row of round, narrowly winged seeds about 1 mm wide.

Habitat: Alpine slopes, often on unstable ground.

Notes: Lemmon's rock cress might be mistaken for the more common species Lyall's rock cress (*Arabis lyallii* S. Wats.), but the leaves of Lyall's rock cress are usually fleshy and hairless, never greyish and felty.

MUSTARD FAMILY **Dicots**

HALIMOLOBOS

Halimolobos virgata (Nutt.) O.E. Schulz
MUSTARD FAMILY (BRASSICACEAE [CRUCIFERAE])

Plants: Biennial (sometimes annual or perennial) herbs; stems usually single (sometimes several), 10–40 cm tall, erect or ascending, freely branched (sometimes unbranched), greyish with long straight, simple or forked hairs mixed with short, branched hairs; from simple root crowns.

Leaves: Mainly in a basal rosette but also alternate on the stem, **densely covered with repeatedly branched (dendritic) hairs**; basal leaves {2} 3–6 cm long, 5–10 {15} mm wide, egg-shaped to lance-shaped, widest above the middle, **wavy-toothed** (sometimes minutely so), tapered to stalks; **stem leaves** smaller (1.5–4 cm long), shorter-stalked upward on the stem, the upper leaves stalkless and **clasping with small basal lobes** (auricles).

Flowers: White with pinkish veins, about 5 mm across, with **4 petals and 4 sepals**; petals spatula-shaped, 3–4 mm long; sepals 2–3 mm long, often purplish, with relatively large, spreading hairs; 6 stamens; styles 0.2–0.5 mm long; numerous, on slender, ascending, 7–11 {5–12} mm long stalks, in loose, elongating clusters (racemes); May–July.

Fruits: Erect (or nearly so), hairless, **cylindrical** (or slightly 4-sided), **linear pods** (siliques), {1.5} **2–4 cm long and 1 mm wide**, strongly nerved; seeds many, crowded in 2 irregular rows, about 1 mm long.

Habitat: Dry prairies; elsewhere, on bushy hillsides, moist meadows and alkali flats; sometimes becoming weedy.

Notes: Halimolobos could easily be mistaken for a rock cress (*Arabis* sp.), but rock cresses have clearly flattened pods, whereas the pods of halimolobos are cylindrical.

Rare *Vascular* Plants of Alberta

Dicots

MUSTARD FAMILY

PURPLE ALPINE ROCKET, PURPLE ROCKET

Erysimum pallasii (Pursh) Fern.
MUSTARD FAMILY (BRASSICACEAE [CRUCIFERAE])

Plants: Biennial or short-lived perennial herbs with **parallel, flat-lying hairs** that are attached at their middle, **low and tufted in flower, 10–20 {3–35} cm tall in fruit**; from swollen, unbranched (rarely branched) crowns on stout taproots.

Leaves: Many, in a prominent **basal rosette, linear to narrowly lance-shaped** and widest above the middle, 5–7 cm long, toothless or wavy-toothed, tapered to a slender stalk; old leaf bases persistent (marcescent).

Flowers: Fragrant, showy, **bright purple**, 2–4 cm across, with **4 sepals** and **4 petals**; sepals oblong, 5–8 {10} mm long, purplish, with translucent edges; petals 10–16 {20} mm long, slender at the base, abruptly flaring to a broad lobe at the tip; styles 1–3 mm long; **50 or more, in compact, short-stalked clusters** (racemes) often nestled among the basal leaves, these clusters greatly elongating in fruit; {July}.

Fruits: Erect to spreading-ascending, **flattened, linear pods** (siliques) {3} 4–10 cm long, 2 {3} mm wide, slightly curved, purplish; lower stalks 7–12 {5–20} mm long; seeds pale brown, oval to oblong, 2 mm long.

Habitat: On shale in alpine areas; elsewhere, on clay banks and sandy, gravelly or rocky slopes. Also noted as a dung-loving calciphile often found near animal burrows, below bird cliffs and around settlements.

Notes: These plants can spend 1 to several years in the rosette stage before flowering and dying. Their long seed pods are well-adapted to the gradual release of seeds. As the pods dry, their sides gradually curve away from the supporting frame or membrane, separating from the top, bottom or occasionally both ends. In this way, a considerable period elapses between the release of the first and last seeds. Released seeds often stick to animal fur, and these plants are often found around ground squirrel burrows and fox or wolf dens.

MUSTARD FAMILY **Dicots**

ALPINE BRAYA, PURPLE-LEAVED BRAYA

Braya purpurascens (R. Br.) Bunge *ex* Ledeb.
MUSTARD FAMILY (BRASSICACEAE [CRUCIFERAE])

Plants: Low, **tufted perennial** (rarely biennial) herbs; stems 5–15 {3.5–20} **cm tall**, single to many, leafless, slightly to densely hairy with simple to 3-branched hairs; from simple or branched woody crowns on thick taproots.

Leaves: **Basal**, tufted, somewhat **fleshy**, green to deep purple, spoon-shaped to narrowly lance-shaped, widest toward the blunt tip, 1–3 {0.6–6} cm long, fringed with hairs, nearly hairless to somewhat hairy on the surfaces, **essentially toothless**.

Flowers: White or purple-tinged, with **4 petals and 4 sepals**; petals 3–4 {4.5} mm long, broad-tipped, tapered to a slender base; sepals purple, egg-shaped, 1.5–2.5 {3.7} mm long; stamens 6 (2+4); stigmas head-like; several, in head-like clusters elongating in fruit (racemes); June–August.

Fruits: **Plump**, **straight pods** (siliques), about 10 mm long and 2–3 mm wide, broadly elliptic to lance-shaped or oblong-cylindrical, **sometimes constricted between seeds** (torulose), minutely hairy, tipped with a stout, 1 {0.5–2} mm long style, erect to spreading on 2–8 mm long stalks; seeds in 2 rows.

Habitat: Calcareous alpine scree slopes; elsewhere, in alpine meadows, on moist clay, sand and gravel barrens and on seashores.

Notes: This species includes *Braya americana* (Hook) Fern. It has also been called *Braya glabella* Richards. • The only other braya species formally reported from Alberta, leafy braya (*Braya humilis* (C.A. Mey.) Robins.), has leafy stems and longer (10–30 mm), narrowly cylindrical, clearly torulose pods. Leafy braya is more common and grows in areas from open woods and gravel bars to alpine slopes.

Dicots   MUSTARD FAMILY

ALPINE BITTER CRESS

Cardamine bellidifolia L.
MUSTARD FAMILY (BRASSICACEAE [CRUCIFERAE])

Plants: Tiny, hairless perennial herbs; stems several, tufted, {2} 3–10 cm tall, essentially leafless; from simple to many-branched root crowns on vertical taproots.

Leaves: In basal rosettes; blades thin, egg-shaped or elliptic, 5–15 mm long, without teeth or with 2–4 vague teeth; stalks slender, 2–4 times as long as the blades.

Flowers: White, about 5 mm across, with 4 sepals and 4 petals; petals 3–4 {5} mm long; 1–5, in short, flat-topped clusters (subumbellate racemes), elongating in fruit; July–August.

Fruits: Linear pods (siliques) 2–4 cm long, flattened, tipped with a short (1–3 mm long) persistent style, opening elastically from the base, held erect on 4–10 mm long stalks; seeds in a single row, without wings.

Habitat: Moist alpine banks and ledges; elsewhere, moist gravelly or rocky sites in subalpine to alpine zones.

Notes: Alpine bitter cress might be confused with some of the small white-flowered whitlow-grasses (*Draba* spp.) or with alpine braya (*Braya purpurascens* (R. Br.) Bunge *ex* Ledeb., p. 107), but those species have smaller (usually less than 1 cm long) pods and their plants are usually at least somewhat hairy. • As the seed pods dry, the sides gradually curl away from the supporting membrane, separating from the top, bottom or occasionally both ends. A considerable amount of time can elapse between the release of the first and last seeds. Under severe climatic conditions, seeds may not be released until the following spring. • Alpine bitter cress is a good example of a species with genetically dwarfed races. Most plants in alpine sites are barely distinguishable from those of high-arctic populations. However, when grown under warmer, temperate conditions, plants from alpine populations are tall and have large, irregular leaves, whereas those from high-arctic regions remain as small as plants on the more desirable sites at 80–82° N, and have small, regular leaves. Alpine plants have retained the ability to produce more robust plants and larger amounts of seed when occasional warm summers occur, but arctic plants never produce flowers far from the ground.

Rare *Vascular* Plants of Alberta

MUSTARD FAMILY — **Dicots**

SMALL BITTER CRESS

Cardamine parviflora L.
MUSTARD FAMILY (BRASSICACEAE [CRUCIFERAE])

Plants: Slender, **hairless** annual or biennial herbs; **stems leafy**, usually solitary, 10–30 {40} cm tall, simple to freely branched; from weak fibrous roots.

Leaves: Alternate on the stem, sometimes also clustered near the base, **pinnately divided** into **3–6 pairs** of **linear to spoon-shaped or oblong** leaflets and tipped with a larger, linear-oblong to somewhat lance-shaped leaflet that is widest above the middle, usually toothless; **leaflet bases not extending along the central stalk** (rachis).

Flowers: White, about **2–3 mm across**, with **4 petals** and **4 sepals**; petals 1.5–3 mm long; 6 stamens, the outer 2 attached lower down than the inner 4; numerous, on spreading stalks in **elongating clusters** (racemes); {March–May} July.

Fruits: Linear, 0.8–1 mm wide, **1–3 cm long pods** (siliques) tipped with a **0.5–0.7 {0.2–1} mm long style**, usually flattened, essentially nerveless, **opening elastically from the base**, pointed upward; stalks 5–8 {4–10} mm long; **seeds in a single row**, wingless, 0.7–0.9 mm long.

Habitat: Sandy ground and dry, open, mixed woodland; elsewhere, in seepage areas and on rocky outcrops, rocky or sandy shores, solifluction soil and scree slopes.

Notes: Mountain bitter cress (*Cardamine umbellata* Greene, also known as *Cardamine oligosperma* Nutt. *ex* Torr. & A. Gray var. *kamtschatica* (Regel) Detling) is another rare Alberta species of moist mountain sites. Most of its leaves are in a rosette at the base of the stem, and their leaflets are relatively wide, ranging from egg-shaped to round or kidney-shaped. Its flowers are slightly larger than those of small bitter cress (3–4 mm long vs. 1.5–3 mm long) and are borne in short (1–2 cm long), open clusters from July to September. • Small bitter cress might be confused with Pennsylvanian bitter cress (*Cardamine pensylvanica* Muhl. *ex* Willd.), but Pennsylvanian bitter cress is usually taller (10–50 cm) with less-branched stems that are hairy at the base. The bases of the leaflets extend down along the central stalk, the uppermost leaflet is much broader (elliptic to kidney-shaped) and the styles at the tips of the seed pods are more conspicuous (0.5–1.5 mm long).

C. UMBELLATA

Rare *Vascular* Plants of Alberta

Dicots MUSTARD FAMILY

MEADOW BITTER CRESS, CUCKOO FLOWER

Cardamine pratensis L.
MUSTARD FAMILY (BRASSICACEAE [CRUCIFERAE])

Plants: Slender, hairless (sometimes sparsely hairy) perennial herbs; stems erect, leafy, 20–40 {8–50} cm tall; from short underground stems (rhizomes).

Leaves: In a basal tuft and alternate on the stem; lower leaves **pinnately divided** into 5–17 small, stalked, toothless or obscurely toothed leaflets; basal leaves long-stalked, with round to lance-shaped leaflets; upper leaves stalkless, with lance-shaped to linear leaflets.

Flowers: White to pinkish or purplish, with **4 petals** and **4 sepals**; petals showy, 8–13 {15} **mm long**, egg-shaped, widest above the middle; sepals erect, ⅓ as long as the petals; 6 stamens, the outer 2 attached lower down than the inner 4; styles up to 1 mm long; several to many, on spreading-ascending stalks in elongating clusters (racemes); {May} June–July.

Fruits: Straight, linear pods (siliques) 20–30 {15–40} mm long and about 1.5 mm wide, tipped with a persistent style (beak) 1–2 mm long, usually flattened, nerveless or nearly so, opening elastically from the base; lowest stalks 12–18 mm long; seeds in a single row, wingless.

Habitat: Moist meadows and swamps; elsewhere, in calcareous shallow water, springs and swampy woods, calcareous meadows and thickets, wet places and along creeks.

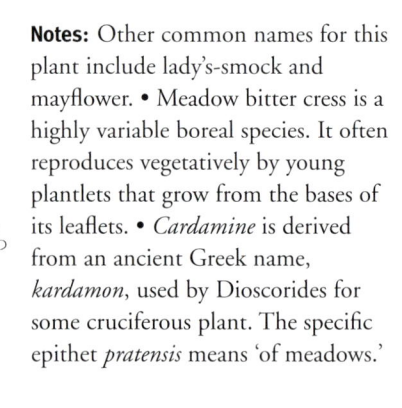

Notes: Other common names for this plant include lady's-smock and mayflower. • Meadow bitter cress is a highly variable boreal species. It often reproduces vegetatively by young plantlets that grow from the bases of its leaflets. • *Cardamine* is derived from an ancient Greek name, *kardamon*, used by Dioscorides for some cruciferous plant. The specific epithet *pratensis* means 'of meadows.'

SUNDEW FAMILY **Dicots**

OBLONG-LEAVED SUNDEW, ENGLISH SUNDEW, GREAT SUNDEW

Drosera anglica Huds.
SUNDEW FAMILY (DROSERACEAE)

Plants: Small, often **reddish, insectivorous** perennial herbs; flowering stems erect, leafless, 6–25 cm tall; from winter resting buds.

Leaves: In a **basal rosette**, covered with reddish, **tentacle-like, gland-tipped hairs**; blades egg-shaped to elongate–spatula-shaped, broadest above middle, 1–3 {3.5} cm long, 3–4 mm wide; stalks 2–6 {7} cm long, with 5 mm long, fringed stipules at base.

Flowers: White, with 4–8 equal petals; petals spatula-shaped, about 6 mm long; sepals oblong, 5–6 mm long; 1–9, in uncoiling clusters (scorpioid cymes); June–August.

Fruits: Capsules, covered with microscopic bumps (tubercles), containing many black, **spindle-shaped, 1–1.5 mm long seeds**.

Habitat: Peaty, usually calcareous sites in swamps and fens, where the water-table is at or just below the surface; elsewhere, also on lakeshores.

Notes: This species has also been called *Drosera longifolia* L. • The edges and upper surfaces of the leaves are covered with stalked, tentacle-like, gland-tipped hairs that secrete a drop of mucilage at their tips. When an insect (usually a midge or another small fly) alights on the surface of the leaf, it becomes stuck on 1 or more of these glands, and nearby tentacles bend inward to entrap the prey. This bending is actually the result of rapid growth along 1 side of the stalk, but it can occur in less than a minute. The more the insect struggles, the more tentacles are activated. An insect may be totally entrapped within 20 minutes, but digestion can take a week. Once digestion has been completed, the tentacles unfurl, poised to entrap another victim. The nitrogen obtained from insects and other prey enables sundews to live in environments that are nitrogen-poor. Small flies, fungus gnats and gall-wasps are the main pollinators of sundew flowers. • The specific epithet *anglica* is Latin for 'English' and was given because this plant was first described from England.

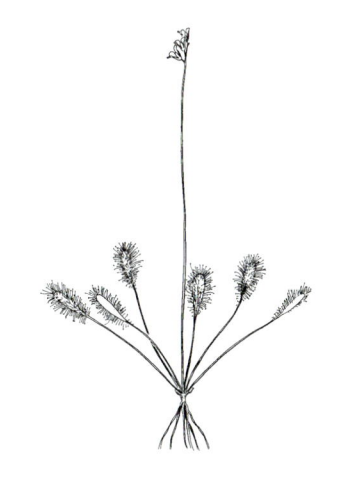

SUNDEW FAMILY

SLENDER-LEAVED SUNDEW, LINEAR-LEAVED SUNDEW

Drosera linearis Goldie
SUNDEW FAMILY (DROSERACEAE)

Plants: Small, often **reddish, insectivorous** perennial herbs; flowering stems erect, leafless, 6–13 cm tall; from winter resting buds.

Leaves: In a **basal rosette**, covered with **reddish, tentacle-like, gland-tipped hairs**; blades linear, 2–5 {6} cm long, about 2 mm wide; stalks with small, fringed stipules at base.

Flowers: White, with 4–8 equal petals; petals egg-shaped, broadest above middle, about 6 mm long; sepals oblong, 5–6 mm long; 1–4, in uncoiling clusters (scorpioid cymes); July {June–August}.

Fruits: Capsules, covered with microscopic bumps (tubercles), containing many black, **oblong to egg-shaped, 0.5–0.8 mm long seeds**.

Habitat: Bogs, often in marly sites; requires alkaline conditions for growth; elsewhere, also on wet, calcareous shores.

Notes: Sundew tea was used in the past for respiratory ailments. Extracts were also said to cure pimples, corns and warts. • The generic name *Drosera* was derived from the Greek *droseros* (dewy), in reference to the dew-like drops of liquid on the leaves. The specific epithet *linearis* refers to the slender, linear leaves.

PITCHER-PLANT FAMILY **Dicots**

PITCHER-PLANT
Sarracenia purpurea L. ssp. *purpurea*
PITCHER-PLANT FAMILY (SARRACENIACEAE)

Plants: Tufted, **insectivorous** perennial herbs; flowering stems leafless, 20–40 cm tall; from stout underground stems (rhizomes).

Leaves: In a **basal rosette, 10–20 {30} cm long**, curved tubes (pitchers) with a **broad wing** on the inner side **and a hood-like flap** over the opening at the top, green to yellowish **green with red or purple veins**, usually containing water; inner surface with **stiff, downward-pointing hairs** near the top, smooth and slippery near the base.

Flowers: Deep reddish purple, 5–7 cm wide, nodding, with a **broad, umbrella-shaped style** surrounded by 5 large, incurved petals and 5 broad sepals; **solitary**, on a long leafless stalk from the centre of the leaf rosette; July.

Fruit: Capsules with 5 cavities, containing many seeds.

Habitat: Wetlands, usually with sphagnum moss (*Sphagnum* spp.).

Notes: Other common names for this unusual plant include side-saddle flower, Indian pipe and frog's trousers. • Young leaves are flat and develop into a pitcher only after they have reached full length. • Pitcher-plants are insectivorous, supplementing their diet with nutrients from animal protein. This helps them to grow in environments where some nutrients are in poor supply. Insects are attracted by the brightly coloured leaves and nectar. Within the pitchers, downward-pointing hairs and a slippery surface direct the insects into the leaf cavity and then prevent escape. The insects drown in the pool of water at the bottom of the tube and are digested by acids and enzymes secreted by the plant. However, some insects, for example the larvae of some species of mosquitoes, are known to live in the leaves of pitcher-plants. • These unusual flowers are pollinated by insects, primarily bumblebees and honey-bees. • The pitcher-plant is the floral emblem of Newfoundland. • The generic name *Sarracenia* commemorates Dr. Michel Sarrasin de l'Etang, an 18th-century physician to the French court who collected some of the first specimens of Canadian plants. The specific epithet *purpurea* means 'purple,' a reference to the purple flowers.

Rare *Vascular* Plants of Alberta

STONECROP FAMILY

SPREADING STONECROP

Sedum divergens S. Wats.
STONECROP FAMILY (CRASSULACEAE)

Plants: Succulent, mat-forming perennial herbs; stems lying on the ground, rooting at joints (nodes) and producing erect non-flowering and flowering branches about 5–15 cm tall.

Leaves: Opposite (or nearly so) along stems, fleshy (succulent), stalkless, hairless, 4–10 mm long, **oval or egg-shaped** (widest above the middle) **to round**, narrower on flowering branches.

Flowers: Yellow, with 4 or 5 petals and sepals; petals longer than the sepals, tipped with a tiny point; sepals triangular, joined at the base; 3–15, in loose to dense, spreading clusters (cymes); {July} August–September.

Fruits: Pod-like fruits that open along one side (follicles), 4 or 5 per flower in star-like clusters, joined at the base for 1.5–2 mm and with tips pointing outward.

Habitat: Moist alpine or subalpine ridges, rocky ledges, crevices and talus slopes.

Notes: Other stonecrops that occur in Alberta have much narrower, alternate leaves. The most common, lance-leaved stonecrop (*Sedum lanceolatum* Torr.), has erect follicles and stems that grow in clumps. Narrow-petalled stonecrop (*Sedum stenopetalum* Pursh) is found only in southwestern Alberta. Its leaves are flattened, becoming brown or dry and translucent, and small bulbs replace some of the flowers. • *Sedum* is an ancient Latin name supposedly derived from *sedeo* (to sit), in reference to the squatty habit of the plants. The specific epithet *divergens* means 'spreading widely,' in reference to the fruits.

SAXIFRAGE FAMILY	**Dicots**

SPIDERPLANT

Saxifraga flagellaris Willd. ssp. *setigera* (Pursh) Tolm.
SAXIFRAGE FAMILY (SAXIFRAGACEAE)

Plants: Perennial herbs; stems 3–20 cm tall, glandular-hairy; from slender underground stems (rhizomes) and also many slender, **arching or trailing stems** (stolons) **up to 10 cm long and sprouting at tips**.

Leaves: Mainly in basal rosettes, egg-shaped to lance-shaped, widest above middle, **spine-tipped, 5–10 {15} mm long, fringed** with stiff white hairs; stem leaves several, gradually smaller upward, hairless above, glandular-hairy beneath.

Flowers: Bright yellow, about **1–1.5 cm across**, with 5 petals and 5 sepals; petals spreading, egg-shaped, widest above middle, pink-spotted near the base; **sepals purple** and **glandular**, in a bell-shaped calyx with lobes 2–5 mm long; 1–3 {5}; July {August}.

Fruits: Pairs of erect pod-like fruits (follicles), {4} 5–10 mm long, **often bright red** with 2 spreading tips.

Habitat: Moist, turfy, limestone slopes and ridges in alpine areas; elsewhere, on streambanks and gravelly to rocky slopes at high to low elevations.

Notes: This species has also been called *Saxifraga setigera* Pursh. • Yellow mountain saxifrage (*Saxifraga aizoides* L.) is another small, yellow-flowered saxifrage, but its plants lack stolons and its linear leaves are succulent, hairless, and borne mainly on the stems (rather than in basal rosettes). • Spiderplant is very decorative in gardens, but it can be difficult to grow. Plants are usually propagated by rooting rosettes at the tips of stolons and then severing from the parent plants once their roots are well established. • Spiderplant has been classified as rare in BC and Alberta, and it is protected by law in New Mexico. • *Saxifraga* is from the Latin *saxum* (rock) and *frangere* (break). This may refer to the tendency of saxifrage plants to grow between rocks, or to the historical use of the plants to treat kidney stones.

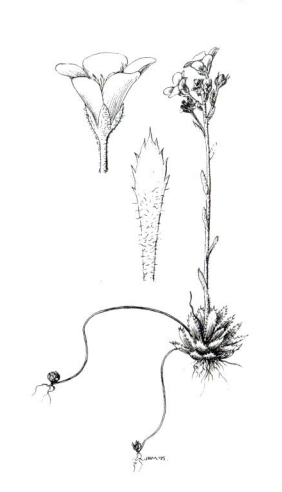

Dicots SAXIFRAGE FAMILY

ALPINE SAXIFRAGE
Saxifraga nivalis L.
SAXIFRAGE FAMILY (Saxifragaceae)

Plants: Perennial herbs, 4–15 {22} cm tall; stems (flowering stalks) single, erect, with silky, gland-tipped hairs, often **purplish**; from short, thick underground stems (rhizomes).

Leaves: In a **basal rosette**, thick, green and hairless on upper surface, **purple on lower surface**, usually with coarse, rusty-coloured hairs along edges; blades 1–3 cm long, round to egg-shaped, tipped with 9–20 coarse, rounded teeth, tapering to broad stalks about as long as the blade.

Flowers: Reddish-tinged, with 5 petals and 5 sepals; petals 3 mm long, greenish white, purplish or reddish on the back, slightly longer than the sepals; sepals purplish; stalks very short, glandular-hairy; 3–12, in **compact, branched clusters** (cymes); July {August}.

Fruits: Pod-like fruits that split open down 1 side (follicles), **purplish, 5–6 {7} mm long**, egg-shaped, 2 per flower, joined at base and with tips curved outward.

Habitat: Moist alpine slopes, ridges and rock crevices.

Notes: Oregon saxifrage (*Saxifraga oregana* J.T. Howell var. *montanensis* (Small) C.L. Hitchc.) is found on alpine slopes in the Crowsnest Pass. It is a much larger plant, 15–120 cm tall, with narrow basal leaves that are 6–20 cm long and widest toward the tip. The fragrant, greenish white flowers are numerous in branched clusters (panicles) with white-hairy stalks. The petals are often unequal in size or with some absent. This species flowers from June to early July. • Rhomboid-leaved saxifrage (*Saxifraga occidentalis* S. Wats.) is a common species that has longer leaves (more than 2 cm long) that lack the purple lower surface, and its flower clusters are more open, flat-topped and spreading. • Red-stemmed saxifrage (*Saxifraga lyallii* Engler) often has purplish stems, but it grows to 30 cm tall and has open flower clusters and fan-shaped leaves. • Another rare saxifrage has been reported to occur in Alberta, but its presence has not been verified. Yellowstone saxifrage (*Saxifraga subapetala* (E. Nels.) Rydb.) is very similar to Oregon saxifrage, and some taxonomists consider it a variety of the same species (*Saxifraga oregana* J.T. Howell var. *subapetala* (E. Nels.) C.L. Hitchc.). It is usually distinguished by its slightly smaller (5–7 mm wide), greenish-tinged flowers in narrow, elongating flower

S. SUBAPETALA

clusters (rather than pyramidal panicles) on sparsely hairy or glandular stalks (rather than on conspicuously long-hairy stalks). • Saxifrages have been used in the treatment of scurvy. The fresh leaves are high in vitamins A and C, and they have been used in salads. Several species are cultivated as ornamentals. • The specific epithet *nivalis* was derived from the Latin *nivea* (snow).

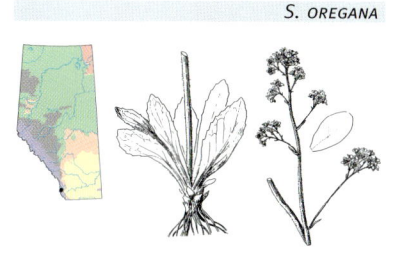

S. OREGANA

STREAM SAXIFRAGE

Saxifraga odontoloma Piper
SAXIFRAGE FAMILY (SAXIFRAGACEAE)

Plants: Perennial herbs; stems erect, 10–75 cm tall, hairless near base, glandular toward tips; from elongated underground stems (rhizomes).

Leaves: Basal; blades **round**, 2.5–10 cm wide, **heart-shaped at the base** and **toothed all along the edges**, hairless or sparsely hairy; **stalks 1–3 times longer than the blades**.

Flowers: White with **2 yellow or green basal spots at the base of each petal**, with 5 petals and 5 sepals; petals 2–5 mm long, broad at tip, **abruptly narrowed to a slender base**; sepals 1.8–2.8 mm long, hairless or glandular-hairy along the edges only; in open, branched clusters (panicles) with spreading to erect, glandular-hairy branches, the hairs often reddish purple; July {June–August}.

Fruits: Pod-like fruits that split open down 1 side (follicles), usually **purplish**, **5–9 mm long**, 2 (sometimes 3 or 4) per flower, fused for ⅓–⅗ of their length and with **firm, slightly spreading tips**.

Habitat: Shady mountain streambanks and wet cliff faces.

Notes: Cordate-leaved saxifrage (*Saxifraga nelsoniana* D. Don ssp. *porsildiana* (Calder & Savile) Hult.) is also rare in Alberta. It is distinguished from stream saxifrage by its narrower (slender to broadly elliptic), often pinkish petals, which taper gradually and lack spots at their bases. It grows in moist rocky sites and along streams in alpine areas. • Alaska saxifrage (*Saxifraga ferruginea* Graham) is another rare Alberta species with open, branched flower

Dicots

SAXIFRAGE FAMILY

clusters (panicles). However, it has much narrower (spatula-shaped) leaves that taper gradually to wedge-shaped bases. Its flowers have 3 broad and 2 narrow petals, and small bulblets usually replace some blooms. Alaska saxifrage grows on moist, often rocky, alpine sites, and blooms in {June} July and August. • Red-stemmed saxifrage (*Saxifraga lyallii* Engler) is a common species whose fan-shaped leaves are toothed only above the middle and have stalks about as long as their blades. • There is some controversy about whether the Alberta plants are hybrids with red-stemmed saxifrage or the true stream saxifrage. • The specific epithet *odontoloma* is from the Greek *odonto* (tooth) and *loma* (fringe, border), presumably referring to the toothed edges of the leaf.

S. NELSONIANA

S. FERRUGINEA

SAXIFRAGE FAMILY   **Dicots**

HEUCHERA-LIKE BOYKINIA, TELESONIX
Boykinia heucheriformis (Rydb.) Rosendahl
SAXIFRAGE FAMILY (Saxifragaceae)

Plants: Perennial herbs; stems 5–20 cm tall, with **glandular hairs**; from thick, scaly, elongated underground stems (rhizomes).

Leaves: Mostly basal; blades **kidney-shaped to round**, 1–6 cm wide, with shallow, rounded or pointed lobes edged with rounded teeth, bearing stalkless glands; **stem leaves** alternate, **fan-shaped**, short-stalked or stalkless.

Flowers: Reddish purple to deep pink; 5 petals, widest toward their tips, equal to or slightly longer than the sepals; 5 sepals, fused at base, 9–13 mm long; 10 stamens; 5–25, in **short, leafy, branched clusters** (panicles), somewhat nodding to 1 side; June–August.

Fruits: Capsules, **split** about halfway to base **into 2 firm, slightly spreading points**; seeds brown, shiny, 1–2 mm long.

Habitat: Rocky outcrops and talus slopes (often limestone) in alpine areas. Alberta populations are disjunct and restricted to glacial refugia.

Notes: This species has also been called *Telesonix jamesii* (Torr.) Raf. and *Telesonix heucheriformis* Rydb. • These plants are eaten by elk and deer. • The generic name *Boykinia* commemorates Samuel Boykin (1786–1848), an American banker, physician and naturalist who lived in Georgia. The specific epithet *heucheriformis* means 'like (in the form of) *Heuchera*,' a similar genus in the saxifrage family.

Rare *Vascular* Plants of Alberta

Dicots SAXIFRAGE FAMILY

BLUE SUKSDORFIA

Suksdorfia violacea A. Gray
SAXIFRAGE FAMILY (Saxifragaceae)

Plants: Delicate perennial herbs, 10–20 cm tall, lightly **hairy and glandular** on lower parts, becoming more densely so in upper portions; from elongated underground stems (rhizomes) bearing **small bulbs**.

Leaves: Basal (1–3) and alternate (2–5); blades **kidney-shaped**, 1–2.5 cm wide with 5–7 **shallow, slightly toothed lobes**, more or less short-hairy; lower leaves long-stalked; stem leaves with large, **deeply cut stipules at base**, becoming smaller, wedge-shaped and almost stalkless upward, with 2–4 teeth at the tip.

Flowers: Light violet (rarely white); **5 petals 5–9 mm long, erect**, narrowed abruptly to a whitish or yellow base; **calyxes narrowly cone-shaped**, 2–3 mm long, with **5 narrowly lance-shaped, erect lobes**; 2–10, borne on 2 or 3 erect branches in elongated, loose, branched clusters (corymbs); May–July.

Fruits: Capsules, 4–6 mm long, with warty, brownish seeds about 0.5 mm long.

Habitat: Among wet rocks along mountain streams and in damp, mossy areas.

Notes: White suksdorfia (*Suksdorfia ranunculifolia* (Hook.) Engl.) is also rare in Alberta, known only from wet rock faces and crevices in Waterton Lakes National Park. It differs from blue suksdorfia in having 5 or 6 basal leaves with 3–10 cm long stalks and round, 2–4 cm wide blades that are deeply cut into 3 wedge-shaped lobes with irregular, rounded teeth. The middle stem leaves have expanded bases that clasp the stem. Its flower clusters are nearly flat-topped with short branches bearing several flowers. The flowers have smaller (2.5–6 mm long), spreading petals that are white (often purplish at the base). The calyxes are broadly bell-shaped, with 5 broadly triangular spreading lobes, and the stamens are attached to a thick disc that is not found in blue suksdorfia flowers. White suksdorfia flowers from June to July. • The distribution of these species may be correlated with the extension of ice during the last glaciation (i.e., they occur beyond the maximum extent of the glaciers).

S. RANUNCULIFOLIA

ALPINE ALUMROOT

Heuchera glabra Willd. *ex* R. & S.
SAXIFRAGE FAMILY (Saxifragaceae)

Plants: Slender perennial herbs; flowering stems single to several, 10–60 cm tall, hairless near the base but **conspicuously glandular and hairy** on the upper parts; from **well-developed, scaly, horizontal to ascending underground stems** (rhizomes).

Leaves: Mainly basal; blades 2–9 cm long, 1.5–4 {9} cm wide, **somewhat maple-leaf-like**, broadly heart-shaped, notched at the base and tipped with {3} **5–7 shallow, irregularly sharp-toothed lobes, essentially hairless**, sometimes with small gland-tipped hairs on the lower surface or with a few tiny, stiff hairs on the upper surface when young; 1 or 2 stem leaves, usually more or less bract-like; **stalks long and slender with 2 broad lobes** (stipules) at the base.

Flowers: White, **cone-shaped, about 5 mm long**, with **5 slender petals**; petals 2–4 times as long as the sepals, with spoon-shaped blades, ¼–½ as long as their slender stalks (claws); **calyxes cone-shaped**, squared at the base, 2–3 {1.5–3.5} **mm long, glandular-hairy**, cut less than halfway to the base into 5 egg-shaped to oblong lobes; **5 stamens, projecting from the flower**, equal to or slightly longer than the petals; pistils with 2 stalkless, head-like stigmas, each at the tip of a hollow, spreading point (beak); numerous, in the axils of small bracts on thread-like branches in **open, branched clusters** (panicles); July–August {June–September}.

Fruits: Oval capsules tipped with 2 firm, spreading points (beaks), protruding from the calyx, 5–6 {9} mm long, **splitting open between the beaks**; seeds brown, somewhat crescent-shaped, 0.7–0.8 mm long and ⅓–¼ as wide, with rows of tiny spines; August.

Habitat: Moist alpine scree slopes; elsewhere, on damp, shady rocks and riverbanks (such as spray zones near waterfalls), rocky meadows and slopes, and grassy hillsides.

Notes: Other common names for this species include alpine heuchera, smooth alumroot, smooth heuchera and mountain saxifrage. • Alumroot plants are commonly used as mordants—substances to make dyes colourfast. • The powdered roots were used to stop bleeding and heal wounds and sores. Root tea was taken as a tonic or as a treatment for diarrhea or stomach flu, and it was also gargled.

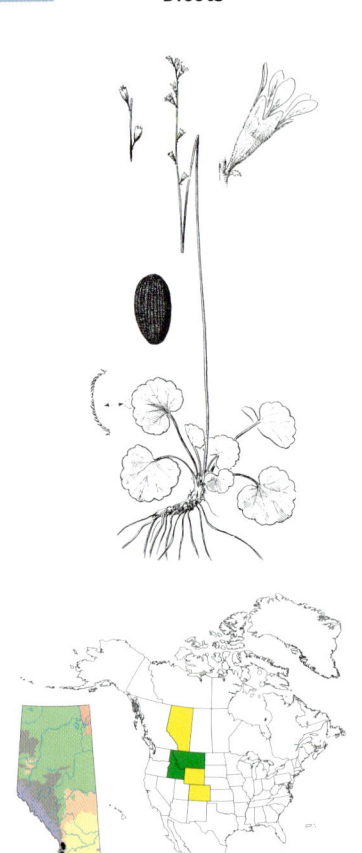

WILLIAMS' CONIMITELLA

Conimitella williamsii (D.C. Eaton) Rydb.
SAXIFRAGE FAMILY (Saxifragaceae)

Plants: Slender perennial herbs; flowering stems leafless, 20–40 {60} cm tall, with fine gland-tipped hairs; from short, ascending underground stems (rhizomes).

Leaves: Basal; blades **round to kidney-shaped**, 1–4 cm wide, leathery, usually purplish on the lower surface, edged with broad, rounded, double teeth and **stiff hairs**; stalks slender, 1.5–6 {1–10} cm long.

Flowers: White, cone-shaped, with **5 small, slender petals**; petals 4–5 mm long, with a blunt, broadened, lance-shaped blade at the tip of a slender base (claw); **calyxes cone-shaped, 4–6 mm long** (up to 9 mm long in fruit), tipped with 5 spreading, egg-shaped lobes about 1 mm long, attached to the ovary for ⅓–½ of its length; 5 stamens, opposite the sepals, with filaments about the same length as the anthers; pistils with 2 stalkless, head-like stigmas, each at the tip of a hollow point; 5–10 {12}, on 1–5 mm long stalks in the axils of slender bracts, forming loose, finely glandular-hairy, elongating clusters (racemes); June.

Fruits: Capsules, **enveloped almost entirely by the enlarged calyxes, splitting open along the inner** (ventral) **side of the 2 small, hollow points at the tips**; seeds numerous, black, slightly roughened, about 1 mm long.

Habitat: Open montane slopes; elsewhere, in rich montane forests and on moist cliffs and rocky slopes, often on limestone.

Notes: The leaves of Williams' conimitella resemble those of alumroots (*Heuchera* spp.) and bishop's-caps (*Mitella* spp.), but the conimitella leaves are broader (more kidney-shaped) and their edges are fringed with stiff hairs. When bishop's-caps are in bloom, they are easily identified by the feathery petals of their flowers. • The generic name *Conimitella* was derived from the Latin *conos* (cone) and *mitella* (bishop's cap), in reference to the cone-shaped flowers of these *Mitella*-like plants. The specific epithet *williamsii* commemorates R.S. Williams (1859–1945), a Montana botanist who built the first cabin where Great Falls, Montana now stands and who was the last rider for the Pony Express.

SAXIFRAGE FAMILY   **Dicots**

SMALL-FLOWERED ROCKSTAR

Lithophragma parviflorum (Hook.) Nutt. *ex* T. & G.
SAXIFRAGE FAMILY (Saxifragaceae)

Plants: Perennial herbs; stems **glandular-hairy, 10–30 cm tall**, often rough to touch; from thin underground stems (rhizomes) with scattered, rice-like bulblets.

Leaves: Mainly **basal**, 1–3 cm long, palmately divided into **3–5 wedge-shaped sections that are once or twice 3-lobed or 3-toothed**, sparsely white-hairy, long-stalked; stem leaves smaller and stalkless upward.

Flowers: White to pinkish or purplish, broadly funnel-shaped or star-like, 5-sided, 12–15 mm wide, with **5 spreading petals** and 5 small calyx lobes; petals **4–10 mm long, deeply 3–5-lobed at the tip** (appearing as if 15 or 25 slender petals), abruptly narrowed to a slender base; 3–9 {11}, in elongating clusters (racemes); {April} May–July.

Fruits: Capsules, splitting open along 3 lines.

Habitat: Moist montane meadows and open woods; elsewhere, often in dry grassland or sagebrush desert.

Notes: Smooth rockstar (*Lithophragma glabrum* Nutt., also called *Lithophragma bulbifera* Rydb.) is another rare Alberta plant. It has nearly hairless leaves, and it often produces small bulblets in the axils of its lower leaves. Smooth rockstar also has fewer (2–5) flowers in more head-like clusters, and its stems are covered with purple-tipped hairs. It grows on dry plains and montane slopes, and flowers in July and August. • Rockstar is sometimes called 'fringe-cup.' The generic name *Lithophragma* was derived from the Greek *lithos* (stone) and *phragma* (wall) because these plants usually grow on rocky slopes.

L. GLABRUM

Rare *Vascular* Plants of Alberta

Dicots

SAXIFRAGE FAMILY

FRINGE-CUPS

Tellima grandiflora (Pursh) Dougl. *ex* Lindl.
SAXIFRAGE FAMILY (Saxifragaceae)

Plants: Perennial herbs; stems 40–80 cm tall, unbranched, **stiff-hairy on the lower portion**, **glandular** and less hairy upward; from short underground stems (rhizomes).

Leaves: Basal and alternate; blades **heart-shaped to kidney-shaped**, **3–10 cm wide**, sparsely stiff-hairy, shallowly 5–7-lobed (sometimes 3-lobed), with irregular, broad, pointed teeth; **basal leaf stalks 5–20 cm long, very hairy**; 1–3 stem leaves, smaller upward.

Flowers: Greenish yellow when young, reddish with age, short-stalked, **fragrant**; **5 petals, fringed** with 5–10 thread-like segments at tips; **calyx greenish, glandular**, 5–8 mm long (longer when fruit appears), with 5 erect, triangular lobes; 10–35, in **narrow, elongated, branched clusters** (panicles); {April–July}.

Fruits: Egg-shaped capsules, approximately 10 mm long, **with 2 spreading beaks**; seeds brown with rows of tiny bumps.

Habitat: Rocky seeps and rich, moist soil.

Notes: The leaves of fringe-cups resemble those of sugar-scoop (*Tiarella unifoliata* Hook.) and certain species of bishop's-cap (*Mitella* spp.) and alumroot (*Heuchera* spp.). However, the flowers of these other species are smaller and not fragrant, and their petals are not tipped with the characteristic fringe. • Fringe-cups was first discovered in Alberta in 1986, in the South Castle River valley. • The Skagit pounded and boiled fringe-cups to make a tea used for any kind of sickness, especially for lack of appetite. It is also said to have been eaten by woodland elves to improve night vision. • Fringe-cups thrives in sites that are high in nitrogen. Its seeds readily germinate and take root, resulting in new, vigorous plants that often crowd out the mother plant. • *Tellima* is an anagram of *Mitella*, the genus under which this plant was first described. *Grandiflora* is derived from Latin meaning 'large-flowered.' The common name 'fringe-cups' refers to the highly divided petals which form a fringe around each flower.

Rare *Vascular* Plants of Alberta

GRASS-OF-PARNASSUS FAMILY

Dicots

SMALL NORTHERN GRASS-OF-PARNASSUS

Parnassia parviflora DC.
GRASS-OF-PARNASSUS FAMILY (PARNASSIACEAE)

Plants: Slender, hairless perennial herbs; stems 5–30 cm tall, with a single, 7–16 mm long, **narrowly egg-shaped leaf** (bract) **usually borne below the middle** and **not clasping** the stem; from short, erect to ascending underground stems (rhizomes).

Leaves: Basal (except for a small, single stem leaf), slender-stalked; **blades 1–2 {2.5} cm long**, elliptic to egg-shaped, **tapered to base**.

Flowers: White with greenish or yellowish veins, with 5 petals, 5 sepals and 5 clusters of specialized, gland-tipped structures (staminodia); petals 5–7-veined, oblong to elliptic, 5–9 {4–10} mm long; sepals usually only slightly shorter than the petals; staminodia slender, borne **in clusters of 5–7 {9} at the tip of small, oblong scales**; single; July–August {September}.

Fruits: Blunt-tipped, ovoid capsules, 7–11 mm long.

Habitat: Bogs and streambanks; elsewhere, in wet meadows.

Notes: This species is sometimes considered a small-flowered variant of northern grass-of-Parnassus (*Parnassia palustris* L.), and these 2 species often appear to merge through *P. palustris* var. *montanensis* (Fern. & Rydb.) C.L. Hitchc., which is almost exactly intermediate in its characteristics. Northern grass-of-Parnassus is usually distinguished by its heart-shaped basal leaves, its clasping stem leaf and its large clusters of 7–15 staminodia. • The white, saucer-shaped flowers are designed to attract flies. The glistening glands at the tips of the staminodia appear to be drops of pure nectar, but they are really false nectaries designed to lure insects to the flowers. When nectar-seeking visitors land at the staminodia, they either touch the stigma (fertilizing the flower) or brush against the stamens (picking up pollen to carry to the next flower). Usually they are so strongly attracted to the false nectaries that they do not even bother eating the pollen on the stamens. The anthers mature first, and later, when the 2-pronged stigma opens, most of the flower's own pollen is gone. This reduces the chances of self-fertilization.

Rare *Vascular* Plants of Alberta

Dicots ROSE FAMILY

Macoun's cinquefoil

Potentilla macounii Rydb.
ROSE FAMILY (Rosaceae)

Plants: Low, **spreading** perennial herbs; stems ascending or spreading on the ground, 4–12 {20} cm long and usually less than half as tall; usually few to several; from stout, sparingly branched root crowns on taproots.

Leaves: Mainly in a basal tuft, but also with 1 or 2 alternate leaves on the stem; blades **pinnately divided into 7–9 {5–11} leaflets**; leaflets about 1 cm long, **cut to the midrib into linear to lance-shaped lobes** about 1 mm wide, **usually greyish with a mixture of flat-lying, stiff hairs and woolly hairs on the upper surface, densely white-woolly beneath** and with overlying straight hairs; stalks with a pair of lance-shaped, 3–10 mm long lobes (stipules) at the base.

Flowers: Yellow, saucer-shaped, 10–15 mm across, with 5 shallowly notched, egg-shaped petals twice as long as the cupped, **silky-hairy calyx**, which has 5 lance-shaped, 3–6 mm long lobes (sepals) alternating with 5 smaller lobes (bractlets); 2–5, in open clusters (cymes); {June–August}.

Fruits: Smooth seed-like fruits (achenes), about 1.5 mm long, with slender, thread-like styles attached below the tip, in compact clusters surrounded by persistent calyxes.

Habitat: Dry rocky slopes on the eastern slopes of the Rocky Mountains.

Notes: This species has also been called *Potentilla concinna* Richards. var. *macounii* (Rydb.) C.L. Hitchc. and var. *divisa* Rydb. • Macoun's cinquefoil could be confused with the more common species sheep cinquefoil (*Potentilla ovina* Macoun), but the leaves of sheep cinquefoil lack woolly hairs (tomentum) on their lower surfaces. • Another small, rare cinquefoil has been discovered in Alberta. Colorado cinquefoil (*Potentilla subjuga* Rydb., also called *Potentilla rubripes* Rydb. and *Potentilla concinna* Richards. var. *rubripes* (Rydb.) C.L. Hitchc.) is very similar to Macoun's cinquefoil, and some taxonomists consider both species to be varieties of *Potentilla concinna*. Colorado cinquefoil is distinguished by its greenish, stiff-hairy upper leaf surfaces (rather than greyish and woolly). It grows on mountain slopes from Alberta to Colorado and flowers in spring and early summer.

P. SUBJUGA

ROSE FAMILY **Dicots**

BRANCHED CINQUEFOIL, STAGHORN CINQUEFOIL

Potentilla multifida L.
ROSE FAMILY (ROSACEAE)

Plants: Densely tufted perennial herbs; stems erect or spreading, 10–20 {40} cm tall, sparsely to densely covered with spreading hairs; from stout, rather woody root crowns.

Leaves: Basal and alternate on the stem, 5–10 cm long; blades **pinnately divided into 5–7 leaflets that are deeply cut into parallel, linear lobes** about 1–1.5 mm wide and arranged **like teeth on a comb** (pectinate), **rolled under along the edges**, green and sparsely covered with flat-lying, stiff hairs on the upper surface, densely **white-woolly beneath**, and silky-hairy on the midribs; stalks either shorter than or much longer than the blade, with a pair of toothless lobes (stipules) at the base.

Flowers: Yellow, saucer-shaped, 6–10 mm across, with 5 egg-shaped petals above an **equally long or slightly longer, silky-hairy, bell-shaped calyx** with 5 triangular to egg-shaped, 3–5 mm long lobes (sepals) alternating with 5 linear lobes (bractlets); 3–15, in open clusters (cymes); July.

Fruits: Dry seed-like fruits (achenes), about 1.2 mm long, with slender, thread-like styles, in compact clusters surrounded by persistent calyxes.

Habitat: Gravel bars and open slopes; elsewhere, on scree slopes and open gravelly or sandy ground, often near streams and lakes or on roadsides.

Notes: This species has also been called *Potentilla virgulata* A. Nels. and *Potentilla bimundorum* Sojak. • Branched cinquefoil could be mistaken for the much more common species prairie cinquefoil (*Potentilla pensylvanica* L.), but the stipules of the upper stem leaves of prairie cinquefoil are toothed (at least near the base) and its stems are usually much taller (20–50 cm tall). • The generic name *Potentilla* is from the Latin *potens* (powerful), in reference to the strong (mainly astringent) medicinal properties of these plants. The specific epithet *multifida* was taken from the Latin *multi* (many) and *fidus* (split), in reference to the many slender lobes of the leaflets.

Dicots ROSE FAMILY

SMOOTH-LEAVED CINQUEFOIL, FEATHER-LEAVED CINQUEFOIL

Potentilla multisecta (S. Wats.) Rydb.
ROSE FAMILY (Rosaceae)

Plants: Sparsely hairy perennial herbs, stems **5–20 cm tall**, with a mixture of short, spreading and long, flat-lying hairs; from a thick, branched root crown.

Leaves: Mostly basal, hairless to sparsely hairy on the upper surface, densely hairy on the veins beneath, 3–4 cm long, pinnately (rarely palmately) **divided into 5–9 leaflets**; leaflets 5–15 mm long, **deeply cut** (to the midrib) **into slender** (0.5–2 mm wide), **linear or oblong segments**; stalks as long as blades; stem leaves alternate.

Flowers: Yellow, 1–1.5 cm across, with **5 sepals and 5 smaller, alternating bractlets** below 5 petals; sepals lance-shaped, 3–5 mm long; petals 5–8 mm long; 2–10, in spreading clusters (cymes); June {July–August}.

Fruits: Dry seed-like fruits (achenes), about 1 mm long.

Habitat: Dry alpine slopes.

Notes: *Potentilla multisecta* is often treated as a variety of mountain cinquefoil (*Potentilla diversifolia* Lehm.), and it may be an alpine ecotype of that species. However, the leaflets of mountain cinquefoil are borne close together, in an almost palmate arrangement (not clearly pinnate), and they are toothed or shallowly lobed (not cut almost to the midrib). • Smooth-leaved cinquefoil could be confused with prairie cinquefoil (*Potentilla pensylvanica* L.), but that is a larger (20–50 cm tall) species of dry prairies and open slopes. • Smooth-leaved cinquefoil often grows with sheep cinquefoil (*Potentilla ovina* Macoun), and in several ways it appears transitional to that species. • The specific epithet *multisecta*, from the Latin *multus* (many) and *sectus* (to cut), refers to the finely divided (feather-like) leaves of this species.

ROSE FAMILY **Dicots**

BUSHY CINQUEFOIL

Potentilla paradoxa Nutt.
ROSE FAMILY (Rosaceae)

Plants: Annual, biennial or short-lived perennial **herbs**; stems **leafy**, spreading to ascending, 20–50 {90} cm long, unbranched or branched, hairless near the base but often stiff-hairy on the upper parts; from simple or branched root crowns on taproots.

Leaves: Alternate, pinnately divided into 2–4 {5} **pairs of leaflets** (upper leaves sometimes with 3 leaflets), **almost hairless**; leaflets 1–3 cm long, oblong to egg-shaped and widest above the middle, tapered to a wedge-shaped base, edged with **blunt teeth**; stalks short, with a pair of well-developed lobes (stipules) at the base.

Flowers: Yellow, saucer-shaped, 5–7 {9} **mm across**, with 5 bractlets, 5 sepals and 5 petals **all about 3–4 mm long**; petals egg-shaped, widest above the middle; sepals stiff-hairy, triangular to egg-shaped, widest above the middle; stamens 20 (sometimes 10–15), often with sterile (abortive) anthers; ovaries numerous, **tipped with an equally long style** that is somewhat **thickened** at the base; **in diffusely branched, leafy, open clusters** (cymes), often bearing flowers over the upper half of the plant; {June–July}.

Fruits: Smooth, dry seed-like fruits (achenes), about 1.2 mm long, **with a corky, wedge-shaped thickening nearly (or equally) as large as the achene on the inner side**, sometimes with obscure, wavy, lengthwise ridges, in compact clusters surrounded by persistent calyxes.

Habitat: Moist flats and sandy lake shores and riverbanks; elsewhere, in damp meadows and prairies.

Notes: Drummond's cinquefoil (*Potentilla drummondii* Lehm. ssp. *drummondii*, sometimes classified as var. *drummondii*) is another rare Alberta cinquefoil, with open clusters of yellow flowers and with sparsely hairy, pinnately divided leaves whose leaflets are lobed only halfway to the middle (or less). It is distinguished from bushy cinquefoil by its few (0–2 rather than many) stem leaves, which are almost palmately divided (rather than clearly pinnately divided) and which are edged with sharp teeth (rather than rounded lobes). Drummond's cinquefoil tends to have showier flowers (10–20 mm across vs. 5–7 mm), with petals that are much longer than the sepals. It grows in moist meadows in subalpine and alpine areas, and flowers

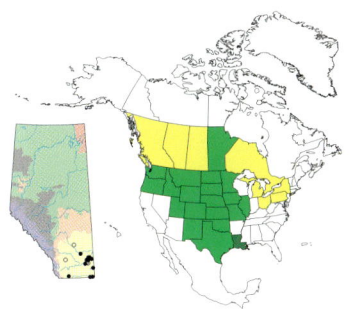

P. DRUMMONDII ssp. *DRUMMONDII*

Rare *Vascular* Plants of Alberta

Dicots

ROSE FAMILY

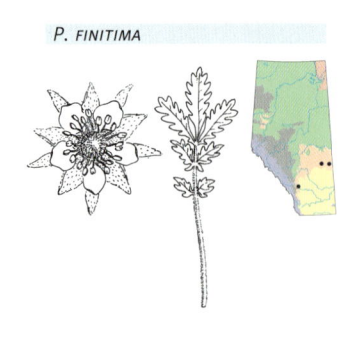

P. FINITIMA

in July {June–August}. • Another rare species in Alberta, sandhills cinquefoil (*Potentilla finitima* Kohli & Packer, also called *Potentilla lasiodonta* Rydb. and *P. pensylvanica* L. var. *arida* B. Boivin), is distinguished from the other 2 species by leaves that are densely woolly-hairy below and by bractlets that are much longer than, and at least as wide as, the sepals.

LOW CINQUEFOIL

Potentilla plattensis Nutt.
ROSE FAMILY (ROSACEAE)

Plants: Low, spreading perennial herbs, with **fat, tapered, flat-lying, white hairs**; stems single to several, slender, branched, usually ascending or spreading on the ground, 10–20 {30} cm long; from thick, rather woody root crowns on taproots.

Leaves: Mostly basal, numerous, **pinnately divided into 7–15 {17} light green, 5–8 mm long leaflets** that are **deeply cut** (almost to the midrib) into oblong to almost linear lobes, stiff-hairy to nearly hairless and **similarly green on both surfaces**.

Flowers: Yellow, **saucer-shaped**, about 1 cm across with 5 broad, slightly heart-shaped petals that are longer than the sepals, 5 stiff-hairy, 4–5 mm long, lance-shaped sepals, and 5 similar bractlets; **ovaries numerous, tipped with slender styles**; 2–10, on slender stalks in open clusters (cymes); June–July {August}.

Fruits: Dry seed-like fruits (achenes), numerous, in compact clusters surrounded by persistent calyxes.

Habitat: Coulees and dry flats in prairie grassland; elsewhere, on moist to dry meadows and hillsides.

Notes: Low cinquefoil is distinguished from other *Potentilla* species in Alberta by its low-spreading stems, thick, white, flat-lying hairs and leaves with 7–15 pinnate leaflets that are green on both surfaces.

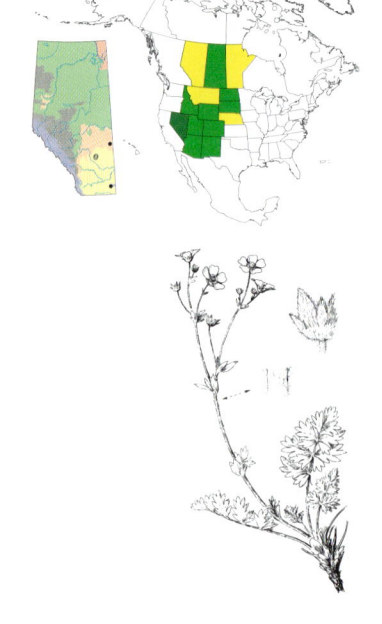

HAIRY CINQUEFOIL

Potentilla villosa Pallas *ex* Pursh
ROSE FAMILY (ROSACEAE)

Plants: Tufted perennial herbs with **long, soft, greyish hairs,** but may be **silky-hairy throughout**; stems 8–12 {2–20} **cm tall, often branched** and curved or bent above the middle, with woolly hairs overlain by spreading hairs; from a stout, branched root crown on thick, ascending underground stems (rhizomes).

Leaves: Basal and alternate, **divided into 3 rather leathery, brownish to olive-green leaflets,** sparsely **silky-hairy on the upper surface, white-woolly with overlying silky hairs and usually prominently veined beneath**; leaflets mostly 0.5–2 cm long, egg-shaped, widest above the middle, wedge-shaped at the base, irregularly cut ⅓–⅔ of the way to the midrib into blunt lobes; 1–3 stem leaves, short-stalked; stalks mostly as long as or longer than the blades, with conspicuous toothless lobes (stipules) about 1 cm long at the base.

Flowers: Yellow, saucer-shaped, **10–15 mm across**, with **5 showy, heart-shaped, 5–8 mm long petals** above a **densely silky-hairy, 7–11 mm long calyx** with 5 triangular, 3–5 {6} mm long lobes (sepals) alternating with 5 elliptic bractlets of similar size or slightly narrower; stamens usually 20; pistils numerous with slightly warty, thick-based styles attached just below the tip and about as long as the fruit; 2 {1–5}, in rather open, spreading clusters (cymes) with conspicuous, leafy bracts; June–August {September}.

Fruits: Dry seed-like fruits (achenes), smooth or with a netted pattern, {1} 1.7–2 mm long, numerous, in compact, head-like clusters.

Habitat: Alpine slopes and ridges; elsewhere, on rocky slopes and ridges, coastal bluffs and arctic tundra, rare inland.

Notes: This species has also been called *Potentilla nivea* L. var. *villosa* (Pallas *ex* Pursh) Regel & Tiling and *Potentilla villosula* Jurtz. • Two common, small alpine cinquefoils could be mistaken for hairy cinquefoil. Snow cinquefoil (*Potentilla nivea* L.) is very similar, but its plants have only cobwebby or woolly hairs, never long, silky hairs like those found on hairy cinquefoil. One-flowered cinquefoil (*Potentilla uniflora* Ledeb.) is usually a smaller (4–8 cm tall) plant with unbranched stems that bear larger (15–20 mm wide), single flowers. The veins on its lower leaf surfaces are hidden beneath dense, woolly hairs. • Hooker's cinquefoil

Dicots

PEA FAMILY

P. HOOKERIANA

(*Potentilla hookeriana* Lehm.) is another small, rare Alberta cinquefoil, with yellow flowers and with leaves divided into 3 leaflets (trifoliate) that are densely woolly on their lower surface. It is distinguished from hairy cinquefoil by its larger flower clusters (usually more than 3 flowers) and by the absence of woolly hairs on its leaf stalks. Hooker's cinquefoil grows on dry, rocky alpine slopes, and flowers in {June} July and August. • Hairy cinquefoil is most common and widespread along the coast. Relative humidity could be a limiting factor in interior sites. • The specific epithet *villosa* is Latin for 'shaggy' or 'hairy,' in reference to the long, soft hairs of these plants.

BODIN'S MILK VETCH

Astragalus bodinii Sheldon
PEA FAMILY (FABACEAE [LEGUMINOSAE])

Plants: Mat-forming perennial herbs; stems numerous, very slender, spreading on the ground, 10–30 {60} cm long; from central taproots.

Leaves: Alternate, pinnately divided into 7–15 {19} leaflets; leaflets lance-shaped to oblong, 4–12 {2–19} mm long, hairless on the upper surface, with scattered, stiff, flat-lying hairs beneath.

Flowers: Pink to bluish purple, pea-like, 7–10 {11} mm long; calyxes tipped with 5 awl-shaped lobes 1.2–2.4 mm long, black-hairy and white-hairy; few, in loose, elongating clusters (racemes) on slender stalks up to 25 cm long; July.

Fruits: Erect, black-hairy pods, **short-stalked** (stalks less than 2 mm long, within the calyx), 5–7 {10} mm long, obliquely egg-shaped to oblong-ellipsoid, abruptly sharp-pointed, single-cavitied.

Habitat: Moist meadows and gravel banks; elsewhere, in thickets and invading disturbed ground along roads.

Notes: This species has also been called *Astragalus yukonis* M.E. Jones. • A more common species, alpine milk vetch (*Astragalus alpinus* L.), is very similar to Bodin's milk vetch, but it has larger (3–4 mm vs. 2.5 mm long) calyxes and its black-hairy pods hang on slender, 3–4 mm long stalks (within the calyxes).

LOW MILK VETCH

Astragalus lotiflorus Hook.
PEA FAMILY (FABACEAE [LEGUMINOSAE])

Plants: **Low, tufted** perennial herbs usually less than 10 cm tall, more or less **silvery with straight, parallel, flat-lying hairs** attached by the middle; stems branched, erect or spreading, {1} 2–10 cm long; from deep taproots.

Leaves: Basal, **5–10 cm long**; pinnately divided into {3} **5–15 leaflets**; leaflets oblong to elliptic, 5–15 {20} mm long, **½–⅕ as wide as they are long, pointed**; with a pair of lance-shaped, 2–5 mm long appendages (stipules) at the base of each stalk.

Flowers: **Yellowish white**, sometimes purplish-tinged, **7–10 mm long**, usually long and narrow, with the upper petal (banner) equal to or longer than the lower and side petals (keel and wings); calyx white-hairy, tipped with long, slender teeth; 10 stamens, 9 united, 1 free (diadelphous); 2–12, on short (1 mm) stalks in compact clusters that elongate in fruit (racemes), **often nestled among the leaves or barely taller than them**, sometimes on elongated stems up to 8 cm tall; May {June}.

Fruits: Densely hairy, rather leathery pods, inflated but slightly flattened from top to bottom, somewhat crescent-shaped, 1–2 {2.5} cm long, 6–8 mm wide, tipped with a long, firm, pointed tip (beak), stalkless (within the calyx), single-celled, covered with an interwoven mixture of woolly and stiff flattened hairs.

Habitat: Dry slopes and prairie, sandy-gravelly dune slacks in active blowout areas; elsewhere, on gravelly or sandy hillsides, pebbly or sandy lakeshores, clay, loam, silt or gravelly soil.

Notes: Pursh's milk vetch (*Astragalus purshii* Dougl. *ex* Hook. var. *purshii*) is a similar species that is also rare in Alberta. Its leaflets are only about twice as long as they are wide, with rounded tips, its hairs are attached by their bases, and its flowers are 20–30 mm long. Pursh's milk vetch grows in dry prairies in Alberta; elsewhere, it is also found in mixed grasslands and on sand plains. It is reported to flower from

A. PURSHII

April to June. • Although there is a considerable amount of potential habitat in the Little Rolling Hills, low milk vetch has not been found there to date. Cultivation of native grasslands on sandier soils is probably the greatest threat to this species. Encroachment of other plants (e.g., Russian thistle [*Salsola kali* L.]) on active dune blowouts may also be a problem. It is not clear whether dunes in native grassland were kept active by periodic drought, fire, grazing or a combination of those factors. • Low milk vetch sometimes produces small, closed, self-fertilized (cleistogamous) flowers near the surface of the ground.

PRICKLY MILK VETCH

Astragalus kentrophyta A. Gray var. *kentrophyta*
PEA FAMILY (FABACEAE [LEGUMINOSAE])

Plants: Prostrate perennial herbs, 5–10 cm tall, with **flat-lying, parallel hairs attached by the middle**; stems spreading on the ground, branched, 10–40 cm long; forming **mats** up to 50 cm across, from strong taproots.

Leaves: Alternate, silky with flat-lying hairs, pinnately divided into 5–7 **stiff, linear to lance-shaped leaflets 5–12 mm long, each tipped with a sharp spine**.

Flowers: Yellowish white, often **tinged with purple**, 4–6 {7} **mm long, pea-like**; upper petal (banner) arched back; flowers **solitary or 2–4**, in small clusters (racemes) **from leaf axils**, on stalks shorter than the leaves; June–July {August–September}.

Fruits: Greyish-hairy, elliptic pods, {3} 4–7 mm long, stalkless, not splitting open.

Habitat: Dry prairies, in hard-packed or gravelly sand in blowout areas and on exposed, eroding soils, often along trails; elsewhere, in low foothills and on sand dunes.

Notes: The needle-like leaves of prickly milk vetch help to minimize water loss in the hot, dry environments where this plant thrives; its long, strong taproots anchor it in shifting sands and reach deep for water in sandy soils. These plants are known to grow in association with mycorrhizal fungi. • Dam construction could flood some populations of this species along the North Milk River.

PEA FAMILY

Dicots

HARE-FOOTED LOCOWEED

Oxytropis lagopus Nutt. var. *conjugens* Barneby
PEA FAMILY (FABACEAE [LEGUMINOSAE])

Plants: Silvery, silky-hairy perennial herbs, densely tufted, often cushion-like, **stemless** except for leafless, hairy flowering stalks; from taproots with branched crowns.

Leaves: Basal, ascending, covered in long silvery hairs, 3–11 cm long, **pinnately divided into 5–17 leaflets**; leaflets opposite or offset, egg-shaped to elliptic, 3–15 mm long, toothless, with edges raised or rolled inward; stalks with a pair of silky-hairy, membranous lobes (stipules) at the base (sometimes almost hairless with age).

Flowers: Pink-purple or blue-purple, pea-like, **13–16 {20} mm long**, with the lower 2 petals joined in a sharp-pointed keel 11–14 mm long; **calyxes** 8–11 mm long, with dense, spreading, silky, white hairs concealing the surface, **swollen to inflated when the flowers open**; 5–18 {20}, in the axils of **dark, shaggy-hairy bracts**, forming head-like to oblong, elongating clusters (racemes) 2–4 cm long on erect to arching or low-spreading stalks 2–13 cm long; late May–early June.

Fruits: Erect or spreading, **papery** or almost membranous **pods**, oblong to egg-shaped, covered **with spreading, silky hairs, broadly heart-shaped in cross-section**, 6–15 mm long, 4–7 mm wide, tipped with a short beak, projecting from the calyx, short-stalked (within the calyx).

Habitat: Sandy or grassy knolls and hillsides; elsewhere, on sagebrush plains.

Notes: Hare-footed locoweed might be confused with some of the smaller milk vetches (*Astragalus* spp.). These 2 genera are most easily separated by the keels of their flowers, which are blunt-tipped in milk vetches and sharply pointed in locoweeds (*Oxytropis* spp.). Milk vetches usually have leaves with stems, whereas most locoweeds are stemless, with all of their leaves arising from a root crown. • Cultivation has eliminated many sites where hare-footed locoweed may have grown, but the steepness of most remaining areas makes further cultivation difficult. Because of the abundance of gravel in suitable habitats, gravel operations now pose the largest single threat. Effects of grazing require further study, although sites are mostly inaccessible to cattle due to the steepness of the hills on which they are found.

Rare *Vascular* Plants of Alberta

PURPLE MOUNTAIN LOCOWEED

Oxytropis jordalii Porsild var. *jordalii*
PEA FAMILY (FABACEAE [LEGUMINOSAE])

Plants: Low perennial herbs; **flowering stems leafless**, green to purplish green, 10–15 cm tall; from branched root crowns.

Leaves: Mainly basal, {4} 6–14 cm long, **pinnately divided into 9–25 leaflets** and with 2 lobes (stipules) at the base of the stalk; leaflets narrowly lance-shaped to oblong-elliptic, curved inward; stipules edged with club-like 'hairs,' otherwise hairless or with soft, spreading hairs.

Flowers: Pink or purple, pea-shaped, 11–14 mm long; keel 10–12 mm long; calyx bell-shaped, 5–7 mm long, with a mixture of black and pale hairs; **6–12**, in compact, head-like clusters that elongate when mature (racemes); June–August.

Fruits: Papery pods, oblong to egg-shaped, 8–15 mm long, thinly covered with mixed black and white hairs, **stalkless or short-stalked** (within the calyx).

Habitat: Alpine and subalpine meadows and dry ridges.

Notes: This species has been reported incorrectly for Alberta. Alberta plants identified as this species are referrable to *Oxytropis campestris* (L.) DC. var. *davisii* Welsh. • This species was named in honour of Louis Henrik Jordal (1919–52), a Norwegian collector who worked in Alaska and was a graduate student in Michigan and an author. • Many species of *Oxytropis* are **poisonous to livestock**, causing a form of insanity called 'locoism' and eventually death. To be lethal, large amounts must be eaten over a long period of time, but these plants are habit-forming, so this often happens. Late yellow locoweed (*Oxytropis monticola* A. Gray), a close relative, probably causes more livestock deaths than all other locoweeds combined. • The generic name *Oxytropis* was taken from the Greek *oxus* (sharp) and *tropis* (keel), in reference to the sharply pointed keels of the flowers.

Rare *Vascular* Plants of Alberta

PEA FAMILY **Dicots**

SILVERLEAF PSORALEA

Psoralea argophylla Pursh
PEA FAMILY (FABACEAE [LEGUMINOSAE])

Plants: **Silvery-hairy perennial herbs** dotted with tiny glands; stems erect, 30–60 cm tall, freely wide-branched; from creeping underground stems (rhizomes).

Leaves: Alternate, densely silvery-hairy, mostly **divided into 3 leaflets** but lower leaves sometimes with 5; **leaflets** egg-shaped and widest above the middle to elliptic, **2–4 {5} cm long, 5–15 {20} mm wide**, toothless; stalks slightly longer or shorter than the blades, with a pair of slender lobes (stipules) 3–8 mm long at the base.

Flowers: Purple or deep blue, pea-like, 7–10 mm long, with the broad upper petal (banner) and 2 slender side petals (wings) all extending forward past the 2 incurved lower petals (keels); **calyxes** silvery with dense, silky hairs, the 2 mm long tube tipped with 4 slender, 2–3 mm long teeth above a lower 6 mm long tooth; several, in 1–3 well-spaced whorls of 2–4 flowers each, forming spike-like clusters 2–5 cm long, at branch tips; June–July.

Fruits: Silky-hairy, egg-shaped **pods**, about 8 mm long, tipped with a short, straight beak, thick, single-seeded, not splitting open, enclosed in a persistent calyx up to 1 cm long.

Habitat: Prairie grassland; elsewhere, on eastern slopes of the Rocky Mountains.

Notes: Silverleaf psoralea lacks the large tuberous roots of its well-known, edible relative, Indian breadroot (*Psoralea esculenta* Pursh). However, it was used by some tribes to make a medicinal wash for cleaning wounds. This plant is very common across the prairies.

Rare *Vascular* Plants of Alberta

PEA FAMILY

ARCTIC LUPINE

Lupinus arcticus S. Wats. ssp. *subalpinus* (Piper & B.L. Robins.) D. Dunn
PEA FAMILY (FABACEAE [LEGUMINOSAE])

Plants: Loosely tufted perennial herbs; stems ascending to erect, 30–50 {13–60} cm tall, few to several; from branched root crowns.

Leaves: Mainly basal, **palmately divided** into 5–9 leaflets; leaflets narrowly elliptic to lance-shaped, often widest above the middle, pointed, 1.3–9 cm long, 4–19 mm wide, hairless on the upper surface, with sparse, stiff, flat-lying hairs beneath; stalks long, mostly 6–21 cm long.

Flowers: Bluish purple (rarely white), **pea-like**, 15–20 mm long; upper petal (banner) bent back from slightly below the middle; numerous, in whorls, in loose (sometimes dense), elongated clusters (racemes) 4–14 cm long on 4–10 cm long stalks; August.

Fruits: Yellowish, silky-hairy pods, 2–4 cm long, containing 5–8 seeds.

Habitat: Open, grassy sites in tundra, heath and woodland in northwest North America.

Notes: *Lupinus* is a large and taxonomically difficult genus, with about 100 species in North America. Many hybridize, producing a wide range of intergrading forms and varieties. Lupines can be very 'plastic'—their size and appearance may vary greatly with different environmental conditions—and this can also make identification difficult. • The generic name *Lupinus* was taken from the Latin *lupus* (wolf) because these plants usually grew on poor land, where they were believed to rob soil of nourishment, like a wolf in the fold. In reality, lupines benefit the soil. They have nitrogen-fixing bacteria in nodules on their roots which allow them to survive on poor sites, and when the plants die and decompose, they add their accumulated nitrogen to the soil.

PEA FAMILY   **Dicots**

Wyeth's lupine

Lupinus wyethii S. Wats.
PEA FAMILY (FABACEAE [LEGUMINOSAE])

Plants: Tufted **perennial herbs**, covered with short, stiff, flat-lying, yellow hairs; flowering **stems unbranched**, **40–50 {20–60} cm tall, solid** (or only slightly hollow).

Leaves: Basal and alternate, numerous, **palmately divided into 9–11 {8–13} leaflets; leaflets narrow**, 2–4 cm long, 2–6 mm wide, widest above the middle, abruptly narrowed to a sharp point, **yellow-hairy on both surfaces**; stalks 3–6 times as long as the blades on lower leaves but much shorter on upper leaves.

Flowers: Deep violet or purple, pea-like, 11–16 mm long; **upper petal** (banner) often reddish, yellowish or white-centred, **hairless, strongly bent backward**; 2 side petals (wings) hairless, curved forward and touching at the tips; 2 lower petals joined in a fringed, sickle-shaped keel with a beak-like tip; **calyx** silky-hairy, deeply **2-lipped**, the upper lip 2-toothed, **cut less than ⅓ of its length**, the lower unlobed; **10 stamens, fused** together **to form a tube** (monadelphous), with larger and smaller anthers alternating; numerous, **scattered or in whorls, forming 15–25 cm long**, elongating **clusters** (racemes) on erect stalks, well above the leaves; {July–August}.

Fruits: Flattened, hairy **pods**, 2.5–4 cm long, splitting in half lengthwise to release 4–8 greyish, mottled seeds.

Habitat: Roadside in subalpine forest and larch–whitebark pine forest; elsewhere, from sagebrush plains and valleys to montane forests to open ridges in the mountains.

Notes: Wyeth's lupine could be confused with least lupine (*Lupinus minimus* Dougl. *ex* Hook., p. 140), but least lupine is generally a smaller (10–40 cm tall), more densely hairy plant with smaller (10–12 mm long) flowers and more deeply cut upper lips on its calyxes. • Another rare species reported for Alberta, large-leaved lupine (*Lupinus polyphyllus* Lindl.), has taller (about 50–100 cm), hollow stems and larger (4–10 cm long) leaflets that are hairless on their upper surface. It grows in moist woods, on shorelines and in meadows; it is reported to flower from mid June to early September.

L. POLYPHYLLUS

Dicots   PEA FAMILY

LEAST LUPINE

Lupinus minimus Dougl. *ex* Hook.
PEA FAMILY (FABACEAE [LEGUMINOSAE])

Plants: Tufted, **densely silky-hairy perennial herbs**; flowering stems 10–40 cm tall, clearly **taller than the leaves**.

Leaves: Mainly **basal** and **less than 10 cm**, plus **1–3 alternate on the stem, palmately divided into 6–8 {5–10} leaflets**; leaflets 1.5–2.5 {1–4} cm long, lance-shaped, widest above the middle, abruptly narrowed to a sharp point, often folded in half lengthwise, densely silky-hairy on both surfaces; stalks 6–12 cm long.

Flowers: Blue, occasionally silvery-brown, **pea-like**, about **10 mm long**, with a **broad, rounded upper petal** (banner), **hairless**, strongly **bent backward** near the middle; 2 side petals (wings) curved forward and touching at the tips; 2 lower petals joined in a fringed, sickle-shaped keel with a sharp, beak-like tip; calyx deeply 2-lipped, the upper lip 2-toothed, the lower 3-lobed or unlobed; **10 stamens, fused** together **to form a tube** (monadelphous), with larger and smaller anthers alternating; several to many, **in whorls forming dense, 2–15 cm long**, elongating **clusters** (racemes) on erect, 3–10 cm long stalks; June {July–August}.

Fruits: Flattened, hairy **pods**, 2–3 cm long, splitting in half lengthwise to release 4–6 seeds.

Habitat: River flats and open gravelly areas to alpine elevations.

Notes: This species could be confused with alpine lupine (*Lupinus lepidus* Dougl. *ex* Lindl.). These 2 species are very similar, but alpine lupine is often matted and its leaves are usually over 10 cm in height, with 4 or more leaves commonly on the stem. The upper petal of its flowers is also elliptic to oblong-elliptic rather than round. Alpine lupine grows on river flats and gravelly habitats from grasslands at low elevations to alpine areas in the mountains. It is reported to flower from March to June.

GERANIUM FAMILY   **Dicots**

Carolina wild geranium

Geranium carolinianum L.
GERANIUM FAMILY (GERANIACEAE)

Plants: Bushy **annual** herbs, 10–50 {70} cm tall; stems usually erect, **freely branched**, with long, soft hairs that are spreading or pointing downward; without rootstocks or branched bases.

Leaves: Alternate, broadly heart-shaped to rounded, 2–7 cm wide, **cut ⅓–½ to the base into 5–7 radiating** (palmate), sharply toothed and lobed **segments** with wedge-shaped bases.

Flowers: Rose-purple, **saucer-shaped**, with 5 broad, **2–8 mm long petals**; petals rounded or with a notched tip; **sepals broadly egg-shaped**, about as long as the petals, tipped with a 1–1.5 mm long bristle, ribbed with 5 silky-hairy nerves; 10 (rarely 5) fertile stamens, in 2 unequal sets; styles joined with the elongated receptacle to form a slender **column** tipped with a short **beak 1–1.5 mm long; stalks shorter than the calyxes**, covered with tiny, downward-pointing hairs; few, in crowded, flat-topped clusters (corymbs); {April–July}.

Fruits: Hairy, awl-shaped capsules (columns) 10–15 mm long, **tipped with a 1–1.5 mm long beak** (to 2.5 mm including the free stigmas), splitting open from the base upward into 5 segments that recoil when dry and fling out the seeds (like a slingshot), borne on **stalks that are shorter than or slightly longer than the calyxes**, containing rounded seeds with a faint netted pattern of about 50 rows of small square or rounded hollows.

Habitat: Clearings and disturbed sites; elsewhere, on granite outcrops and in dry, rocky woods, often on sandy soil.

Notes: Western plants of this species have also been called *Geranium sphaerospermum* Fern. or *Geranium carolinianum* L. var. *sphaerospermum* (Fern.) Breitung. • Carolina wild geranium might be confused with the common weedy species Bicknell's geranium (*Geranium bicknellii* Britt.), but the fruits of Bicknell's geranium are borne on longer stalks (much longer than the calyxes) and the fruits are about 2 cm long, with 3–5 mm long beaks. • The sticky, glandular hairs and stiff, downward-pointing hairs on the upper stems and calyxes help to deter insects from climbing up to steal nectar from the flowers.

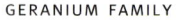

GERANIUM FAMILY

WOOLLY GERANIUM
Geranium erianthum DC.
GERANIUM FAMILY (Geraniaceae)

Plants: Perennial herbs; stems erect or ascending, 20–80 cm tall, with downward-pointing hairs; from thick, scaly, branched underground stems (rhizomes).

Leaves: Few, alternate, 1 or 2 near the base and 1 or 2 on the stem, **wider than they are long**, usually **3–10 cm long** from the hollow between 2 lowest lobes (sinus) to the tip, **cut almost to the base into {3} 5–7 radiating** (palmate), sharply toothed and lobed **segments**, sometimes with the edges of the 2 lowest segments overlapping, **uniformly hairy on the lower surface**, long-stalked, but stalkless on the upper stem.

Flowers: Rose to blue or violet (rarely white), very showy, saucer-shaped, with 5 broad petals; **petals 16–20 mm long, fringed near the base and soft-hairy on the lower half of the upper surface**; sepals 8–12 mm long, bristle-tipped, glandular-hairy; 10 stamens (rarely 5), in 2 unequal sets; styles joined with the elongated receptacle to form a slender, beak-like column; 2–several, on slender stalks from leaf axils, **usually not as tall as the leaves**; {June–August}.

Fruits: Cylindrical capsules tapered to **long, slender beaks**, 25–32 mm long (including the 6–9 mm long beak), splitting open from the base upward into 5 segments that recoil when dry and fling out the seeds (like a slingshot).

Habitat: Moist open woods and meadows in prairie, montane and subalpine zones; elsewhere, in willow thickets and rich meadows, from forests to above timberline.

Notes: This species has also been called *Geranium pratense* L. var. *erianthum* (DC.) Boivin. • Woolly geranium might be confused with the common species sticky purple geranium (*Geranium viscosissimum* Fisch. & Mey.), but the undersides of the leaves of sticky purple geranium are hairy only along the veins, and the flowers are much more conspicuous, clearly overtopping the leaves. • The generic name *Geranium*, from the Greek *geranos* (crane), refers to the long, slender beak of the fruits, which was thought to resemble the bill of a crane. The specific epithet *erianthum*, from the Greek *eryon* (wool) and *anthos* (flower), refers to the woolly anthers of the flowers.

FRINGED MILKWORT

Polygala paucifolia Willd.
MILKWORT FAMILY (POLYGALACEAE)

Plants: Perennial herbs, 1–14 cm tall; stems hairless, branched toward the tip; from slender runners (stolons).

Leaves: Alternate, **crowded near stem tip**, short-stalked or stalkless; blades **egg-shaped to oval**, pointed, 1–5 cm long, 1–2 cm wide; lower stem leaves small to scale-like.

Flowers: Pink to rose-purple, 10–20 mm long, pea-like, with **3 petals (middle petal with a fringed crest** about 6 mm long) and 5 sepals (3 small sepals 3–6 mm long and **2 large, petal-like sepals** 10–17 mm long); 1–4, at stem tips; May–early July.

Fruits: Egg-shaped, 2-seeded capsules, 6–7 mm long, splitting open lengthwise; seeds about 3.5 mm long, egg-shaped, tipped with a small, whitish 2-lobed appendage.

Habitat: Moist coniferous or mixed woods.

Notes: Fringed milkwort produces 2 types of flowers. If the showy pink, springtime flowers (chasmogamous) are not fertilized, the plants produce small, inconspicuous flowers (cleistogamous) on short side stems near the ground surface. These self-fertilize without opening and produce seed for the following year. The leaves of fringed milkwort continue to grow to the end of the growing season. • The generic name *Polygala* was derived from the Greek *poly* (much or many) and *gala* (milk), in reference to the milky secretions of some plants of this genus, and the belief that a diet high in milkwort would increase the milk production of both domestic cows and nursing mothers. The small number of leaves of fringed milkwort is reflected in the specific epithet *paucifolia*, which means 'few-leaved.' Some people call this the 'Snoopy flower,' because it resembles a tiny, dancing Snoopy (of *Peanuts* cartoon fame).

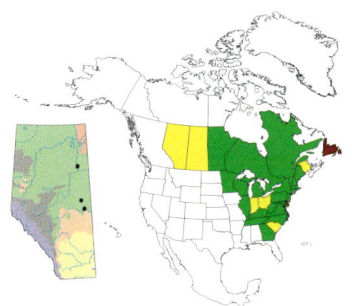

MOUNTAIN HOLLYHOCK

Iliamna rivularis (Dougl. *ex* Hook.) Greene
MALLOW FAMILY (Malvaceae)

Plants: Stout perennial herbs with **sparse, star-shaped hairs**; stems leafy, 0.5–2 m tall; from underground stems (rhizomes).

Leaves: Alternate, **maple-leaf-shaped**, 5–15 cm wide with {3} 4–7 triangular lobes, coarsely round-toothed.

Flowers: Rose-purple to pink or nearly white, cup-shaped, about 3–4 cm across, with 5 broad petals each about 2 cm long; sepals 3–5 mm long, blunt-tipped, above 3 slender bracts; **on short** (less than 1 cm), **stout stalks**; stamen with filaments fused into a tube tipped with a cluster of separate anthers; styles slender, tipped with head-like stigmas; in long, loose clusters (racemes); July {August–September}.

Fruits: Hairy, oblong, wedge-shaped segments (carpels) about 8 mm long, arranged like wedges on a wheel of cheese, splitting to release 2–4 seeds.

Habitat: Moist foothill and montane slopes and meadows, often along roads and streams.

Notes: This species has also been called *Sphaeralcea rivularis* (Dougl. *ex* Hook.) Torr. *ex* Gray. • These showy flowers look like miniature garden hollyhocks. Their fruits break into sections, like the segments of an orange, and these are easily collected and planted in a garden. But be forewarned: some people find that the stiff hairs **irritate their skin**.
• In some areas, mountain hollyhock is the first species to appear after a forest fire.

ST. JOHN'S-WORT FAMILY　　　　**Dicots**

WESTERN ST. JOHN'S-WORT

Hypericum scouleri Hook.
ST. JOHN'S-WORT FAMILY (CLUSIACEAE [HYPERICACEAE])

Plants: Perennial herbs, 20–40 {10–80} cm tall; stems many, erect, hairless, with reddish buds; from spreading, horizontal underground stems (rhizomes).

Leaves: Opposite, oblong to egg-shaped, blunt-tipped, 1–3 cm long, black-dotted along edges, stalkless, slightly clasping.

Flowers: Bright yellow, about 2–2.5 cm across, with 5 broad petals edged with black dots or teeth and 5 black-dotted sepals ½ as long as the petals; 75–100 stamens gathered in 3 bundles; few, in leafy, open clusters (cymes); late July–early August.

Fruits: Membranous capsules, 6–9 mm long, slightly 3-lobed, containing yellowish, net-veined seeds up to 1 mm long.

Habitat: Moist slopes, ledges and shores in alpine and subalpine areas.

Notes: This species has also been called *Hypericum formosum* H.B.K. var. *scouleri* (Hook.) C.L. Hitchc. • Large St. John's-wort (*Hypericum majus* (Gray) Britt.), another rare Alberta plant, is an annual species with small, inconspicuous flowers in which the petals and sepals are both about 4–5 mm long and the petals have no black dots. Its leaves are blunt-tipped, and they have 5–7 strong, parallel nerves. It grows on wet sites in the plains, foothills and boreal forest, and flowers from late June to September. • Two compounds found in St. John's-wort plants (hypericin and pseudohypericin) have been found to be potent antiviral agents with no serious side-effects, so St. John's-wort is being studied for use in the treatment of AIDS. Hypericin, extracted from the dark dots on the leaves and flowers, is used as an antidepressant.

H. MAJUS

Rare *Vascular* Plants of Alberta　　145

THREE-STAMENED WATERWORT, MUD CRUD
Elatine triandra Schk.
WATERWORT FAMILY (ELATINACEAE)

Plants: Small, tufted, **semi-aquatic** annual herbs; stems slender, spreading on the ground, rooting at the joints, sending up erect 2–15 cm tall branches; **often forming mats**.

Leaves: Opposite, essentially toothless, hairless, **linear to spatula-shaped, 3–8 mm long**, squared or notched at the tip, tapered to short stalks at the base.

Flowers: Very small and inconspicuous, about 2 mm long, with {2} 3 sepals and {2} 3 petals attached at base of a single ovary; 3 stamens; **solitary, in leaf axils**; {early summer–fall}.

Fruits: Membranous, **2- or 3-celled capsules, splitting lengthwise along the partitions** to release many straight or slightly curved, cylindrical seeds, honeycombed with 8–10 lengthwise rows of 9–25 {27} angular pits.

Habitat: Shallow water and mud flats, mostly by ponds and slow-moving streams.

Notes: Some taxonomists divide this species into 3 varieties: var. *americana* (Pursh) Fassett, var. *brachysperma* (A. Gray) Fassett (sometimes called *Elatine brachysperma* A. Gray) and var. *triandra*. The last 2 varieties have been found in Alberta. • Ducks enjoy eating these tender semi-aquatic plants. • The origin of the generic name *Elatine* is unclear. Some say it was derived from the Greek *elatino* (fir-like), and others report that it is the ancient name of some obscure herb. The specific epithet *triandra* comes from the Greek *tri* (three) and *andros* (male), referring to the 3 stamens in the tiny flowers.

Macloskey's violet, Northern white violet

Viola pallens (Banks *ex* DC.) Brainerd
VIOLET FAMILY (VIOLACEAE)

Plants: Small perennial herbs, usually 3–6 cm tall; **stems absent**, leaves and flower stalks all growing from the base of the plant; with slender, **creeping underground stems** (rhizomes) and white, **thread-like runners** (stolons).

Leaves: Basal, **broadly heart-shaped**, longer than they are wide, 1–3 cm long, rounded or blunt-pointed, **hairless**, edged with indistinct to prominent, rounded teeth; stipules thin, lance-shaped, with tiny glandular teeth; stalks 2–4 cm long.

Flowers: 5-petalled, **white** (even on the back) **with purple lines** (pencilling) on the 3 lower petals, bilaterally symmetric, 5–14 mm long; **side petals usually with a few short hairs** at the base, lowest petal projecting back from its base in a short, fairly prominent spur; solitary, nodding, on leafless stalks, usually taller than the leaves; May–July {August}.

Fruits: Rounded to cylindrical, **hairless capsules**, greenish, splitting into 3 parts; seeds numerous, black at maturity, with a small swelling near the base, 1–1.3 mm long, ejected by force.

Habitat: Wet ground in moist woods; elsewhere, most common in coniferous stands and bogs; often on sphagnum hummocks, but also in swamps and wet thickets, hollows, crevices and meadows.

Notes: This taxon is often treated as a subspecies of *Viola macloskeyi* Lloyd (*Flora of Alberta* [Moss 1983]). True *Viola macloskeyi* is known only from Oregon and California. • Kidney-leaved violet (*Viola renifolia* A. Gray) could be confused with Macloskey's violet, but its plants do not have stolons, its side petals always lack hairs, and its leaves are often hairy and larger (2–6 cm wide). • Macloskey's violet is also very similar to marsh violet (*Viola palustris* L.), but marsh violet usually has blue to violet-coloured petals, and its leaves are usually wider (2.5–3.5 cm wide) and heart-shaped to kidney-shaped. • Another rare Alberta species, broad-leaved yellow prairie violet (*Viola praemorsa* Dougl. *ex* Lindl. ssp. *linguifolia* (Nutt.) M.S. Baker & J.C. Klausen *ex* M.E. Peck, also called *Viola nuttallii* Pursh var. *linguifolia* (Nutt.) Jepson), has yellow flowers, and its slender, narrowly

Dicots VIOLET FAMILY

V. PRAEMORSA

egg-shaped to lance-shaped leaves have wedge-shaped bases that taper to their stalks (rather unusual for a violet). It grows in dry, open forests at low elevations and flowers in July. It is distinguished from the common yellow prairie violet (*Viola nuttallii* Pursh) by its long-hairy, slightly broader blades (usually less than 3 times as long as wide), which have rounded tips. The leaf blades of common yellow prairie violet are usually at least 3 times as long as they are wide, and they have pointed tips. • Violets produce 2 kinds of flowers. The showy flowers that we recognize as violets appear in the spring. Their nectar-filled spurs and delicately lined petals attract insects, which carry pollen from one bloom to the next. Normally only a small percentage of these flowers are pollinated, so most plants produce inconspicuous (cleistogamous) flowers in the summer. These flowers have small (or no) petals, and they never open, but most produce seed through self-fertilization within the closed calyx.

CROWFOOT VIOLET

Viola pedatifida G. Don
VIOLET FAMILY (VIOLACEAE)

Plants: Low perennial herbs; **stems absent**, leaves and flower stalks all growing from base of plant; with **short, vertical underground stems** (rhizomes) up to 5 mm thick.

Leaves: Basal, **palmately divided into 3 parts and each part cut into 2–4 linear lobes**, usually **wedge-shaped at the base**, up to 10 cm wide, hairy to nearly hairless, edged with tiny hairs.

Flowers: Showy, **bright violet**, bilaterally symmetric, 1–2 cm long, with 5 showy petals and 5 persistent, lobed sepals; 3 lower petals hairy (bearded) at base; lowest petal projecting back from its base in a short, fairly prominent spur; **solitary, nodding**, on leafless stalks from rhizomes, taller than the leaves; {April} May–June.

Fruits: Yellowish grey, hairless capsules, 10–15 mm long; seeds numerous, 2 mm long, light brown, with a small swelling near the base, ejected by force.

Habitat: Dry gravelly hills and exposed banks in prairie grassland.

Notes: Some violets disperse their seeds explosively. The walls of the drying capsules build up pressure as they fold inward, and eventually the seeds are catapulted into the air. Some violet seeds have swellings or 'oil bodies,' which attract ants. The ants often carry the seeds some distance before eating the oil body.

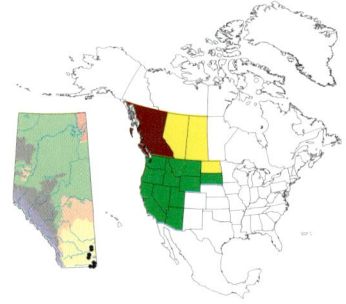

SMOOTH BOISDUVALIA

Boisduvalia glabella (Nutt.) Walp.
EVENING-PRIMROSE FAMILY (ONAGRACEAE)

Plants: **Low-growing annual** herbs, pale greenish, hairless or with sparse, flat-lying hairs; stems spreading and lying on the ground, 10–30 cm long, usually **branched from near the base**; from slender taproots.

Leaves: Alternate, numerous and fairly evenly spaced; blades **lance-shaped, 10–15 {5–18} mm long**, 3–6 mm wide, **edged with small teeth** or toothless, stalkless.

Flowers: Pinkish or purplish red, **inconspicuous**, with **4 sepals, 4 petals and 8 stamens on a receptacle** (hypanthium) **at the tip of an elongating, stalk-like ovary**; sepals erect; petals 2–4 mm long, deeply 2-lobed; styles single; **anthers erect, attached near their base**; single in upper leaf axils, crowded near the stem tips, forming elongating, leafy clusters (racemes); {June–July}.

Fruits: Slender, 4-ribbed capsules, 6–8 mm long, 1–1.5 mm wide, slightly curved and pointed, finely hairy, with **4 cavities** each containing 6–14 seeds; **seeds** about 1 mm long, brownish, hairy but **lacking a tuft of hairs** (not comose); mid to late August.

Habitat: Mud flats, especially on alkaline clays, in the prairies; elsewhere, on dried streambanks and slough bottoms and in moist depressions or mud flats.

Notes: Some taxonomists now classify this species as *Epilobium pygmaeum* (Speg.) P. Hoch & Raven. • The generic name *Boisduvalia* commemorates Jean Alphonse Boisduval (1801–79), a French naturalist and author of a flora of France.

EVENING-PRIMROSE FAMILY   **Dicots**

SHRUBBY EVENING-PRIMROSE
Calylophus serrulatus (Nutt.) Raven
EVENING-PRIMROSE FAMILY (ONAGRACEAE)

Plants: Erect perennial sub-shrubs, sparsely hairy to pale grey with dense, fine, flat-lying hairs; stems slender, 10–50 {60} cm tall, usually **many-branched from the base**, leafy; from branched, **somewhat woody root crowns**.

Leaves: Alternate, with **bunches of small leaves in the axils of larger leaves**; blades linear to oblong or spatula-shaped, {1} 2–5 cm long, 3–10 mm wide, edged with few to many fine sharp teeth, sometimes toothless.

Flowers: Yellow, with 4 sepals, 4 petals and 8 stamens on a funnel-shaped, 4-angled, 6–15 mm long receptacle (hypanthium) **at the tip of an elongating, stalk-like ovary**; **sepals separate**, spreading or **bent backward**, ridged lengthwise on the back; petals 8–25 mm long, edged with fine, rounded teeth; stamens scarcely ½ as long as the petals, with **anthers attached near their middle**; styles shorter than the stamens, tipped with a shallowly 4-lobed, disc-shaped stigma; stalkless in upper leaf axils, forming elongating clusters (racemes); {May} June–July.

Fruits: Slender capsules, 15–25 {30} mm long, greyish white with fine hairs, splitting lengthwise into **4 compartments**; **seeds numerous, lacking a tuft of hairs** (not comose).

Habitat: Sandy prairies and dunes; elsewhere, in moist depressions in grasslands, gravel flats, disturbed sites and dry fields.

Notes: This species has also been called *Oenothera serrulata* Nutt.

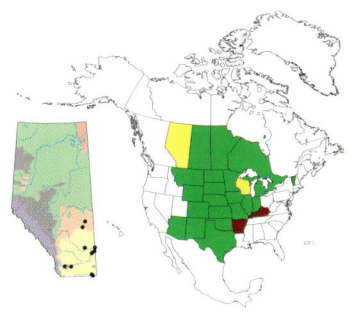

Rare *Vascular* Plants of Alberta   151

UPLAND EVENING-PRIMROSE

Camissonia andina (Nutt.) Raven
EVENING-PRIMROSE FAMILY (ONAGRACEAE)

Plants: Small **annual herbs**, often wider than they are tall, **greyish** with fine, close hairs; **stems slender, 3–10 {15} cm long**, **branched** from near the base, **ascending and leafy near the tips**, spreading and essentially leafless near the base; from weak taproots.

Leaves: Alternate; blades **linear to narrowly lance-shaped** and widest above the middle, {5} 10–30 mm long, 0.5–2 mm wide, toothless; stalks short.

Flowers: Lemon-yellow, **tiny** and inconspicuous, with **4 petals, 4 sepals and 8 stamens on a receptacle** (hypanthium) **less than 2 cm long at the tip of an elongating, stalk-like ovary**, stalkless; **sepals bent backward**; petals 1–1.5 mm long, egg-shaped; stamens short, with oval **anthers, 0.5 mm long and attached near their middle**; stigma head-like, projecting past the petals on an elongated style; single in the upper leaf axils, forming narrow, crowded, leafy clusters (spikes); May {June–July}.

Fruits: Spindle-shaped, 4-sided **capsules**, thickened at the base, 6 {4–8} mm long, greyish white with fine hairs, splitting lengthwise into **4 compartments** to release many **seeds without tufts of hairs** (not comose).

Habitat: Dry prairie slopes, flats and depressions, on moist sandy soil; elsewhere, on exposed sandy soil, in moist swales on south-facing hillsides, and in sagebrush areas.

Notes: This species has also been called *Oenothera andina* Nutt. • Upland evening-primrose cannot tolerate heavy grazing; however, minor disturbance from moderate grazing may be needed to provide a suitable seedbed of exposed soil. • A Washington variety, var. *hilgardii* (Greene) Munz (sometimes called *Camissonia hilgardii* (Greene) Raven), has much larger flowers, with petals up to 4.5 mm long.

EVENING-PRIMROSE FAMILY **Dicots**

TARAXIA, SHORT-FLOWERED EVENING-PRIMROSE
Camissonia breviflora (Torr. & Gray) Raven
EVENING-PRIMROSE FAMILY (ONAGRACEAE)

Plants: Low, tufted **perennial** herbs, 5–10 cm tall, more or less densely covered with fine, close hairs, **without flowering stems**; from long taproots.

Leaves: In a basal rosette; blades lance-shaped, often widest above the middle, 5–10 {3–15} **cm long** and 5–15 mm wide, deeply cut into many **pinnate lobes**; stalks slender, ½–⅔ as long as the blades.

Flowers: Yellow (not purplish with age), cupped, about 1 cm across, with **4 sepals, 4 petals and 8 stamens on a flared, 10–25 mm long receptacle (hypanthium) at the tip of an elongating, stalk-like ovary**; sepals bent backward; petals 5–8 {10} mm long; stamens unequal, the longer 4 slightly more than ½ as long as the petals; style about the same length as the petals, tipped with a **round, slightly lobed stigma**; few to several, **nestled among the leaves**, arising **from the root crown**; {June} July.

Fruits: Hairy, leathery, **narrowly egg-shaped to oblong to spindle-shaped capsules** tapered toward the tip, 4-sided, 15 {10–25} **mm long**, roughened with many rounded bumps, **splitting lengthwise into 4 compartments; seeds numerous, lacking a tuft of hairs** (not comose); May–July.

Habitat: Clay flats, including dry slough bottoms and alkaline shores; elsewhere, in dry meadows and on streambanks.

Notes: This species has also been called *Oenothera breviflora* T. & G. and *Taraxia breviflora* (T. & G.) Nutt. • Another rare Alberta species, low yellow evening-primrose (*Oenothera flava* (A. Nels.) Garrett), is similar to taraxia but is distinguished by its stigma, which has 4 linear, 3 mm long lobes, and by its larger, 12–18 {10–20} mm long petals. The flowers of low yellow evening-primrose turn purple with age, and the 2–3 cm long capsules are winged on all 4 angles. It grows on clay flats and slough edges and is reported to flower in July and August. • Low yellow evening-primrose could be confused with the common butte-primrose (*Oenothera caespitosa* Nutt.), but butte-primrose has white (pink with age) petals that are more than 2 cm long, and its capsules are not winged.

OENOTHERA FLAVA

Rare *Vascular* Plants of Alberta

Dicots

EVENING-PRIMROSE FAMILY

HALL'S WILLOWHERB

Epilobium halleanum Hausskn.
EVENING-PRIMROSE FAMILY (ONAGRACEAE)

Plants: Slender perennial herbs; stems erect, 2–60 cm tall, usually simple, with **lines of stiff hairs running down from the leaf bases**, usually with stiff and glandular hairs on upper parts; from slender underground runners (rhizomes) with **clusters of small, round, compact bulbs** (turions) near the stem base.

Leaves: Opposite; blades lance-shaped, 5–47 mm long, 3–14 mm wide, edged with short, stiff hairs but otherwise mostly hairless, toothless or edged with small, conspicuous teeth; **stalks present**, sometimes short.

Flowers: White, often fading to pink, with **4 sepals, 4 petals and 8 stamens on a receptacle** (hypanthium) about 2 mm long at the tip of an elongating, stalk-like ovary; sepals green, 1.2–2.8 mm long, hairless or sparsely glandular-hairy; petals 1.6–5.5 mm long, notched; stamens of 2 lengths, with anthers 0.3–0.9 mm long; ovary 10–14 mm long, usually densely covered with glandular and low, stiff hairs (rarely hairless); styles 0.8–5 mm long tipped with a **club-shaped stigma**; few to several, on slender stalks in elongating clusters (racemes), nodding in bud; July.

Fruits: Slender capsules, 24–60 mm long, hairless to hairy (not glandular), **on stalks 8–38 mm long**, splitting lengthwise into 4 compartments; **seeds** numerous, narrowly egg-shaped, 1.1–1.6 mm long, less than 0.8 mm wide, rough **with tiny bumps** (papillae), tipped with a **3–6 mm long tuft of white hairs** (coma), which **easily breaks away**; July.

Habitat: Moist ground; elsewhere, in moist coniferous forests.

Notes: This species has also been included in *Epilobium glandulosum* Lehm. var. *macounii* (Trel.) C.L. Hitchc. • Hall's willowherb is similar to another rare species, Rocky Mountain willowherb (*Epilobium saxmontanum* Hausskn.), which differs by having stalkless, clasping leaves that are often wider (4–24 mm) than those of Hall's willowherb. The underground bulbs (turions) of Rocky Mountain willowherb are longer and fleshier than those of Hall's willowherb, the capsules are stalkless (rather than stalked), and the seeds have a conspicuous collar (inconspicuous on seeds of Hall's willowherb). Rocky Mountain willowherb

E. SAXIMONTANUM

grows in moist montane and subalpine meadows and on streambanks; it flowers in July and August. • Another small willowherb has falsely been reported to occur in Alberta. Oregon willowherb (*Epilobium oreganum* Greene, also called *Epilobium exaltatum* Drew) is very similar to Rocky Mountain willowherb, and some taxonomists consider both species part of the same variety (*Epilobium ciliatum* Lehm. var. *glandulosum* (Lehm.) Hoch & Raven). It is distinguished by its thicker (almost as wide as long) flower tube and spreading (rather than almost erect) sepals, and by the many spreading branches on its slender stems. It grows in similar moist sites and flowers in July and August. • The generic name *Epilobium* comes from the Greek *epi* (upon) and *lob* (pod), in reference to the pod-like fruit that develops from the base of the fading flower.

SLENDER-FRUITED WILLOWHERB

Epilobium leptocarpum Hausskn.
EVENING-PRIMROSE FAMILY (ONAGRACEAE)

Plants: Delicate perennial herbs, often loosely clumped, **upper leaf axils sometimes with bud-like growths** (gemmae), which fall off and grow into new plants; stems erect to ascending, 8–30 cm tall, branched, hairless except for **lines of stiff hairs running down from the leaf bases**; from slender underground runners (rhizomes) with **small, round bulbs** (turions) underground and at the base of the stem.

Leaves: Opposite; blades narrowly lance-shaped to elliptic, 8–40 mm long, 4–13 mm wide, sparsely hairy along the edges but otherwise hairless, **toothed toward the tip**; stalks winged and 3–5 mm long on lower leaves, absent on upper leaves.

Flowers: White (fading to pink), with **4 sepals, 4 petals and 8 stamens on a receptacle** (hypanthium) **at the tip of an elongating, stalk-like ovary**; sepals lance-shaped, 2.9–4 mm long, with straight, stiff, flat-lying hairs; petals 4–6.5 mm long, notched; stamens of 2 sizes with anthers 0.4–0.8 mm long; ovaries 12–19 mm long, densely covered with straight, stiff hairs; styles 3.2–3.8 mm long, tipped

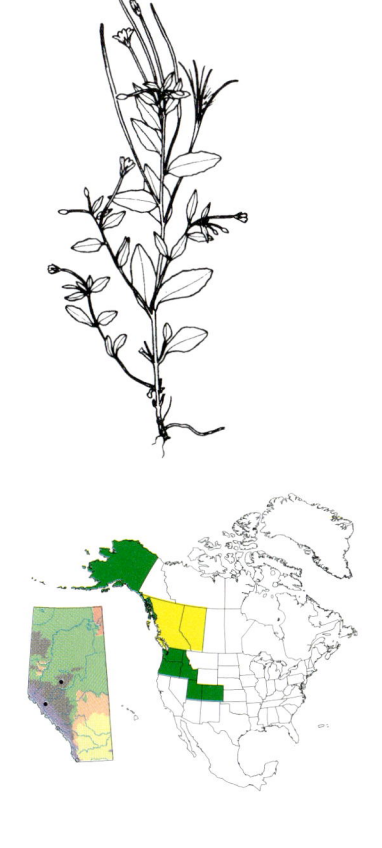

Dicots EVENING-PRIMROSE FAMILY

E. GLABERRIMUM

E. MIRABILE

with a **broadly club-shaped stigma**; several, in nodding to somewhat erect, elongated clusters (racemes); July–August.

Fruits: Slender capsules, 25–55 mm long, 1–2 mm wide, sparsely hairy, split lengthwise to reveal 4 compartments; **seeds numerous, 0.8–1.2 mm long, rough with tiny bumps** (papillae), tipped with a **persistent tuft of tawny, reddish, 3–6 mm long hairs** (coma); {August}.

Habitat: Moist, open, stony slopes.

Notes: This species has also been included in *Epilobium glandulosum* Lehm. var. *macounii* (Trel.) C.L. Hitchc.
• Another rare species, pale willowherb (*Epilobium glaberrimum* Barbey ssp. *fastigiatum* (Nutt.) Hoch & Raven, also called *Epilobium platyphyllum* Rydb., *Epilobium fastigiatum* Nutt. var. *glaberrimum* (Barbey) Piper and *Epilobium affine* Bong. var. *fastigiatum* Nutt.), has hairless stems, often with a greyish, waxy appearance (glaucous), and stalkless leaves. Its flowers are pink to rose-purple, and its seeds are 0.8–1 mm long with distinct rows of small bumps (papillae). Pale willowherb grows on rocky mountain slopes and streambanks and in moist forests and meadows. It flowers in {July} August and produces fruit in August and September.
• A species of subalpine scree slopes, wonderful willowherb (*Epilobium mirabile* Trel., also called *Epilobium clavatum* Trel. var. *glareosum* (G.N. Jones) Munz) has also been reported from Alberta. It is distinguished from pale willowherb by its white flowers, densely close-hairy stems and larger (1.7–2.2 mm long) seeds, which are covered with netted veins.

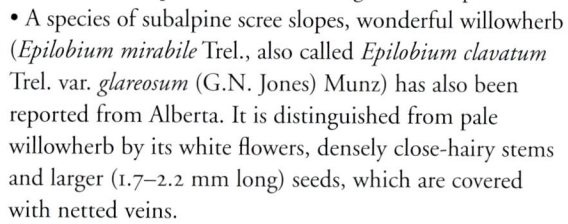

EVENING-PRIMROSE FAMILY   **Dicots**

CLUB WILLOWHERB

Epilobium clavatum Trel.
EVENING-PRIMROSE FAMILY (ONAGRACEAE)

Plants: Clumped perennial herbs; **stems ascending, 5–22 cm tall, with lines of stiff, flat-lying hairs running down the stem from the leaf bases**, also with stiff and glandular hairs on the upper stems; forming large clumps from **scaly underground stems** (rhizomes) that send out **numerous wiry shoots and suckers** (soboles).

Leaves: Opposite; blades egg-shaped to elliptic, **12–28 mm long**, 6–16 mm wide, hairless or with sparse hairs on the midrib and edges, **toothless** (sometimes finely toothed); stalks 0–3 mm long.

Flowers: Pink to rose-purple, with 4 sepals, 4 petals and 8 stamens on a receptacle (hypanthium) **at the tip of an elongating, stalk-like ovary**; sepals 2.5–4.2 mm long, hairless or sparsely glandular-hairy; petals 36–60 mm long, notched; stamens of 2 sizes, with anthers 0.4–0.9 mm long; ovaries 8–20 mm long; styles 1.4–3.2 mm long, tipped with a narrowly club-shaped to head-like stigma; few, in **erect, usually glandular-hairy, elongating clusters** (racemes); July–August {June–September}.

Fruits: Club-shaped capsules, **2–4 cm long**, hairless or sparsely hairy, splitting lengthwise into 4 compartments; **seeds** numerous, narrowly egg-shaped, widest above the middle, **1.3–2 mm long**, net-veined or rough with tiny bumps (papillae), tipped with a **5–15 mm long tuft of white hairs** (coma) which **breaks away easily**.

Habitat: Moist alpine slopes.

Notes: A similar species, white willowherb (*Epilobium lactiflorum* Hausskn.), is also rare in Alberta. White willowherb is a larger plant, with taller (15–50 cm) stems, larger (20–55 mm long), toothed leaf blades, and longer (5–10 cm) capsules. As its name suggests, it has white flowers, though these are occasionally pink-tinged or marked with red veins. White willowherb grows at slightly lower elevations, on moist montane and subalpine slopes, and flowers from June to August.

E. LACTIFLORUM

EVENING-PRIMROSE FAMILY

YELLOW WILLOWHERB

Epilobium luteum Pursh
EVENING-PRIMROSE FAMILY (ONAGRACEAE)

Plants: Robust perennial herbs; stems erect, several, sometimes branched above, 15–75 cm tall, with **lines of hairs running down from the leaf bases**; from scaly, underground runners (rhizomes) that send out numerous wiry shoots and suckers (soboles); occasionally with condensed bulbs (turions) at the stem base.

Leaves: Mostly opposite; blades egg-shaped (sometimes elliptic), 25–78 mm long, 12–35 mm wide, edged with fine, translucent teeth, hairless except for sparse, straight hairs on the midrib and edges; stalks short or absent.

Flowers: Yellow or cream-coloured, with **4 sepals, 4 petals and 8 stamens on a 1–2 mm tall, cupped receptacle** (hypanthium) **at the tip of an elongating, stalk-like ovary**; sepals narrowly lance-shaped, 10–12 {8–13} mm long, densely covered with glands; petals 12–22 mm long, heart-shaped with shallowly notched tips; stamens of 2 lengths, with anthers 22–30 mm long; ovaries 20–35 mm long, densely covered with glands; **styles erect**, 15–22 mm long, **extending beyond the petals, tipped with a broadly 4-lobed stigma**; 2–10, on 5–30 mm long stalks, borne singly in upper leaf axils, forming elongating clusters (racemes), nodding in bud; {July} August–September.

Fruits: Erect, linear capsules, 35–75 mm long, splitting lengthwise into **4 compartments**; **seeds** numerous, **1–1.2 mm long**, tapered to a long, slender point, net-veined, tipped **with a persistent, reddish tuft of 6.5–8 mm long hairs** (coma); {July} August–September.

Habitat: Moist woods and streambanks in the mountains.

Notes: This is the only willowherb in Alberta that has yellow flowers. The specific epithet *luteum* means 'yellow.'

EVENING-PRIMROSE FAMILY

Dicots

LOW WILLOWHERB, GROUNDSMOKE

Gayophytum racemosum Nutt. *ex* T. & G.
EVENING-PRIMROSE FAMILY (ONAGRACEAE)

Plants: Small, slender **annual** herbs; stems erect, 10–30 {5–40} cm tall, hairless or sparsely hairy, usually **freely branched mostly from near the base**, appearing very leafy above; from weak taproots.

Leaves: Alternate (the lower sometimes opposite); **blades linear to narrowly lance-shaped** and widest above the middle, 1–3 cm long, only gradually reduced in size up the stem, usually hairless (sometimes sparsely hairy), stalkless or short-stalked.

Flowers: White, sometimes reddish, **tiny** and inconspicuous; with **4 sepals, 4 petals and 8 stamens at the tip of an elongating, stalk-like ovary**; sepals bent backward, hairless or sparsely hairy, soon shed; petals about 1 mm long; stamens of 2 types, those opposite the petals shorter and sterile; ovaries hairless or sparsely hairy, tipped with a rounded stigma; borne singly in upper leaf axils, crowded, forming narrow, elongating clusters (racemes); {June–August}.

Fruits: Erect, slender, hairless **capsules** 8–14 {7–15} mm long, essentially stalkless, **splitting lengthwise into 2 compartments**; few **seeds** (fewer than 10 per cavity), about 1 mm long, **lacking a tuft of hairs** (not comose).

Habitat: Disturbed ground on open montane slopes; elsewhere, on sandy or gravelly soil and snow beds at foothill to alpine elevations.

Notes: This species has also been called *Gayophytum helleri* Rydb. var. *glabrum* Munz. • The generic name is a combination of the name J. Gay (author of a flora of Chile) and the Greek *phyton* (plant)—hence 'Gay's plant.'

Rare *Vascular* Plants of Alberta

Dicots   MARE'S-TAIL FAMILY

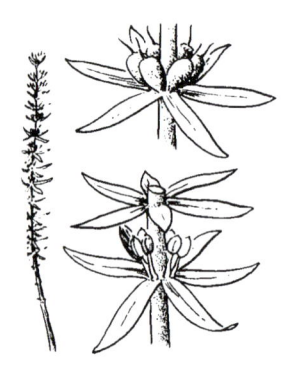

MOUNTAIN MARE'S-TAIL
Hippuris montana Ledeb.
MARE'S-TAIL FAMILY (HIPPURIDACEAE)

Plants: Aquatic or amphibious perennial herbs, hairless, with delicate, unbranched stems 1–10 cm tall, 0.2–0.5 mm thick; from slender, creeping underground stems (rhizomes).

Leaves: Linear, pointed, 5–10 mm long, 0.5–1 mm wide, stalkless, **in whorls of 5–8**, not toothed, lobed or divided.

Flowers: Green, stalkless; **tiny, lacking petals**, with 1 short-stalked anther and 1 ovary about 1 mm long and bearing a thread-like style, **usually male or female** and with female flowers above male flowers (sometimes with both male and female parts); **single in upper leaf axils**, thus forming whorls; July–August {September}.

Fruits: Single-seeded, nut-like, not splitting open.

Habitat: Moist open sites along streams and on mossy banks.

Notes: Mountain mare's-tail could be confused with common mare's-tail (*Hippuris vulgaris* L.), but common mare's-tail is a larger plant, with leaves over 1 mm wide and 1 cm long and stems more than 1 mm thick. Most of its flowers have both male and female parts. Mare's-tails also resemble horsetails (*Equisetum* spp.), but horsetails have hollow, jointed stems, and they reproduce by spores rather than flowers and seeds. • Hippuridaceae is a small family, consisting of a single genus, and some taxonomists include it in the closely related family Haloragidaceae (the water-milfoil family). The generic name *Hippuris* was derived from the Greek *hippos* (horse) and *oura* (tail).

CARROT FAMILY **Dicots**

BISCUIT-ROOT

Lomatium cous (S. Wats.) Coult. & Rose
CARROT FAMILY (APIACEAE [UMBELLIFERAE])

Plants: Delicate, essentially **hairless** perennial herbs; flowering stalks **usually leafless, 5–35 cm tall; from thick, round to elongated taproots.**

Leaves: Basal, sometimes with 1 stem leaf, parsley-like, **deeply cut 3 or more times** into narrowly elliptic to lance-shaped segments.

Flowers: Yellow, small, with 5 petals, 5 stamens and 1 **hairless ovary** tipped with 2 styles; many tiny flowers on 1–3 mm long stalks in dense, **round clusters** (umbellets), each cluster above a whorl of small (2–5 mm long), broad-tipped, egg-shaped, elliptic or lance-shaped bracts and borne at the tip of a slender stalk; 5–20 unequal (1.5–10 cm long) stalks (rays) forming an umbrella-shaped cluster (umbel); May {April–July}.

Fruits: Broadly elliptic pairs of flattened seed-like fruits (schizocarps), 5–12 mm long and 3–5 mm wide, with 5 prominent ribs (sometimes with alternating smaller ribs), edged with a broad **wing almost as wide as the body,** splitting apart when mature and usually hanging from the tip of the central stalk (carpophore), hairless or granular-roughened.

Habitat: Dry, open, rocky slopes and flats in the montane zone; elsewhere, on dry hillsides, often with sagebrush, in the foothills and mountains.

Notes: This species has also been called *Lomatium montanum* Coult. & Rose. • All *Lomatium* species are edible. The large, starchy roots were cooked or dried and ground into flour for making bread, cakes and biscuits—hence the name 'biscuit-root.' The roots and meal were important trade items among some tribes. Large, flat cakes of meal were carried on long journeys for food. The leaves add a strong parsley taste to salads. • The generic name *Lomatium* was taken from the Greek *loma* (a border), in reference to the flattened edges (wings) on the fruits.

Dicots CARROT FAMILY

PURPLE SWEET CICELY

Osmorhiza purpurea (Coult. & Rose) Suksd.
CARROT FAMILY (APIACEAE [UMBELLIFERAE])

Plants: Slender perennial herbs; stems 20–60 cm tall, rather leafy, hairy or hairless; from **thick, aromatic roots**.

Leaves: Basal and **alternate on the stem**, 1–3 times divided in 3s; leaflets lance-shaped, 2–6 cm long, sharply toothed or lobed; lower leaves stalked, upper leaves nearly stalkless.

Flowers: Purple or pinkish, sometimes greenish white, small, with 5 petals, 5 stamens and a broad, thickened disc with 2 styles each less than 0.5 mm long; few, on erect or widely spreading stalks in small, **umbrella-shaped clusters** (umbellets), these clusters borne on slender stalks (rays) in larger umbrella-shaped clusters (umbels); bracts at the base of the umbellets very small, often absent; {June} July.

Fruits: Bristly-hairy, club-shaped pairs of seed-like fruits (mericarps), **8–13 mm long, tipped with a small, thick, 2-styled disc** (stylopodium) that is **0.5–2 mm high** and **wider than it is long**, prominently 5-ribbed (sometimes with alternating smaller ribs), splitting apart from the base when mature and usually hanging from the tip of the central stalk (carpophore).

Habitat: Moist coniferous woods and slopes in montane and subalpine regions; elsewhere, in meadows, on streambanks and in redwood forests.

Notes: This species has also been called *Osmorhiza chilensis* Hook. & Arn. var. *purpurea* (Coult. & Rose) Boivin.
• Purple sweet cicely is easily confused with spreading sweet cicely (*Osmorhiza depauperata* Philippi) and blunt-fruited sweet cicely (*Osmorhiza berteroi* DC., also called *Osmorhiza chilensis* Hook. & Arn.), with which it is thought to hybridize. However, both of these species have hairy leaves. Blunt-fruited sweet cicely usually has white or greenish (sometimes purple) flowers, and its 12–22 mm long fruits are narrowed (constricted) below the relatively narrow (longer than wide) tip. Spreading sweet cicely has pink-purplish flowers and wide-spreading or reflexed rays tipped with club-shaped fruits that curve abruptly to broad-pointed tips. These more common species grow in moist woods to middle elevations in the mountains. • Smooth sweet cicely (*Osmorhiza longistylis* (Torr.) DC.) is also rare in Alberta. Unlike purple sweet cicely, it has persistent bracts

O. LONGISTYLIS

WINTERGREEN FAMILY **Dicots**

O. LONGISTYLIS

at the base of the small flower clusters, whitish flowers, longer styles (2–3 mm) and longer fruits (18–22 mm). Like purple sweet cicely, its fruits are bristly-hairy and taper into long, slender tails. Smooth sweet cicely grows at lower elevations, in moist woods in the parkland and prairies, rather than in mountain forests as does purple sweet cicely. It flowers in June. • The roots have a licorice-like odour when crushed. Several tribes enjoyed their delicate sweet flavour and ate them steamed, boiled or in stews.

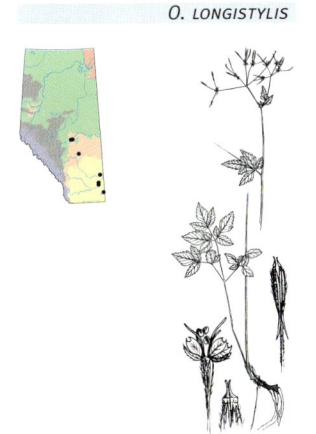

ARCTIC WINTERGREEN

Pyrola grandiflora Radius
WINTERGREEN FAMILY (PYROLACEAE)

Plants: Low, **evergreen** perennial **herbs**, hairless; stems 5–15 {25} cm tall; from slender, creeping underground stems (rhizomes).

Leaves: Basal, **firm** and leathery, shiny green on the upper surface, often reddish beneath; blades 2–4 {1–4.5} cm long, **oval to round**, thickened at edges, with obscure, rounded teeth or toothless; stalks 1–8 cm long, longer than or equal to their blades.

Flowers: Creamy or greenish white, often pinkish-tinged, about **1.5–2 cm across**, with 5 showy, 5–10 {12} mm long, spreading, rounded petals, 5 smaller, {2} 3–4 mm long sepals, small (at most 2.3 mm long) lemon-yellow anthers and a **long, curved style** with a collar below the tip; 4–11, on 2–8 mm long stalks in elongated clusters (racemes); July {June–August}.

Fruits: Round capsules, splitting open from base to tip; segments edged with cobwebby hairs at first; seeds numerous, tiny, with a loose, fleshy seed coat.

Habitat: Alpine slopes; elsewhere, in woods and heathlands.

Notes: The leaves of arctic wintergreen are very similar to those of common pink wintergreen (*Pyrola asarifolia* Michx.), but common pink wintergreen has smaller (5–7 mm long), pink petals, smaller (2–3 mm long) sepals and large (up to 3.5 mm long), reddish purple anthers.

Rare *Vascular* Plants of Alberta 163

Dicots   WINTERGREEN FAMILY

WHITE-VEINED WINTERGREEN
Pyrola picta J.E. Smith
WINTERGREEN FAMILY (PYROLACEAE)

Plants: Low, evergreen perennial herbs, hairless; stems reddish brown, 10–20 {25} cm tall; from slender, creeping underground stems (rhizomes).

Leaves: Basal, firm and leathery, **deep green, mottled with white or grey** along the larger veins on the upper surface, usually somewhat purplish on lower surfaces; blades 1–6 cm long, **narrowly to broadly egg-shaped,** commonly **pointed** at the tip and tapered at the base, thickened at edges, with small teeth or toothless, longer than or equal to their stalks.

Flowers: Yellowish to greenish white (or purplish), mostly **nodding,** about **1 cm across,** with 5 showy, spreading, rounded petals, 5 small (1.5 mm long), fused, reddish sepals and a **long** (5 mm), **curved style**; several, on spreading, 4–8 mm long stalks in elongated clusters (racemes); {June} July–August.

Fruits: Round capsules, splitting open from base to tip; segments edged with cobwebby hairs at first; seeds numerous, tiny, with a loose, fleshy seed coat.

Habitat: Moist to dry, coniferous forests.

Notes: This species is easily distinguished from other wintergreens by the whitish mottling along its leaf veins. The leaves could be confused with those of rattlesnake-plantain (*Goodyera* spp.), but the leaves of that orchid are narrower and softer (not leathery). • The generic name *Pyrola* was taken from the Latin *pyrus* (pear) because the leaves of some species are often pear-shaped. The specific epithet *picta* means 'brightly marked or painted'—a reference to the white veins on the leaves. • Although these plants have green leaves and are capable of producing food by photosynthesis, they also take nutrients from decaying plant material using specialized fungi (mycorrhizae) that are intimately associated with their roots.

INDIAN PIPE FAMILY | **Dicots**

PINE-SAP

Monotropa hypopitys L.
INDIAN PIPE FAMILY (MONOTROPACEAE)

Plants: Fleshy, saprophytic herbs, **waxy white, yellowish white or pinkish**, lacking green pigment, often drying black, commonly fragrant; stems stout, unbranched, 10–30 cm tall, often clustered, usually more or less **downy**; from dense masses of matted roots.

Leaves: Alternate, **scale-like**, smooth-edged or somewhat fringed, 1–1.5 cm long on upper stem, thicker and smaller toward stem base.

Flowers: Waxy white to yellowish or pinkish, **urn-shaped, 10–14 mm long**, with 2–5 (usually 4) lance-shaped sepals and **4 or 5 overlapping petals; petals 10–12 mm long, scale-like**, somewhat **pouched** at the base, **fuzzy** on 1 or both sides; uppermost flowers largest; few to many, short-stalked, in dense, elongating clusters (racemes), **nodding at first but soon erect**; {May–June} July.

Fruits: Erect, egg-shaped to round **capsules**, 5–8 mm long, opening from tip by 4 lengthwise slits; seeds numerous, tiny.

Habitat: Rich, shady coniferous forests, on humus.

Notes: This species has also been called *Hypopitys monotropa* Crantz. It is found around the world in the northern hemisphere. • Pine-sap resembles the more common species Indian pipe (*Monotropa uniflora* L.), but Indian pipe is white (rarely pinkish) and its stems are tipped with single nodding flowers. • In colour and texture, the nodding flowers of pine-sap resemble congealed pine resin, and these plants often grow under pines, so it was believed that they 'sapped' the strength of these trees. This may be true, to some extent. The roots of pine-sap share mycorrhizal fungi with some trees, and nutrients from the conifers may pass to the pine-sap via the fungi. The interrelationships between mycorrhizal fungi, pine-sap and trees are complex, and studies suggest that substances from the fungi help trees to fight some diseases. • The generic name comes from the Greek *monos* (one) and *tropos* (direction), referring to the flowers which are turned to 1 side. The specific epithet *hypopitys* was derived from the Greek *hypo* (beneath) and *pitys* (pine tree), referring to habitat.

Dicots

INDIAN PIPE FAMILY

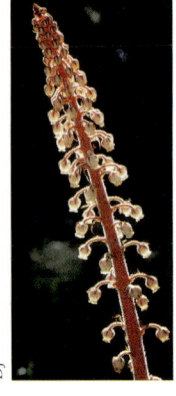

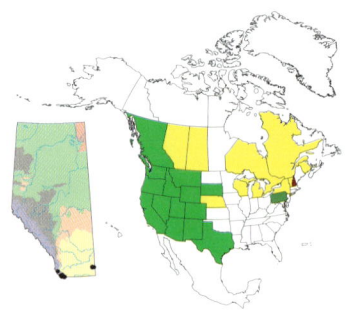

PINE-DROPS

Pterospora andromedea Nutt.
INDIAN PIPE FAMILY (MONOTROPACEAE)

Plants: Fleshy, clammy-fuzzy herbs, covered with glandular hairs, purplish or reddish brown to pinkish, lacking green pigment; stems erect, unbranched, 20–80 {90} cm tall, stout (often more than 2 cm wide at the base), remaining as fibrous dried stalks for at least 1 year; from dense masses of mycorrhizal roots.

Leaves: Numerous, alternate, **scale-like**, lance-shaped, 1–3.5 cm long, thick, crowded near the stem base.

Flowers: Pale yellowish, urn-shaped, {5} 6–8 mm long, hairless, with 5 short, spreading lobes at tip and 5 **glandular-hairy, reddish brown sepals** at base, **nodding**, on glandular-hairy stalks about 1 cm long, from the axils of slender bracts; numerous, in **erect, elongating clusters** (racemes) **10–30 cm long** (usually equal to the rest of the stem); {June} July–September.

Fruits: **Nodding**, round, **pincushion-like capsules**, 8–12 mm broad, purplish red; seeds numerous, tiny, tipped with a broad, netted wing many times longer than the egg-shaped seed.

Habitat: Dry coniferous woods, on deep humus.

Notes: Pine-drops was once thought to be a parasite, feeding on the roots of coniferous trees. It was later classified as a saprophyte, believed to derive its food from dead organic matter on the forest floor through a symbiotic association with soil fungi (mycorrhizae). It is now thought to be a parasite of fungi that feed on the roots of coniferous trees. Because of this dependence on soil fungi, it is extremely sensitive to disturbance. • Some tribes pounded the stems and fruits of these plants to make a medicinal tea for treating bleeding from the lungs. The dry powder was used as a snuff for nosebleeds. Europeans used pine-drops as a sedative and to stimulate perspiration. • The generic name *Pterospora* was derived from the Greek *pteros* (a wing) and *sporos* (seed), alluding to the lovely translucent wing borne by each seed. The specific epithet *andromedea* alludes to the resemblance of these flowers to those of the genus *Andromeda*. The plants grow under pines and their flowers resemble drops of resin—hence the common name 'pine-drops.'

PRIMROSE FAMILY **Dicots**

Greenland primrose

Primula egaliksensis Wormskj.
PRIMROSE FAMILY (Primulaceae)

Plants: Tiny perennial herbs; flowering stalks leafless, usually {5} 6–18 cm high.

Leaves: In a basal rosette, thin, 1.5–5.5 cm long (including stalks); blades egg-shaped to spatula-shaped, often widest above middle, flat or wavy along the edges; **stalks often as long as the blades or longer**.

Flowers: Violet or deep lilac (sometimes white) with a yellow eye, 5–9 mm across, with 5 petals fused in a slender, 6–8 mm long tube (longer than the sepals) but each ending in a broad (1.6–4 mm wide), abruptly spreading, notched lobe (limb); sepals 4–9 mm long, fused into a tube at the base, cut about ¼–½ of the way to the base into 5 lobes, fringed with **gland-tipped hairs**; 5 stamens, not extending past the mouth of the corolla; styles thread-like (filiform), with head-like (capitate) stigmas; 1–6, in a flat-topped cluster (umbel) above a whorl (involucre) of lance-shaped, 3–6 mm long bracts with enlarged, cupped bases; June–July {August}.

Fruits: Slender, cylindrical capsules, **2–3 times as long as the sepals**, splitting into 5 parts at the tapered tip; seeds pale, smooth.

Habitat: Marshy ground, wet meadows and shores in alpine and subalpine areas; elsewhere, in wet meadows and on wet, calcareous lakeshores and riverbanks.

Notes: Erect primrose (*Primula stricta* Hornem.) is another slender primrose that is also rare in Alberta. It is a larger plant (10–30 cm tall) with round-toothed leaves, and its showy lilac flowers, 8–12 mm wide, appear in June and July. Erect primrose grows on moist alpine slopes and in wet areas in the boreal forest. It prefers saline soil. • The generic name *Primula* was derived from the Latin *primus* (early or first) because many species flower early in spring.

P. STRICTA

Dicots

PRIMROSE FAMILY

CHAFFWEED, FALSE PIMPERNEL

Anagallis minima (L.) Krause
PRIMROSE FAMILY (Primulaceae)

Plants: Small, **hairless annual** herbs, **2–10 cm tall**; branches erect or with bases spreading on the ground and tips ascending, often rooting at the joints.

Leaves: Alternate, oblong, spoon-shaped or egg-shaped and widest above the middle, 4–8 {10} mm long; almost stalkless, with the lower edges extending down the stem.

Flowers: Greenish, **inconspicuous, about 2 mm long**, with 4 sepals and 4 petal lobes; corolla white or pinkish, scarcely half as long as the sepals, with 4 spreading lobes at the tip of a broad tube, withered and persistent on the capsule lid; sepals narrowly lance-shaped, 2–3 mm long, edged with small sharp teeth; 4 stamens opposite the corolla lobes; solitary in leaf axils on short (1 mm long) stalks; {May–July} August.

Fruits: Rounded capsules about 2 mm long, splitting in 2 just above the middle, **shedding the top as a lid and leaving the cupped lower ½ on the plant**; seeds brown, pitted.

Habitat: Moist soil and mud by ponds; elsewhere, occasionally in flooded areas and in damp places along paths in woods, usually in acid soils.

Notes: This species has also been called *Centunculus minimus* L. • The generic name *Anagallis* was the ancient Greek name for pimpernel. The specific epithet *minima* means 'small.'

PRIMROSE FAMILY **Dicots**

MOUNTAIN DWARF PRIMROSE, MOUNTAIN DOUGLASIA
Douglasia montana A. Gray
PRIMROSE FAMILY (PRIMULACEAE)

Plants: Low, **cushion-forming or matted** herbs, about 3 cm tall; **stems slender, branched, slightly woody**, spreading, tipped with tufts of short leaves; flowering stems erect, leafless; from persistent **root crowns** and underground stems (rhizomes).

Leaves: Crowded **at branch tips** in **compact rosettes**, linear to lance-shaped, about 4 {8} **mm long**, fringed with tiny, stiff hairs that are rough to the touch.

Flowers: Purple, lilac or bright pink, 6–8 mm long and about 1 cm across, with **5 showy corolla lobes and 5 slender-pointed calyx lobes**; corolla with erect or spreading, oblong to egg-shaped lobes about 4 {5} mm long at the tip of a narrow tube, also bearing 5 small ridges around the mouth of the tube; calyx reddish, hairy or hairless, equal to or slightly longer than the corolla tube, cut about halfway to the base into 5 ridged lobes; stamens opposite the corolla lobes and contained in the flower tube; styles thread-like; single (sometimes 2), on 1–4 leafless, 5–25 mm long, hairy stalks from the centres of the leaf rosettes, usually with 1 or 2 bracts at the base of the flower; {June–July} August.

Fruits: Capsules, slightly longer than the calyx tube, containing 1 or 2 brown, finely pitted seeds.

Habitat: Dry alpine slopes; elsewhere, in dry places from foothills to mountain tops.

Notes: Mountain dwarf primrose is superficially similar to the common plant moss phlox (*Phlox hoodii* Richards.), but moss phlox has paired hairy leaves, hairy sepals, nearly stalkless flowers and broader, more rounded petals.
• Mountain dwarf primrose grows with white dryad (*Dryas octopetala* L.) on a single alpine summit in Waterton Lakes National Park. It is extremely rare, with a population of about 10,000 plants at this single site. • The genus *Douglasia* was named in honour of David Douglas (1798–1834), a famous early botanical explorer in the northwest. The specific epithet *montana* means 'of the mountains.'

PRIMROSE FAMILY

LANCE-LEAVED LOOSESTRIFE

Lysimachia hybrida Michx.
PRIMROSE FAMILY (Primulaceae)

Plants: Erect perennial herbs; stems leafy, 40–100 cm tall, almost hairless; from **short** (never extensive) **underground stems** (rhizomes).

Leaves: Opposite or whorled, linear to **lance-shaped, gradually tapered to both ends**, 3–10 cm long, 1–3 cm wide, rough-edged but rarely fringed with hairs; **leaf stalks hairless or inconspicuously fringed**, the lower stalks 10–30 mm long.

Flowers: Bright yellow, about **15 mm across**, with 5 {6} **broadly egg-shaped, flat-spreading petal lobes** at the tip of a short, inconspicuous tube; each lobe usually tipped with a short, stiff point and edged with fine, irregular teeth; **calyxes deeply cut into 5 {6} firm, pointed segments**; 5 stamens, alternating with tiny sterile stamens; styles slender; **single, on thread-like stalks in upper leaf axils**, often appearing in whorls; July {June–August}.

Fruits: Round or egg-shaped capsules, splitting open lengthwise into 5 parts.

Habitat: Moist meadows and shores; elsewhere, in thickets, dry to moist open woods, and swamps.

Notes: This species has also been called *Lysimachia lanceolata* Walt. ssp. *hybrida* (Michx.) J.D. Ray, *Steironema lanceolatum* (Walt.) A. Gray var. *hybridum* (Michx.) A. Gray and *Steironema hybridum* (Michx.) Raf. • Fringed loosestrife (*Lysimachia ciliata* L.) is a similar, much more common species distinguished by its more extensive, elongated underground stems (rhizomes) and broader (3–6 cm wide), egg-shaped to lance-shaped leaves, with shorter (5–20 mm) stalks that are conspicuously fringed with hairs. The flowers of fringed loosestrife are also larger, with lobes about 1 cm long. • The generic name *Lysimachia* honours King Lysimachus, a Macedonian general who ruled Thrace 323–281 BC. Some sources report that it was taken from the Greek *lysis* (a losing or ending) and *mache* (strife). There was an ancient belief that these plants would calm savage beasts and that a piece hung from the yokes of oxen would be enough to appease any strife or unruliness. Loosestrife does appear to repel gnats and flies, so this might help to keep bothersome pests away from the faces of such beasts of burden.

GENTIAN FAMILY **Dicots**

NORTHERN FRINGED GENTIAN, RAUP'S FRINGED GENTIAN

Gentianopsis detonsa (Rottb.) Ma ssp. *raupii* (Porsild)
A. Löve & D. Löve
GENTIAN FAMILY (GENTIANACEAE)

Plants: Hairless **annual or biennial** herbs; stems erect, 5–40 {60} cm tall, branched in the upper sections, **often purplish**.

Leaves: Basal (often in rosettes) and **opposite** on the stem; basal leaves **egg-shaped to spoon-shaped**, 5–35 {60} mm long; stem leaves widest at or above the middle, 15–65 mm long, 1–7 mm wide; **uppermost leaves blunt-tipped**.

Flowers: Blue (rarely whitish), **broadly goblet-shaped**, **1–4 cm long**, 0.8–1.5 cm wide, with a corolla tube cut about halfway to the base into **4 elongate, rounded lobes fringed with slender teeth**; 4 calyx lobes, **smooth or wrinkled** but without tiny bumps (papillae), **not prominently ridged**, glossy and purplish at the centre, thin and translucent along the edges; 1–few, on erect, long stalks; late June–early August.

Fruits: Cylindrical capsules, as long as the persistent flower tube, containing many small, dark seeds; {August}.

Habitat: Wet meadows and saline flats.

Notes: This species has also been called *Gentianella detonsa* (Rottb.) G. Don ssp. *raupii* (Porsild) J. Gillett and *Gentiana detonsa* Rottb. • Northern fringed gentian resembles the common fringed gentian (*Gentianopsis crinita* (Froel.) Ma), but the calyxes of fringed gentian have tiny bumps (papillae, visible only with magnification) at their bases and on the dull green or purplish ridges (keels) of at least 2 of their lobes. The upper leaves of fringed gentian are pointed rather than blunt. • Subspecies *raupii* is the only subspecies in Alberta. In Canada, it is found only in the Mackenzie Basin drainage and the Hudson Bay areas. • The specific epithet *detonsa* was taken from the Latin *de* (concerning) and *tonsus* (shaven). The subspecies name *raupii* honours H.M. Raup, an American botanist born in 1901, who spent considerable time in the Northwest Territories.

MARSH GENTIAN, LOWLY GENTIAN

Gentiana fremontii Torr.
GENTIAN FAMILY (Gentianaceae)

Plants: Low biennial herbs, 1–10 cm high; stems lying on the ground to erect-ascending, branched at base; from weak, slender roots.

Leaves: In basal rosettes and opposite on the stem, pale green with **broad, white edges**; basal leaves round to egg-shaped, 2–13 mm long, 1.5–1.8 mm wide; stem leaves pressed to the stem, **scarcely bent backward**, not toothed, oblong to lance-shaped, less than 6 mm long, with **bases of pairs fused into a sheath ¼–½ as long as the blades**.

Flowers: Greenish purple to whitish, tubular, 7–10 {15} mm long, tipped with 5 lobes; calyx tubes, 4–8 {12} mm long, cut less than ¼ of the way to the base into 4 or 5 faintly ridged lobes; solitary, erect; June {July–August}.

Fruits: Slender, papery capsules, splitting in half to release many small seeds; August.

Habitat: Moist grassy meadows; elsewhere, in moist sandy areas and meadows.

Notes: This species has also been called *Gentiana aquatica* auct., non L. • Marsh gentian is similar to moss gentian (*Gentiana prostrata* Haenke *ex* Jacq.), and many floras (including the second edition of *Flora of Alberta*) include it in that species. The leaves and sepals of moss gentian do not have wide, white edges, the stem leaves are bent backward and the basal and stem leaves are very similar in shape and size. Moss gentian grows in alpine areas. • The generic name *Gentiana* refers to Gentius, king of Illyria (an ancient region along the east coast of the Adriatic Sea) in the second century BC, who was said to have discovered the medicinal properties of *Gentiana lutea*. The specific epithet *fremontii* honours Major-General John Charles Fremont (1813–90), a plant collector of the western US.

ALPINE GENTIAN, GLAUCOUS GENTIAN, PALE GENTIAN

Gentiana glauca Pallas
GENTIAN FAMILY (Gentianaceae)

Plants: Small, somewhat fleshy perennial herbs, often **bluish green** with a thin, waxy coating (glaucous); stems erect, **3–10 {2–15} cm tall**; from creeping underground stems (rhizomes).

Leaves: In **basal rosettes** (winter rosettes present) and in **2 or 3 {4} pairs on the stem**, shiny, bluish green; basal leaves oval, widest at or above the middle, 1 {0.5–2} cm long, 1 cm wide; stem leaves oval, widest at or below the middle, 0.5–1.5 cm long.

Flowers: Dark to greenish blue, tubular, **12–18 {10–20} mm long**, tipped with **5 small** (about 2.5 mm long), blunt-tipped, broadly triangular to **egg-shaped, 3-veined lobes, alternating with 5 vertical folds** (pleats); calyx 5–7 mm long, green or purplish, tubular, tipped with 5 lance-shaped, blunt or pointed, unequal lobes, about ½ the length of the tube; **5 stamens**, contained within the corolla tube; 3–5 {7}, stalkless, in closely crowded clusters (cymes), sometimes with short-stalked lower flowers in upper leaf axils; June–August {September}.

Fruits: Erect capsules projecting slightly from the mouth of the persistent flower tube, containing many flattened, net-veined seeds.

Habitat: Moist subalpine and alpine meadows and slopes.

Notes: The compact basal rosettes of leaves and clusters of small, blue-green flowers distinguish this species from other gentians in Alberta. • The tubular flowers of the gentians are adapted for pollination by bumblebees. • The specific epithet *glauca* refers to the blue-green (glaucous) plants and flowers.

GENTIAN FAMILY

MARSH FELWORT

Lomatogonium rotatum (L.) Fries *ex* Nyman
GENTIAN FAMILY (Gentianaceae)

Plants: Slender **annual** herbs; stems erect, 10–35 {5–45} cm tall, unbranched or with strongly upward-pointing branches; from poorly developed taproots.

Leaves: Basal and opposite; lower leaves spatula-shaped, soon withered; middle and upper leaves **opposite**, mostly **linear to lance-shaped, 1–3** {0.5–5} **cm long**, 0.1–0.3 cm wide, slender-pointed, stalkless.

Flowers: White or bluish with purple veins, saucer-shaped, about 2 cm across, with **4 or 5 lobes widely spreading** at the tip of a **very short tube**; each lobe 10–12 {6–15} mm long, 8–20 mm wide, with 2 small, fringed scales at the base; calyxes with 4 or 5 slender lobes with prominent midveins, alternating with and equal to or longer than the petals; stamens 0.5 cm long, with flattened filaments, attached to the base of the petals; ovaries to 1 cm long, with 2 **stigmas** attached to the surface and **extending down the sides** as fine lines; few, on slender stalks from upper leaf axils, in elongated clusters (raceme-like); {late July} August–early September.

Fruits: Oblong to egg-shaped capsules, up to 1.5 cm long, splitting open from the tip, the segments curving strongly back; seeds light to dark brown, longer than they are wide, 0.5–0.75 mm long; late August–early September.

Habitat: Wet meadows and flats, often on saline soils.

Notes: A common relative, felwort (*Gentianella amarella* (L.) Börner), sometimes resembles marsh felwort, but felwort has many blue, funnel-shaped flowers that are borne in the axils of the leaves over much of the stem. • Marsh felwort has spread rapidly across North America since deglaciation. • The generic name *Lomatogonium* was derived from the Greek *lomato* (fringed) and *gone* (a seed), a reference to the stigma, which runs in lines down the side of the ovary. The specific epithet *rotatum* is from the Latin *rotatus* (wheel-shaped), in reference to the broad, almost flat, round flowers.

GREEN MILKWEED

Asclepias viridiflora Raf.
MILKWEED FAMILY (ASCLEPIADACEAE)

Plants: Perennial herbs with white, **milky juice** (latex) and tiny, downy hairs; stems 20–60 {120} cm long, often sprawling at the base; from deep taproots.

Leaves: Opposite or alternate, 2–8 {10} cm long, egg-shaped to oblong–lance-shaped, leathery with age, wavy-edged.

Flowers: Pale green and purple, about 1 cm long, with **5 greenish white, spreading to down-turned petal lobes around a projecting tube of fused stamens tipped with a broad, purplish, 5-hooded structure lacking horns** (corona); sepals persistent but inconspicuous; stalks hairy, short; numerous, in dense, round clusters (umbels) at stem tips and in leaf axils; {June–July}.

Fruits: Pairs of soft, spindle-shaped, pod-like fruits that split open down 1 side (follicles), finely hairy but otherwise smooth (not bumpy), stalked; seeds broad, flat, tipped with a tuft of long, white, silky hairs.

Habitat: Dry hillsides, often on sandy soil.

Notes: Green milkweed is very similar to low milkweed (*Asclepias ovalifolia* Dcne., p. 176). Both are relatively small species (usually less than 60 cm tall) with downy pods that lack the numerous small bumps found on the fruits of many other milkweed species. Both have greenish white flowers, but in low milkweed the tips of the 5 horns on each flower are curved inward, whereas in green milkweed these points are straight. • At the slightest damage, these plants ooze quantities of milky sap, which **contains toxic alkaloids**. Some cases of **livestock poisoning** have been attributed to milkweeds, but some insects eat milkweed with impunity, and many have developed the ability to incorporate alkaloids from the plants into their own bodies. This makes them poisonous and less tasty to predators. • Milkweed flowers produce large amounts of fragrant nectar, and their pollen grains stick together in masses called pollinia. When an insect lands, its feet slip into little slits on the sides of the flower and catch on wire-like filaments with a mass of pollen on each end. When the insect pulls its foot out, it carries tiny saddlebags of pollen. Sometimes an insect cannot free its foot and is trapped on the flower.

LOW MILKWEED

Asclepias ovalifolia Dcne.
MILKWEED FAMILY (ASCLEPIADACEAE)

Plants: Downy perennial herbs with white, **milky juice** (latex); stems {15} 20–60 cm tall; from deep taproots.

Leaves: Mostly opposite, 3–7 {8} cm long, egg-shaped to lance-shaped, rounded or slightly pointed at the tip, tapered to a short stalk at the base.

Flowers: Greenish white (sometimes purplish tinged), less than 1 cm long, with **5 spreading to down-turned petal lobes around a projecting tube of fused stamens tipped with a broad yellowish, 5-hooded structure with 5 incurved horns** (corona); sepals persistent but inconspicuous; stalks long and slender; 10–18, in loose, round clusters (umbels); June–August.

Fruits: Hanging pairs of soft, stalked, **spindle-shaped, pod-like fruits** that split open down 1 side (follicles), downy-hairy but otherwise smooth (not bumpy); seeds broad, flat, tipped with a tuft of long, white, silky hairs.

Habitat: Open woods and slopes; elsewhere, in moist prairie.

Notes: The young shoots, unopened flower buds and immature seed pods of several milkweed species were eaten raw or cooked in soups and stews. The **poisonous alkaloids** are destroyed by heat. It has been suggested that milkweed has a tenderizing effect on meat. Milkweed seeds have also been used for food occasionally. • The Cheyenne collected the dried juice from milkweed stems and chewed it like gum. The juice was also used to temporarily brand livestock because it stained the hides on animals for several months. The fibrous older milkweed plants were sometimes used to make string and rope.

Dicots

COMMON DODDER

Cuscuta gronovii Willd.
MORNING-GLORY FAMILY (CONVOLVULACEAE)

Plants: Leafless, parasitic annual herbs; stems rather coarse, **orange or yellow**, often **climbing high, twining** around other plants and drawing nutrients from them by means of **suckers**; producing aerial roots but not rooted in the ground when mature.

Leaves: Reduced to tiny scales.

Flowers: Whitish, 2.5–3 {2–4} mm long, **bell-shaped**, tipped **with 4 or 5 blunt, broadly egg-shaped lobes** that are shorter than the tube, bearing **copiously fringed scales in the throat**, which are much shorter than the tube; calyx short, with rounded lobes; stigmas head-like; stalks short or absent; several, in **dense, head-like clusters** (cymes); August {July–October}.

Fruits: **Ovoid capsules**, 3–6 mm long, tipped with styles about half as long as the capsule, **not splitting open**, containing 2–4 seeds, each 2–3 mm long.

Habitat: Growing on a variety of coarse plants (usually shrubs) in moist, shady areas.

Notes: This is a difficult genus to identify because species are often differentiated on the basis of microscopic flower characteristics. • When a dodder seed germinates, it produces roots and a slowly rotating shoot. This shoot wraps around the first plant it contacts and soon fastens itself to the host using aerial roots and suckers. Once the shoot has joined with a suitable host, the roots and lower stems wither and decay. Dodders lack chlorophyll and are unable to produce food through photosynthesis, so they depend entirely on their host for sustenance. Their stems produce specialized suckers that penetrate the host tissue and draw out nutrients. Most plants produce many flowers and abundant seed as the season progresses, ensuring continuation of the cycle the following year.

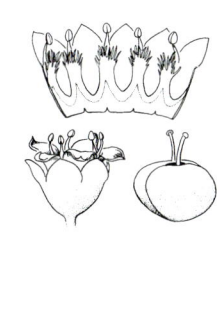

Rare *Vascular* Plants of Alberta

PHLOX FAMILY

NORTHERN LINANTHUS

Linanthus septentrionalis H.L. Mason
PHLOX FAMILY (POLEMONIACEAE)

Plants: Slender **annual** herbs; stems erect, **freely branched**, 5–30 cm tall; from taproots.

Leaves: Opposite, palmately divided into 5–7 thread-like segments, 5–20 mm long, hairless or with short hairs, stalkless.

Flowers: White, pale blue or lavender, funnel-shaped, 2–5 {6} **mm long** and wide, with 5 broad, spreading lobes at the tip of a **short tube with a ring of hairs at the mouth**; calyx tubular, ½–⅔ as long as the petals, **tipped with 5 equal, triangular lobes** with slender points, dry and translucent below the hollows (sinuses); styles projecting from the flowers, tipped with a 3-lobed stigma; **single, on slender, spreading stalks from upper leaf axils**, forming spreading, branched clusters (cyme-like panicles); late May–June {July}.

Fruits: Capsules, with 2–4 {8} seeds in each of 3 cavities.

Habitat: Dry hillsides and plains; elsewhere, in forest openings and dry meadows.

Notes: This species has also been called *Linanthus harknessii* (Curran) Greene var. *septentrionalis* (Mason) Jepson & V. Bailey. • Linanthus seeds become gummy when wet. • The generic name *Linanthus* was taken from the Greek *linon* (flax) and *anthos* (a flower) because these blooms were thought to resemble tiny flax flowers. The specific epithet *septentrionalis* is Latin for 'northern.'

PHLOX FAMILY **Dicots**

SLENDER PHLOX

Phlox gracilis (Hook.) Greene ssp. *gracilis*
PHLOX FAMILY (POLEMONIACEAE)

Plants: **Slender annual** herbs, not tufted; stems single, erect, rarely spreading on the ground, 10–40 cm tall, usually **branched and with fine glandular hairs** (at least on the upper stem); from weak taproots.

Leaves: Mainly opposite, **toothless**, elliptic to oblong, 2–5 cm long; **upper leaves lance-shaped to linear, alternate.**

Flowers: **Pink or lavender, trumpet-shaped,** {9} 10–14 mm long, with **5 spreading lobes** at the tip of a **slender, 8–12 mm long, yellowish tube**; calyx glandular-hairy, 5–6 mm long, cut about halfway to the base into **5 slender lobes**, not enlarging in fruit, eventually **ruptured by the growing capsule**; **5 stamens**; stigmas 3-lobed; single and stalkless or in pairs (1 stalkless and 1 stalked), borne in the axils of alternate leaf-like bracts, forming branched, cyme-like clusters; {March–May} June.

Fruits: Capsules, splitting open along 3 vertical lines.

Habitat: Dry to moist open ground; elsewhere, in thickets and bottomlands.

Notes: This species has also been called *Microsteris gracilis* (Hook.) Greene. • The generic name *Phlox* was taken from the Greek *phlox* (flame) in reference to the brightly coloured flowers of some species. The specific epithet *gracilis* is Latin for 'slender' or 'graceful.'

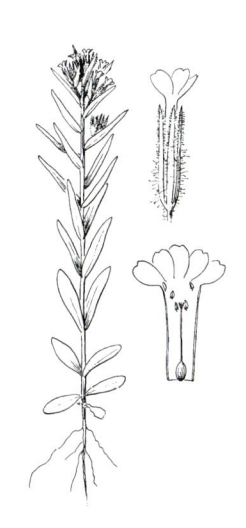

Rare *Vascular* Plants of Alberta

Dicots   WATERLEAF FAMILY

Sitka Romanzoffia, Mistmaiden, Cliff Romanzoffia, Sitka Mistmaiden

Romanzoffia sitchensis Bong.
WATERLEAF FAMILY (Hydrophyllaceae)

Plants: Delicate perennial herbs, usually almost hairless; stems slender, 10–30 cm tall, with a **bulbous base**; from underground stems (rhizomes).

Leaves: Mainly basal, **round to kidney-shaped, shallowly 5–7-lobed to coarsely toothed**, 1–3 {4} cm wide, rather fleshy, long-stalked.

Flowers: White or cream-coloured with a golden eye, broadly funnel-shaped, 7–10 mm across, with 5 broad petal lobes and 5 small, slender, hairless sepals; few to several, in **loose, uncoiling clusters** (raceme-like cymes); {June} July–August.

Fruits: Oblong, 2-chambered capsules with many seeds.

Habitat: Moist, rocky slopes, cliffs and ledges, in subalpine to alpine zones, often in spray zones near waterfalls; elsewhere, in wet places in the montane zone.

Notes: This little plant could be mistaken for a saxifrage (*Saxifraga* spp.) at first glance, but its flowers have broad, 5-lobed funnels (rather than separate petals) and form 1-sided, uncoiling clusters (rather than branched, panicle-like clusters). The capsules of Sitka romanzoffia are oblong and blunt-tipped, rather than 2-beaked. • The generic name *Romanzoffia* commemorates Nikolai Rumiantzev, Count Romanzoff (1754–1826), a Russian patron of botany who sponsored Kotzebue's expedition to the Pacific Northwest coast.

WATERLEAF FAMILY **Dicots**

WATERPOD

Ellisia nyctelea L.
WATERLEAF FAMILY (HYDROPHYLLACEAE)

Plants: Delicate, more or less hairy annual herbs; stems {5} 10–40 cm tall, **usually freely branched**, sometimes unbranched.

Leaves: Opposite on lower stem, alternate above, 2–6 {10} cm long, **deeply pinnately cut** into 7–13 lance-shaped or oblong lobes; **stalks fringed** with stiff hairs.

Flowers: Whitish, often tinged blue, 6–12 mm across, **narrowly bell-shaped**, with tiny appendages in the throat, equal to or slightly longer than the calyx; calyx cut almost to the base into 5 lobes 3–5 mm long; stalks usually less than 1 cm long; **single, from leaf axils**; May–June {July}.

Fruits: Round, bristly-hairy capsules about 6 mm in diameter, **above 5 large** (usually about 1 cm long), **net-veined, wide-spreading sepals**, on long (up to 5 cm) stalks, splitting into 2 parts to release 4 round, 2.5–3 mm long, net-veined seeds.

Habitat: Moist shady woods and streambanks; elsewhere, on lake edges.

Notes: The generic name *Ellisia* commemorates the English botanist John Ellis, who lived 1710–76.

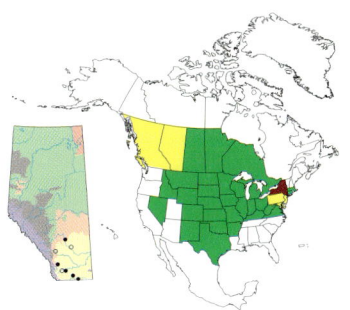

SMALL BABY-BLUE-EYES
Nemophila breviflora A. Gray
WATERLEAF FAMILY (Hydrophyllaceae)

Plants: Delicate annual herbs; stems weak, usually ascending or loosely erect, sometimes spreading on the ground, 10–20 {30} cm long, generally branched, angled, with tiny, downward-pointing prickles; from weak taproots.

Leaves: Mostly alternate, egg-shaped in outline, **pinnately divided into 3–5 oblong to lance-shaped, pointed lobes** up to 2 cm long and 7 mm wide, thin, with coarse, stiff, often flat-lying hairs, conspicuously **fringed** with hairs (especially on the stalk).

Flowers: White or purplish, **cupped**, with a **5-lobed, 2 mm long corolla** and a longer, **deeply 5-lobed calyx**; calyxes **fringed with long, stiff hairs** and bearing a slender, 1–2 mm long, backward-bent lobe (auricle) in each of the 3–5 mm deep hollows (sinuses) between the lobes (sepals), growing with age to equal or surpass the capsule; **5 stamens**, shorter than the sepals; styles 0.5–1 mm long; **single, short-stalked**, from leaf axils or stem tips; {April–May} June–July.

Fruits: Rounded capsules, 3–5 mm long, **spreading or nodding on 5–15 mm long stalks**, cupped in the persistent calyxes, splitting in 2 to release a **single seed** tipped with an appendage that is soon shed.

Habitat: Moist montane woods and open slopes, in protected places, usually aspen groves, from foothills to moderate elevations in the mountains; elsewhere, in shrub thickets and coniferous woods, often on moist shale.

Notes: The appendage at the tip of the seed attracts ants, which feed on it. The ants also carry the seeds to their nests, thus helping to disperse and plant the seeds. • The generic name *Nemophila* was taken from the Greek *nemos* (a grove) and *philein* (to love) in reference to the wooded habitat of some species. The specific epithet *breviflora* means 'with short flowers.'

WATERLEAF FAMILY **Dicots**

LINEAR-LEAVED SCORPIONWEED

Phacelia linearis (Pursh) Holzinger
WATERLEAF FAMILY (HYDROPHYLLACEAE)

Plants: Slender **annual** herbs, hairy with tiny and bristly hairs; stems erect, 10–40 {50} cm tall, unbranched to freely branched; from weak taproots.

Leaves: Alternate, **linear to lance-shaped**, 2–6 {1.5–11} cm long, 1.5–12 mm wide (excluding the lobes), **undivided or some deeply cut** into 3–7 linear to lance-shaped lobes, mostly stalkless; lower leaves soon withered.

Flowers: Bright blue to blue-lavender (rarely white), **broadly bell-shaped**, showy, about 1 cm long, 8–18 mm across, with **5 broad petal lobes**, 5 linear, bristly fringed sepals, and **5 hairless stamens extending only slightly past petals**; several, in **compact clusters** (scorpioid cymes) **that uncoil as the flowers develop** and elongate in fruit; {April–May} June–July.

Fruits: Capsules containing about 6–15 black, oblong, coarsely pitted seeds.

Habitat: Dry, open slopes and plains.

Notes: Many phacelias make excellent ornamental plants in rock gardens. They are best grown from seed. • The common name 'scorpionweed' likens the coiled branches of the flower cluster to the tail of a scorpion. The generic name *Phacelia* was derived from the Greek *phakelos* (a fascicle or bundle), in reference to the dense flower clusters of some species.

LYALL'S SCORPIONWEED

Phacelia lyallii (A. Gray) Rydb.
WATERLEAF FAMILY (HYDROPHYLLACEAE)

Plants: Dwarf **perennial** herbs; stems 10–20 {25} cm tall, sparsely hairy on the upper parts; from stout, branched crowns on taproots.

Leaves: Mainly basal but also alternate on the stem, **lance-shaped**, widest above the middle, **5–10 cm long**, up to 3 cm wide, **coarsely pinnately divided** into many oblong or lance-shaped lobes, with scattered hairs, sometimes glandular-hairy.

Flowers: Deep blue to bluish purple, bell-shaped, 8 {5–9} cm long and wide, with **5 petal lobes**, 5 slender, stiff-hairy sepals and **5 long, slender stamens 1½–2½ times as long as the petals** and projecting from the mouth of the flower, numerous, in **short, dense clusters that uncoil** as the flowers develop and elongate in fruit (scorpioid cymes); July–August.

Fruits: Capsules containing about 8–18 oblong, densely pitted seeds.

Habitat: Rocky alpine slopes, on talus and in rock crevices.

Notes: Lyall's scorpionweed is closely related to the more common species silky scorpionweed (*Phacelia sericea* (Graham) A. Gray) and was once considered a variety of that species. However, silky scorpionweed has more finely divided leaves (cut almost to the midrib into slender, linear segments) and longer (up to 20 cm long), spike-like flower clusters, and its stamens are 3 times as long as their corollas.

BORAGE FAMILY **Dicots**

LARGE-FLOWERED LUNGWORT, LONG-FLOWERED BLUEBELLS

Mertensia longiflora Greene
BORAGE FAMILY (BORAGINACEAE)

Plants: Bluish green perennial herbs, 10–30 cm tall; stems erect or ascending, **hairless**, solitary or in clusters of 2–5; from **short, dark tuberous roots**, shallow and easily detached.

Leaves: Alternate, few; blades **elliptic to egg-shaped, usually widest above the middle**, blunt-tipped, **lacking distinct side veins**, 2–8 {10} cm long, **mostly 1½–4 times as long as wide**, hairless or stiff-hairy, sometimes fringed with hairs (ciliate), stalked on the lower stem to stalkless on the upper stem; **lower leaves soon withered**.

Flowers: Blue (occasionally white), **15–20 mm long**, with 5 rounded lobes at the tip of a long **tube 2–3 times as long as the flared upper part** (throat and limb), smooth inside; calyx hairless, 3–6 mm long, deeply cut into 5 lance-shaped lobes; stamens with broad, flat, 1.5–3 mm long filaments; styles extending past the anthers but included in or projecting slightly from the flower; numerous, **in compact, nodding clusters, elongating in fruit** (raceme-like), often with small flowering branches in the axils of nearby leaves; {April} May–June.

Fruits: Clusters of 4 small, erect nutlets, somewhat ridged, attached above the base to a convex receptacle.

Habitat: Open woods, moist slopes and meadows; elsewhere, on mesic, basaltic or sandy soils with sagebrush or ponderosa pine and along subalpine streams.

Notes: The common lungwort in Alberta, tall lungwort (*Mertensia paniculata* (Ait.) G. Don), has taller (usually 40–80 cm) plants with hairy stems, and its leaves have strong side veins.
• Lance-leaved lungwort (*Mertensia lanceolata* (Pursh) A. DC., also known as prairie lungwort) is a similar species that is also rare in Alberta. Some taxonomists include *M. lanceolata* in *M. longiflora*. These species are often difficult to distinguish, and both grow on slopes in the mountains in southwestern Alberta. However, lance-leaved lungwort has narrower (mostly

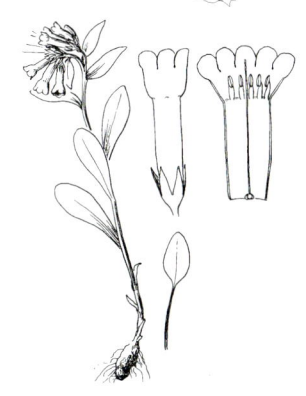

M. LANCEOLATA

Rare *Vascular* Plants of Alberta

oblong to lance-shaped) stem leaves, and the corolla tubes of its flowers are only slightly, if at all, longer than the throat and limb. It flowers from {May} June to July. • Most first-year plants produce only basal leaves. • The generic name *Mertensia* commemorates F.C. Mertens (1764–1831), a German botanist.

WESTERN FALSE GROMWELL

Onosmodium molle Michx. var. *occidentale* (Mack.) I.M. Johnston
BORAGE FAMILY (BORAGINACEAE)

Plants: Large, **loosely rough-hairy perennial herbs** with stiff, flat-lying and spreading hairs; **stems several, stout**, {30} 40–80 cm tall, branched above; from woody roots.

Leaves: Alternate, egg-shaped to broadly lance-shaped, 4–6 {3–8} cm long, with **5–7 prominent veins**, toothless, stalkless; lower leaves soon withered.

Flowers: Dull **yellowish white or greenish, tubular**, {10} **12–20 mm long**, tipped with 5 small (2–4 mm long), **erect, triangular lobes**; calyx 7–12 mm long, with 5 slender, densely stiff-hairy lobes; numerous, in leafy clusters with **dense, 1-sided, uncoiling branches** (scorpioid racemes or spikes); June–July.

Fruits: Shiny white or brownish nutlets, 2.5–3.5 {5} mm long, smooth or with a few scattered pits.

Habitat: Gravelly banks and dry, open woods.

Notes: This variety has also been called *Onosmodium occidentale* Mack. • People are reported to have been **poisoned** after drinking herbal teas containing this plant. • The generic name *Onosmodium* was taken from *Onosma*, a similar Eurasian genus. The specific epithet *molle* means 'soft,' perhaps in reference to the soft appearance of these hairy plants.

SMALL CRYPTANTHE, TINY CRYPTANTHE

Cryptantha minima Rydb.
BORAGE FAMILY (BORAGINACEAE)

Plants: Dwarf **annual herbs, 10–20 cm tall** in flower; stems branched, erect, bristly-hairy.

Leaves: Mostly **alternate, spatula-shaped, 5–15 mm long** and 2–3 mm wide, gradually becoming smaller higher up the stem.

Flowers: White, tiny, 2.5–3 mm long and a little over 1 mm across; **calyxes 2–2.5 mm long** in flower, enlarging to about 5 mm long in fruit, eventually shed, **covered with long, bristly hairs**; numerous, **in the axils of small bracts** on 1 side of slender, uncoiling branches, forming spreading clusters (cymes); {May–June}.

Fruits: Small clusters of 4 whitish nutlets of different shapes (heteromorphic), 1 smooth, 1.6–2 mm long and about 1.2 mm wide, the other 3 covered with tiny bumps, 1.2–1.5 mm long and 0.7–0.9 mm wide.

Habitat: Dry, eroded prairie slopes.

Notes: Another species that may occur in Alberta, slender cryptanthe (*Cryptantha affinis* (A. Gray) Greene), differs from small cryptanthe in having somewhat larger plants, 2–3 cm long linear to oblong leaves, slightly larger flowers and 4 nearly identical, egg-shaped nutlets. It is also reported to flower slightly later, in June and July. • A more common Alberta species, Fendler's cryptanthe (*Cryptantha fendleri* (A. Gray) Greene), produces clusters of 4 smooth, lance-shaped nutlets. It grows on sandy soil and dunes and blooms from May to July {August}. It is locally abundant in Alberta, but its populations seem to fluctuate greatly with changes in moisture conditions. It also requires eroding surfaces, so stabilization of dunes is a threat to some populations. • The generic name *Cryptantha* was taken from the Greek *krypto* (to hide) and *anthos* (flower) because several species have flowers that never open and appear hidden. The specific epithet *minima* is Latin for 'very small or least.'

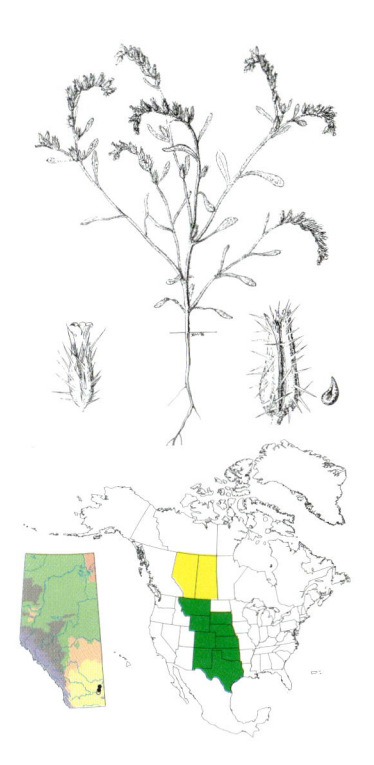

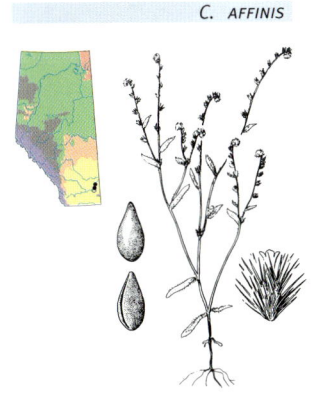

C. FENDLERI

C. AFFINIS

Dicots   BORAGE FAMILY

HOUND'S-TONGUE, WILD COMFREY

Cynoglossum virginianum L. var. *boreale* (Fern.) Cooperrider
BORAGE FAMILY (Boraginaceae)

Plants: Coarse-hairy perennial herbs; stems erect, unbranched, leafy, 40–80 cm tall; from taproots.

Leaves: Basal and alternate, with **many stiff, flat-lying hairs**; lower leaves tapered to long, slender stalks; upper leaves smaller, stalkless and clasping; blades elliptic, 7–20 cm long, 2–7 cm wide.

Flowers: Blue (sometimes white), **funnel-shaped**, with 5 spreading lobes at the tip of a short tube, {5} **6–8 mm across**; sepals {1} 2–3 mm long; numerous, in **leafless clusters** with 3–several **elongating, 1-sided**, spreading **branches** (false racemes); {June–July}.

Fruits: Clusters of **4 nutlets** {3.5} 4–5 mm long, covered with short, **barbed prickles**, borne on 5–15 mm long, down-turned stalks.

Habitat: Dry woods; elsewhere, sometimes in moist woods.

Notes: This species has also been included in *Cynoglossum boreale* Fern. • A somewhat similar plant, fringed stickseed (*Hackelia ciliata* (Dougl. *ex* Lehm.) I.M. Johnston), has larger (8–12 mm wide) blue flowers in more compact clusters and leaves that do not clasp the stem. Fringed stickseed is best identified by its nutlets, which are edged with broad prickles that are joined for at least ⅓–½ of their length and that curve upward to form a shallow cup. It grows on dry, open slopes, often in sagebrush and pine woodlands, and has been reported to occur in Alberta. It flowers from May to June. • The hooked spines on the nutlets catch on the fur, feathers or clothing of passers-by, effectively dispersing the seeds to new sites. • There have been **reports of deaths of horses** after feeding on hay containing the related species common hound's-tongue (*Cynoglossum officinale* L.), and cattle grazing on pasture with hound's-tongue are said to have been poisoned. It is also said to be **mildly poisonous to people**, and **the stiff hairs on its fuzzy leaves can irritate sensitive skin**. • According to early herbalists, a poultice of bruised hound's-tongue leaves was the only treatment needed to cure lesions from the bites of mad dogs. Hair loss from high fevers could also be controlled by massaging a salve, made from bruised leaves and swine's grease, into the scalp. • The bruised fresh roots of hound's-tongue have a

Rare *Vascular* Plants of Alberta

BORAGE FAMILY **Dicots**

heavy, unpleasant odour that was said to repel insects and rodents. • The rough surface of the nutlets, or possibly of the leaves of some species, was thought to resemble the surface of a dog's tongue—hence the common name 'hound's-tongue' and the generic name *Cynoglossum,* from the Greek *cynos* (dog) and *glossa* (tongue).

SPATULATE-LEAVED HELIOTROPE
Heliotropium curassavicum L.
BORAGE FAMILY (BORAGINACEAE)

Plants: Low, **succulent**, short-lived, **hairless** perennial herbs (usually annual elsewhere); stems with **spreading bases** (often on the ground) and ascending tips, 10–40 cm long, usually **coated with a whitish bloom** (glaucous); from weak, slender roots.

Leaves: Alternate, **fleshy, lance-shaped to spatula-shaped**, widest above middle, 2–4 {6} cm long, short-stalked, toothless; basal leaves reduced to scales.

Flowers: **White** (sometimes faintly bluish) **with a yellow eye**, 6–8 {5–9} mm across with 5 widely spreading lobes at the tip of a short tube; calyx 2–3 mm long; numerous, in branched clusters with **short, densely flowered, 1-sided, uncoiling branches** (scorpioid spikes) on **leafless stalks** up to 6 {10} cm long; June–July {August–September}.

Fruits: Clusters of **4 nutlets** 1.5–2 mm long, tardily splitting apart.

Habitat: Saline flats, often on dried lake beds; elsewhere, most common along the seacoast and in salt marshes.

Notes: The generic name *Heliotropium* was taken from the Greek *helios* (the sun) and *tropos* (to turn) because some species were said to open in the morning facing east and to gradually turn to the west, following the sun as the day progresses. Then, during the night, the same flower was said to turn east again to meet the rising sun the following morning. The specific epithet *curassavicum* means 'of Curaçao' (in the Caribbean Sea).

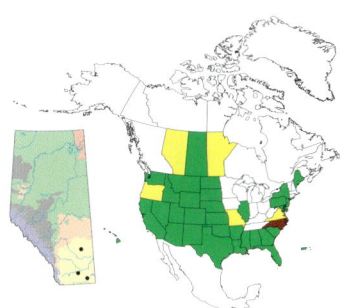

Rare *Vascular* Plants of Alberta

Dicots MINT FAMILY

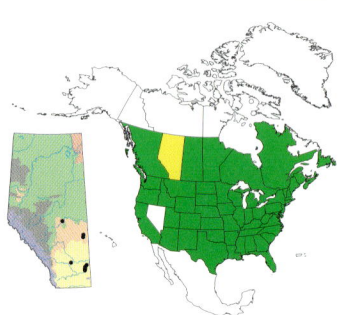

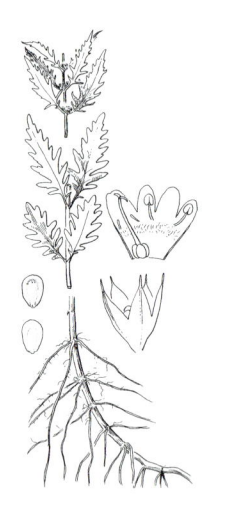

AMERICAN WATER-HOREHOUND

Lycopus americanus Muhl. *ex* W.C. Barton
MINT FAMILY (LAMIACEAE [LABIATAE])

Plants: Slender perennial herbs; stems erect, 20–80 cm tall, 4-sided, nearly hairless (hairy at the joints); from trailing runners (stolons) and slender underground stems (rhizomes).

Leaves: Opposite, spreading horizontally, lance-shaped, gradually tapered at the tip, abruptly narrowed at the base, 2–7 {8} cm long, **deeply pinnately cut**, short-stalked; upper leaves smaller, toothed to toothless.

Flowers: Bluish white, about 3 mm across, with 4 petals (lobes) at the tip of a short tube, hairy within; sepals fused into a cupped **calyx tipped with 4 equal, slender-pointed lobes**; numerous, **in dense whorls from upper leaf axils**; July {June–August}.

Fruits: Nutlets, 3-sided, **1–1.5 mm long**, 0.8–0.9 {1.2} mm wide, with a thickened, corky ridge along the side angles and across the tip, shorter than the calyx lobes.

Habitat: Marshy sites and moist, low ground along streams.

Notes: American water-horehound might be confused with a mint (*Mentha* spp.), but water-horehounds do not have the characteristic minty aroma. • Plants of this genus are called water-horehounds because they grow in wet, marshy environments and resemble true horehound (*Marrubium vulgare* L.), a European mint that is also a common introduced weed in much of North America. • The generic name *Lycopus* was taken from the Greek *lycos* (wolf) and *pous* (foot) as a translation of the French common name for these plants, *patte du loup*.

FALSE DRAGONHEAD

Physostegia ledinghamii (Boivin) Cantino
MINT FAMILY (LAMIACEAE [LABIATAE])

Plants: Perennial herbs, **hairless or with very fine hairs**; stems **4-sided**, erect, 30–90 {20–100} cm tall, unbranched or branched at the base of the flower cluster; from an underground stem (rhizome) or rhizome-like base with fibrous roots.

Leaves: Opposite, 3–10 cm long, 5–20 mm wide, **elliptic–lance-shaped to linear-oblong**, pointed, **stalkless**, edged with **blunt or hooked teeth** (sometimes almost toothless); lower leaves soon withered.

Flowers: **Rose-pink to lavender-purple**, showy, 8–16 mm long, **funnel-shaped**, enlarged at the mouth and tipped with an **erect, hooded upper lip** and a **spreading, 3-lobed lower lip**; calyx somewhat inflated, 4–6 mm long, 5-toothed, 10-nerved, **with short, fine, gland-tipped hairs**; 4 stamens, under the upper lip; several to many, almost stalkless, in the axils of **small, egg-shaped bracts**, in **elongated, spike-like clusters** (racemes) at the stem tip and sometimes also from the axils of upper leaves; {July–September}.

Fruits: Smooth, ovoid nutlets.

Habitat: Moist woods and streambanks; elsewhere, on lake shores and in marshes.

Notes: This species was once included with *Physostegia parviflora* Nutt. *ex* A. Gray, *Physostegia virginiana* (L.) Benth. var. *ledinghamii* Boivin and *Dracocephalum nuttallii* Britt. It is not synonymous with *Dracocephalum parviflorum* Nutt. (also known as *Moldavica parviflora* (Nutt.) Britton), which is American dragonhead.

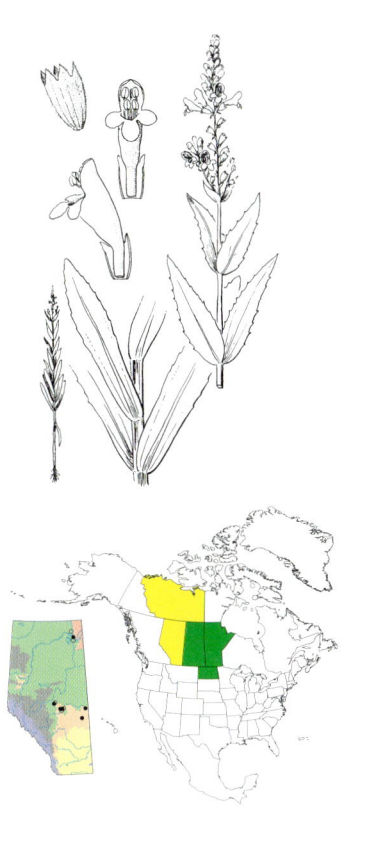

Dicots FIGWORT FAMILY

SHRUBBY BEARDTONGUE, SHRUBBY PENSTEMON

Penstemon fruticosus (Pursh) Greene var. *scouleri* (Lindl.) Cronq.
FIGWORT FAMILY (SCROPHULARIACEAE)

Plants: Low, bushy semi-shrub, {5} 10–40 cm tall; stems freely branched, often reddish and brittle, glandular-hairy near top, forming **dense clumps from a woody base.**

Leaves: **Opposite** and in basal clusters, narrowly **lance-shaped**, 2–5 {6} cm long (much smaller upward on flowering stems), **dark green and hairless above**, saw-toothed to nearly toothless.

Flowers: **Bluish lavender to pale purple**, 3.5–4.5 {3–5} **cm long, tubular, 2-lipped**, long-white-hairy on lower lip; calyx with dense, glandular hairs on slender, pointed lobes; anthers long-woolly; borne **in pairs in short clusters** (racemes) at branch tips; June–July {August}.

Fruits: Capsules, 8–12 mm long.

Habitat: Dry, rocky slopes and open woods in subalpine and alpine zones.

Notes: Shrubby beardtongue could be confused with elliptic-leaved beardtongue (*Penstemon ellipticus* Coult. & Fisher), a low (5–15 cm tall), mat-forming (rather than bushy) species with broader (elliptic to egg-shaped) leaves. • The leaves and flowers of shrubby beardtongue were added to cooking pits with wild onions (*Allium* spp.) and the roots of arrow-leaved balsamroot (*Balsamorhiza sagittata* (Pursh) Nutt.). • The stems, flowers and leaves were soaked or boiled to make an eyewash and a bath for relieving rheumatism. The boiled tea was taken to relieve sore backs, ulcers and kidney problems and to 'clean out your insides.' It was also used as a wash for treating arthritis, aches and sores (on both humans and horses). • Bees and hummingbirds pollinate these long, tubular flowers. • The generic name *Penstemon* was derived from the Latin *pente* (five) and *stemon* (thread), referring to the 5 (4 fertile plus 1 sterile) stamens of these flowers. Penstemons are called beardtongues because of their bearded sterile stamen.

WATER SPEEDWELL

Veronica catenata Pennell
FIGWORT FAMILY (SCROPHULARIACEAE)

Plants: **Semi-aquatic** perennial herbs; stems stout, **succulent**, **hairless** or with a few fine, glandular hairs in the flower cluster, more or less erect, 40–80 {20–100} cm tall, with ascending branches; from creeping underground stems (rhizomes) with fibrous roots.

Leaves: **Opposite, narrowly elliptic to oblong**, 2–7 {10} cm long, {2.5} 3–5 times as long as wide, toothless or sharply toothed, stalkless, **clasping** the stem.

Flowers: **White to pink or pale bluish**, about 5 mm across, with a short, inconspicuous tube tipped with **4 broad, nearly equal, widely spreading lobes** and with 2 stamens projecting from the centre; **calyxes** deeply cut into **4 lobes**, highly variable; numerous, on slender 3–8 mm long stalks in loose, elongating, 4–12 cm long, branched clusters (racemes) **from upper leaf axils**, essentially hairless; July {June–September}.

Fruits: Slightly heart-shaped capsules, 2.5–4 mm long, mostly wider than high, shallowly notched at the tip, borne on slender, widely spreading stalks; seeds numerous, up to 0.5 mm long.

Habitat: Marshy, muddy or gravelly ground and shallow water, by ponds and streams and in ditches; elsewhere, by springs and spring-fed streams.

Notes: This species includes *Veronica comosa* Richt. var. *glaberrima* (Pennell) Boivin and *Veronica salina* auct. non Schur. It is sometimes included in *Veronica anagallis-aquatica* L. • Water speedwell might be confused with marsh speedwell (*Veronica scutellata* L.), but marsh speedwell has linear leaves that do not clasp the stem with their bases. • Another Alberta speedwell, thyme-leaved speedwell (*Veronica serpyllifolia* L. var. *humifusa* (Dickson) Vahl), has slightly smaller (5–8 mm wide) flowers in elongated clusters at the stem tips only (not from leaf axils). Its lower leaves are stalked and its capsules are deeply notched and covered with glandular hairs. Thyme-leaved speedwell blooms in June

V. SERPYLLIFOLIA

{May–August}. It grows on moist, montane slopes in Alberta and in moist meadows and roadside ditches elsewhere. • The generic name *Veronica* comes from the Latin *vera iconica* (true likeness). The flowers of some species were said to bear markings similar to those that were left on the cloth used by St. Veronica to wipe the face of Jesus as he carried the cross.

WATER HYSSOP

Bacopa rotundifolia (Michx.) Wettst.
FIGWORT FAMILY (SCROPHULARIACEAE)

Plants: **Succulent, aquatic** perennial herbs; stems branched, 10–40 cm long, floating, bearing spreading hairs at the tips; from fibrous roots.

Leaves: **Opposite**, hairless, with several **nerves in a fingerlike pattern**, nearly round to broadly egg-shaped and widest above the middle, 1–3 cm long, stalkless.

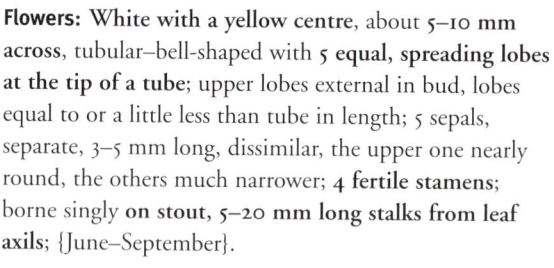

Flowers: **White with a yellow centre**, about 5–10 mm across, tubular–bell-shaped with **5 equal, spreading lobes at the tip of a tube**; upper lobes external in bud, lobes equal to or a little less than tube in length; 5 sepals, separate, 3–5 mm long, dissimilar, the upper one nearly round, the others much narrower; **4 fertile stamens**; borne singly **on stout, 5–20 mm long stalks from leaf axils**; {June–September}.

Fruits: **Capsules**, splitting open lengthwise, containing many seeds.

Habitat: Mud-bottomed pools.

Notes: The generic name *Bacopa* is the Latin form of an aboriginal name that was used for a plant of this genus by natives of French Guiana.

FIGWORT FAMILY　　　　**Dicots**

FIELD TOADFLAX, BLUE TOADFLAX

Nuttallanthus canadensis (L.) D.A. Sutton
FIGWORT FAMILY (SCROPHULARIACEAE)

Plants: Slender, bluish green annual or winter annual herbs, essentially hairless, 10–60 cm tall, with **a single to several erect**, unbranched, sparsely leafy **stems** and also a **spreading rosette of short, low-lying stems** with ascending tips; from short taproots.

Leaves: Mostly **paired or in 3s on lower stems**, mostly **alternate on upper erect stems, linear**, 1–3.5 cm long and 1–2.5 mm wide on erect stems, shorter and wider on low-lying stems.

Flowers: **Light blue, funnel-shaped**, with a small, 2-lobed upper lip, a **broad, paler, 3-lobed lower lip** and a **slender tube** (spur) extending down and **curved forward** from the base, 8–12 mm long (excluding the 2–9 mm long spur); lower lip paler, scarcely raised; several, in leafless, elongating clusters (racemes); {April–June}.

Fruits: Broadly ellipsoid capsules, 2.5–4 mm long, with many angled seeds.

Habitat: Moist, sandy sites.

Notes: This species has also been called *Linaria canadensis* (L.) Dum.-Cours.

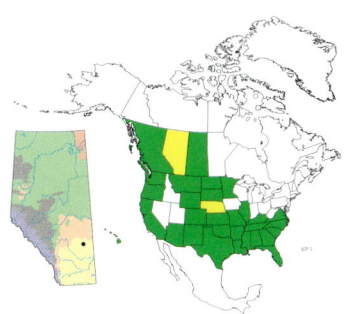

FIGWORT FAMILY

SMOOTH MONKEYFLOWER

Mimulus glabratus Kunth
FIGWORT FAMILY (SCROPHULARIACEAE)

Plants: Low, **diffusely spreading** perennial herbs; stems weak, hairless or inconspicuously hairy, **creeping or floating**, often **rooting at the joints** (nodes), 5–60 cm long, leafy; from underground stems (rhizomes).

Leaves: **Opposite**, broadly egg-shaped to **round** to somewhat kidney-shaped, 1–4 cm long, **veiny**, with 3–7 fingerlike (palmate) nerves, toothless or edged with small, irregular teeth, **nearly stalkless**.

Flowers: **Bright yellow**, sparingly (if at all) red-spotted, {10} **15–25 mm long**, **tubular**, with an **open throat and 2 broad lips**, the upper lip 2-lobed and the lower lip fuzzy and 3-lobed; **calyx** 5-angled, 5-toothed, irregular, **with a large upper tooth** and short, blunt side and lower teeth, **inflated in fruit**, 5–12 mm long; 4 anther-bearing stamens; few, long-stalked, in the axils of upper leaves; {May–August}.

Fruits: Many-seeded capsules, inside inflated calyxes.

Habitat: Wet places, often in water and around springs.

Notes: Smooth monkeyflower might be confused with yellow monkeyflower (*Mimulus guttatus* DC., p. 197), but yellow monkeyflower is a larger, more erect plant with larger flowers (25–45 mm long) and larger mature calyxes (12–17 mm long). • Small yellow monkeyflower (*Mimulus floribundus* Dougl. *ex* Lindl.) is another rare Alberta species. It is a small (5–25 cm tall), usually sticky-hairy annual plant with small (7–15 mm long), only slightly 2-lipped flowers, which have sharply angled calyxes with 5 nearly equal teeth. Small yellow monkeyflower grows on moist banks in montane areas, and it flowers in July {May–October}. • The generic name *Mimulus* was derived from the Latin *mimus* (a mimic or actor) because these flowers were thought to resemble small, grinning faces.

M. FLORIBUNDUS

FIGWORT FAMILY **Dicots**

YELLOW MONKEYFLOWER
Mimulus guttatus DC.
FIGWORT FAMILY (SCROPHULARIACEAE)

Plants: Highly variable annual or perennial herbs; stems stout and erect or weak and trailing, 20–60 {5–80} cm tall, often rooting at joints (nodes); from fibrous roots (annual) or stolons (perennial).

Leaves: Opposite, oval or egg-shaped, widest above the middle, 1–5 cm long, **coarsely and unevenly wavy-toothed, 3–7-veined**; lower leaves stalked; upper leaves stalkless and clasping.

Flowers: Bright yellow, snapdragon-like, trumpet-shaped and 2-lipped, {1} 2–4 cm long but usually less than 2 cm when few in number, slender-stalked; lower lip larger than upper, **hairy and spotted with crimson** or reddish brown; sepals quite unequal, fused, 1–2 cm long; few to several (usually more than 5), in loose clusters (racemes) on slender stalks; {April–June} July–August.

Fruits: Broadly oblong capsules, 1–2 cm long, contained **in inflated 'balloons' of fused sepals**.

Habitat: Wet meadows, springs and streambanks.

Notes: Large mountain monkeyflower (*Mimulus tilingii* Regel) is also rare in Alberta. It is very similar to yellow monkeyflower (some taxonomists include these taxa in the same species), but its plants are low-growing (less than 20 cm tall) with many sod-forming underground stems (rhizomes) (sometimes with stolons too), and its flowers are few (usually 1–5) and large (2–4 cm long). It grows in wet alpine sites, where it flowers in August {July–September}.
• Yellow monkeyflower plants were eaten raw or cooked by native peoples and early settlers in the Rocky Mountains, but they are bitter when raw. The stems and leaves were also applied to burns and wounds as poultices, to speed healing and stop bleeding.
• The 2 rounded lobes of the stigma are initially widespread, but when they are brushed by a pollen-laden insect, they immediately fold together (like the pages of a book), holding the pollen firmly. This prevents self-pollination when the same insect backs out of the flower.

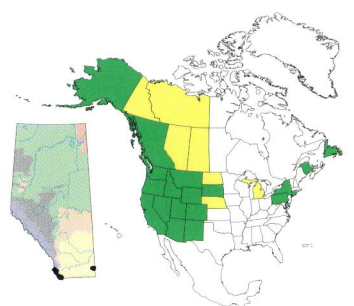

M. TILINGII

Dicots FIGWORT FAMILY

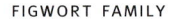

Brewer's monkeyflower

Mimulus breweri (Greene) Coville
FIGWORT FAMILY (SCROPHULARIACEAE)

Plants: Delicate, often reddish annual herbs with copious glandular hairs; stems slender, **less than 15 cm tall**, unbranched or sparingly branched; from weak taproots.

Leaves: Opposite, long and narrow, 1–2 cm long, 1–4 mm wide, faintly 3-nerved, entire, stalkless.

Flowers: Red to purple, often with **yellow markings in the throat**; funnel-shaped, **5–10 mm long**, somewhat 2-lipped, with 5 small, nearly equal petal lobes at the tip of a slender tube; calyx tube glandular-hairy, tipped with 5 short, nearly equal teeth; {June–August}.

Fruits: Small, ascending capsules, splitting in half to release numerous seeds.

Habitat: Moist slopes and meadows; elsewhere, on moderately dry slopes.

Notes: Red monkeyflower (*Mimulus lewisii* Pursh) is the only other red-flowered monkeyflower that occurs in Alberta. It is a 30–70 cm tall perennial with large (3–5 cm long) flowers and grows in wet places in the mountains.
• Brewer's monkeyflower was first discovered in Alberta in 1997 and is known only from Waterton Lakes National Park. • The specific epithet *breweri* commemorates W.H. Brewer, an American naturalist who lived 1828–1910.

FIGWORT FAMILY **Dicots**

LEAFY LOUSEWORT, SICKLETOP LOUSEWORT

Pedicularis racemosa Dougl. *ex* Hook.
FIGWORT FAMILY (Scrophulariaceae)

Plants: Perennial herbs, essentially hairless; stems 15–50 cm tall, **unbranched**, clumped; from fibrous roots.

Leaves: Alternate, **lance-shaped to linear**, 3–8 cm long, largest near middle of stem, **coarsely round-toothed**, often reddish.

Flowers: White, sometimes tinged pink, 12–15 mm long, 2-lipped, with a **broad, 3-lobed lower lip and an arching upper lip tapered to a sickle-shaped, down-curved beak**; calyx 2-lobed; few flowers, from leaf axils and in long, loose clusters (racemes); {June} July–August.

Fruits: Smooth, flattened, curved capsules.

Habitat: Dry, open subalpine slopes; elsewhere, in coniferous montane stands and on alpine slopes.

Notes: A more common species, coil-beaked lousewort (*Pedicularis contorta* Benth. *ex* Hook.), is very similar to leafy lousewort, but its calyx is 5-lobed and its leaves are deeply cut into narrow, toothed leaflets, giving them a fern-like appearance. • Large-flowered lousewort (*Pedicularis capitata* J.E. Adams) is a small (5–15 cm tall) species with compact, head-like clusters of relatively large (25–30 mm long), white to faintly pinkish flowers with broad, square-tipped (beakless) hoods. Its small leaves are finely pinnately divided (rather fern-like), and its flowering stems are usually leafless. Large-flowered lousewort is also rare in Alberta, where it grows on calcareous alpine slopes and flowers from June to August. • Louseworts are said to be edible in an emergency, raw or cooked, but they contain enough glycosides to **cause severe illness** if they are eaten in quantity. • The generic name *Pedicularis* from the Latin *pedis* (a louse), and the common name lousewort both come from the early belief that animals that ate these plants were more likely to be infested with lice. Most animals do not eat these plants, and therefore it is more likely that such livestock were grazing in poor pastures, where they were more likely to become weakened and louse-ridden.

P. CAPITATA

Dicots

FIGWORT FAMILY

PURPLE RATTLE

Pedicularis sudetica Willd. ssp. *interioides* Hult.
FIGWORT FAMILY (Scrophulariaceae)

Plants: Slender perennial herbs, **hairless to sparsely soft-hairy**; stems erect, unbranched, {5} 10–40 cm tall, with 0–2 {3} small leaves, single or in small clumps; from **short, branched** underground stems (rhizomes).

Leaves: Mainly basal, linear to lance-shaped, **feather-like**, **pinnately divided** into narrow, toothed lobes, long-stalked.

Flowers: Rose-pink to crimson-purple, showy, **15–24 mm long**, strongly 2-lipped, with a **slender, purple, arched, 6–13 mm long upper lip** (galea) **tipped with 2 tiny, slender teeth** above a pink or spotted, **broad, 3-lobed lower lip**; calyxes 8–12 mm long, tipped with 5 slender, pointed lobes, woolly to hairless; **numerous, in woolly to soft-hairy spikes**, compact and head-like at first but elongating in fruit, often with 1 or 2 leafy bracts at the base; July {August}.

Fruits: Slightly flattened, **oblong capsules**, 9–14 mm long, tipped with a **short point**, containing many rough-textured seeds.

Habitat: Fens; elsewhere, in wet, calcareous tundra and on lakeshores.

Notes: Two other rare species, arctic lousewort (*Pedicularis langsdorfii* Fisch. *ex* Stev. ssp. *arctica* (R. Br.) Pennell, also called *Pedicularis arctica* R. Br.) and woolly lousewort (*Pedicularis lanata* Cham. & Schlecht., p. 201), have leafy stems and stout, yellowish taproots (rather than rhizomes). • Louseworts have typical bee-pollinated flowers, and their range coincides with that of bumblebees. All louseworts are partial parasites, capable of producing their own food, but also of stealing nutrients from their neighbours through sucker-like attachments on their roots. Once this relationship is established, the plants are very difficult to transplant.

P. LANGSDORFII

FIGWORT FAMILY　　　　**Dicots**

WOOLLY LOUSEWORT

Pedicularis lanata Cham. & Schlecht.
FIGWORT FAMILY (SCROPHULARIACEAE)

Plants: Stout, **woolly** perennial herbs; stems erect, unbranched, 5–15 {40} cm tall, usually densely white-woolly and leafy; from **bright lemon-yellow taproots**.

Leaves: Alternate, **crowded**, linear to lance-shaped, feather-like, **pinnately divided** into narrow, toothed lobes.

Flowers: Rose-pink to lavender (rarely white), showy, faintly scented, **18–20** {15–25} **mm long**, strongly 2-lipped, with a narrow, **straight or slightly arched, 3–5** {8} **mm long upper lip** (galea) above a **broad, 2–5 mm long, 3-lobed lower lip**; calyx 4–5 mm long, tipped with 5 pointed, triangular lobes 0.5–1 mm long; **numerous, in dense, copiously woolly, elongating spikes**; June–July.

Fruits: Slightly flattened, pointed, egg-shaped capsules, 8–13 mm long; seeds large, with loose, ashy-grey, honeycombed seed coats.

Habitat: Rocky alpine slopes; elsewhere, in moist, stony tundra.

Notes: Woolly lousewort might be confused with another rare Alberta species, arctic lousewort (*Pedicularis langsdorfii* Fisch. *ex* Stev. ssp. *arctica* (R. Br.) Pennell, also called *Pedicularis arctica* R. Br., p. 200), but arctic lousewort is much less woolly (often almost hairless) and the upper lip of its flower is tipped with 2 slender teeth which are not found on woolly lousewort flowers. Arctic lousewort grows on moist alpine slopes and flowers in July {June–August}. • Purple rattle (*Pedicularis sudetica* Willd., p. 200) has flowers very similar to those of woolly lousewort, but its stems are hairless and almost leafless. • Northern native peoples ate the shoots and young flowering stems raw, cooked or mixed in fat and stored. The roots have been likened to carrots, and they were eaten raw, roasted or boiled. Children enjoyed sucking nectar from the base of the long flower tubes. • Some tribes used the roots as a tobacco substitute or boiled them to make a yellow dye.

P. LANGSDORFII

Dicots

FIGWORT FAMILY

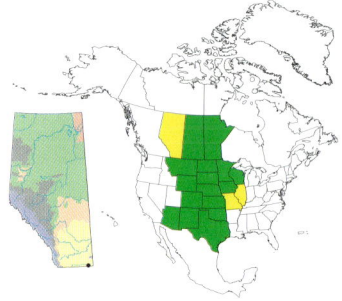

DOWNY PAINTBRUSH, DOWNY PAINTED-CUP

Castilleja sessiliflora Pursh
FIGWORT FAMILY (SCROPHULARIACEAE)

Plants: Loosely clumped perennial herbs, **ashy-grey with silky and woolly hairs**; stems erect or ascending, 10–40 cm tall, sometimes branched near the base; from branched root crowns.

Leaves: Alternate, 2–4 cm long, linear and undivided on the lower stem, **broader and tipped with 3–5 slender, spreading lobes on the upper stem**, densely covered with short, often woolly hairs.

Flowers: Pale yellow, sometimes pinkish or purplish, {4} **4.5–5.5 cm long**, projecting well past the bracts, with a **long, arched tube tipped with a 1 cm long upper lip** (galea) **and a prominent, 3-lobed lower lip** 3–5 mm long, covered with tiny, often gland-tipped hairs; calyx yellowish, minutely glandular-hairy, tubular, 2.5–4 cm long, tipped with 4 slender, almost equal lobes 5–15 mm long; bracts leafy, green, mostly cut to below the middle into 3 slender lobes, short-hairy or woolly; **4 stamens**, enclosed in the upper lip; styles slender, usually protruding from the tip of the upper lip; numerous, in the axils of bracts in dense spikes that elongate in fruit; June {May–July}.

Fruits: Capsules, splitting open lengthwise, containing many seeds with netted veins; {May–July}.

Habitat: Dry mixed prairie grasslands, on dry, sandy soils.

Notes: Downy paintbrush resembles stiff yellow paintbrush (*Castilleja lutescens* (Greenm.) Rydb., included in *Castilleja septentrionalis* Lindl. in the first edition of the *Flora of Alberta*, but not a synonym of that species), but stiff yellow paintbrush has straighter (not arched), shorter (less than 2 cm) flower tubes, branched stems, spreading leaves and short, stiff hairs. It grows on grassy slopes in Alberta (also in open coniferous woods elsewhere) and is reported to bloom from May to August.

FIGWORT FAMILY   **Dicots**

PALE GREENISH PAINTBRUSH

Castilleja pallida (L.) Spreng. ssp. *caudata* Pennell
FIGWORT FAMILY (SCROPHULARIACEAE)

Plants: Slender perennial herbs; stems 20–40 cm tall, single or in small clumps, hairless or sometimes finely hairy on the upper parts and rather coarsely hairy in the flower cluster; from short, weak taproots.

Leaves: Alternate, **narrowly lance-shaped** (not lobed), 3–9 cm long, hairless on the upper surface, finely hairy beneath; upper leaves up to 1 cm wide at the base, **tapered to slender tails** at the tip.

Flowers: Yellowish to yellowish green, slender, **tubular, 16–26 mm long, 2-lipped, almost hidden in the axils of showy greenish to yellowish bracts**; bracts simple (rarely cut into 3–5 slender lobes at the tip); **calyx tubular, 4-lobed, cut at least halfway to the base** on the sides; **corolla tubular**, slightly longer than the calyx, with a slender, greenish, **3.5–8 mm long upper lip** and a broad, toothed, yellowish lower lip about ⅓ as long as the upper lip; 5–12, in elongating clusters (racemes); June–July.

Fruits: Capsules, 7–10 mm long, splitting lengthwise; seeds numerous, wedge-shaped, 1.5–2 mm long, with coarse, netted ridges.

Habitat: Moist tundra, lakeshores; in Alaska, on streambanks, terraces and bars, and in woods and thickets.

Notes: Pale greenish paintbrush might be confused with stiff yellow paintbrush (*Castilleja lutescens* (Greenm.) Rydb., included in *Castilleja septentrionalis* Lindl. in the first edition of the *Flora of Alberta*, but not a synonym of that species), but the flower clusters of stiff yellow paintbrush have 3–7-lobed bracts and the leaves are very slender (linear), densely short-hairy to short-bristly and rough to the touch.
• Paintbrush plants commonly join roots with neighbouring plants to absorb nutrients.

Dicots   BROOMRAPE FAMILY

Russian ground-cone

Boschniakia rossica (Cham. & Schlecht.) Fedtsch.
BROOMRAPE FAMILY (OROBANCHACEAE)

Plants: Fleshy, **brown, cone-like parasitic herbs,** hairless (or nearly so); stems 10–40 cm tall, 4–15 mm thick, usually yellowish or purplish, single or few together; from 15–25 mm thick, **tuber-like underground stems** (rhizomes).

Leaves: Alternate, overlapping, **yellowish to purplish or brown, scale-like,** 3–10 mm long, triangular to egg-shaped, smooth to fringed along edges.

Flowers: Brownish red to purplish, 8–13 {15} mm long, with 5 lobes forming **2 unequal fringed lips**, stalkless, in the axils of fringed bracts; calyxes 3–6 mm long, irregularly lobed, fringed with short hairs; numerous, in **dense spike-like clusters** (racemes) **6–25 cm long**; {June} July.

Fruits: Capsules, 6–7 mm long, splitting open irregularly; seeds very small (0.001 mg) and numerous (over 300,000 per plant).

Habitat: Beside, and parasitic on, green alder in open, wooded or shrubby areas.

Notes: Russian ground-cone lacks chlorophyll and cannot produce food through photosynthesis. Instead, it parasitizes green alder shrubs, drawing all of its nutrients from their roots. The tiny seeds are swept into cracks in the soil, where they grow quickly, in search of a host plant. If they are lucky enough to come into contact with an alder root, they join with it and begin to develop a tuber and eventually a cone-like flowering stalk. The drain of reserves from year to year makes the end of the host root die off at an early age, so established tubers are usually attached to root tips. Plants may grow underground for 4–5 years before sending up a flower spike. • The tubers are often eaten by bears.

BROOMRAPE FAMILY · Dicots

ONE-FLOWERED CANCER-ROOT

Orobanche uniflora L.
BROOMRAPE FAMILY (OROBANCHACEAE)

Plants: Delicate parasitic herbs; 1–3 flowering stalks per plant, slender, **5–10 {3–20} cm tall**, covered with fine, glandular hairs; from **soft, 1–5 cm long underground stems.**

Leaves: Inconspicuous, scale-like, alternate near base, lance-shaped, up to 10 mm long.

Flowers: Purplish to yellowish white, 15–25 {35} mm long, 2-lipped, with rounded, minutely fringed lobes at the tip of a curved tube; calyx 4–12 mm long, with 5 slender, tapered lobes; single, on erect, slender stalks; {April–May} June–August.

Fruits: Capsules, splitting lengthwise when mature.

Habitat: Moist woods, with stonecrops (*Sedum* spp.), saxifrages (*Saxifraga* spp.), species of the aster family (Asteraceae) and other vascular plants; elsewhere, in open, moist to dry sites or woods.

Notes: Another rare Alberta species, Louisiana broomrape (*Orobanche ludoviciana* Nutt.), produces dense, 10–25 cm tall spikes of purplish (sometimes yellowish), stalkless flowers. These sticky, glandular-hairy plants are parasitic on sageworts (*Artemisia* spp.) and other members of the aster family (Asteraceae) in the prairies, and they flower in July–August {September}. • Broomrapes lack chlorophyll and cannot produce their own food through photosynthesis. Instead, they join to the roots of host plants, which supply all of their nutrients. • The generic name *Orobanche* was taken from the Greek *orobos* (a clinging plant) and *ancho* (to strangle), in reference to the parasitic roots of these plants.

O. LUDOVICIANA

Rare *Vascular* Plants of Alberta

Dicots

BLADDERWORT FAMILY

SMALL BUTTERWORT, HAIRY BUTTERWORT
Pinguicula villosa L.
BLADDERWORT FAMILY (LENTIBULARIACEAE)

Plants: Small, **insectivorous** perennial herbs; flowering stems 2–5 cm tall, with long, soft hairs near the base and **glandular hairs above**; forms winter buds.

Leaves: Basal, egg-shaped to elliptic, up to 1 cm long, fleshy, upper surface covered with **greasy-slimy glands**, **yellowish green**, **rolled upward** along the edges.

Flowers: Purple to bluish violet, violet-like, 7–8 {6–10} **mm long** (including spur), funnel-shaped, with a 2-lobed upper lip and a broad, 3-lobed lower lip at the mouth and a straight, nectar-producing spur extending back from the base; calyxes with a 3-lobed upper lip and a 2-lobed lower lip; **single**, nodding on erect, slender stalks; mid June–July.

Fruits: Erect, round capsules, 3–5 mm long, splitting into 2 parts, containing many small seeds.

Habitat: Sphagnum hummocks in peatlands.

Notes: There is only 1 other species of butterwort in Alberta, common butterwort (*Pinguicula vulgaris* L.). Common butterwort is easily distinguished by its hairless flowering stalks and larger (15–20 mm long) flowers.
• Butterwort leaves are greasy to the touch because of the droplets of mucilage produced by stalked glands on the upper surface. When small insects land on the leaf, they are restricted by this glue-like substance, and their movement results in the production of more mucilage. This leads to the eventual death of the trapped insect by suffocation and its digestion by acid and enzymes that are also secreted by the leaves. The leaves also change shape to entrap their prey. Their edges roll over the prey, and the surface becomes dished to hold secreted liquids. These changes are slow (inward rolling of leaves can take as long as 2 days), but they help to cover the insect in glands and fluids while preventing it from being washed off the leaf during periods of rain. • The common name 'butterwort' and the scientific name *Pinguicula,* from the Latin *pinguis* (fat), both refer to the greasy or 'buttery' feel of the leaves. The specific epithet *villosa* is Latin for 'hairy' or 'shaggy,' in reference to the hairy flowering stalks.

BLADDERWORT FAMILY

Dicots

HORNED BLADDERWORT

Utricularia cornuta Michx.
BLADDERWORT FAMILY (LENTIBULARIACEAE)

Plants: Semi-aquatic perennial herbs; **stems mainly underground**, root-like, delicate, finely branched, bearing specialized leaves; **flowering stems erect**, above ground, **10–25 {3–30} cm tall**, leafless, straight, wiry.

Leaves: Alternate, **thread-like**, small, **with tiny bladders** (specialized leaflets) along the edges, mostly **underground** so seldom seen.

Flowers: **Bright yellow**, fragrant, **15–20 mm long** and wide, with an egg-shaped upper lip (widest toward the tip), a **large, protruding, helmet-shaped lower lip** at the mouth, and a **slender, 10–12 mm long spur** projecting back and down from the base; calyx 2-lipped, much smaller than the corolla; 1–3 {6}, short-stalked, each in the axil of 3 small bractlets, forming small, elongating clusters (racemes); June–July {August}.

Fruits: Capsules covered by persistent, beaked calyxes.

Habitat: Poor fens, muddy shores and calcareous wetlands.

Notes: Horned bladderwort is seldom seen, except when in flower. Its inconspicuous leaves, buried in the moist soil, and its yellow, long-spurred flowers distinguish it from other Alberta species. • Bladderworts are carnivorous plants. They obtain nutrients by trapping and digesting tiny animals. The small bladders have a trapdoor, surrounded by stiff hairs. When a small invertebrate brushes against these sensitive hairs, the trapdoor springs open and water rushes in, carrying the prey with it. Once the bladder is full, the door closes, and digestive enzymes and acids are secreted, beginning the process of digestion. As the prey is broken down, specialized cells absorb water from the bladder, sending nutrients to the plant. This process takes about 30–120 minutes. When it is completed, a partial vacuum is recreated in the bladder, and the trap is reset. • In the east, horned bladderwort has been observed growing in water with a pH of 5.6–6.9. • The generic name *Utricularia* comes from the Latin *utriculus* (a small bottle), in reference to the tiny bladders, which trap small aquatic animals. The specific epithet *cornuta* is Latin for 'horned,' a reference to the stiff spur at the base of each flower.

Rare Vascular Plants of Alberta

WESTERN RIBGRASS

Plantago canescens J.E. Adams
PLANTAIN FAMILY (Plantaginaceae)

Plants: **Tufted perennial herbs**, soft-hairy with **segmented** (jointed) **hairs**; flowering **stems leafless**, **woolly-hairy**, 10–15 {8–30} cm tall, longer than the leaves; from stout crowns on thick, elongated taproots.

Leaves: In **basal clusters**, **lance-shaped** (sometimes oblong to linear), 5–15 {25} cm long, 3–5 nerved, toothless or irregularly toothed.

Flowers: White or greenish, inconspicuous, **about 2 mm long**, with 4 spreading, 2 mm long, minutely fringed lobes, stalkless, in the axils of broadly egg-shaped, 2 mm long, scale-like bracts with translucent edges; stamens sometimes conspicuous, up to 7 mm long; numerous, in **dense, 3 (to 12) cm long spikes**, elongating in fruit, lower flowers sometimes more widely spaced; June.

Fruits: **Oblong capsules**, splitting horizontally below the middle (circumscissile), freeing a blunt, **cupped lid** and releasing 2–4 black, finely pitted seeds, each 1–1.5 mm long.

Habitat: Grassy and gravelly slopes, on non-alkaline soil; elsewhere, on steep open slopes, riverbanks, eskers and scree slopes.

Notes: This species has also been called *Plantago septata* Morris *ex* Rydb. • Western ribgrass could be confused with the more common species saline plantain (*Plantago eriopoda* Torr.), but saline plantain has wider leaves, yellowish brown or reddish wool on its root crowns (around the leaf bases) and sparingly hairy stems and leaves. • The seed coat of most plantains is high in mucilage, and it swells and becomes slimy when wetted. Plantain seeds have been used for many years as a bulk laxative, and the clear, sticky gum they exude has been used in the manufacture of lotions and hair-wave products.

SEASIDE PLANTAIN

Plantago maritima L.
PLANTAIN FAMILY (PLANTAGINACEAE)

Plants: Perennial herbs, essentially hairless; **stems** (flowering stalks) **leafless**, 5–25 cm tall, usually slightly taller than the leaves; from **long, thick taproots** with branched crowns.

Leaves: Basal, fleshy, linear to lance-shaped, up to 22 cm long and 1.5 cm wide, with **3–5** (sometimes 6 or 7) **parallel nerves**, smooth-edged or with a few small teeth.

Flowers: White or greenish, with **4 dry, translucent**, 1–1.5 mm long, **spreading lobes** at the tip of a **hairy tube**; calyx with 4 lobes rounded on the back, all alike or 2 long and 2 short, fringed with hairs; bracts broadly egg-shaped, 1.5–6 mm long, about as long as the sepals; numerous, in **1–10 cm long spikes**; June {July–August}.

Fruits: Egg-shaped or cone-shaped **membranous capsules**, with the **top coming off like a lid** (circumscissile); seeds 1.5–3 mm long; seed coat mucilaginous when wet.

Habitat: Saline marshes and alkaline soil.

Notes: Linear-leaved plantain (*Plantago elongata* Pursh) grows on alkaline prairie. It is a smaller (up to 8 cm tall) annual plant, with very slender leaves (about 1 mm wide).
• Plantain and *Plantago* were derived from the Latin *planta*, or 'sole of the foot,' because the leaves were thought to resemble a footprint. Common plantain (*Plantago major* L.) was sometimes called 'white man's foot,' because it seemed to follow the white man wherever he settled. • The specific epithet *maritima* is Latin for 'of the sea,' because this species is often found in coastal areas.

MADDER FAMILY

THIN-LEAVED BEDSTRAW

Galium bifolium S. Wats.
MADDER FAMILY (RUBIACEAE)

Plants: Slender, **hairless annual herbs**, 5–20 cm tall; stems mostly erect, unbranched or moderately branched, usually 4-sided; from short, weak taproots.

Leaves: Opposite (upper leaves) **and in circles** (whorls) **of 4** (lower leaves) that often have **1 pair smaller than the other**, linear to elliptic or a little wider, **blunt-tipped or slightly pointed, mostly 1–2 {2.5} cm long**, toothless.

Flowers: White, inconspicuous, **tiny, 3-lobed, soon shed**; calyx essentially absent; 4 stamens (rarely 3); 2 styles; solitary, short-stalked in leaf axils, erect at first, but eventually widely spreading on stalks up to 3 cm; {May–August}.

Fruits: Nodding pairs of round, 1-seeded **nutlets 2.5–3.5 mm long, bristly with short, hooked hairs**.

Habitat: Dry, open areas, often on disturbed sites; elsewhere, in moist meadows and open coniferous forests.

Notes: The 2 other small bedstraws in Alberta that have leaves in whorls of 4—small bedstraw (*Galium trifidum* L.) and Labrador bedstraw (*Galium labradoricum* Wieg.)—are both perennial species with creeping rhizomes and hairless fruits. • The generic name *Galium* was derived from the Greek *gala* (milk) because a common European species, *Galium verum* L., was used to curdle milk. The specific epithet *bifolium,* from the Latin *bi* (two) and *folium* (leaf), refers to the paired leaves on the upper stem.

LONG-LEAVED BLUETS
Hedyotis longifolia (Gaertn.) Hook.
MADDER FAMILY (RUBIACEAE)

Plants: Delicate, **tufted perennial herbs**; stems numerous, 10–25 cm tall, usually 4-sided, hairless and purplish, unbranched or branched in upper parts; from branched root crowns.

Leaves: Opposite, linear to narrowly oblong, 1–3 cm long, tapered to a stalkless base with **2 small, whitish or purplish appendages** (stipules), 1-nerved, smooth or minutely roughened along the edges.

Flowers: Purplish to pink or white, {3–5} **6–9 mm long, funnel-shaped**, tipped **with 4 spreading lobes**, hairy within the tube; **calyx persistent, with 4 awl-shaped, 1–2** {2.5} **mm long lobes**, often extending past the tip of the mature capsule; 4 stamens, often projecting out past the petals; numerous, in crowded or loose, leafy, spreading clusters (cymes); June–July {May–September}.

Fruits: Ovoid capsules, splitting open across the tip to release small, pitted seeds.

Habitat: Sandy soil in open woods and on dunes; elsewhere, in grasslands.

Notes: This species has also been called *Houstonia longifolia* Gaertn. • These flowers are dimorphous—that is, they have 2 forms. In one case, the style is short and hidden in the corolla tube, while the stamens project out from the mouth of the flower; in the other case, the styles project from the flowers and the stamens are contained within. Pollen from long-stamened flowers is usually transferred to the stigmas of long-styled flowers, and pollen from short-stamened flowers usually pollinates short-styled flowers. This is one way the flowers can reduce the chance of self-pollination and therefore increase the exchange of genes among local populations.

Dicots   BLUEBELL FAMILY

ALPINE HAREBELL, ARCTIC HAREBELL, ARCTIC BELLFLOWER

Campanula uniflora L.
BLUEBELL FAMILY (CAMPANULACEAE)

Plants: Small, hairless perennial herbs; stems solitary or few, usually **less than 10 cm tall** (elsewhere, up to 30 cm); from taproots.

Leaves: Alternate, simple; lower leaves spatula-shaped, 2–3 cm long, **toothless**; upper leaves linear.

Flowers: Pale blue, funnel-shaped, with 5 fused petals, **5–10 mm long**, 4–8 mm wide; **5 sepals, hairy, slender**, joined at base; 5 stamens, attached at the base of the corolla; 3 stigmas; **solitary**, nodding when young, erect when mature; July {August}.

Fruits: Erect, hairy capsules, 1 cm long, containing many smooth seeds.

Habitat: Exposed, stony alpine slopes; elsewhere, on calcareous cliffs and scree slopes and in moderately moist to dry meadows and ridges in the montane, subalpine and alpine zones.

Notes: Alaska harebell (*Campanula lasiocarpa* Cham.) is the only other *Campanula* species in Alberta with solitary flowers. It is easily distinguished by its toothed leaves, broader (bell-shaped) flowers and toothed calyx lobes.
• The generic name *Campanula* is derived from the Latin *campana* (bell), in reference to the shape of the flower. The specific epithet *uniflora* means 'one-flowered.'

LOBELIA FAMILY | **Dicots**

DOWNINGIA

Downingia laeta (Greene) Greene
LOBELIA FAMILY (Lobeliaceae)

Plants: Small, hairless annual herbs; stems 5–20 cm tall, slender, with or without branches, erect or spreading and ascending, sometimes rooting at the joints; from fibrous roots.

Leaves: Alternate, lance-shaped, 5–20 mm long, toothless, stalkless; lower leaves smaller, narrower and soon shed.

Flowers: White to light blue or purplish with white or yellow markings, inconspicuous, bearing **5 petal lobes and 5 leaf-like sepals at the tip of a long, slender, stalk-like tube** (an elongated, inferior ovary); **corolla 4–7 {8} mm long, slightly 2-sided, with 5 similar lobes** at the tip of a short tube; **sepals** 4–7 {3–8} mm long, **linear** to linear-elliptic, joined only slightly at the base; **5 stamens, fused into a tube** around the style at the centre of the flower and tipped with 2 bristles; stalkless (though ovaries may resemble stalks) in the axils of leaf-like bracts, forming elongating clusters (racemes) at the stem tip; {June} July–August.

Fruits: Slender, pod-like capsules, 2–4.5 cm long, 1–2 mm thick, containing many seeds.

Habitat: Shallow water and banks near ponds; elsewhere, in wet meadows.

Notes: With its long, stalk-like ovaries tipped with small, inconspicuous petals, downingia might be mistaken, at first glance, for one of the small willowherbs (*Epilobium* spp.). However, small willowherbs have opposite leaves and their flowers have 4 equal, separate petals.

Rare *Vascular* Plants of Alberta   213

Dicots

LOBELIA FAMILY

WATER LOBELIA
Lobelia dortmanna L.
LOBELIA FAMILY (Lobeliaceae)

Plants: **Emergent aquatic** perennial herbs; hairless; stems 10–100 cm tall, slender, hollow, essentially leafless; from fibrous roots.

Leaves: In a **basal rosette**, **submerged**, fleshy, **tubular**, **2–8 cm long**, somewhat curved; stem leaves reduced to a few thin scales.

Flowers: **Lilac to pale blue or white, showy**, 10–20 mm long, with a broad, **deeply 3-lobed lower lip** and **2 linear, backward-curved upper lobes** at the tip of a **slender tube** that is **split almost to the base** along its upper surface; calyx about 5 mm long, tipped with 5 blunt lobes; 5 stamens, forming a tube around the style inside the flower tube, with some or all anthers hairy at the tip; few, in elongating clusters (racemes) on slender, leafless stems; July {June–August}.

Fruits: Nodding, oblong **capsules**, **5–10 mm long**, 3–5 mm wide, with **2 chambers** (locules), splitting open from the tip to release many rough seeds less than 1 mm long.

Habitat: Shallow water near the edges of ponds, lakes and on sandy shores; elsewhere, also on gravelly, silty or muddy bottoms, sometimes deeply submersed but with flower stalks above water.

Notes: Water lobelia often grows with quillworts (*Isoetes* spp.), and the sterile rosettes of these plants might be confused at first glance. However, quillwort leaves have broad, flattened bases and grow from thick, corm-like stems. • Another rare Alberta species, spiked lobelia (*Lobelia spicata* Lam.), is a terrestrial plant that grows on dry to moist sandy soil on the prairie. It is easily distinguished by its leafy stems, broad, lance-shaped to egg-shaped leaves and many flowers in slender, almost spike-like clusters. • The generic name *Lobelia* commemorates Matthias de L'Obel, a Belgian botanist who lived 1538–1616.

L. SPICATA

ASTER FAMILY　　　**Dicots**

BUR RAGWEED

Ambrosia acanthicarpa Hook.
ASTER FAMILY (ASTERACEAE [COMPOSITAE])

Plants: Erect annual herbs; stems 20–60 {10–80} cm tall, leafy, much branched, with stiff, white hairs; from slender taproots.

Leaves: Alternate, numerous, 2–8 cm long, once or twice deeply pinnately divided into slender lobes, mostly stiff-hairy on both surfaces (sometimes hairless on the upper side) and stalked (upper leaves sometimes stalkless).

Flowerheads: Yellowish, either **male or female**, but with **both sexes on the same plant** (monoecious), including **disc florets only** with **inconspicuous** (sometimes absent) **petals**; male (staminate) flowerheads 2–4 mm wide, several-flowered, with involucres of fused oval bracts, borne in long, nodding, leafless clusters (racemes) at stem tips; female (pistillate) flowerheads about 7–8 mm long, single-flowered, the developing seed (pistil) enclosed by an **involucre** of fused bracts armed **with 8–15 spreading, straight spines**, solitary or clustered in the upper leaf axils; July {August–October}.

Fruits: Dry seed-like fruits (achenes) without bristles (pappus) at the tip, **enclosed in a ridged, bur-like, spiny involucre**.

Habitat: Sand dunes; elsewhere, in open, usually sandy places.

Notes: This species has also been called *Franseria acanthicarpa* (Hook.) Coville. It can be distinguished from other ragweeds (*Ambrosia* spp.) by its involucres, which have several rows of spines rather than just 1. • The specific epithet *acanthicarpa* comes from *Acanthus* (a subtropical genus of plants with seeds enclosed in spiny bracts) and *carpa* (fruit), hence 'fruit like *Acanthus*.' • The pollen of ragweeds (*Ambrosia artemisiifolia* L., *Ambrosia trifida* L. and *Ambrosia psilostachya* DC.) is a common cause of hay fever.

ASTER FAMILY

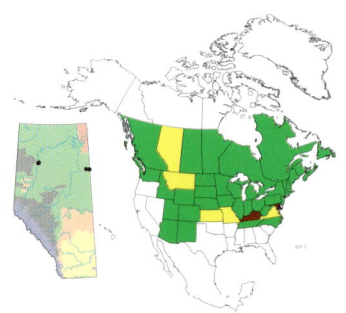

SPOTTED JOE-PYE WEED

Eupatorium maculatum L.
ASTER FAMILY (ASTERACEAE [COMPOSITAE])

Plants: Robust perennial herbs; stems stout, erect, 60–150 cm tall, **purple or purple-spotted**; from fibrous roots.

Leaves: In **whorls of 3–6**, lance-shaped to egg-shaped, 6–20 cm long, 2–7 cm wide, sharply pointed at the tip, coarsely sharp-toothed, stalked.

Flowerheads: **Pink to purple**, with **disc florets only**; involucres 6–9 mm high, often pinkish, in overlapping rows of bracts; numerous, in showy, **flat-topped clusters** (corymbs) **15–20 cm across**; August {July–September}.

Fruits: Gland-dotted seed-like fruits (achenes), 3–4.5 mm long, tipped with a tuft of white, hair-like bristles (pappus).

Habitat: Wet to moist meadows and open woods.

Notes: This species has also been included in *Eupatorium purpureum* L. or called *Eupatorium bruneri* A. Gray. • The Navajo boiled this plant to make a tea for treating arrow wounds. • Joe Pye was a famous medicine man who is said to have used this plant in New England to induce sweating and thereby break fevers during a typhus outbreak. The generic name *Eupatorium* commemorates Eupator Mithridates, a first-century king of Pontus, who was said to have used a plant of this genus as an antidote for poison.

ASTER FAMILY **Dicots**

LARGE-FLOWERED BRICKELLIA

Brickellia grandiflora (Hook.) Nutt.
ASTER FAMILY (ASTERACEAE [COMPOSITAE])

Plants: **Perennial** herbs; stems 20–80 cm tall, minutely hairy, often branched above, somewhat woody at the base; from long, thick, spindle-shaped roots.

Leaves: Mostly alternate, but sometimes opposite, usually somewhat hairy, faintly dotted with glands; blades **triangular to egg-shaped or lance-shaped**, 2–10 {12} cm long and 1–6 cm wide, notched to broadly wedge-shaped at the base, coarsely toothed; stalks 0.3–7 cm long.

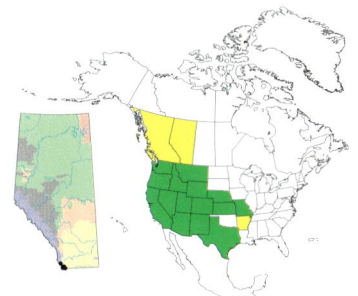

Flowerheads: Cream-coloured to greenish yellow, 8–12 mm wide and high, with 5-lobed **disc florets only**; involucres 7–11 mm high, with several overlapping rows of **striped bracts**, the outer bracts **hairy** with slender, elongated tips; receptacles flat, without hairs or scales (naked); **florets with both male and female parts**; few to several, **nodding**, in spreading clusters (panicle-like cymes); July–September.

Fruits: Seed-like, 10-ribbed fruits (achenes), about 5 mm long, thickened at the base, **tipped with a tuft of white, hair-like bristles** (pappus).

Habitat: Eroded slopes and rocky banks.

Notes: Large-flowered brickellia may be confused with another rare species, purple rattlesnakeroot (*Prenanthes sagittata* (A. Gray) A. Nels.), but rattlesnakeroot has milky juice and winged leaf stalks. Its flowerheads have hairless involucres without stripes and are borne in narrow, elongated clusters. Purple rattlesnakeroot grows in moist, often shaded sites. • The generic name *Brickellia* honours John Brickell (1749–1809), a physician and botanist who lived in Savannah, Georgia. The specific epithet *grandiflora*, from the Latin *grandis* (large) and *floreo* (flower), refers to the large flowerheads.

PRENANTHES SAGITTATA

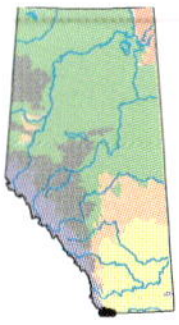

Dicots

ASTER FAMILY

ALPINE TOWNSENDIA

Townsendia condensata Parry *ex* A. Gray
ASTER FAMILY (ASTERACEAE [COMPOSITAE])

Plants: Low, **cushion-like** perennial herbs **without flowering stems**; from taproots.

Leaves: In rather erect, basal clusters, 1.2–3.5 cm long, 2–4 mm wide, **spatula-shaped**, with long, white hairs; blades short, without teeth, abruptly narrowed to long stalks.

Flowerheads: White, pink or lavender, about 4–5 cm across, with numerous 8–16 mm long ray florets around a 1–4 cm wide, yellow button of disc florets; involucres 8–18 mm high, **densely woolly**, bracts overlapping, lance-shaped to linear, pointed, with whitish, translucent edges fringed with soft hairs; receptacles flat, without hairs or scales (naked); ray florets with female parts only; disc florets with both male and female parts; **single to few, stalkless, nestled among the leaves**, early June–early August.

Fruits: Flattened, 2-nerved seed-like fruits (achenes), 3–4.4 mm long, covered with forked or barbed hairs, tipped with a **tuft of rough, hair-like bristles** (on disc florets) **or a crown of small scales** (on ray florets).

Habitat: Exposed shale ridges in alpine areas.

Notes: Another rare species, low townsendia (*Townsendia exscapa* (Richards.) Porter), is distinguished from alpine townsendia by its slightly hairy (not densely woolly) involucral bracts and its linear to lance-shaped (rather than spatula-shaped) leaves. Low townsendia grows on dry hillsides and prairies rather than on alpine ridges, and flowers in May. • The generic name *Townsendia* commemorates David Townsend (1787–1858), an amateur botanist of West Chester, Pennsylvania. The specific epithet *condensata* means 'compressed,' in reference to the compact, cushion-like plants.

T. EXSCAPA

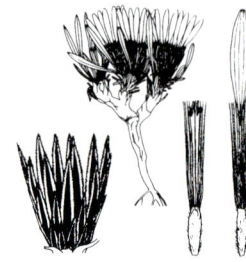

ASTER FAMILY — **Dicots**

FEW-FLOWERED ASTER

Aster pauciflorus Nutt.
ASTER FAMILY (ASTERACEAE [COMPOSITAE])

Plants: Perennial herbs; **stems** {10} 20–50 cm tall, **hairless near the base**, branched and glandular near the top; from slender underground stems (rhizomes).

Leaves: Alternate, much smaller upward on the stem; **lower leaves** linear to lance-shaped, widest above the middle, **4–8 cm long and 2–3 mm wide**, rather fleshy, **hairless**, without teeth, stalked; upper leaves linear, stalkless.

Flowerheads: Blue or whitish, 15–25 mm across, **usually with more than 15, 4–6 mm long ray florets** around a yellow cluster of 5-lobed disc florets; involucres **rather fleshy, hairless, glandular**, 5–8 mm high, with **unequal** linear to lance-shaped **bracts in 2 or 3 loose, overlapping rows**; ray florets with female parts only; disc florets with both male and female parts; styles with flattened branches more than 0.5 mm long and hairy appendages; **few**, borne singly on spreading branches in open, somewhat rounded to flat-topped clusters; July–August.

Fruits: Several-nerved seed-like fruits (achenes), **tipped with a tuft of hair-like bristles** (pappus).

Habitat: Alkaline flats; elsewhere, also in salt marshes.

Notes: Another rare species, meadow aster (*Aster campestris* Nutt.), resembles few-flowered aster, but it is found in dry, open areas, often on alkaline soil, at low to moderate elevations in the mountains. Meadow aster has finely hairy stems and numerous purplish flowerheads with rays up to 12 mm long. Its leaves are {2} 3–5 cm long, about 5 mm wide and only slightly smaller on the upper stem. Meadow aster flowers in July and August. • The specific epithet *pauciflorus* means 'few-flowered'; *campestris* means 'of the fields.'

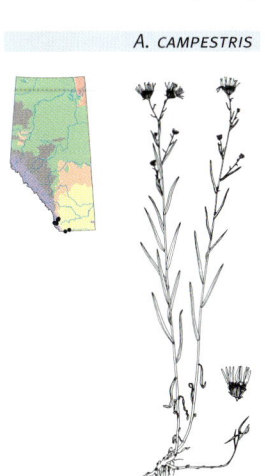

A. CAMPESTRIS

Rare *Vascular* Plants of Alberta

Dicots ASTER FAMILY

Eaton's aster

Aster eatonii (A. Gray) J.T. Howell
ASTER FAMILY (Asteraceae [Compositae])

Plants: Perennial herbs; stems 30–80 {100} cm tall, often reddish, minutely hairy on the upper parts; from creeping underground stems (rhizomes).

Leaves: Alternate, linear to lance-shaped, 6–12 {5–15} cm long, 4–20 mm wide (usually **7–13 times as long as wide**), hairless or rough with short, stiff hairs, usually **toothless and stalkless**.

Flowerheads: White or pink, with 20–40 ray florets **5–12 mm long** around a **yellow button of disc florets**; involucres 5–9 {4.5–10} mm high; **involucral bracts leafy** or completely green (at least some of the outer bracts), loosely overlapping or curved outward; numerous, in **long, narrow, leafy clusters** (raceme-like); {July} August–September.

Fruits: Dry, hairy seed-like fruits (achenes), tipped with a tuft of **white, hair-like bristles** (pappus).

Habitat: Moist montane woodland and streambanks.

Notes: Eaton's aster often hybridizes with leafy-bracted aster (*Aster subspicatus* Nees). These species are very similar, but leafy-bracted aster has only a single to few flowerheads, and its lower leaves are usually stalked and sharply toothed. • Another rare annual aster reported to occur in Alberta, tansy-leaved aster (*Machaeranthera tanacetifolia* (Kunth) Nees., also called *Aster tanacetifolius* Kunth), is easily identified by its taproot and finely dissected, almost feathery leaves in combination with large (about 3 cm wide) blue to bluish purple flowerheads with yellow centres. It grows on dry, open plains and hillsides, and is reported to flower in July and August. • The generic name *Aster* was taken from the Greek *astron* (star), in reference to the star-like appearance of these radiate flowerheads.

MACHAERANTHERA TANACETIFOLIA

ASTER FAMILY

Dicots

FLAT-TOPPED WHITE ASTER

Aster umbellatus Mill.
ASTER FAMILY (ASTERACEAE [COMPOSITAE])

Plants: Perennial herbs; stems 30–200 cm tall, finely hairy, leafy near the top; from creeping underground stems (rhizomes).

Leaves: Alternate, 5–15 cm long, lance-shaped to egg-shaped or elliptic, gradually narrowed to the base and to the tip, smooth or rough to the touch on the upper surface, finely hairy beneath, toothless, short-stalked to almost stalkless.

Flowerheads: White, about 12–20 mm across, with 4–7 wide-spreading, 4–8 mm long ray florets around a yellow cluster of 5-lobed disc florets; receptacles without hairs or scales (naked); involucres hairy, 3–5 mm high, with few, pointed bracts pressed together and ascending; ray florets with female parts only; disc florets with male and female parts; styles with flattened branches more than 0.5 mm long and hairy appendages; **numerous, in flat-topped clusters** (corymbs); July–September.

Fruits: Several-nerved seed-like fruits (achenes), **more or less hairy**, tipped with a tuft of **hair-like bristles** (pappus) in **2 rows**, the **outer row tiny** (less than 1 mm long), the **inner row** longer, stiff, and **often thick-tipped**.

Habitat: Moist woodlands and swampy sites; elsewhere, in moist thickets and meadows.

Notes: This species has an interesting distribution. It is generally an eastern species, with populations in Saskatchewan limited to the east-central region (Duck Mountain and the Porcupine Hills), so the Alberta populations are rather disjunct from the main range.
• The specific epithet *umbellatus* comes from the Latin *umbella* (a parasol), in reference to the flat-topped, umbrella-shaped flower clusters.

Rare *Vascular* Plants of Alberta

WILD DAISY FLEABANE

Erigeron hyssopifolius Michx.
ASTER FAMILY (ASTERACEAE [COMPOSITAE])

Plants: Tufted perennial herbs; stems slender, leafy, mostly **10–30 cm tall**, sparsely hairy or nearly hairless, branched, often with **short, leafy shoots in leaf axils**; from a short, branched woody base with many fibrous roots.

Leaves: Alternate, numerous, **crowded** (overlapping on the stem), **largest at mid-stem, linear or linear–lance-shaped**, 2–3 cm long and **1–3 mm wide**, thin, soft, nearly hairless but with scattered hairs along the edges, not toothed, stalkless or short-stalked.

Flowerheads: White, sometimes pink or purplish, about 1.5 cm across, with 20–50 small (**4–8 mm long**, about 1.5 mm wide) ray florets around a button of yellow disc florets; involucres 4–6 mm high, the **bracts equal, thin**, soft, pointed, **with a few somewhat sticky hairs**; receptacles flat, without hairs or scales (naked); 1–5, on long, mostly leafless stalks; July–August.

Fruits: Hairy, nerved seed-like fruits (achenes), tipped with a **tuft of yellowish, hair-like bristles** (pappus).

Habitat: Shores, banks and ledges; elsewhere, on calcareous ledges, talus and gravelly shores, in fens and in wet openings in coniferous forests.

Notes: The specific epithet *hyssopifolius* means 'with leaves like hyssop.' Hyssop (*Hyssopus officinalis* L.) is a Eurasian mint species that has toothless, lance-shaped leaves less than 3 cm long and that produces leafy shoots in the leaf axils, very similar to those of wild daisy fleabane.

ASTER FAMILY — Dicots

TRIFID-LEAVED FLEABANE, THREE-FORKED FLEABANE
Erigeron trifidus Hook.
ASTER FAMILY (ASTERACEAE [COMPOSITAE])

Plants: Small, **densely tufted** perennial herbs; flowering stems 4–8 {3–10} cm tall, leafless (sometimes with a small bract near the middle), densely glandular-hairy and usually with long, spreading, many-celled hairs, often white-woolly below the flowerhead; from compact, many-branched crowns on taproots.

Leaves: Basal; blades lance-shaped to narrowly egg-shaped or oblong, 1–2 {3} **cm long**, usually **tipped with 3** (sometimes 2) lance-shaped to oblong **lobes**, each 3–5 {8} mm long and **more than 1 mm wide**, sparsely glandular-hairy on both surfaces and fringed with stiff hairs, abruptly narrowed to a 1–2 mm wide 'stalk' (basal segment).

Flowerheads: White (sometimes pinkish), drying purple, about 1–1.2 cm across, with 20–30 {40}, linear to spatula-shaped, 10–15 mm long ray florets around a yellow button of disc florets; involucres 8–10 mm high, glandular-hairy, crimson or crimson-tipped, with 25–30 narrowly lance-shaped, 5–12 mm long, **equal bracts**; receptacles flat, without hairs or scales (naked); solitary; {May–July}.

Fruits: Densely hairy, 2-nerved seed-like fruits (achenes), 2–2.5 mm long, tipped with a **tuft of 15–20 hair-like bristles** (pappus) 3–4 mm long.

Habitat: Alpine slopes.

Notes: Trifid-leaved fleabane produces seed asexually and probably originated as a hybrid between *Erigeron compositus* Pursh and *Erigeron lanatus* Hook. • The specific epithet *trifidus* means 'split into 3 parts,' in reference to the 3-lobed leaves.

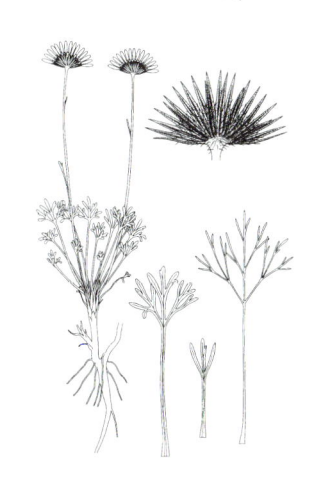

Rare *Vascular* Plants of Alberta

FRONT-RANGE FLEABANE

Erigeron lackschewitzii Nesom and Weber
ASTER FAMILY (ASTERACEAE [COMPOSITAE])

Plants: Perennial herbs; stems with **long, soft hairs, without glands**; from long, thick taproots.

Leaves: Basal and alternate on the stem; basal leaves linear to broadly lance-shaped, not toothed; 5–10 stem leaves.

Flowerheads: Blue, with **30–68 ray florets**; involucral bracts **woolly with interwoven hairs** that often have thin, purple 'bands' (cross-walls); **solitary**, on woolly stalks.

Fruits: Dry seed-like fruits (achenes), 1.5–2.5 mm long, tipped with a tuft of 15–24 hair-like bristles (pappus).

Habitat: Gravelly slopes in *Dryas*-dominated tundra; elsewhere, in exposed mountain sites on limestone and dolomite rock.

Notes: Another rare Alberta species, yellow alpine fleabane (*Erigeron ochroleucus* Nutt.), closely resembles front-range fleabane. It is distinguished by its smaller (3–3.5 {2.5–3.8} mm long) disc corollas, its 4–8 mm high flowerheads with 40–80 {30–88} ray florets and its smaller (1.1–1.5 mm long) achenes. Yellow alpine fleabane grows in dry exposed areas, primarily in the montane zone, and flowers from June to August. Two varieties are recognized, var. *ochroleucus* and var. *scribneri* (Canby) Cronq., the latter being a shorter, more slender form. • These small fleabanes might be confused with the more common species large-flowered fleabane (*Erigeron grandiflorus* Hook.). However, the lower leaves of large-flowered fleabane are broadly lance-shaped or spatula-shaped with rounded tips, the flowering involucres are 8–10 mm tall with conspicuous glandular hairs and the flowerheads have 100–125 ray florets. Large-flowered fleabane grows on alpine slopes. • Front-range fleabane is endemic to southwestern Alberta and Montana. It is a species that has evolved only recently and appears to have been derived from yellow alpine fleabane. • The specific epithet *lackschewitzii* honours Klaus Lackschewitz (1911–95), the Montana botanist who first collected this species. The common name refers to the range of this species, in the front ranges of the Rocky Mountains.

E. OCHROLEUCUS

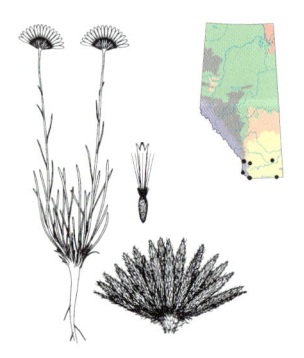

ASTER FAMILY — **Dicots**

PALE ALPINE FLEABANE

Erigeron pallens Cronq.
ASTER FAMILY (ASTERACEAE [COMPOSITAE])

Plants: Small, **loosely matted** perennial herbs, sparsely to moderately **sticky-glandular and hairy** with long, multi-celled hairs, some hairs with thin, purplish black bands (cross-walls); flowering **stalks leafless**, 2–3 {6} cm tall; from slender-**branched, woody bases** and underground stems (rhizomes).

Leaves: Basal; blades **spatula-shaped to lance-shaped, widest above the middle, some tipped with 3 rounded lobes**, up to 2.5 cm long and 4 mm wide.

Flowerheads: White or pinkish to purplish, about 1.5–2 cm across, with 50–60 {90} small (4–5 mm long and 0.5–0.9 mm wide) ray florets around a yellow button of 3.5–5 mm long disc florets; involucres 6–8 mm high, green or sometimes purplish, with few thin bracts, woolly and sticky, with long, flattened, gland-tipped hairs; receptacles flat, without hairs or scales (naked); **solitary**; June–early August.

Fruits: Short-hairy, 2-nerved seed-like fruits (achenes), tipped with a **tuft of yellowish white, hair-like bristles** (pappus) in 2 rows, the inner row with 35–40 bristles, longer than the corolla, and the outer row with short, inconspicuous bristles.

Habitat: Rocky slopes in alpine areas; elsewhere, in sandy or gravelly areas including riverbanks.

Notes: This species has also been called *Erigeron purpuratus* Greene ssp. *pallens* (Cronq.) G. Douglas. • The specific epithet *pallens* is from the Latin *pallere* (to become pale), perhaps referring to the pale colour of the flowerheads.

Dicots — ASTER FAMILY

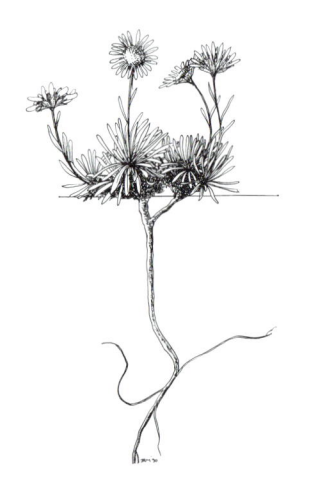

DWARF FLEABANE

Erigeron radicatus Hook.
ASTER FAMILY (ASTERACEAE [COMPOSITAE])

Plants: Small perennial herbs; flowering **stems erect, 3–5 {10} cm tall**, with coarse or soft hairs (especially near the top), almost leafless; from **branched root crowns** on taproots.

Leaves: Basal, sometimes also with 2 or 3 small, linear stem leaves; **basal leaves slender, lance-shaped, widest above the middle, 1–2 {3.5} cm long** and 2.5 mm wide, fringed with hairs, hairless or finely hairy on the surfaces, **lacking teeth.**

Flowerheads: White, 1–2 cm across, with 20–50 small (5–8 mm long and 2 mm wide) **ray florets** around a yellow button of 2.3–3 mm long disc florets; involucres about 5 mm high and 7–10 mm wide, greenish, glandular-sticky, **with short, soft hairs**, and **equal**, narrowly lance-shaped to oblong **bracts**; receptacles flat, without hairs or scales (naked); **solitary**; late May–July.

Fruits: Ribbed seed-like fruits (achenes), **tipped with 6–12 fragile bristles** and some short, narrow, outer scales (together making up the pappus).

Habitat: Dry, open ridges, rocky slopes, hilltops and grasslands, in sites considered to have escaped Wisconsinan glaciation.

Notes: The total Alberta population of dwarf fleabane is estimated at 1,000 plants and the Canadian population at 5,000 plants. Populations of this species are very local and infrequent and therefore easily eradicated by disturbance. However, populations commonly are found in sites that are difficult to access. • Another rare species, yellow alpine fleabane (*Erigeron ochroleucus* Nutt., see p. 224), resembles dwarf fleabane, but yellow alpine fleabane is tufted (rather than branched at the base) and its leaves are larger (more than 2 cm long) with enlarged, translucent, whitish or purplish bases. Its flowering stems have evident leaves and its involucral bracts are relatively broad and dark yellowish green and sometimes have a brown midrib and purple tips. The ray florets can be blue or purple, as well as white, and up to 12 mm long and the disc florets are 2.8–4.3 mm long. Its achenes have as many as 20 pappus bristles, and their outer bristles are larger than those of dwarf fleabane. Yellow alpine fleabane grows on dry, open slopes and flowers from June to August.

226 Rare *Vascular* Plants of Alberta

ASTER FAMILY

Dicots

> **SPREADING FLEABANE**
>
> *Erigeron divergens* T. & G.
> ASTER FAMILY (ASTERACEAE [COMPOSITAE])

Plants: Biennial or perennial herbs; **stems leafy, 10–40 cm tall** (sometimes to 70 cm), often **freely branched**, with spreading hairs and glands; from taproots.

Leaves: Basal and alternate, gradually smaller upward on the stems, numerous, covered with **spreading white hairs on both surfaces**; basal leaves lance-shaped, widest above the middle, or spatula-shaped, 1–4 cm long and 1–3 mm wide, **without teeth**, stalked, withered and shed first; stem leaves linear, **toothless**, stalkless or short-stalked.

Flowerheads: Blue, pink or white (drying pink or purplish), about 2 cm across, with 75–150 slender, 5–6 {10} mm long, 0.5–1.2 mm wide ray florets around a 7–11 mm wide yellow button of 2–3 mm long disc florets; involucres **hairy and glandular**, 4–5 mm high, made up of narrow, greenish, equal or overlapping bracts, with thin, **translucent edges and tips**; receptacles flat, without hairs or scales (naked); **several to many**, in branched, open clusters; {May–July}.

Fruits: Seed-like fruits (achenes), 2–4-nerved, tipped with **5–12 fragile bristles and several short, narrow outer scales** (together constituting the pappus).

Habitat: Dry, open areas at lower elevations in the mountains.

Notes: This species has also been called *Erigeron divaricatus* Nutt. • The generic name *Erigeron* was taken from the Greek *eri* (early) and *geron* (old man), probably referring to the early flowering and fruiting of most fleabanes. The specific epithet *divergens,* from the Latin *dis* (apart) and *verger* (to turn), refers to the spreading hairs that cover these plants.

Dicots

ASTER FAMILY

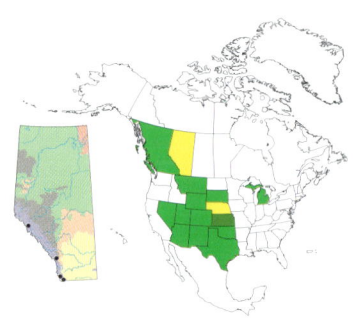

CREEPING FLEABANE

Erigeron flagellaris A. Gray
ASTER FAMILY (ASTERACEAE [COMPOSITAE])

Plants: Biennial or short-lived perennial herbs; stems 5–40 cm tall, hairy; **erect** (flowering) **to trailing and rooting at the tips** (vegetative runners); from taproots.

Leaves: Basal and alternate on stems; basal leaves lance-shaped, widest above the middle, without teeth, stalked; stem leaves few, small, linear to narrowly lance-shaped and widest above the middle.

Flowerheads: White (sometimes pink or blue), with many 5–10 mm long, 1 mm wide ray florets around a yellow button of 2.5–3.5 mm long disc florets; involucres with short, spreading hairs and small glands, 3.5–5 mm high, **the bracts greenish with translucent edges and tips**; receptacles flat, without hairs or scales (naked); solitary; June–August.

Fruits: Nerved seed-like fruits (achenes), **tipped with 10–15 bristles and several short, narrow outer scales** (together comprising the pappus).

Habitat: Dry, open woods, lakeshores and disturbed or poorly vegetated areas; elsewhere, in sagebrush meadows, disturbed grasslands and rocky streambanks.

Notes: The specific epithet *flagellaris* is from the Latin *flagellare* (to whip), referring to the whip-like, trailing stems.

WOOLLYHEADS

Psilocarphus elatior A. Gray
ASTER FAMILY (ASTERACEAE [COMPOSITAE])

Plants: **Small**, **loosely greyish-woolly annual** herbs; **stems erect or spreading at the base**, {1} 3–15 cm tall, branched; from weak taproots.

Leaves: **Opposite** (mostly), **linear to linear-oblong**, 5–18 {35} mm long, **without teeth**, **stalkless**; uppermost leaves surrounding the flowerheads.

Flowerheads: **White**, round, 4–8 mm wide, with 2 types of **disc florets**; outer florets numerous, **thread-like**, loosely **enveloped by a small** (about 3 mm long), **sac-like, woolly bract with a small translucent appendage just below the tip**, with female parts only; central florets few, tubular, without a bract, with only the male parts functioning; **true involucres absent**; **receptacles round**; numerous, in **compact clusters** at branch tips and in branch axils; May {June–July}.

Fruits: Oblong seed-like fruits (achenes), 1–1.7 mm long, brown, smooth, **without bristles** (pappus) **at the tip**, loose in a woolly, bladder-like bract.

Habitat: Dried beds of pools from spring runoff.

Notes: This species has also been called *Psilocarphus oregonus* Nutt. var. *elatior* A. Gray. • The generic name *Psilocarphus* comes from the Greek *psilos* (bare) and *karphos* (chaff), in reference to the small, translucent appendage near the tip of each female flower bract.

ASTER FAMILY

AROMATIC EVERLASTING

Antennaria aromatica Evert
ASTER FAMILY (ASTERACEAE [COMPOSITAE])

Plants: Mat-forming perennial herbs with a **citronella-like odour** (when fresh); flowering stems 2–6 cm tall, **woolly and glandular**; from **short trailing stems or runners** (stolons) and branched, woody root crowns.

Leaves: Mostly in flattened, **basal rosettes**, but also alternate on the stems, densely woolly and glandular on both surfaces; **basal leaves** 5–10 mm long and 3–8 mm wide, **spatula-shaped**, tapered to a wedge-shaped base and tipped with a short, sharp point; **stem leaves small**, linear to lance-shaped, widest above the middle, 3–7 mm long and 0.5–2 mm wide, the uppermost often with brown, translucent tips; withered leaves persistent (marcescent).

Flowerheads: Green or brownish, 8–20 mm wide, including **disc florets only**, **male or female**, with sexes **on separate plants** (dioecious); receptacles without hairs or scales (naked); female flowerheads with thread-like florets, 2-lobed styles and 5–7 mm high involucres; male flowerheads with tubular florets, undivided styles and 4–6 mm high involucres; **involucral bracts** loosely woolly, usually glandular and light green or light brown at the base, becoming **translucent brown or green** toward the blunt or pointed, irregularly cut tips; 2–5, in fairly compact clusters; July–early August.

Fruits: Dry seed-like fruits (achenes), 1.5–2 mm long, tipped with a tuft of hair-like bristles (pappus) that are joined at the base and **shed as a unit**.

Habitat: Limestone talus from treeline to alpine regions, in areas that were not covered by ice during the Wisconsinan glaciation.

Notes: This species has also been called *Antennaria pulvinata* Greene. • It appears that aromatic everlasting has not migrated very far since the last glaciation. • The specific epithet *aromatica* refers to the fragrant living plants.

ASTER FAMILY Dicots

CORYMBOSE EVERLASTING

Antennaria corymbosa E. Nels.
ASTER FAMILY (ASTERACEAE [COMPOSITAE])

Plants: Thinly woolly perennial herbs; flowering stems erect, slender, 10–30 cm tall; forming loose mats from **leafy trailing stems or runners** (stolons).

Leaves: **Mostly in flat basal rosettes**, but also alternate on the flowering stems; basal leaves 2–4 cm long, **narrowly lance-shaped**, widest above the middle, tapered to a narrow base, **greyish and thinly woolly on both surfaces**; **stem leaves small**, linear.

Flowerheads: Whitish with **dark spots**, about 4 mm across, with disc florets only, male or female, with sexes on separate plants; receptacles without hairs or scales (naked); involucres 4–5 mm high, their **bracts woolly at the base** and with a **dark brown or blackish spot below the whitish tip**; male flowerheads with tubular florets and undivided styles; female flowerheads with thread-like florets and 2-lobed styles; several, on short stalks in **compact, flat-topped clusters**; {June–July} August.

Fruits: Dry seed-like fruits (achenes), tipped with a tuft of hair-like bristles (pappus) that are joined at the base and **shed as a unit**.

Habitat: Moist open woods and meadows; elsewhere, in moist or wet subalpine and alpine meadows.

Notes: Corymbose everlasting is similar to another rare species, one-headed everlasting (*Antennaria monocephala* D.C. Eaton), but one-headed everlasting has only a single flowerhead per stem, its involucral bracts have blackish tips and its leaves are green and hairless (or sparsely hairy) on their upper surface. One-headed everlasting grows on alpine slopes and ledges, and flowers in July and August.
• The specific epithet *corymbosa* refers to the arrangement of the flowerheads in flat-topped, corymb-like clusters.

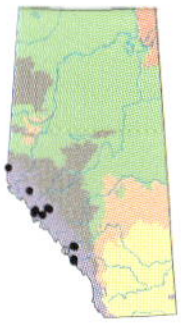

A. MONOCEPHALA

Rare *Vascular* Plants of Alberta 231

Dicots ASTER FAMILY

SILVERY EVERLASTING

Antennaria luzuloides T. & G.
ASTER FAMILY (ASTERACEAE [COMPOSITAE])

Plants: Clumped perennial herbs; flowering stems erect, 10–50 {70} cm tall, clustered, grey-woolly; from branched, rather woody root crowns.

Leaves: Basal and alternate on the stems; **basal leaves erect**, narrowly **lance-shaped**, widest above the middle, 3–8 cm long and 2–8 mm wide, usually with **conspicuous nerves**, thinly woolly, tapered to a short stalk; **stem leaves well developed**, mostly **linear**, gradually smaller upward.

Flowerheads: Greenish brown with whitish-tipped bracts, about 5 mm across, with disc florets only, male or female with sexes on separate plants; receptacles without hairs or scales (naked); **involucres 4–5 mm high, mostly hairless to the base**, with pale greenish brown, whitish-tipped, **translucent** bracts; male flowerheads with tubular florets and undivided styles; female flowerheads with thread-like florets and 2-lobed styles; several to many, in **compact, often flat-topped clusters**; {May–June} July.

Fruits: Dry seed-like fruits (achenes), tipped with a tuft of hair-like bristles (pappus) that are joined at the base and **shed as a unit**.

Habitat: Gravelly slopes and open places at low to moderate elevations in the mountains.

Notes: The generic name *Antennaria* refers to the fancied resemblance of the thick pappus hairs in the male florets of some species to the antennae of some insects. In silvery everlasting, these hairs are strongly club-shaped, flattened at their tips and more or less minutely toothed.

ASTER FAMILY   **Dicots**

COMMON CUDWEED, TALL CUDWEED
Gnaphalium microcephalum Nutt.
ASTER FAMILY (ASTERACEAE [COMPOSITAE])

Plants: Tufted, short-lived **perennial** herbs, **white-woolly**; stems several, clumped, erect, 20–70 cm tall, unbranched or moderately branched; from woody **taproots**.

Leaves: Basal and alternate on stems, **numerous, mostly linear** (lower leaves sometimes lance-shaped and widest above the middle), 3–10 cm long and 2–10 mm wide, stalkless, with smooth (toothless) **edges** that sometimes **extend slightly down the stem as wings**.

Flowerheads: Yellow or whitish, about 6 mm high, **with disc florets only**; involucres 4–7 mm high, with many **translucent white to tan bracts**, woolly only at the base (if at all); receptacles flat, without hairs or scales (naked); **outer florets** very slender, with **female** parts only; **central florets** with **both male and female** parts; numerous, on short stalks in small, compact heads forming open clusters; August {July–September}.

Fruits: Hairless, nerveless seed-like fruits (achenes), **with a tuft of hair-like bristles** (pappus) that is **soon shed**.

Habitat: Dry, open sites; elsewhere, on burned sites and streambanks, and around hot springs.

Notes: This species has also been called *Gnaphalium thermale* E. Nels. • Common cudweed resembles another rare species, clammy cudweed (*Gnaphalium viscosum* Kunth), but clammy cudweed has single stems and glandular stem and upper leaf surfaces. Clammy cudweed grows in open woods and flowers from July to September. • The cudweeds (*Gnaphalium* spp.) may be confused with pearly everlasting (*Anaphalis margaritacea* (L.) Benth. & Hook. f. *ex* C.B. Clarke), but the leaves of pearly everlasting have dark green upper surfaces and their edges do not run down the stem. Its plants spread by means of underground stems (rhizomes), and its pappus bristles remain attached to the seeds. • The cudweeds also resemble some species of everlasting (*Antennaria* spp.), but everlasting has male and female flowerheads on separate plants and lacks extended leaf edges.

G. VISCOSUM

Dicots

ASTER FAMILY

COMMON BEGGARTICKS, TALL BEGGARTICKS

Bidens frondosa L.
ASTER FAMILY (ASTERACEAE [COMPOSITAE])

Plants: Annual herbs; stems {20} 30–150 cm tall, usually reddish, furrowed, hairless or sparsely hairy, branched toward the top; from weak, slender roots.

Leaves: Opposite, numerous, finely hairy or (usually) hairless, 5–10 cm long; **blades pinnately divided into 3–5** lance-shaped, coarsely toothed **leaflets** up to 10 cm long and 3 cm wide, **the tip leaflet** (and sometimes lower leaflets) **stalked**; leaf stalks slender, 1–6 cm long.

Flowerheads: Orange (sometimes pale yellow), bell-shaped to bowl-shaped, 12–20 mm across, with **0–8 inconspicuous** (less than 3.5 mm long) **ray florets** around a cluster of 5-lobed disc florets; **involucral bracts in 2 distinct overlapping rows**, with 4–8 {16} green, leafy, fringed, **10–20 mm long outer bracts** around the broader, translucent (at least in part), 7–9 mm long inner bracts; **receptacle surface with dry scales**; disc florets with both male and female parts; several, in open, branched clusters; {June–October}.

Fruits: Flat seed-like fruits (achenes), {5} 6–12 mm long, strongly 1-nerved on each face, brown or blackish, **tipped with 2 stiff, barbed bristles** (awns).

Habitat: Moist ground and ditches; elsewhere, in wet or occasionally rather dry waste places.

Notes: The generic name *Bidens,* from the Latin *bi* (two) and *dens* (tooth), refers to the 2 barbed bristles at the tip of each achene. The specific epithet *frondosa* is Latin for 'leafy.'

COMMON TICKSEED

Coreopsis tinctoria Nutt.
ASTER FAMILY (ASTERACEAE [COMPOSITAE])

Plants: Erect, hairless **annual** herbs; stems slender, branched, 40–100 {30–120} cm tall; from fibrous roots.

Leaves: Usually opposite, fairly scattered, 5–10 cm long, once or twice pinnately divided into slender lobes, short-stalked to almost stalkless.

Flowerheads: Orange-yellow, about **2.5 cm across**, with 8 {6–10} **ray florets around a dark, red-purplish button of fertile disc florets**; ray florets 8–12 mm long, often reddish brown or purplish at the base, lacking stamens and pistils; involucres 7 {6–10} mm high and 10–15 mm wide, with **2 unequal rows of bracts**, all somewhat **joined at the base**; outer bracts slender, about 2 mm long; inner bracts egg-shaped, 5–8 mm long, orange or brown with translucent edges; several to many, **on long, slender stalks** in branched, open clusters; July {June–September}.

Fruits: Black, oblong to wedge-shaped seed-like fruits (achenes), flattened parallel to the outer bracts, **1–4 mm long**, curved inward, tipped **with or without 2 tiny bristles** (pappus).

Habitat: Clay flats and slough edges; elsewhere, in irrigation ditches.

Notes: This showy wildflower is often grown in flower gardens, and it frequently escapes to grow in the wild. It can occur very sporadically, growing abundantly one year and then disappearing for several seasons. • The common name 'tickseed' refers to the small, dark achenes, which were thought to resemble tiny insects.

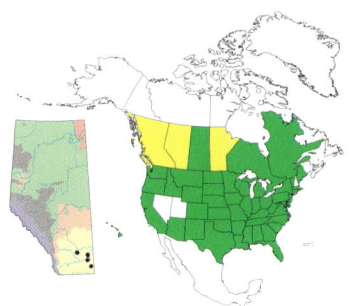

Dicots

ASTER FAMILY

GREENTHREAD, TICKSEED

Thelesperma subnudum A. Gray var. *marginatum* (Rydb.) T.E. Melchert *ex* Cronq.
ASTER FAMILY (ASTERACEAE [COMPOSITAE])

Plants: Essentially hairless perennial herbs; stems usually single, 10–20 cm tall; from creeping underground stems (rhizomes).

Leaves: Opposite (sometimes alternate on the upper stem), mostly crowded near the stem base; **lower leaves irregularly pinnately divided into slender**, 1.5–5 cm long and {1} 1.5–3 mm wide **lobes**; upper leaves linear, gradually reduced to bracts on the upper stem.

Flowerheads: Yellow, about 1 cm across, with **disc florets only**; **involucres 6–9 mm high**, with 2 overlapping rows of bracts, the outer bracts leafy and 2.5–5 mm long, **the inner bracts larger**, translucent, fused together for the lower ⅓–⅔; **receptacles** flat, covered **with thin scales**; florets with both male and female parts; 1–3, on 7–10 cm long stalks from upper leaf axils; {May–June}.

Fruits: Flattened, linear–oblong seed-like fruits (achenes), hairless but sometimes covered with tiny bumps (papillae), tipped with **2 inconspicuous** (sometimes absent), **barbed bristles** (awns).

Habitat: Dry eroded hills, on sandy soil; elsewhere, in mixed grasslands dominated by western wheatgrass (*Elymus smithii* (Rydb.) Gould).

Notes: This species has also been called *Thelesperma marginatum* Rydb. • The generic name *Thelesperma* comes from the Greek *thele* (nipple) and *sperma* (seed), referring to the tiny, nipple-like bumps on the seeds of some species.

ASTER FAMILY — Dicots

PICRADENIOPSIS

Picradeniopsis oppositifolia (Nutt.) Rydb. *ex* Britt.
ASTER FAMILY (ASTERACEAE [COMPOSITAE])

Plants: Tufted perennial herbs, **greyish with fine, flat-lying hairs**; stems leafy, erect, 10–20 {50} cm tall, **usually freely branched** from near the more or less woody base (sometimes unbranched); from tough, slender, creeping underground stems (rhizomes).

Leaves: Opposite (sometimes a few alternate), **numerous**, 1–4 cm long, with **tiny, pitted glands**; blades once or twice palmately divided into 3 {5} linear segments; stalks short.

Flowerheads: Yellow, 7–10 {12} mm across, with **few** (5 or 6), **inconspicuous** (2–4 {5} mm long) **ray florets** around a dense cluster of disc florets; involucres finely greyish-hairy, 5–7 mm high, with **2 overlapping rows of bracts**, the outer bracts ridged lengthwise, the inner bracts with thin, almost translucent edges; receptacles flat, without hairs or scales (naked); ray florets with female parts only; **disc florets glandular** on the lower part of their tubes, with both male and female parts; several, borne at the tips of upper branches, forming flat-topped clusters; July {August–September}.

Fruits: Slender, 4-sided seed-like fruits (achenes), 4 mm long, glandular, **tipped with a few blunt, translucent scales** (pappus).

Habitat: Badlands and roadsides; elsewhere, as a rather persistent weed in cultivated land and on saline flats and dry plains.

Notes: This species has also been called *Bahia oppositifolia* (Nutt.) A. Gray. • The specific epithet *oppositifolia* refers to the opposite leaves.

ASTER FAMILY

TUFTED HYMENOPAPPUS

Hymenopappus filifolius Hook. var. *polycephalus* (Osterh.) B.L. Turner
ASTER FAMILY (ASTERACEAE [COMPOSITAE])

Plants: Perennial herbs; stems 20–40 {10–90} cm tall, few to several, almost hairless or sparsely covered with soft, woolly tufts; from stout, woody root crowns on deep, woody taproots.

Leaves: Mostly basal, but also alternate on the stem, 2–7 cm long; **blades once or twice pinnately divided into linear or thread-like segments**, covered with soft, white, woolly hair when young, almost hairless and purplish with age, stalked (lower leaves) to stalkless (upper leaves); stem leaves few and smaller.

Flowerheads: Yellowish, 12–20 mm across, with tubular **disc florets only**; involucres 5–7 mm high, their **bracts** pressed together, with broad, blunt, **yellowish**, **translucent** tips; receptacles small, without hairs or scales (naked); florets with both male and female parts; several to many, in open, flat-topped clusters (corymbs); May–August.

Fruits: Silky-hairy seed-like fruits (achenes), **4- or 5-sided, widest toward the tip**, 15–20 nerved, tipped **with a crown of tiny scales** (pappus) less than 1 mm long.

Habitat: Dry gravelly or sandy sites on valley slopes and at the edges of coulees and badlands.

Notes: This species has also been called *Hymenopappus polycephalus* Osterh. • The generic name *Hymenopappus* comes from the Greek *hymen* (membrane) and *pappos* (old man), in reference to the translucent scales and greyish, silky hairs (like those of an old man) on the fruits. The specific epithet *filifolius*, from the Latin *filum* (thread) and *folium* (leaf), refers to the thread-like segments of the leaves; *polycephalus*, from the Greek *polys* (many) and *kephale* (the head), refers to the many flowerheads.

ASTER FAMILY **Dicots**

Indian tansy

Tanacetum bipinnatum (L.) Schultz-Bip. ssp. *huronense* (Nutt.) Breitung
ASTER FAMILY (ASTERACEAE [COMPOSITAE])

Plants: Perennial herbs; stems 10–80 cm tall, **with long, soft hairs** (at least when young); from slender, elongated underground stems (rhizomes).

Leaves: Alternate, **2 or 3 times pinnately divided** into small leaflets with pointed lobes, **5–20 cm long**, 2–8 cm wide, stalkless or short-stalked, covered with long, soft hairs.

Flowerheads: Yellow, with a **few small ray florets** around a **1–2 cm wide button of disc florets**; involucral bracts overlapping, dry, translucent at the tips and on the edges; **1–15, in flat-topped clusters**; May–July.

Fruits: Dry seed-like fruits (achenes), angled or ribbed, commonly with glands, **lacking bristles at the tip** but sometimes with a small crown of short scales.

Habitat: Sandy or gravelly shores, sand dunes and gravel bars.

Notes: This species has also been called *Tanacetum huronense* Nutt. var. *bifarium* Fern. and *Chrysanthemum bipinnatum* L. var. *huronense* (Nutt.) Hult. • Common tansy (*Tanacetum vulgare* L.) is Alberta's only other tansy species. It is an introduced species that is stouter and larger than Indian tansy, with stems 40–180 cm tall. The leaves of common tansy are hairless and dotted with tiny pits, and the flowerheads are more numerous (20–200) and smaller (5–10 mm wide). Common tansy is a noxious weed.

Dicots   ASTER FAMILY

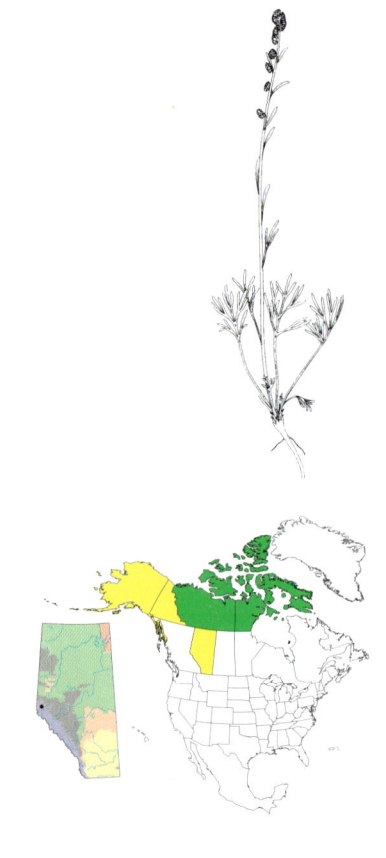

FORKED WORMWOOD

Artemisia furcata Bieb. var. *furcata*
ASTER FAMILY (ASTERACEAE [COMPOSITAE])

Plants: Perennial, usually aromatic **herbs; stems mostly 10–20 {5–30} cm tall**, greyish white with fine, silky hairs; from branched root crowns and short, stout underground stems (rhizomes).

Leaves: Mostly basal, but also alternate on the stem, **greyish on both surfaces** with flat-lying, silky hairs; **basal leaves 2–5 {1–12} cm long, once or twice palmately divided into 3 or 5 slender** (about 1 mm wide), linear to narrowly elliptic **segments**, long-stalked; stem leaves few, sometimes not divided, short-stalked or stalkless.

Flowerheads: Yellow, 6–10 mm across, with 5-lobed, glandular and sometimes hairy disc florets; involucral bracts overlapping, 3–6 mm long, egg-shaped to elliptic, with **broad, brownish black, translucent edges**, irregularly cut or fringed tips, and woolly outer surfaces; receptacles without hairs or scales (naked); outer florets with female parts only; **central florets with both male and female parts**; about 5–12, stalked or stalkless, **in narrow, elongated clusters** (racemes or spikes); {July} August–September.

Fruits: Sparsely long-hairy seed-like fruits (achenes) without bristles (pappus) at the tip.

Habitat: Rocky **alpine** slopes; elsewhere, on ledges and rocky or sandy slopes at low to high elevations.

Notes: This species has also been called *Artemisia trifurcata* Steph. *ex* Sprengel and *Artemisia hyperborea* Rydb. • The generic name *Artemisia* honours Artemis, the Greek goddess of the moon, wild animals and hunting. The specific epithet *furcata*, from the Latin *furca* (fork), refers to the divided (forked) leaves.

ASTER FAMILY — Dicots

Herriot's sagewort, mountain sagewort
Artemisia tilesii Ledeb. ssp. *elatior* (T. & G.) Hult.
ASTER FAMILY (ASTERACEAE [COMPOSITAE])

Plants: Erect, **aromatic** perennial herbs; stems usually finely **woolly**, 60–120 {15–150} cm tall; from woody root crowns, strong underground stems (rhizomes) and fibrous roots.

Leaves: **Alternate**, elliptic to lance-shaped, sometimes widest above middle, {2} 5–20 cm long, highly variable, **toothless to sharply toothed or pinnately lobed**, green and almost hairless on the upper surface, **densely white-woolly beneath**.

Flowerheads: Yellowish, sometimes tinged reddish, small, with 20–40 **disc florets**; involucres {3} **4–5 mm high**, **woolly**, with blunt-tipped, often dark-edged bracts; numerous, short-stalked or stalkless, in erect, **elongating, branched clusters** (panicles); July–October.

Fruits: Dry, **hairless** seed-like fruits (achenes), **without bristles** (pappus) at the tip.

Habitat: Open woods and river flats; elsewhere, on open, rocky or gravelly alpine slopes or in heathlands.

Notes: This species has also been called *Artemisia herriotii* Rydb. • This aromatic plant was widely used as a medicine by native peoples. It was brewed to make medicinal teas for treating infections, fevers, cancer, rheumatism, tuberculosis, stomach aches, sprains, sore limbs, athlete's foot, itching, earaches, toothaches, chest ailments and colds. Modern herbalists give the tea to treat colds, constipation, kidney problems and internal bleeding. It is also used as a gargle for sore throats and applied to sore eyes and cuts. Raw leaves are chewed to relieve colds, flu, fever, headaches and ulcers, and they are also recommended as a mosquito repellent. Boiled leaves have been used as poultices on arthritic joints, and they were sometimes mixed with dog food as a supplement.

ASTER FAMILY

LONG-LEAVED ARNICA

Arnica longifolia D.C. Eat.
ASTER FAMILY (ASTERACEAE [COMPOSITAE])

Plants: Densely tufted perennial herbs; flowering stems several, erect, 30–60 cm tall, hairy, with smaller, sterile, leafy stems from the thickened base; forming large patches from **branched root crowns and short, stout underground stems** (rhizomes).

Leaves: Opposite, in 5–7 **pairs**, largest near mid-stem; blades narrowly lance-shaped to elliptic, gradually tapered to a pointed tip, 5–12 cm long and 1–2 cm wide, somewhat rough to the touch (finely stiff-hairy) and often slightly sticky (finely glandular-hairy), **toothless**; stalks short (lower leaves) to absent (upper leaves), often joining at the base to encircle the stem.

Flowerheads: Yellow or orange, bell-shaped, about 2 cm across, with 8–11 {13} ray florets around a dense cluster of disc florets; involucres **evenly glandular-hairy**, 7–10 mm high, with mostly equal, **sharp-pointed bracts**; receptacles without hairs or scales (naked); ray florets 1–2 cm long, with female parts only; disc florets 5-lobed, with both male and female parts; single to several, in open, branched clusters; July–August {September}.

Fruits: Slender, 5–10-nerved seed-like fruits (achenes), glandular-hairy, tipped with a tuft of **finely barbed, straw-coloured, hair-like bristles** (pappus).

Habitat: Open, rocky subalpine slopes and cliffs; elsewhere, on well-drained soil or rock by springs and seeps and along cliffs and riverbanks at moderate to high elevations in the mountains.

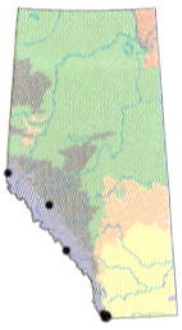

A. AMPLEXICAULIS

Notes: Another rare species, stem-clasping arnica (*Arnica amplexicaulis* Nutt.), resembles long-leaved arnica but has toothed leaves, longer underground stems (usually) and fewer stems that are seldom, if at all, tufted. Its flowerheads have pale-yellow rays, the leaves can be up to 6 cm wide and the pappus bristles are light brown and somewhat feathery. Stem-clasping arnica grows in moist woods and along streambanks (usually at lower elevations than long-leaved arnica); it flowers in July and August.

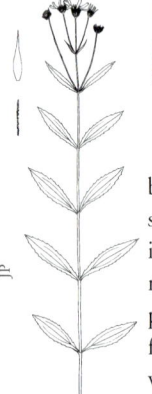

ASTER FAMILY **Dicots**

NODDING ARNICA

Arnica parryi A. Gray
ASTER FAMILY (ASTERACEAE [COMPOSITAE])

Plants: Perennial herbs; stems solitary, 20–50 {60} cm tall, usually silky-woolly near the base, glandular (at least on the upper parts); from slender underground stems (rhizomes).

Leaves: Opposite, basal and **in 2 or 3 {4} pairs on the stem**, much smaller upward on the stem; blades lance-shaped to egg-shaped, 5–20 cm long and 1.5–6 cm wide, toothless or finely toothed, more or less glandular and silky-hairy, tapered to a short stalk (lower leaves) or sometimes stalkless (upper leaves).

Flowerheads: Yellow, bell-shaped (sometimes narrowly so), about 1–2 cm across, **with disc florets only**; involucres mostly 10–14 mm high, glandular and finely hairy, with almost equal, sharply pointed bracts; receptacles without hairs or scales (naked); **florets** 5-lobed, **with both male and female parts**; several, often **nodding in bud**, in open, branched clusters; July–August.

Fruits: Slender, 5–10-nerved seed-like fruits (achenes), hairless, hairy or glandular, tipped with a tuft of **straw-coloured to light brown, barbed to almost feathery bristles** (pappus).

Habitat: Open woods at lower elevations in the mountains; elsewhere, on grassy hillsides and scree slopes in the foothills and at moderate elevations in the mountains.

Notes: This is the only *Arnica* in Alberta that lacks ray florets.

Dicots   ASTER FAMILY

ALPINE MEADOW BUTTERWEED

Packera subnuda D.K. Trock & T.M. Barkley
ASTER FAMILY (ASTERACEAE [COMPOSITAE])

Plants: Small, **hairless perennial** herbs; **stems** usually single and **5–15 {30} cm tall**; from creeping underground stems (rhizomes) with fibrous roots.

Leaves: Basal and alternate on the stems; basal leaves **2–4 {6} cm long**, with **broadly egg-shaped to round blades** that are much shorter than their stalks, edged with rounded teeth or more or less toothless; 2 or 3 stem leaves, smaller upward on the stem, lance-shaped, often widest above the middle, pinnately lobed.

Flowerheads: Yellow, **2–3 cm wide**, with few, 7–14 mm long ray florets around an 8–15 mm wide button of many disc florets; **involucres** {5} 6–8 mm high, **green or crimson**, **not black-tipped**, with **equal**, linear **bracts**; receptacles without hairs or scales (naked); ray florets with female parts only; disc florets with both male and female parts; solitary (sometimes 2); {June–September}.

Fruits: Seed-like fruits (achenes), 5–10-ribbed, tipped with a **tuft of soft, white, hair-like bristles** (pappus).

Habitat: Moist meadows in alpine and subalpine zones; elsewhere, in forest clearings.

Notes: This species has also been called *Packera buekii* D.K. Trock and T.M. Barkley, *P. cymbalarioides* (Buek) Weber & Löve and *Senecio cymbalarioides* Buek. It is *not* synonymous with *Senecio cymbalarioides* (T. & G.) Nutt. non Buek or *Senecio streptanthifolius* Greene. • Alpine meadow butterweed is distinguished from other small, hairless butterweeds (*Senecio* spp.) in Alberta by its single flowerheads. It is most similar to arctic butterweed (*Packera cymbalaria* (Pursh) W.A. Weber & A. Löve, previously called *Senecio cymbalaria* Pursh, *Senecio conterminus* Greenm. and *Senecio resedifolius* Less.), but alpine meadow butterweed has stems arising from a slender, creeping rhizome and its plants are either hairless or with permanent woolly hairs only in and near the leaf axils or at the base of the head. Arctic butterweed, on the other hand, has stems that arise from a short, branching, horizontal to ascending root crown, and its plants are usually hairless when mature but they are often lightly cobwebby-woolly, especially on the lower half of the plant. Arctic butterweed also grows in slightly different habitats, preferring dry, rocky alpine or subalpine sites. • Another rare, single-headed species, large-flowered ragwort (*Senecio megacephalus* Nutt.), is distinguished by its larger plants (20–50 cm tall) with narrower and longer lower leaves (10–20 cm long) and larger flowerheads (3–5 cm wide). It grows on rocky alpine and subalpine slopes in a few localities in the Waterton area. • The generic name *Packera* honours Dr. John Packer (1929–), a professor of botany at the University of Alberta who produced the second edition of the *Flora of Alberta* (Moss 1983).

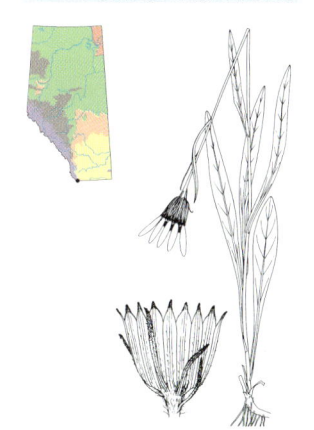

SENECIO MEGACEPHALUS

AMERICAN SAW-WORT

Saussurea americana D.C. Eat.
ASTER FAMILY (ASTERACEAE [COMPOSITAE])

Plants: Perennial herbs, **cobwebby-hairy** (at least when young); stems stout, 30–100 cm tall, unbranched except in the flower clusters; from short, stout underground stems (rhizomes).

Leaves: Alternate, numerous; blades **narrowly triangular**, up to 15 cm long and 8 cm wide, progressively smaller upward on the stem (becoming lance-shaped), sharply toothed, bright green on upper surface, **cobwebby-hairy beneath**; stalks up to 3 cm long (lower leaves) to absent (upper leaves).

Flowerheads: Purple, with **disc florets only**, 10–14 mm high; florets with both male and female parts, 11–12 mm long, 5-lobed; bracts hairy, green with dark edges; outer bracts broad; inner bracts narrow, 7–8 mm long, progressively longer than outer bracts; numerous, short-stalked, in **crowded, erect, sometimes flat-topped clusters**; {July–August} September.

Fruits: Light brown, hairless seed-like fruits (achenes), 4–6 mm long, longer than wide, tipped with 2 rows of bristles (pappus); outer bristles short, hair-like, soon shed; **inner bristles feathery**, joined at the base, shed together.

Habitat: Moist open slopes and meadows.

Notes: Brook ragwort (*Senecio triangularis* Hook.) has similar large, triangular, toothed leaves but its flowers are yellow.

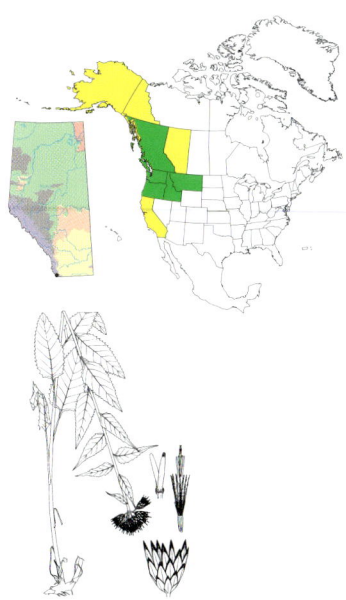

Rare *Vascular* Plants of Alberta

Dicots

ASTER FAMILY

ELK THISTLE

Cirsium scariosum Nutt.
ASTER FAMILY (ASTERACEAE [COMPOSITAE])

Plants: Erect, **spiny biennial** herbs; stems 30–80 cm tall, fleshy, ribbed, hairy, mostly unbranched; from taproots.

Leaves: Alternate, gradually smaller toward the top of the stem, stalkless, **with spiny edges**; lower leaves narrowly egg-shaped (wider above the middle), tapering at the base, **hairless on the upper surface**, woolly beneath, **lobed only slightly to halfway to the midrib**; upper leaves narrowly elliptic to linear, shallowly lobed.

Flowerheads: White to pink or reddish purple, with **disc florets only**; involucres 2–3 cm high, **without cobwebby hairs or glands**, the inner bracts often longer than the outer; outer bracts broadly lance-shaped, 2.5–5 mm wide at the base and tipped with a slender spine 2–4 mm long; **receptacle densely bristly**; disc florets with both male and female parts, **5 lobes of the corolla elongated**, styles with a thickened, hairy ring, anthers 6–10 mm long; 5–15, short-stalked, clustered at the stem tip and sometimes in upper leaf axils; June–September.

Fruits: Oblong, somewhat flattened seed-like fruits (achenes), 5.5–6.5 mm long, hairless, brown with a narrow, yellowish band at the top, tipped with a **tuft of feathery bristles** (pappus) 4–5 mm long, **shorter than the corolla**.

Habitat: Moist meadows, streambanks, open woodlands and slopes from low to high elevations.

Notes: Some taxonomists have included this species with *Cirsium hookerianum* Nutt. and others with *Cirsium foliosum* (Hook.) DC. It has also been called *Cnicus scariosus* (Nutt.) A. Gray and *Carduus scariosus* (Nutt.) Heller. The taxonomy is still problematic. • The generic name *Cirsium* is from the Greek *kirsos* (a swollen vein), for which thistles, called *kirsion*, were a reputed remedy.

ASTER FAMILY   **Dicots**

RUSH-PINK

Stephanomeria runcinata Nutt.
ASTER FAMILY (ASTERACEAE [COMPOSITAE])

Plants: Perennial herbs with **milky juice**; stems slender, 10–20 cm tall, **freely branched**; from taproots.

Leaves: Alternate, **linear to narrowly lance-shaped**, 1–7 cm long, 4–15 mm wide, with **sharp, backward-pointing teeth**, stalkless; upper leaves smaller and toothless.

Flowerheads: Pink or white, 10–15 mm long, **tubular**, usually **5-flowered**, with **ray florets only**; florets with both male and female parts; involucres 9–12 mm high, with 5 main bracts; **numerous**, at branch tips; June–July.

Fruits: Pitted or bumpy seed-like fruits (achenes), about 5 mm long, tipped with a tuft of white, **feathery bristles** (pappus).

Habitat: Dry hills and plains, often on eroded sites.

Notes: Skeletonweeds (*Lygodesmia* spp.) resemble rush-pink, but none of their leaves are toothed and the pappus bristles at the tips of their achenes are hair-like (not feathery). • Major modifications to waterways pose the greatest threat to the survival of this species in Alberta. • Although it is an attractive plant, rush-pink is not useful for horticultural applications because it prefers rather extreme climatic and habitat conditions. • The generic name *Stephanomeria* is derived from the Greek *stephanus* (a crown or wreath) and *mereia* (a division). The specific epithet *runcinata* is Latin for 'a large saw,' referring to the downward-pointing, saw-like teeth of the leaves.

Rare *Vascular* Plants of Alberta

Dicots

ASTER FAMILY

PRAIRIE FALSE DANDELION

Nothocalais cuspidata (Pursh) Greene
ASTER FAMILY (Asteraceae [Compositae])

Plants: Perennial herbs **with milky juice; stems single, leafless,** erect, {5} 10–30 cm tall, hairless or with woolly hairs on the upper parts; from stout taproots.

Leaves: Basal, several, 7–30 cm long and 3–20 mm wide, **linear,** tapered to a slender point; **fringed with long, soft hairs; edges** toothless, usually **irregularly curled** and crinkled.

Flowerheads: Yellow, about 2 cm high and wide, **with ray florets only;** florets with both male and female parts; involucres 17–25 mm high, often speckled, with **almost equal** (slightly overlapping), lance-shaped to linear, often long-tapered **bracts; solitary;** {April–June}.

Fruits: Dry seed-like fruits (achenes), 8–10 mm long, gradually tapered to both ends, **tipped with a tuft of hair-like bristles** (pappus) **mixed with slender, needle-like scales** (not much wider than the hairs).

Habitat: Prairies; elsewhere, in dry, open places, often on gravelly soil and in parkland.

Notes: This species has also been called *Microseris cuspidata* (Pursh) Schultz-Bip. and *Agoseris cuspidata* (Pursh) Raf. • The specific epithet *cuspidata* (pointed) refers to the long, slender points on the tapered bracts and leaves. • Prairie false dandelion could be confused with the common yellow false dandelion (*Agoseris glauca* (Pursh) Raf.), but that species does not have long, white hairs on its leaf edges and does not have slender scales in the pappus (at the tip of its achenes).

ASTER FAMILY | **Dicots**

ANNUAL SKELETONWEED
Shinneroseris rostrata (A. Gray) S. Tomb
ASTER FAMILY (ASTERACEAE [COMPOSITAE])

Plants: **Annual** herbs with **white milky juice**; **stems stiff, slender**, erect or spreading, 30–60 {10–100} cm tall, striped, hairless and **often nearly leafless**, with **strongly ascending branches**; from tough, thin taproots.

Leaves: Alternate, **5–20 cm long** on lower stems (smaller and awl-shaped on the upper stems), narrowly linear, sharp-tipped, 3-nerved, stalkless.

Flowerheads: **Pink or rose**, about 12 mm across, with 7–9 {6–10} **ray florets**; florets with both male and female parts; **involucres cylindrical, 10–16 mm high**, with 7–9 main bracts and a few small basal ones; receptacles flat, without hairs or scales (naked); numerous, at the tips of short, scaly branches in elongated clusters toward the main branch tips (racemes); August {July–September}.

Fruits: Slender, spindle-shaped seed-like fruits (achenes), 8–10 mm long, 4–8-ribbed, tipped with a **tuft of whitish, soft, hair-like bristles** (pappus).

Habitat: Sandy banks and dunes, where there is considerable loose sand; elsewhere, in canyons and on sandy plains.

Notes: This species has also been called *Lygodesmia rostrata* A. Gray. • Annual skeletonweed is sometimes confused with common skeletonweed (*Lygodesmia juncea* (Pursh) D. Don), a much more widespread perennial plant that has creeping underground stems (rhizomes), yellowish milky juice and smaller (seldom over 5 mm long), usually scale-like leaves.

SLENDER HAWK'S-BEARD
Crepis atribarba Heller
ASTER FAMILY (Asteraceae [Compositae])

Plants: Perennial herbs with milky juice; 1 or 2 stems, **15–70 cm tall**; from branched, somewhat woody root crowns on **taproots**.

Leaves: Mostly basal, but also alternate on the stem; **lower leaves** 10–35 cm long, pinnately cut into **linear to narrowly lance-shaped segments**, hairless to woolly, mostly **without teeth**; upper stem leaves much shorter, linear, toothless.

Flowerheads: Yellow, with about 10–35 {40} **ray florets**; involucres grey-woolly, sometimes with bristly, black hairs, **lacking glands**, 8–14 {15} mm high and 3–5 mm wide, with **2 overlapping rows of bracts**, the 5–10 outer bracts less than half as long as the 8–10 inner bracts; florets 10–18 mm long, with both male and female parts; 3–30, in open, branched, almost leafless clusters; {May} June–July.

Fruits: Cylindrical, 10–20-ribbed seed-like fruits (achenes), gradually tapered to a slender point, **usually greenish**, tipped with a tuft of **whitish, hair-like bristles** (pappus).

Habitat: Dry, open, grassy slopes at moderate elevations in the mountains.

Notes: This species has also been called *Crepis exilis* Osterh. and *Crepis occidentalis* Nutt. var. *gracilis* DC., and has been misspelled *Crepis atrabarba* Heller. • Small-flowered hawk's-beard (*Crepis occidentalis* Nutt.) is also

C. OCCIDENTALIS

ASTER FAMILY — **Dicots**

C. INTERMEDIA

rare in Alberta. It is distinguished from slender hawk's-beard by its broader (5–10 mm wide), glandular-hairy involucres, its smaller plants (mostly less than 35 cm tall) and its brownish seeds. The lower leaves of small-flowered hawk's-beard may have broader segments that are less deeply lobed and toothed. It grows on dry, eroding slopes in the prairies and flowers in May and June {July}.

• Another rare species, intermediate hawk's-beard (*Crepis intermedia* A. Gray), is intermediate in form between slender hawk's-beard and small-flowered hawk's-beard. Like slender hawk's-beard, it lacks gland-tipped hairs and has involucres less than 5 mm wide, but like small-flowered hawk's-beard, it has brownish (or yellowish) seeds and its leaf segments are usually toothed. Intermediate hawk's-beard grows in dry, open areas and flowers in {May–July} August. • The generic name *Crepis* was originally used by the Roman naturalist Pliny (AD 23–79) to refer to another plant. It was taken from the Greek *krepis* (a boot or sandal).

WESTERN WHITE LETTUCE, WING-LEAVED RATTLESNAKEROOT

Prenanthes alata (Hook.) D. Dietr.
ASTER FAMILY (ASTERACEAE [COMPOSITAE])

Plants: Perennial herbs with **milky juice**; stems single, leafy, erect, 30–70 {15–80} cm tall, hairy near the tip; from slightly thickened roots.

Leaves: Alternate, several; blades **arrowhead-shaped or heart-shaped** (lower and middle leaves) to oblong or lance-shaped (upper leaves), up to 12 cm long and 11 cm wide, hairless, thin, with or without irregular teeth; stalks with narrow, flattened sections (wings) along the edges.

Flowerheads: White, about 1 cm across, **with 10–15 ray florets only**; involucres 10–13 mm high, with **about 8, sparsely glandular-hairy bracts**; ray florets with both male and female parts; few to several, long-stalked, from upper

Rare *Vascular* Plants of Alberta

Dicots

ASTER FAMILY

leaf axils, **forming broad, branched, flat-topped clusters** (corymbs), with long side branches spreading or **nodding in fruit**; August {July–September}.

Fruits: Ribbed, cylindrical seed-like fruits (achenes), tipped with a **tuft of dirty-white, hair-like bristles** (pappus).

Habitat: Edges of moist woods and thickets; elsewhere, on streambanks and in other moist, often shaded places.

Notes: This species has also been called *Prenanthes hastata* Jones. • Another rare Alberta species, purple rattlesnakeroot (*Prenanthes sagittata* (A. Gray) A. Nels.), also called arrow-leaved rattlesnakeroot, resembles western white lettuce but is distinguished by its narrow, elongated cluster of ascending (not nodding) flowerheads (though these may become spreading when in fruit). It flowers in July and August. The ranges of the 2 species are widely separated. Purple rattlesnakeroot grows in moist montane and foothills woodlands and thickets in extreme southwestern Alberta, whereas western white lettuce is a west coast species with a widely disjunct population in the Swan Hills. • These 2 plants could be confused with yet another rare species, large-flowered brickellia (*Brickellia grandiflora* (Hook.) Nutt., p. 217), but brickellia does not have milky juice or winged leaf stalks, and it has more distinctly toothed leaves and striped involucral bracts.
• The generic name *Prenanthes*, from the Greek *prenes* (drooping) and *anthe* (blossom), refers to the nodding mature flowerheads. The specific epithet *alata* means 'winged,' referring to the flattened edges of the leaf stalks.

P. SAGITTATA

Rare *Vascular* Plants of Alberta

ASTER FAMILY   **Dicots**

WOOLLY HAWKWEED

Hieracium cynoglossoides Arv.-Touv. *ex* A. Gray
ASTER FAMILY (ASTERACEAE [COMPOSITAE])

Plants: Perennial herbs with **milky juice**, bristly with **long, spreading, white or yellowish hairs**, sometimes **also with star-shaped hairs**; stems single, erect, 20–100 cm tall, covered with long, white or yellowish bristles; from **short underground stems** (rhizomes).

Leaves: Alternate, mainly on the lower ⅓ of the stem, {5} 8–20 cm long, **linear to lance-shaped**, tapered to winged stalks, toothless or with indistinct small teeth; upper leaves much smaller, stalkless.

Flowerheads: Yellow, about **1 cm across** with 15–50 ray florets 7–14 mm long; florets with both male and female parts; involucres 7–12 mm high, with **1 row of bracts** covered **with long, dark-based, white or yellowish bristles**, sometimes with short, black, gland-tipped hairs; several, in flat-topped, often compact clusters (corymbs); {June} July–September.

Fruits: Slender, ribbed seed-like fruits (achenes), tapered to the base, blunt and with a tuft of **white, hair-like bristles** (pappus) at the tip.

Habitat: Open woods and montane slopes.

Notes: This species includes *Hieracium albertinum* Farr. • Hawkweeds are often difficult to identify. Many can produce large amounts of seed without fertilization, and when this happens each of the offspring is genetically identical to the parent. A single successful plant can produce large populations of identical offspring, each capable of reproducing and spreading to other areas. Hundreds of species (microspecies) have been identified in this genus. • The genus name *Hieracium* was taken from the Greek *hieros* (hawk). The ancient Romans (e.g., Pliny, AD 23–79) believed that hawks used these plants to maintain their excellent vision, and consequently hawkweeds were also used to make salves and drops for improving human eyesight.

Rare *Vascular* **Plants** of Alberta

Dicots

ASTER FAMILY

PINK FALSE DANDELION

Agoseris lackschewitzii D. Henderson & R. Moseley
ASTER FAMILY (ASTERACEAE [COMPOSITAE])

Plants: Perennial herbs **with milky juice**; 1–3 flowering stems, leafless, 6–35 {50} cm tall, hairy along their entire length; from slender taproots.

Leaves: Basal; **blades** lance-shaped, widest above the middle, 6–20 {4–27} cm long and 0.7–2.5 {3.1} cm wide, **thin, hairless**, often with purple flecks, toothless or with a few small teeth along the edges; stalks edged with thin, flattened sections (wings) and fringed with hairs.

Flowerheads: Pink in bud and in full flower, with 50–70 5-toothed **ray florets**, each **5–10 mm long** and 1.5 mm wide; florets with both male and female parts; involucral bracts in several overlapping rows, light green with a dark purple central stripe and purple mottling, 11–25 mm long; inner bracts lance-shaped, pointed, with transparent edges; **outer bracts** broader and blunt, **hairy at the base; solitary**; July–August {September}.

Fruits: Rounded, 10-ribbed seed-like fruits (achenes), 6–8 mm long, with a slender, **gradually tapered point** (beak) that is **shorter than the body** (4.2–6.6 mm long), tipped with a tuft of 2 rows of white, hair-like bristles (pappus) 6–12 mm long.

Habitat: Moist subalpine and alpine meadows.

Notes: Other species of false dandelions (*Agoseris* spp.) have orange or yellow flowerheads that can age or dry to a pinkish colour; therefore, care should be taken not to use flower colour as the only differentiating characteristic. The achenes of orange false dandelion (*Agoseris aurantiaca* (Hook.) Greene) have abrupt edges at the tip of the body that form clearly visible stair-steps narrowing from the body to the tip. The leaves of yellow false dandelion (*Agoseris glauca* (Pursh) Raf.) are usually thicker, wider and greyer. • This species is named in honour of Klaus Lackschewitz (1911–95), a Montana botanist who made outstanding contributions to the knowledge of mountain flora in that state.

ASTER FAMILY **Dicots**

TALL BLUE LETTUCE

Lactuca biennis (Moench) Fern.
ASTER FAMILY (ASTERACEAE [COMPOSITAE])

Plants: Coarse **annual or biennial** herbs with **milky juice**, somewhat weedy, essentially hairless; stems leafy, 50–200 {250} cm tall, unbranched (except in the flower cluster); from taproots.

Leaves: Alternate, deeply pinnately divided into sharply toothed lobes (sometimes merely toothed), 10–40 cm long, 4–20 cm wide, hairless (sometimes with hairs on the lower side of the veins); stalks winged, more or less clasping the stem.

Flowerheads: Bluish to creamy white (rarely yellowish), about 5 mm across, with 13–34 {55} **ray florets**; florets with both male and female parts; involucres narrowly cone-shaped to cylindrical, 10–14 mm high; few to many, in large, fairly narrow, branched clusters (panicles); July–August.

Fruits: Thin, flattened, many-nerved seed-like fruits (achenes), 4–5.5 {7} mm long, tipped with a tuft of **light-brown, soft, hair-like bristles** (pappus) at the end of a **short, stout point** (without a slender beak).

Habitat: Moist woods and clearings; elsewhere, in swampy sites and by hot springs.

Notes: This species has also been called *Lactuca spicata* (Lam.) A.S. Hitchc. • **Caution: the milky sap of these plants can cause a skin reaction and internal poisoning**. Some tribes used the sap as a treatment for rashes, pimples and other skin problems. They also applied the leaves to insect stings. The leaf tea was used as a sedative and nerve tonic, and the root tea was taken to treat diarrhea, heart and lung problems, bleeding, nausea, and general aches and pains. • Tall blue lettuce is related to cultivated lettuce (*Lactuca sativa* L.).

Grass-like Plants

RUSH FAMILY

Grass-like Plants

PARRY'S RUSH

Juncus parryi Engelm.
RUSH FAMILY (JUNCACEAE)

Plants: Perennial herbs; stems slender, round, 10–30 cm tall; densely tufted, sometimes forming mats.

Leaves: Most reduced to **brown bladeless sheaths**; uppermost leaf usually with a **slender, cylindrical blade** 3–6 cm long.

Flower clusters: Small, irregular, widely branched (cymes), with **1–3 single flowers from** the base of a slender, cylindrical, 2–3 {8} cm long **bract that appears to be a continuation of the stem**; 6 tepals, scale-like, 6–7 mm long, lance-shaped, with dry, translucent edges; the 3 outer tepals slightly longer and more slenderly pointed than the 3 inner tepals; July {August–September}.

Fruits: Lance-shaped cylindrical capsules, pointed at the tip, equalling or slightly longer than the tepals; **seeds** 0.6 mm long, **with a slender tail at each end** (at least as long as the body).

Habitat: Wet meadows and slopes in the mountains.

Notes: Thread rush (*Juncus filiformis* L.) is a rare species of Alberta fens and marshes, and on lakeshores and streambanks elsewhere. Its flower cluster, which appears in June and July, also seems to be growing from the side of the stem, because the cylindrical bract looks like a continuation of the stem. In thread rush, this bract is at least as long as the stem below, and usually longer. The round, 10–60 cm tall stems are very slender (rarely more than 1 mm wide), with fine, lengthwise grooves, and they grow in rows or small clumps from slender, elongated underground stems (rhizomes). The leaves of thread rush are reduced to tight sheaths, sometimes with tiny, bristle-like blades. The flower clusters are 2–4 cm long and sparingly branched, with 5–15 flowers. Each has 6 small (2.5–4.5 mm long), greenish, lance-shaped tepals, which are slightly longer than the egg-shaped, abruptly sharp-pointed capsules. The tiny (0.4 mm) seeds are pointed but not tailed.

J. FILIFORMIS

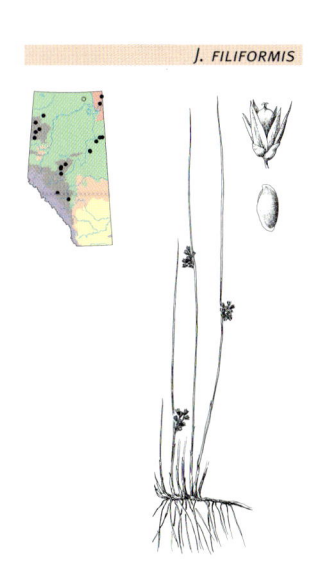

Rare *Vascular* Plants of Alberta

Grass-like Plants

RUSH FAMILY

FEW-FLOWERED RUSH

Juncus confusus Coville
RUSH FAMILY (JUNCACEAE)

Plants: Tufted perennial herbs; stems erect, 30–50 cm tall; from dense fibrous roots.

Leaves: On the lower stem; blades **thread-like**, less than 1 mm wide, **with 2 conspicuous, membranous lobes** (auricles) at the base (at the top of the sheath).

Flower clusters: Pale yellowish brown, **0.5–2 cm long, with a few flowers** on unequal, short stalks **at the base of a slender, erect bract several times as long as the cluster**; 6 tepals, 3.5–4 mm long, yellowish with a broad, yellowish green stripe and dry, translucent edges; {June–August}.

Fruits: Egg-shaped cylindrical capsules, 3-sided and notched at the tip, 3–3.5 mm long, slightly shorter than the tepals; **seeds oblong, minutely pointed**, about 0.5–1 mm long.

Habitat: Moist soil in open woods, thickets, and grassland.

Notes: The generic name *Juncus* is the old Latin name for rushes.

RUSH FAMILY | **Grass-like Plants**

Regel's rush

Juncus regelii Buch.
RUSH FAMILY (JUNCACEAE)

Plants: Perennial herbs; stems slightly flattened, 10–60 cm tall, never reaching above the leaves, single or clustered; from stout, elongated underground stems (rhizomes).

Leaves: Blades flat, grass-like, 10–15 cm long, 2–4 mm wide, **lacking basal lobes** (auricles) or with lobes less than 1 mm long; sheaths with narrow, membranous edges.

Flower clusters: Small, irregular, widely branched (cymes), with 1–3 {5} **round, chestnut-brown heads** on unequal, 0–8 cm long branches, each head with **10–30 flowers** and 8–20 mm in diameter; **6 tepals**, scale-like, 4–6 mm long, **with a broad, greenish midstripe**, broadly lance-shaped, rough with tiny bumps near tips; the outer 3 tepals with slender-pointed tips; the inner 3 tepals slightly shorter and broader with wide, translucent edges and less pointed tips; basal bract 1–4 cm long; June {July–August}.

Fruits: Egg-shaped cylindrical capsules, rounded at the tip, about the same length as the tepals; **seeds** 1.2–1.8 mm long **with a long tail at each end** (each about as long as the body).

Habitat: Wet montane, subalpine and alpine meadows.

Notes: Long-styled rush (*Juncus longistylis* Torr.) is a much more common species that is similar to Regel's rush, but its leaf blades have 2 prominent rounded lobes at their bases, its flowerheads have 3–8 flowers, its seeds are smaller (0.4–0.5 mm long) and abruptly pointed (without tails), and its capsules are tipped with abrupt, long beaks.

Grass-like Plants

RUSH FAMILY

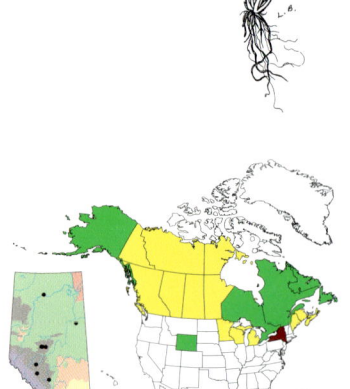

MARSH RUSH

Juncus stygius L. ssp. *americanus* (Buch.) Hult.
RUSH FAMILY (JUNCACEAE)

Plants: Perennial herbs; **stems thread-like, erect, 6–35 cm tall**, single or in small tufts; from slender underground stems (rhizomes).

Leaves: Basal and on the stem, {1} 2 or 3, very slender and **thread-like**, cylindrical.

Flower clusters: Small, **single head** (sometimes 2) **of 1–4 flowers** above a single, erect **bract of equal length** (or slightly longer); 6 tepals, scale-like, 3–6 mm long, lance-shaped to awl-shaped, pale brown with a reddish midvein; August.

Fruits: Light yellowish brown, egg-shaped **capsules**, abruptly sharp-pointed, 6–8.5 mm long, **projecting past the tepals**; seeds 3–3.5 mm long, **with a conspicuous tail at each end**.

Habitat: Fens; elsewhere, in mossy areas around springs and seepages.

Notes: Two-glumed rush (*Juncus biglumis* L.) is rare in moist alpine areas of western Alberta. Its erect stems form loose clumps 2–30 cm high, and its leaves, on the lower stem, are very slender and cylindrical with blades 2–7 cm long. The lowermost bract is leaf-like and extends beyond a solitary flowerhead of 1–4 flowers, with dark brown to blackish, 2.5–4.5 mm long, blunt-tipped tepals. The capsules are egg-shaped with notched tips and dark purplish streaks or mottles, and the seeds are about 1.5 mm long, with a short, white tail at each end. Two-glumed rush flowers in July.

J. BIGLUMIS

RUSH FAMILY   **Graminoids**

NEVADA RUSH

Juncus nevadensis S. Wats.
RUSH FAMILY (JUNCACEAE)

Plants: Perennial herbs; stems 10–70 cm tall; from slender, elongated underground stems (rhizomes).

Leaves: Blades cylindrical, 5–20 cm long, 1–2 mm thick, divided by cross-partitions, with 2 small (1–3 mm), membranous lobes (auricles) projecting from the base of the blade.

Flower clusters: Light brown to dark purplish brown, open, irregularly branched cymes, 1–12 cm long, with 3–30 few-flowered, 4–10 mm wide heads on unequal branches; **6 tepals, lance-shaped, slenderly pointed**, 3–5.5 mm long, with dry, translucent edges, the outer 3 segments sometimes slightly longer than the inner segments; **lowermost bract much shorter than the flower cluster**; {July–August}.

Fruits: Dark brown, egg-shaped to cylindrical capsules, abruptly narrowed to a short, firm tip, equal to or slightly shorter than the tepals; **seeds 0.4 mm long, minutely pointed**.

Habitat: Shorelines and other wet sites from lowland to alpine elevations.

Notes: Short-tailed rush (*Juncus brevicaudatus* (Engelm.) Fern.) is also rare in Alberta, but it grows in marshes and on shores in the northeastern corner of the province. It forms tufts of 10–50 {70} cm tall stems, each with 2–4 cylindrical leaves that are divided by cross-partitions. The flower cluster is narrow and 3–12 cm long, with few to many, 2–7-flowered heads. The greenish to light brown or reddish, 2–3 mm long tepals are narrowly lance-shaped with blunt or pointed tips and dry, translucent edges. The reddish brown capsules of short-tailed rush are longer than the tepals, reaching 2.5–4.5 mm in length, with abruptly narrowed, short, firm tips. Unlike Nevada rush, the seeds of short-tailed rush have short appendages (tails) at each end. • Nevada rush provides good to excellent cattle feed, especially early in the year, and later, in hay. It is less palatable to sheep, particularly when dry.

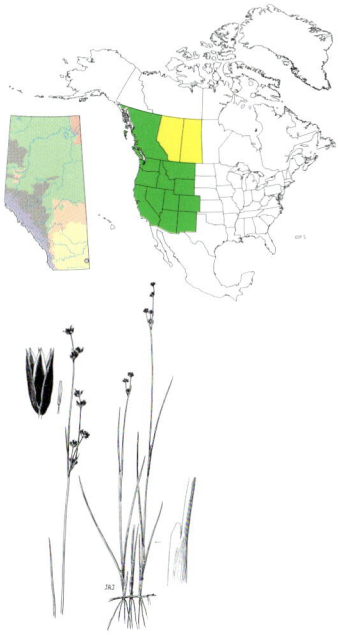

J. BREVICAUDATUS

Rare *Vascular* Plants of Alberta

Grass-like Plants

RUSH FAMILY

SHARP-POINTED WOOD-RUSH

Luzula acuminata Raf.
RUSH FAMILY (JUNCACEAE)

Plants: Loosely tufted perennial herbs; stems slender, erect, 10–40 cm tall; from fibrous roots.

Leaves: Mainly basal, up to 30 cm long, 3–12 mm wide; stem leaves smaller; sheath forming a closed cylinder (fused down 1 side).

Flower clusters: Open, flat-topped (umbels), with **single flowers** (rarely pairs or small clusters of flowers) on spreading, **unbranched, 3–6 cm long, thread-like stalks**; each flower with 6 scale-like tepals, 2.5–4.5 mm long, reddish brown with thin, translucent edges; {April–May}.

Fruits: Abruptly sharp-pointed capsules, **3–4 mm long, projecting past the tepals**; seeds 1.5–2.5 mm long (including a **conspicuous appendage** (caruncle) almost as long as the body of the seed).

Habitat: Moist woodland, often on disturbed sites; elsewhere, on rocky sites and clearings in hardwood, coniferous and mixedwood forests.

Notes: This species has also been called *Luzula saltuensis* Fern. • Another rare Alberta species, reddish wood-rush (*Luzula rufescens* Fisch. *ex* E. Mey.), is very similar to sharp-pointed wood-rush, but it has narrower (less than 5 mm wide) basal leaves, smaller seeds (about 1.5 mm long), and smaller flowers (up to 3.5 mm long) on shorter (up to 3 cm long) stalks. Reddish wood-rush grows on damp grassy slopes, on the edges of bogs and marshes and on moist sand and gravel bars.

L. RUFESCENS

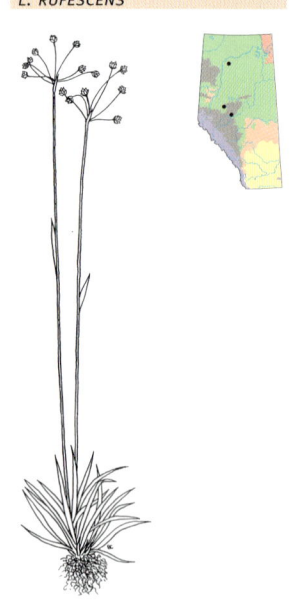

RUSH FAMILY   **Grass-like Plants**

Greenland wood-rush

Luzula groenlandica Bocher
RUSH FAMILY (JUNCACEAE)

Plants: Densely tufted perennial herbs; stems leafy, erect, 10–50 cm tall; from fibrous roots.

Leaves: Basal and alternate, **flat, 3–7 mm wide**, often reddish brown, shorter than the flowering stems, with **blunt, thickened** (callused) **tips**; 2 or 3 stem leaves, short; sheath forming a closed cylinder (fused down 1 side).

Flower clusters: Compact, 3–5 cm long, **head-like**, sometimes with 1 or 2 additional flowerheads on short side branches, above a **conspicuous, thick-tipped, leafy bract** that sticks out at an angle from the main stalk; each flower with 6 scale-like **tepals, 2–2.5 mm long**, reddish brown with broad, translucent edges; inner 3 tepals clearly shorter than the outer 3.

Fruits: Dark brown capsules, 2.5–4 mm long, abruptly sharp-pointed, projecting past the tepals; **seeds 0.8–1.1 mm long** (including the **short appendage** [caruncle]).

Habitat: Moist, turfy tundra.

Notes: Greenland wood-rush is easily confused with the more common species field wood-rush (*Luzula multiflora* (Retz.) Lej.). Field wood-rush is distinguished by its slightly larger tepals (2.5–3.5 mm long) and larger seeds (1.1–1.4 mm long) with relatively long caruncles.

Rare *Vascular* Plants of Alberta

Grass-like Plants

SEDGE FAMILY

AWNED NUT-GRASS

Cyperus squarrosus L.
SEDGE FAMILY (CYPERACEAE)

Plants: Sweet-smelling, tufted **annual** herbs; **stems slender, 1–15 cm tall**, 3-sided; from fibrous roots.

Leaves: Mostly basal, 2 or 3 per stem, about as long as the stems; blades 2–10 cm long, 1–2 mm wide.

Flower clusters: Crowded and head-like or more open with 3 {1–5} similar branches (rays) each 3–7 cm long; 2–4 leaf-like bracts at the base, much longer than the flower cluster; spikelets 2–10 mm long, flattened, lance-shaped and widest above the middle, 5–20-flowered, greenish to pale brown; flowers with **1 stamen** and 3 stigmas, **lacking petals and sepals**, borne in 2 vertical rows in the axils of 1–2 mm long, several-nerved, egg-shaped to lance-shaped **scales** that **taper to a slender, outwardly curved point 0.5–1.2 mm long**; {June–July}.

Fruits: Dry, 3-sided seed-like fruits (achenes), egg-shaped and widest above the middle, 1 mm long, brown, shed with the scales; July–August.

Habitat: Moist soil, usually on sandy river flats; elsewhere, on muddy lakeshores.

Notes: This species has also been called *Cyperus aristatus* Rottb. and *Cyperus inflexus* Muhl. • Another rare species, sand nut-grass (*Cyperus schweinitzii* Torr.), has larger (10–70 cm tall), perennial plants, with rough stems bearing hard, corm-like thickenings at the base. Its flowers have 3 stamens and larger (about 3.5 mm long), light-green to light brown scales with short, erect tips, its spikelets are borne on very uneven stalks in irregular clusters and its achenes are 2.5–3.5 mm long, elliptic and pointed at both ends. Sand nut-grass grows on dry sandy soil, including active sand dunes. • The specific epithet *squarrosus* refers to the scales of the spikelets, which have long, curved tips that bend outward at right angles to the main stalk, an unusual characteristic for species of this genus in Canada.

C. SCHWEINITZII

SEDGE FAMILY — Grass-like Plants

ENGELMANN'S SPIKE-RUSH

Eleocharis engelmannii Steud.
SEDGE FAMILY (CYPERACEAE)

Plants: Tufted **annual** herbs; stems several, ascending, 10–40 {5–50} cm tall, 0.5–2 mm thick, ribbed; from fibrous roots.

Leaves: Reduced to bladeless sheaths; upper sheaths pale brown, tipped with a tiny, tooth-like blade.

Flower clusters: Single, erect spikelets at stem tips, **cylindrical to egg-shaped**, 5–15 mm long, with many **spirally arranged flowers**; flowers **with 3 stamens and 1 ovary** with a **2** {3}**-pronged style**, borne in the axil of a 1.7–2.5 mm long, brownish scale with translucent edges; {June–September}.

Fruits: Smooth, shining, **lens-shaped** seed-like fruits (achenes), 1–2 mm long, **broadly egg-shaped**, widest above the middle, **abruptly tipped with a low** (about 0.3 mm high), broad, **pyramid-shaped projection** (tubercle) **above a tiny ridge**, also with **5–7** brownish, **hair-like bristles** (modified petals and sepals) from the base.

Habitat: Wet places; elsewhere, in moist sandy soils, wet sand, peat, mud and marshes.

Notes: This species has also been called *Eleocharis ovata* (Roth) R. & S. var. *engelmannii* Britt. and *Eleocharis obtusa* (Willd.) Schultes var. *engelmannii* (Steud.) Gilly. • The generic name *Eleocharis* comes from the Greek *helos* (marsh) and *charis* (grace). The specific epithet *engelmannii* honours St. Louis physician and botanist George Engelmann (1809–84).

Grass-like Plants **SEDGE FAMILY**

SLENDER SPIKE-RUSH

Eleocharis elliptica Kunth
SEDGE FAMILY (CYPERACEAE)

Plants: Scattered or loosely clustered perennial herbs; **stems 5–40 {90} cm tall**, very slender, compressed to almost cylindrical, {4} **6–8 {10}-angled**, with {4} 6–8 vascular bundles; from stout, creeping, **reddish to purplish black underground stems** (rhizomes).

Leaves: Reduced to **bladeless sheaths**, conspicuously reddish purple near the base.

Flower clusters: Single spikelets, **oval to egg-shaped** and widest above the middle, {3} **5–12 mm long**, with 10–30 spirally arranged flowers; flowers with 3 stamens and 1 ovary tipped with {2} **3 stigmas**, borne in the axil of an **oval**, 2–3 mm long, **reddish purple to brown or blackish scale** with pale, translucent edges and a notched tip with a small, sharp point in the centre of the notch; {May–August}.

Fruits: Unequally 3-sided seed-like fruits (achenes), 0.6–1 {1.5} mm long, **yellow to dull orange**, with a **honeycomb-netted** surface, tipped with a **short**, pyramid-shaped projection (tubercle) that is **ringed by a tiny ridge**, usually bearing very small, hair-like bristles at the base.

Habitat: Wet places, usually neutral or calcareous conditions.

Notes: This species has also been called *Eleocharis tenuis* (Willd.) Schultes var. *borealis* (Svenson) Gleason and *E. compressa* Sullivant var. *borealis* Drepalik & Mohlenbrock. • Quill spike-rush (*Eleocharis nitida* Fern.) has been reported to occur in Alberta. It is distinguished from slender spike-rush by its achenes, which are minutely wrinkled or roughened, but lack honeycombed ridges. It is also a smaller plant (2–15 cm tall), with few-ribbed, cylindrical stems, and the scales in its spikes have entire (not notched) tips. Quill spike-rush grows in damp, rocky or sandy sites.

E. NITIDA

SEDGE FAMILY Grass-like Plants

SLENDER BEAK-RUSH

Rhynchospora capillacea Torr.
SEDGE FAMILY (CYPERACEAE)

Plants: Tufted perennial herbs; stems slender, 10–40 cm tall, 3-sided, solid; from very short underground stems (rhizomes) and fibrous roots; also with long, thin, bulb-like outgrowths (turions) about 1 mm in diameter and 1 cm long, formed at the base of the stem in late summer.

Leaves: Mostly basal with bristle-like blades (at most 0.5 mm wide) rolled inward at the edges, equalling or slightly longer than the stems; **stem leaves** with **bristle-like** blades; sheaths fused in a tube (closed).

Flower clusters: 1 or 2 rather loose, ascending, ellipsoid heads (glomerules), each with 1–4 {10} **spikelets**; spikelets ellipsoid, 5–7 mm long, reddish brown, on erect stalks, with spirally arranged scales, the upper 2 or 3 scales enclosing flowers; flowers with **3 stamens** bearing 1–2.5 mm long anthers, a **single ovary** (the upper flowers sometimes with stamens only) and **6** (rarely 9 or 12) **small, hair-like bristles** (modified sepals and petals), borne in the axils of thin, brown, lance-shaped scales with a midrib extending beyond the tip; **the lowermost scale without a flower**; {July}.

Fruits: Granular-textured, oblong to **egg-shaped** seed-like fruits (achenes), widest above the middle, abruptly narrowed at the base, {1.7} 2.5–4.2 mm long, tipped with an egg-shaped to **lance-shaped projection** (tubercle) **about as long as the achene body**, surrounded by 6 {12}, **downward-barbed bristles** from the base.

Habitat: Calcareous fens; elsewhere, in calcareous sites in meadows and swamps and on shores.

Notes: This is the only species of this genus that has been found in Alberta. Another species, white beak-rush (*Rhynchospora alba* (L.) Vahl), has been found very close to the southern and western boundaries of Alberta and may also grow here. It is distinguished by the top-shaped clusters of spikelets, pale brown to whitish scales and more numerous bristles (9–12), which have soft hairs at the base. • The generic name *Rhynchospora* comes from the Greek *rhynchos* (beak) and *spora* (seed), referring to the characteristic projection (tubercle) at the tip of each achene.

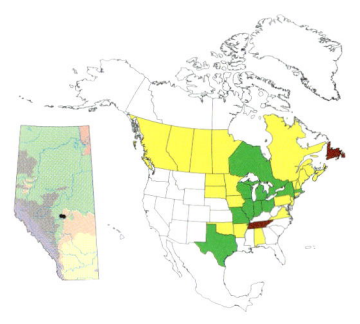

R. ALBA

Grass-like Plants

SEDGE FAMILY

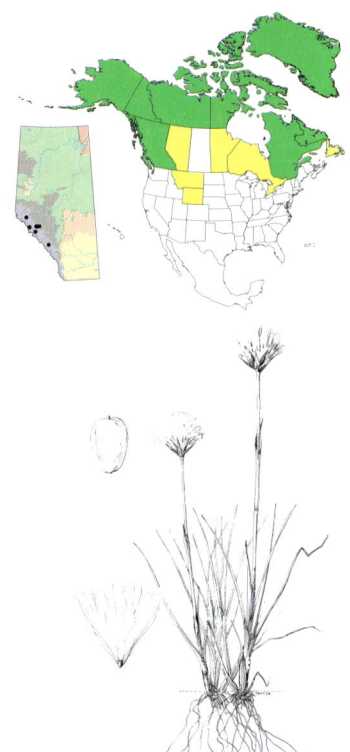

BEAUTIFUL COTTON-GRASS

Eriophorum callitrix Cham. *ex* C.A. Mey.
SEDGE FAMILY (CYPERACEAE)

Plants: **Tufted** perennial herbs with conspicuous persistent leaf bases; stems slender, **6–25 cm tall**, 1–8 per tuft; from very short underground stems (rhizomes) and fibrous roots.

Leaves: Mostly basal, stiff, linear, 1–2 mm wide; **stem leaves usually single** (sometimes none or 2), attached **below mid-stem**, with conspicuously **inflated sheaths** and short or absent blades.

Flower clusters: Single, erect, egg-shaped spikelets at the stem tips, fluffy-white when mature, **widest above the middle**, 1.5–2.5 cm long, many-flowered; **flowers with 1–3 stamens** bearing 0.6–1 mm long anthers, **a single ovary** tipped with a 3-pronged style **and many shiny, white bristles** (modified sepals and petals), borne in the axils of **erect, black or greenish black scales**, the **lower 10–15 scales without flowers**; June–July.

Fruits: Dry, 3-sided seed-like fruits (achenes), 1.8–2 mm long, surrounded by a fluffy tuft of 2–3 cm long, white, hair-like bristles; July–August.

Habitat: Wet alpine turf; elsewhere, in calcareous turfy tundra.

Notes: Beautiful cotton-grass has been confused with 2 common species. One is close-sheathed cotton-grass (*Eriophorum brachyantherum* Trautv. & Mey.), which is taller (30–60 cm), with softer, longer basal leaves and with

E. SCHEUCHZERI

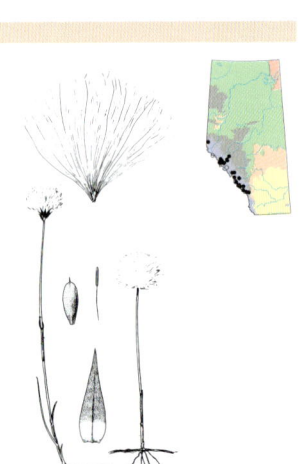

stem leaves that lack inflated sheaths and are attached above the middle of the stem. The other is sheathed cotton-grass or dense cotton-grass (*Eriophorum vaginatum* L. ssp. *spissum* (Fern.) Hult.), a species of dry, peaty bogs, which has white-edged scales that bend sharply downward.
• Another rare species, one-spike cotton-grass (*Eriophorum scheuchzeri* Hoppe), is not tufted but instead has stems that arise singly or in small groups from creeping stems. It is also distinguished from beautiful cotton-grass by its channelled or triangular basal leaves, its pale-edged scales (of which fewer than 7 of the lower ones are empty) and its beaked achenes, which are tipped with a firm point 0.1–0.2 mm long. One-spike cotton-grass grows on marshy ground in Alberta and on damp tundra and wet peat elsewhere. It is reported to flower from July to August.
• The generic name *Eriophorum* comes from the Greek *erion* (wool or cotton) and *phoros* (bearing), in reference to the long, cottony hairs of the seed heads.

DWARF BULRUSH, DWARF CLUB-RUSH

Trichophorum pumilum (Vahl) Schinz & Thellung
SEDGE FAMILY (CYPERACEAE)

Plants: Perennial herbs, 5–17 {20} cm tall; **stems smooth**, stiff and wiry, **round**, with remains of dead leaves and stems at the base; in loose clumps from short underground stems (rhizomes).

Leaves: Mainly **blackish to brown sheaths**; upper sheaths with a slender **blade 5–15 mm long**.

Flower clusters: Single, egg-shaped spikelets at stem tips, 2–3 {5} mm long, 3–5-flowered; scales chestnut-brown with a green midvein and translucent edges; lowermost scale (bract) blunt-tipped, usually less than ½ the length of the spikelet; **no bristles**; 3 stigmas; June {July}.

Fruits: Dry seed-like fruits (achenes), 1–1.5 mm long, **dark brown to blackish**, smooth, with a tiny point at the tip.

Habitat: Calcareous fens.

Grass-like Plants

SEDGE FAMILY

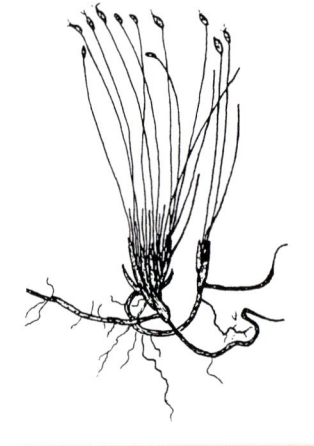

Notes: This species has also been called *Scirpus rollandii* Fern. and *Scirpus pumilus* Vahl ssp. *rollandii* (Fern.) Raymond. • Clinton's club-rush (*Trichophorum clintonii* (A. Gray) S.G. Smith, also called *Scirpus clintonii* A. Gray and Clinton's bulrush) is also rare in Alberta. It also has a single terminal spikelet, but its rough, 3-sided stems are 10–30 cm tall, growing in dense clumps from short rhizomes. Its leaves are flat and much shorter than the stems, and there is a single narrow bract beneath each spikelet. The spikelets are lance-shaped to egg-shaped, 4–5 {3–6} mm long and 4–7-flowered. Their scales are egg-shaped and pale brown, and the lowermost scale has a midvein that is prolonged into a blunt bristle about as long as the spikelet. The achenes are brown, smooth, 3-sided and 1.5–2 mm long, with 3–6, upwardly barbed, 2–3 mm long bristles from the base and 3 stigmas at the tip. Clinton's club-rush is reported to flower from May to June. • Tufted club-rush (*Trichophorum cespitosum* (L.) Hartman, also called tufted bulrush and *Scirpus cespitosus* L.) is a more common species that also has a single spikelet at the stem tip, but it grows in dense clumps (without rhizomes). The spikelets of tufted club-rush are slightly shorter and wider than those of dwarf bulrush, and their lowermost scale is tipped with a bristle at least 1 mm long that sometimes extends past the end of the spikelet. • *Pumilum* is Latin for 'dwarf' or 'small,' in reference to the small stature of dwarf bulrush. The species was originally named for its discoverer, Louis Roland, Frère Rolland-Germain.

T. CLINTONII

SEDGE FAMILY — Grass-like Plants

RED BULRUSH

Blysmus rufus (Hudson) Link
SEDGE FAMILY (CYPERACEAE)

Plants: Perennial herbs; stems erect, **5–60 cm tall**, smooth and nearly round; growing in **clumps** from extensively creeping underground stems (rhizomes).

Leaves: Mainly basal, lowermost reduced to bladeless sheaths, but upper 1–3 with stiffly erect blades 1–3 mm wide and up to 15 cm long, **blunt-tipped, shorter than the stems**.

Flower clusters: Reddish brown spikes, 1–2 cm long, 5–10 mm wide, with few to several spikelets in 2 vertical rows (ranks) and a single, erect, scale-like or leaf-like bract at the base; **spikelets 2–5-flowered**, 5–10 mm long; scales lance-shaped, shiny brown, lacking stiff bristles (awns) at tips; 2 stigmas; {June} July.

Fruits: Dry seed-like fruits (achenes), 4–5 mm long, 1–1.7 mm wide, pointed at both ends, **light brown** with a **dark, elongated tip**, smooth, **stalked**, with **3–6 fine, upwardly barbed bristles from base,** all shorter than the achene and soon falling off.

Habitat: Saline marshes.

Notes: This species has also been called *Scirpus rufus* (Huds.) Schrad. • The specific epithet *rufus* is Latin for 'reddish,' referring to the colour of the spikelets.

Grass-like Plants

SEDGE FAMILY

RIVER BULRUSH

Bolboschoenus fluviatilis (Torr.) J. Sojak.
SEDGE FAMILY (CYPERACEAE)

Plants: Perennial herbs 50–150 cm tall; **stems stout, 3-sided,** sharply angled, erect, regularly spaced along elongated underground stems (rhizomes) with **fleshy thickenings at the joints.**

Leaves: Mainly on the **middle and upper stem**; long-tapering, flat, with a prominent midvein, 5–15 mm wide, pale green.

Flower clusters: Flat-topped, with 5–12 often nodding, head-like, **with 1–5 spikelets** on flat, long, unequal stalks **above a whorl of 3–5 leaf-like bracts**; spikelets egg-shaped or broadly cylindrical, 1–4 cm long, 6–10 mm thick; scales brown, tipped with a stiff, curved, 2 mm long bristle (awn); **3 stigmas**; {June–July}.

Fruits: Dry, egg-shaped seed-like fruits (achenes), widest near the pointed tip, **3-sided,** dull grey-brown, 4–5 mm long, with **6 fine, downward-barbed bristles from the base** that are approximately as long as the achene.

Habitat: Margins of ponds, lakes and rivers.

Notes: This species has also been called *Scirpus fluviatilis* (Torr.) A. Gray and *Schoenoplectus fluviatilis* (Torr.) M.T. Strong. • Prairie bulrush (*Bolboschoenus maritimus* (L.) Pallas ssp. *paludosus* (A. Nels.) T. Koyama, also called *Scirpus paludosus* A. Nels.) also has underground rhizomes with swollen joints (nodes), but it is shorter (20–70 cm), with lens-shaped achenes bearing 2 stigmas (rather than 3). • River bulrush is less alkali tolerant than prairie bulrush. It often forms dense colonies in calm water up to 1 m deep. • 'Bulrush' is a corruption of 'pole-rush' or 'pool-rush,' referring to the habit or habitat of these plants. The specific epithet *fluviatilis* is Latin for 'of rivers,' referring to the common habitat of river bulrush.

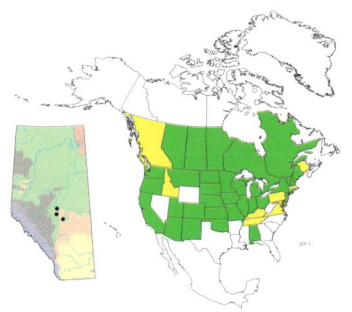

SEDGE FAMILY — Grass-like Plants

PALE BULRUSH

Scirpus pallidus (Britt.) Fern.
SEDGE FAMILY (CYPERACEAE)

Plants: **Robust perennial** herbs; stems erect, pale green, 30–100 {150} **cm tall, 3-sided** with blunt edges, loosely clustered; from short, stout underground stems (rhizomes) and runners (stolons).

Leaves: Mostly **on the lower half of the stem**; blades flat, **5–15 mm wide**; sheaths green, pale brown when dry, not reddish.

Flower clusters: Irregular open, flat-topped (umbels or occasionally compound cymes), with 4–8 dense heads of spikelets on unequal stalks above 3 or 4 short, leaf-like bracts; spikelets greenish brown, 2–8 mm long; scales 2–3 mm long, hairless, pale brown, with a pale midvein extending from the tip as a short bristle (awn) about 0.5 mm long; **3 stigmas**; June–July.

Fruits: Dry seed-like fruits (achenes), pale brown, about 1 mm long, **unequally 3-sided**, pointed at tip, with **6 fine, whitish bristles from base, each downward-barbed above the middle** and almost as long as the achene.

Habitat: In marshes and wet meadows.

Notes: Small-fruited bulrush (*Scirpus microcarpus* C. Presl, also called *Scirpus rubrotinctus* Fern.) is a common species that somewhat resembles pale bulrush, but it has reddish leaf sheaths and lens-shaped achenes with 2 stigmas.
• Late in the season, the flower clusters of pale bulrush are often replaced by leafy shoots, and the plants reproduce vegetatively. • When abundant, this species has been used to weave light-duty baskets and to decorate clothing. The leaves were laid over and under food in steaming pits. In the spring, the tender young stems were eaten raw when the shoots were about 20 cm tall. • The small bristles growing from the base of each achene are modified petals and sepals. • *Pallidus* means 'pale' in Latin and refers to the colour of the spikelets.

Rare *Vascular* Plants of Alberta

Grass-like Plants

SEDGE FAMILY

CAPITATE SEDGE

Carex capitata L.
SEDGE FAMILY (CYPERACEAE)

Plants: Tufted perennial herbs; stems wiry, 10–30 {60} cm tall; from very short, ascending underground stems (rhizomes).

Leaves: Mainly basal, shorter than stems; blades thread-like, up to 1 mm wide, hairless, with somewhat roughened edges; sheaths chestnut-brown or purplish brown; remains of old leaves persistent (marcescent).

Flower clusters: Solitary, round to egg-shaped spikes, {4} 5–10 **mm long**, with **male flowers at the tip** and 6–25 **female flowers at the base** (androgynous), bractless; scales of female flowers dark brown to tan, egg-shaped to nearly round, blunt, with broad, translucent edges, much shorter than the perigynia; **2 stigmas**; June–August.

Fruits: Lens-shaped seed-like fruits (achenes), 1.5 mm long, 1 mm wide, enclosed in perigynia; **perigynia** pale green or yellowish brown, **spreading, 2–2.5 {3.8} mm long**, 1–2 mm wide, broadly egg-shaped, abruptly tapered to a slender beak 0.4–0.9 mm long, flattened, with sharp edges, **stalkless, nerveless**, hairless.

Habitat: Wet sites, often in calcareous fens; elsewhere, in moist meadows and shrubby open woods, often above treeline.

Notes: The specific epithet *capitata*, from the Latin *caput* (head), means 'with a head,' in reference to the head-like spikes. No other species closely resembles this distinctive sedge.

SEASIDE SEDGE

Carex incurviformis Mack. var. *incurviformis*
SEDGE FAMILY (CYPERACEAE)

Plants: Low, creeping perennial herbs; stems solitary, 2–10 {30} cm tall, erect or curved; **from long, scaly, tough, cord-like underground stems** (rhizomes) and runners (stolons).

Leaves: Tufted near the base, 4–7 per stem, often longer than the stems; blades greyish green, erect or spreading, 2–10 {15} cm long, 1–2 mm wide, usually smooth and with the edges rolled upward and inward; lower sheaths often lacking blades.

Flower clusters: Dense, round or egg-shaped, with **3–5 crowded spikes, appearing like a single spike** 5–15 {20} mm long and 8–15 mm wide, with a small, brown, papery bract at the base; **spikes stalkless, with male flowers at the tip** and female flowers at the base (androgynous); scales of female flowers brownish, with broad, papery, translucent edges, egg-shaped, about 3 mm long and 2 mm wide, shorter than the perigynia; **2 stigmas**; June {July}.

Fruits: Lens-shaped seed-like fruits (achenes), about 1.5 mm long, enclosed in perigynia; **perigynia** dark brown, spreading at maturity, 3.5–5 mm long, 1–2 mm wide, broadly egg-shaped, tipped with a slender beak about 1 mm long, faintly to conspicuously nerved, **stalked**, not winged.

Habitat: Gravelly alpine areas; elsewhere, in salt marshes and turfy tundra and on sand dunes and river flats.

Notes: This variety has also been called *Carex maritima* Gunn. var. *incurviformis* (Mack.) Boivin. • The spikes of Hood's sedge (*Carex hoodii* Boott) and Hooker's sedge (*Carex hookerana* Dewey) resemble those of seaside sedge, but these species are much larger (15–80 cm tall), their stems are tufted rather than solitary and they grow in drier habitats. Hooker's sedge is also rare in Alberta. It is distinguished by its elongated heads of more widely separated spikes, its well-developed lower bract (at times exceeding the head) and its scales, which are tipped with strong bristles (awns). Hooker's sedge grows on prairies and dry banks, and in open woods at lower elevations; it flowers in June. • When it is abundant, seaside sedge provides a reasonable seed source for small birds.

C. HOOKERANA

Grass-like Plants

SEDGE FAMILY

FOX SEDGE

Carex vulpinoidea Michx.
SEDGE FAMILY (CYPERACEAE)

Plants: Densely clumped perennial herbs; stems 20–90 {100} cm tall, stiff, 3-sided with sharp angles, very rough to the touch near the top; from short, tough underground stems (rhizomes).

Leaves: Mostly flat, 4 or 5 per stem, 2–5 {6} mm wide, slender-pointed, usually longer than stems; lowest leaves typically smallest; **sheaths red-dotted, with horizontal wrinkles on the inner side.**

Flower clusters: Dense, greenish to yellowish or dull brown, cylindrical heads, 3–12 {15} cm long, 5–20 mm wide, with **numerous stalkless, often compound, spikes**, sometimes with lower spikes somewhat separate; spikes with inconspicuous **male flowers at the tip and female flowers below** (androgynous); lowermost **bract bristle-like**, up to 5 cm long; **scales** yellowish brown with a green midvein, about as long as but narrower than the perigynia, egg-shaped, tipped **with a long** (1–5 mm) **bristle**; 2 stigmas; {May–July}.

Fruits: Dry, lens-shaped seed-like fruits (achenes) enclosed in perigynia; **perigynia straw-coloured to greenish** (sometimes becoming black), pointed upward to outward, egg-shaped, 2–2.5 {3} mm long, 1–1.8 mm wide, rounded on one side and flat on the other, thick-edged (not winged), tapered to a **flattened beak ½–¾ as long as the body edged with tiny sharp teeth and tipped with 2 distinct teeth.**

Habitat: Swamps and wet meadows; rather exacting in its habitat, requiring non-saline, non-acid soils that are permanently wet but with some drainage.

Notes: This species has also been called *Carex multiflora* Muhl. and *Carex setacea* Dewey. • Awl-fruited sedge (*Carex stipata* Muhl. *ex* Willd.) superficially resembles fox sedge, but its perigynia taper gradually to a beak that is as long as the body, and its scales are tipped with a short bristle.
• *Vulpinoidea* is derived from the Latin *vulpes* (fox), because the flower cluster was thought to resemble a fox's tail.

SEDGE FAMILY   Grass-like Plants

BROWNED SEDGE

Carex adusta Boott
SEDGE FAMILY (CYPERACEAE)

Plants: Clumped perennial herbs; **stems stiffly erect, 20–80 cm tall**, rough to the touch near the top only; from short underground stems (rhizomes).

Leaves: Firm, 4–7 per stem, **shorter than the stems**; blades flat, 2–4 {5} mm wide.

Flower clusters: Stiff, erect, 2–3 {1–4} cm long, **egg-shaped to cylindrical heads** of 4–15 **olive-coloured** spikes; spikes with **male flowers at the base** and female flowers at the tip (gynaecandrous), 6–12 mm long, nearly round; **lowermost bract** broad at the base, prolonged into a **stiff, narrow blade** up to 2.5 cm long; scales reddish brown with narrow, white, translucent edges, broadly egg-shaped, **concealing the perigynia**; 2 stigmas; July {May–August}.

Fruits: Lens-shaped seed-like fruits (achenes), 1.8–2.1 mm broad, enclosed by perigynia; **perigynia** brown, **4–5 mm long**, short-stalked, **shiny**, finely nerved on the back below the middle, edged with a narrow **wing bearing tiny teeth above the middle**, tipped with a **flat, 2-toothed beak**; July {August}.

Habitat: Dry, acidic, usually sandy soil, often under pine trees; generally confined to sandy, disturbed areas in the boreal forest.

Notes: This species has also been called *Carex pinguis* Bailey. • A more common species, white-scaled sedge (*Carex xerantica* Bailey), resembles browned sedge but has narrower, more egg-shaped spikes that are slightly tapered at the base above an inconspicuous lower bract. Its light-coloured perigynia are not shiny, and they remain closely pressed to the spike when mature. White-scaled sedge grows in grasslands on the plains. • The crowded spikes of the widespread wetland species Bebb's sedge (*Carex bebbii* (Bailey) Olney *ex* Fern.) resemble those of browned sedge, but their scales are shorter and narrower and do not conceal the perigynia, and their lowermost bracts are only slightly prolonged (not prominent). • When abundant, browned sedge provides a good source of seeds for small birds. • *Adusta* comes from the Latin word for 'burned' or 'scorched,' probably referring to the colour of the scales covering the perigynia.

Grass-like Plants

SEDGE FAMILY

PASTURE SEDGE

Carex petasata Dewey
SEDGE FAMILY (Cyperaceae)

Plants: Tufted perennial herbs; stems slender, stiff, smooth, 30–80 {90} cm tall, brown at the base; from short, fibrous underground stems (rhizomes).

Leaves: Alternate, 2–5 per stem, clustered near the base, much shorter than the stems; blades firm, flat or nearly so, usually 2–4 mm wide; **dried leaves of previous years persistent and conspicuous** (marcescent).

Flower clusters: Erect, head-like, with **3–6 overlapping spikes**, 2–4 {6} cm long and 1–1.5 cm wide; **spikes with female flowers at the tip** and male flowers at the base (gynaecandrous), **stalkless** (or nearly so), 9–18 mm long; lowermost **bract inconspicuous**, usually scale-like, sometimes tipped with a small bristle; **scales of female flowers reddish brown to brown** or straw-coloured, with broad, translucent edges, broadly egg-shaped, **largely covering the perigynia; 2 stigmas**; {May–July}.

Fruits: Dry, **lens-shaped** seed-like fruits (achenes), 2–3 mm long and about 1.5 mm wide, enclosed in perigynia; **perigynia** ascending, pale when young, brown at maturity, finely ribbed on both sides, flattened to a **narrow, finely toothed wing** along the edges, **oblong to lance-shaped, 6–8 mm long** and about 2 mm wide, tapered to a slender, nearly round beak about 2 mm long.

Habitat: Dry grassland and open woods; elsewhere, in grassland, sagebrush scrub, and dry or even wet meadows, sometimes to timberline.

Notes: Pasture sedge could be confused with 2 common species, meadow sedge (*Carex praticola* Rydb.) and head-like sedge (*Carex phaeocephala* Piper), but both of these sedges have smaller perigynia (3.5–6.5 mm long) that are covered by their scales. Head-like sedge is a smaller (10–30 cm tall) plant that grows in large clumps, usually at higher elevations. It usually has only 3 or 4 spikes, and its perigynia are broader than those of pasture sedge and tend to retain their green colour longer. Meadow sedge is similar to pasture sedge in appearance, but it usually has more numerous (2–7) and more widely spaced spikes.
• Plants growing in shaded situations tend to have much paler scales, without the brownish colour of those from more open areas.

SEDGE FAMILY — Grass-like Plants

SMALL-HEADED SEDGE

Carex illota Bailey
SEDGE FAMILY (CYPERACEAE)

Plants: **Tufted** perennial herbs; stems slender, stiff, **10–40 cm tall**, 3-sided, with sharp, rough edges; from short underground stems (rhizomes).

Leaves: Flat, 2–5 per stem, 1–3 mm wide, shorter than the stems.

Flower clusters: Dense, blackish heads of 3–6 stalkless spikes, 6–15 mm long, 5–15 mm wide; spikes with **male flowers at the base** and female flowers at the tip (gynaecandrous), 5–15-flowered, 4–7 mm long; **bracts inconspicuous**, hair-like; scales dark brown to greenish black with narrow, translucent edges, shiny, broadly egg-shaped, blunt, shorter and often narrower than the perigynia; **2 stigmas**; {June} July.

Fruits: Dry, lens-shaped seed-like fruits (achenes), 1.3–1.5 mm long, enclosed in perigynia; **perigynia** blackish brown toward the tip, green to straw-coloured at the spongy thickened base, **pointed outward**, broadly lance-shaped, 2.5–3.2 {3.5} mm long, **flat on the inner side, bulging on the outer side**, sharp-angled but neither flattened nor toothed along the edges, tapered to a beak with 2 teeth.

Habitat: Moist mountain slopes and alpine meadows.

Notes: This species has also been called *Carex bonplandii* var. *minor* Boott and *Carex dieckii* Boeck. • Two other rare Alberta sedges, Presl's sedge (*Carex preslii* Steud.) and Hayden's sedge (*Carex haydeniana* Olney, also called *Carex macloviana* D'Urv. ssp. *haydeniana* (Olney) Taylor & MacBryde), resemble small-headed sedge but are distinguished by their perigynia, which are edged with finely toothed, thin wings and which lack spongy, thickened bases. Presl's sedge is a tufted, 20–80 cm tall species of dry, open slopes in southwestern Alberta. It produces dense, 1–2 cm long heads of 2–8 spikes. Each spike is 6–10 mm wide and has 10–25 flowers. The plump, 3–4.5 mm long, obscurely veined perigynia have dark green tips, and contain rectangular, lens-shaped achenes at least 1.75 mm long. The scales are reddish brown with a green midvein and translucent edges. Presl's sedge flowers in July {June–August}. Hayden's sedge grows in moist, open, subalpine and alpine areas in the mountains. It is distinguished from Presl's

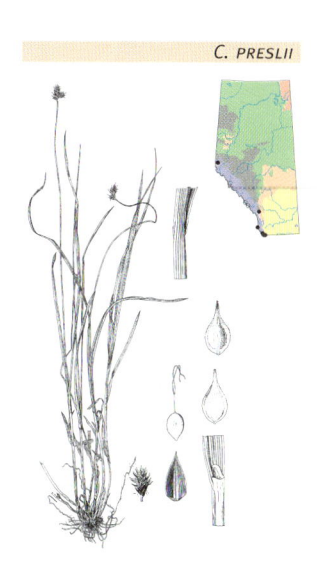

C. PRESLII

Grass-like Plants   SEDGE FAMILY

C. HAYDENIANA

sedge by its larger (4.5–6 mm long), very flat perigynia containing relatively small (1.4–1.7 mm) achenes. • Inland sedge (*Carex interior* Bailey) is a widespread sedge that resembles small-headed sedge, but it grows at lower elevations, its spikes are less crowded, its perigynia are very finely toothed along the upper edges and its scales tend to be brownish. Its upper spike has a very distinct zone of male flowers at the base, and its mature perigynia are widely spreading. • Small-headed sedge is moderately palatable to sheep and cattle and, where abundant, is an important forage plant. • *Illota* is a Latin word meaning 'dirty' or 'unwashed.'

BROOM SEDGE

Carex scoparia Schk. *ex* Willd.
SEDGE FAMILY (CYPERACEAE)

Plants: Densely tufted perennial herbs; stems slender, erect, 20–80 {100} cm tall, 3-sided with sharp edges, rough to the touch near the top; from short underground stems (rhizomes).

Leaves: Firm, 2–6 per stem, 15–90 cm long, 1–3 mm wide, flat or channelled lengthwise, shorter than the stem.

Flower clusters: Shiny, often **straw-coloured**, 2–6 cm long, with 3–10 {12} clearly defined, **stalkless spikes**; spikes 4–10-flowered, with **male flowers at the base and female flowers at the tip** (gynaecandrous), egg-shaped, 6–16 mm long; lowermost bract bristle-like or lacking; scales brownish with a green midvein and narrow translucent edges, egg-shaped, pointed, narrower and shorter than the perigynia; 2 stigmas; {June–July}.

Fruits: Brownish, lens-shaped seed-like fruits (achenes) 1.3–1.8 mm long, enclosed in perigynia; **perigynia** straw-coloured to brown, **egg-shaped to lance-shaped, 4–7 mm long**, 1.2–2.6 mm wide, veined on both sides, **flattened, edged with a thin wing** from base to beak, tapered to a flat, finely toothed, 1–2 mm long beak tipped with 2 shallow teeth.

Grass-like Plants

Habitat: Moist to wet sites; elsewhere, moist open woods.

Notes: Tinged sedge (*Carex tincta* (Fern.) Fern., also known as *Carex mirabilis* Dewey var. *tincta* Fern.) is another rare Alberta species that resembles broom sedge, but it has broader, more egg-shaped perigynia (at least half as wide as long) with conspicuous nerves on the outer side, but fine, less visible nerves on the inner side. Tinged sedge has relatively small perigynia that are 3–4.5 {5} mm long and 1.5–2 mm wide. The spikes form oblong heads 1–4 cm long and 1–1.5 cm thick, with light reddish brown scales as long as the perigynia. Tinged sedge is reported to flower from May to July, and grows in meadows and open woodlands in central and southwestern Alberta. • Broad-fruited sedge (*Carex tenera* Dewey) and Bebb's sedge (*Carex bebbii* (Bailey) Olney *ex* Fern.) are more widespread species that resemble both broom sedge and tinged sedge. The spikes of broad-fruited sedge are more widely spaced (like beads on a string), their perigynia have nerves on the outer side only, and their scales are greenish translucent to tawny. Bebb's sedge has smaller perigynia (3–3.5 mm long) that are nerveless or obscurely nerved and brownish at maturity, and its flower clusters are squared (rather than tapered) at the base. • Crawford's sedge (*Carex crawfordii* Fern.) is also similar to broom sedge but has much narrower (lance-shaped to awl-shaped) perigynia on which the flattened edge (wing) almost disappears at the base. • Broom sedge produces abundant seed, but its plants have low nutritional value and are seldom grazed. • The specific epithet *scoparia*, from the Latin *scopa* (a broom), refers to the general broom-like appearance of this sedge.

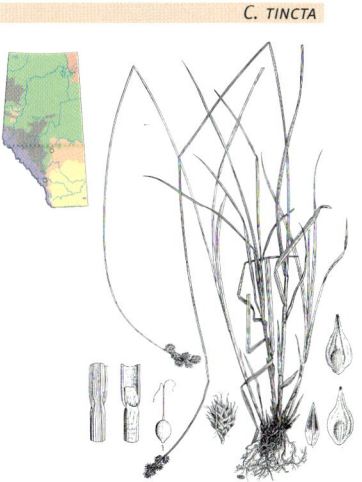

C. TINCTA

Grass-like Plants

SEDGE FAMILY

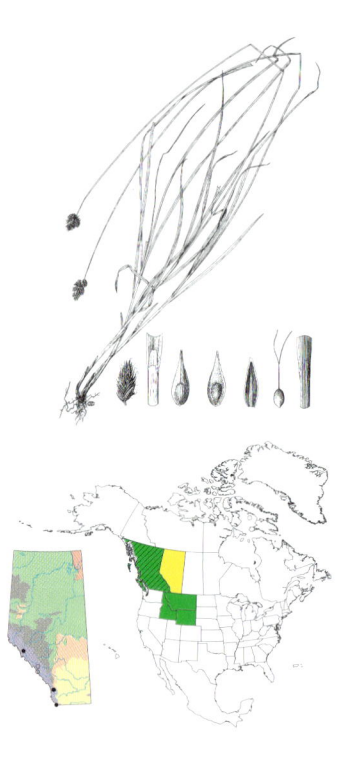

BROAD-SCALED SEDGE

Carex platylepis Mack.
SEDGE FAMILY (CYPERACEAE)

Plants: Tufted perennial herbs; **stems coarse**, 40–70 cm tall; from short underground stems (rhizomes).

Leaves: 3–7 scattered on the lower third of the stem, much shorter than the stems; **blades** 2.5–5 mm wide, **spreading, prominently wrinkled and shallowly pitted on the upper side**.

Flower clusters: 5–8 spikes in a head 1.5–3.5 cm long, clustered or the lower ones slightly separate; spikes with female flowers at the tip, male flowers at the base (gynaecandrous); bracts scale-like; scales egg-shaped, dull reddish brown with translucent edges and a green midvein; 2 stigmas; {May–August}.

Fruits: Lens-shaped seed-like fruits (achenes), about 2 mm long and 1 mm wide, enclosed in perigynia; **perigynia** dull green to yellowish brown, appressed-ascending, oblong to lance-shaped, 4–4.5 mm long, 1.5–2 mm wide, **about as long and wide as the scales**, thin, edged with a wing nearly to base, toothed above the middle, tapered to a beak 1–1.5 mm long.

Habitat: Dry, open coniferous woods.

Notes: Broad-scaled sedge can be confused with several other species. Meadow sedge (*Carex praticola* Rydb.) is distinguished by its coarser growth form, with broader, spreading leaves, leafy stems, clustered spikes and shorter (4 mm long) perigynia. Thick-headed sedge (*Carex pachystachya* Cham.) and small-winged sedge (*Carex microptera* Mack.) have heads of crowded spikes with flatter perigynia containing distinctly smaller achenes. They are also plants of moister habitats. • Another species reported for Alberta, Piper's sedge (*Carex piperi* Mack.), is similar, but its spikes are usually less densely clustered (almost zigzagged), and its perigynia have slender, more rounded beaks and fewer, less conspicuous veins. Piper's sedge has sometimes been included in meadow sedge or broad-scaled sedge, but those species have dull, reddish brown flower scales, without the broad, silvery-white edges found in Piper's sedge, and their flowering heads tend to be shorter, with more densely clustered spikes. Piper's sedge grows in damp meadows.

C. PIPERI

SEDGE FAMILY **Grass-like Plants**

TAPER-FRUIT SHORT-SCALE SEDGE

Carex leptopoda Mack.
SEDGE FAMILY (CYPERACEAE)

Plants: Loosely tufted perennial herbs; stems 20–65 {80} cm tall, sharply triangular; smooth or roughened below the head; from slender, **elongate underground stems** (rhizomes).

Leaves: Light green to yellowish green with a bloom, 2.5–6 mm wide, flat, shorter than the stems.

Flower clusters: Oblong to egg-shaped **heads of 5–7 spikes**, 3.5–4.5 cm long, lower 1–3 spikes usually separate; spikes stalkless or on a short stalk, 9–13 mm long, **13–25-flowered**, usually with female flowers at the tip and male flowers at the base (gynaecandrous); lowest bract 16–26 mm long, with awn 13–23 mm long; scales pointed to short-awned, greenish with white to straw-coloured edges, about as long as the perigynia; 2 stigmas; {May–July}.

Fruits: Dry, lens-shaped seed-like fruits (achenes) enclosed in a perigynia; **perigynia stalked**, ascending to erect, green to pale brown, **prominently** to weakly **veined**, egg-shaped to narrowly egg-shaped, 3.2–3.9 mm long, 1.1–1.5 mm wide, tapering to a toothed beak 1–1.7 mm long, the tip entire or slightly 2-toothed.

Habitat: Moist woods and thickets in extreme southwestern Alberta.

Notes: This species has also been called *Carex deweyana* Schwein. ssp. *leptopoda* (Mack.) Calder & R.L. Taylor and *C. deweyana* var. *leptopoda* (Mack.) B. Boivin. It can be distinguished from the more common and widespread species in Alberta, Dewey's sedge (*Carex deweyana* Schwein.), in having more spikes per stem (5–7 vs. 2–5), more female flowers in the lowest spike (13–25 vs. 5–12), and short-stalked perigynia with more prominent venation. • The specific epithet *leptopoda* is derived from the Greek *lept* (slender or small) and *pod* (foot), in reference to the stalked perigynia.

Rare Vascular Plants of Alberta

Grass-like Plants

SEDGE FAMILY

LACHENAL'S SEDGE

Carex lachenalii Schk.
SEDGE FAMILY (CYPERACEAE)

Plants: **Tufted** perennial herbs; stems slender, stiff, **5–30 cm tall**, sometimes a little rough to the touch near the top; often from short underground stems (rhizomes).

Leaves: Basal and on the lower stem, 1–2.5 mm wide, **flat with the edges rolled under**, equalling or shorter than stems.

Flower clusters: Compact heads of **2–4 stalkless spikes**, 0.5–2 {3.5} **cm long**; spikes with **female flowers at the tip** and male flowers at the base (gynaecandrous) (male flowers sometimes inconspicuous), 5–15 mm long, brownish, 15–30-flowered; bracts inconspicuous; **scales brownish**, sometimes with pale, translucent edges, egg-shaped, **blunt-tipped**, hiding most of the perygynia; 2 stigmas; {July} August.

Fruits: Dry, lens-shaped seed-like fruits (achenes) enclosed in perygynia; **perigynia golden-brown to reddish, densely dotted with tiny pits**, convex on the outer side and flat on the inner, widest at or above the middle, 2–3.5 mm long, 1–1.5 mm wide, tapered to a short, more or less flattened beak with a **translucent tip**.

Habitat: Moist alpine slopes and snow beds; elsewhere, in shallow ponds and open woods, usually in calcareous areas.

Notes: This species has also been called *Carex bipartita* All. (misapplied) and *Carex lagopina* Wahl. • Hudson Bay sedge (*Carex heleonastes* Ehrh. *ex* L.f.) is a rare Alberta species that grows in wet, calcareous sites such as fens and marshes. Its flower clusters resemble those of Lachenal's sedge, with 2–4 crowded spikes, but the perigynia of Hudson Bay sedge are light grey (sometimes turning brown) with short, reddish brown (not translucent) tips, and the upper stems are very rough. • Narrow sedge (*Carex arcta* Boott) is a rare species that grows in moist woods in Alberta and also in

C. ARCTA

SEDGE FAMILY

Grass-like Plants

C. HELEONASTES

wet meadows and on streambanks elsewhere. It is very similar to Hudson Bay sedge and was included in that species at one time, but its perigynia are broadest near the base (rather than at or above the middle) and are tipped with conspicuous, flattened, sharply toothed beaks. Narrow sedge has larger, more compact, head-like flower clusters, with 5–15 crowded spikes. It flowers in July. • The Alberta populations of Lachenal's sedge are widely separated (disjunct) from the main range of this species.

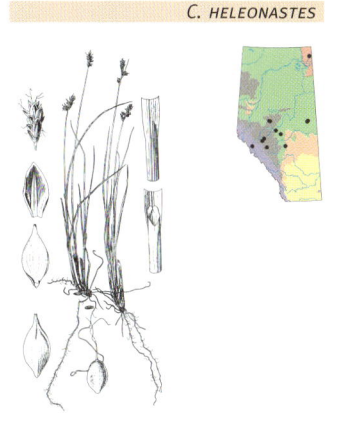

RYE-GRASS SEDGE

Carex loliacea L.
SEDGE FAMILY (CYPERACEAE)

Plants: Loosely tufted perennial herbs; **stems slender, weak, 20–40 {60} cm tall**, rough to the touch near the top, indistinctly 3-sided; from slender creeping stems.

Leaves: Soft, yellowish green, 4–8 per stem, shorter than or as long as stems, 0.5–2 mm wide, flat or with lengthwise grooves, long-pointed.

Flower clusters: Narrow, **1–2.5 {3} cm long heads of 2–5 {8} stalkless spikes**, with the upper spikes touching and the lower ones well separated; **spikes 3–5 mm long**, with **male flowers at the base** and female flowers at the tip (gynaecandrous); lowermost bract bristle-like, 2–8 mm long; **scales white, translucent with a green midvein**, egg-shaped with lengthwise ridge on the back, blunt-tipped, shorter than the perigynia; 2 stigmas; {May–June} July.

Fruits: Dry, brownish, lens-shaped seed-like fruits (achenes) enclosed in perigynia; **perigynia light green**, 3–8 per spike, pointed upward or spreading, widest at or below the middle, **blunt-tipped** (not beaked), {2} 2.5–3 mm long, flat on one side and rounded on the other, spongy, thickened and short-stalked at the base, **distinctly many-ribbed, densely dotted with tiny white pits**.

Habitat: Marshes, moist banks; elsewhere, in sphagnum and wet black spruce bogs.

Rare *Vascular* Plants of Alberta   287

Grass-like Plants

SEDGE FAMILY

Notes: Three-seeded sedge (*Carex trisperma* Dewey) is a rare species of bogs, swamps and wet woods of central Alberta. It resembles rye-grass sedge, but its flower clusters are longer (3–6 cm long) with widely spaced spikes, its spikes have only 1–5 perigynia, its lowermost bracts are bristle-like and 2–7 cm long (many times longer than the lowest spike), and its greenish brown perigynia are 3–4 mm long. Three-seeded sedge has been reported to flower from May to June. • A similar but more common species, two-seeded sedge (*Carex disperma* Dewey), is easily distinguished from rye-grass sedge and three-seeded sedge by its spikes, which have female flowers at the base and male flowers at the tip (androgynous). • Thin-flowered sedge (*Carex tenuiflora* Wahlenb.) is also similar to, but more common than, rye-grass sedge. It is distinguished by its dense, egg-shaped to rounded heads of spikes, and its larger, whitish silvery, green-centred scales that almost hide the faintly nerved perigynia. • *Loliacea* means 'lolium-like' in Latin. *Lolium* is the scientific name for darnel or rye-grass, a genus of grass that this sedge superficially resembles.

C. TRISPERMA

SEDGE FAMILY | **Grass-like Plants**

WEAK SEDGE

Carex supina Willd. ssp. *spaniocarpa* (Steud.) Hult.
SEDGE FAMILY (CYPERACEAE)

Plants: Loosely tufted perennial herbs; **stems** slender, erect, 5–30 cm tall, 3-sided with sharp, rough edges, **reddish at the base**; from slender, brown underground runners (stolons).

Leaves: Crowded, 0.5–2 mm wide, channelled lengthwise, **shorter than the stems**.

Flower clusters: Compact, **2–3 cm long heads** of 1–4 stalkless spikes; **uppermost spike with male flowers only**, pale, 5–15 mm long, **1–2 mm wide**; **lower spikes** with female flowers only, **rounded**, 5 mm long, 3–6 mm wide; lowermost bract short, with a bristle-like blade; **scales reddish brown** with a light-coloured centre and translucent edges, broadly egg-shaped, pointed, **about as long as the perigynia**; 3 stigmas; {May–June} July.

Fruits: Dry, yellowish brown, 3-sided seed-like fruits (achenes), 2 mm long, enclosed in perigynia; **perigynia 3–5 {10} per spike, smooth and shiny, rust-coloured, plump**, 2.5–3.5 mm long, 1–1.5 mm wide, tapered to a short beak.

Habitat: Dry, gravelly, eroding slopes and sandy sites.

Notes: This species has also been called *Carex spaniocarpa* Steud. • Glacier sedge (*Carex glacialis* Mack.) is a northern tundra species that is rare in Alberta; disjunct populations have been discovered on dry, calcareous mountain slopes. This small, densely tufted sedge has smooth, wiry stems that are 4–15 cm tall and reddish purple at the base. Its stiff, curved leaves are flat at the base but channelled toward the tip. They are narrow (0.5–1.5 mm wide) and shorter than the stems, and the remains of old leaves persist at the stem base. Each loose, head-like flower cluster is 5–15 mm long, with 2–4 small (2–7 mm long), stalkless (upper) to short-stalked (lower) spikes. The lowermost bract has a short, tubular sheath and a bristle-like blade 5–15 mm long. The uppermost spike is male and the lower spikes are female, with a few (1–6) small (1.5–2.5 mm long), round perigynia that are slightly longer than their scales. The perigynia are greenish at the base, brownish at the tip and faintly veined, with an abrupt, short, cylindrical beak. The broadly egg-shaped scales are reddish brown to purplish black with a pale midvein and broad translucent edges. Glacier sedge is reported to flower from June to July.

Rare *Vascular* Plants of Alberta

Grass-like Plants

SEDGE FAMILY

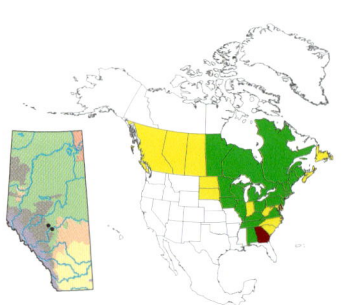

STALKED SEDGE

Carex pedunculata Muhl.
SEDGE FAMILY (CYPERACEAE)

Plants: Tufted, bright green perennial herbs; stems weak, slender, {5} 10–30 cm tall, barely exceeding the leaves, rough to the touch near the top, **strongly purple-tinged at the base**; from short underground stems (rhizomes).

Leaves: Mostly basal, thick, ascending to curved, blades 5–25 cm long, 2–4 {5} mm wide, flat near the base, becoming pleated or with the edges curved upward at the tip, finely toothed on the upper ⅓.

Flower clusters: 4–10 cm long, with 3 or 4 {5} spikes; uppermost spike usually with male flowers only (sometimes with a few female flowers at the base), elliptic, 5–15 mm long; **lower** 2 or 3 {4} **spikes** separate, ascending or spreading, usually with female flowers only, elliptic, 7–10 {30} mm long, 3–8-flowered, **on long, curved, thread-like stalks, some often hidden among the basal leaves; lowermost bract with a short blade**; scales purplish brown, oblong to egg-shaped, with translucent edges and a green **midvein prolonged into a short bristle** (awn), smaller than the perigynia; 3 stigmas; May–June.

Fruits: Brown, 3-sided seed-like fruits (achenes), filling the perigynia; **perigynia** pale green, egg-shaped, 3.5–5 mm long, nerveless, stalked, spongy at the base, **hairless** or slightly hairy above, tipped with a minute, toothless beak.

Habitat: Rich, relatively dry woods, frequently with poplars; elsewhere, also in alder swamps and white spruce and balsam fir forests.

Notes: Stalked sedge is most likely to be confused with the more common beautiful sedge (*Carex concinna* R. Br.), but that species has stalkless lower female spikes, the lowermost bract is bladeless, the scales do not have awns, and the perigynia are more evidently hairy. • Stalked sedge is an eastern species. The Alberta localities are widely disjunct from the nearest records in eastern Saskatchewan.
• *Pedunculata* means 'with peduncles' (the stalk of a cluster of flowers), referring to the long stalks of the female spikes.

SEDGE FAMILY — Grass-like Plants

BALD SEDGE, SHAVED SEDGE

Carex tonsa (Fern.) Bickn. var. *tonsa*
SEDGE FAMILY (CYPERACEAE)

Plants: Small, densely tufted, **almost stemless** perennial herbs; stems 4–16 cm tall, of unequal lengths, **fertile stems hidden among the basal leaves**; from short, stout, brown to reddish brown underground stems (rhizomes).

Leaves: All basal, **stiff**, flat to channelled, pale green, smooth or slightly roughened above; **blades** 5–15 cm long, 2–4 {5} **mm wide**; old leaves numerous at base of plant.

Flower clusters: Compact, with 3 or 4 {5} stalkless or short-stalked spikes, 10–15 mm long; uppermost spike with male flowers only, 5–8 {4.5–11} mm long on a stalk 0.8–15 mm long; lower 2 or 3 {4} spikes with 5–10 female flowers; lowermost bract small, shorter than the flower cluster; **scales** reddish brown with a green midrib, egg-shaped, pointed, **as long as or longer than the perigynia**; 3 stigmas; {May–June}.

Fruits: Dry, 2-ridged, brown, seed-like fruits (achenes), enclosed in perigynia; **perigynia 2-ribbed, hairless, except for lines of hairs on the ribs along the beak**, 3.5–4 {3.2–4.7} mm long, {1.1} 1.3–1.6 mm wide, abruptly narrowed to a straight, 0.9–1.5 {1.9} mm long beak, tipped with 2 teeth 0.2–0.5 mm long.

Habitat: Open woods (particularly pine) and sandy areas in the boreal forest, especially in disturbed areas.

Notes: This species has also been called *Carex umbellata* Schkuhr *ex* Willd. var. *tonsa* Fern. and *Carex rugosperma* Mack. var. *tonsa* (Fern.) Voss. • Umbellate sedge (*Carex umbellata* Schkuhr *ex* Willd.) is identified by its softer, narrower (1–3 mm wide) leaves and distinctly hairy, smaller (2.2–3.3 mm long) perigynia, with shorter (0.5–1 mm long) beaks. It grows in habitats similar to those of bald sedge, but its distribution extends into the Rocky Mountains. • Two other closely related sedges are more common in Alberta, bent sedge (*Carex deflexa* Hornem.) and Ross' sedge (*Carex rossii* Boott). Both have stalkless male spikes, lowermost bracts longer than the flower clusters and scales shorter than the perigynia. Bent sedge is slender, loosely tufted, with soft, spreading leaves less than 2 mm wide. Its male spikes are 2–5 mm long and its perigynia are 2.5–3 mm long, with short (0.4–0.7 mm long),

C. UMBELLATA

shallowly toothed beaks. Ross' sedge is stout, densely tufted, with stiff, erect leaves more than 2 mm wide. Its male spikes are 12–15 mm long and its perigynia are 3–4.5 mm long, with long (0.8–1.7 mm long), deeply toothed beaks.

BACK'S SEDGE

Carex backii Boott
SEDGE FAMILY (CYPERACEAE)

Plants: Small tufted herbs; **stems weak**, narrowly winged, up to 25 cm tall, but much **shorter than the leaves**; forming dense mats.

Leaves: Flat, dark green, 3–6 mm wide.

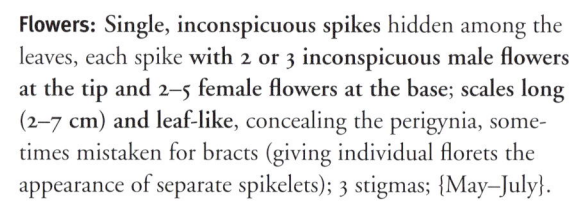

Flowers: Single, inconspicuous spikes hidden among the leaves, each spike **with 2 or 3 inconspicuous male flowers at the tip and 2–5 female flowers at the base**; scales long (2–7 cm) and leaf-like, concealing the perigynia, sometimes mistaken for bracts (giving individual florets the appearance of separate spikelets); 3 stigmas; {May–July}.

Fruits: Dry, 3-sided seed-like fruits (achenes) enclosed in perigynia; **perigynia 5–6 mm long, with the upper ⅓ of the body empty**, tipped with stout, toothless beaks about 2 mm long.

Habitat: Dry (to moist), shady woods; elsewhere, in riparian woodland.

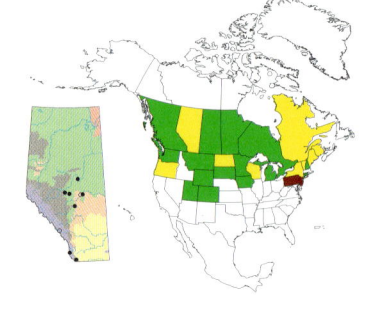

Notes: Rocky Mountain sedge (*Carex saxmontana* Mack., also called *Carex backii* Boott var. *saxmontana* (Mack.) Boivin) has shiny green leaves with down-curled edges, smaller (4–5 mm long) perigynia in which the achene fills the upper part of the body, and conical, slightly toothed beaks, about 1 mm long. It is reported from the Waterton-Crowsnest area of Alberta; Back's sedge occurs in the northern parkland and southern boreal forest in the central part of the province.

SEDGE FAMILY   **Grass-like Plants**

CRAWE'S SEDGE
Carex crawei Dewey
SEDGE FAMILY (CYPERACEAE)

Plants: Perennial herbs; stems slender, stiffly erect, 5–30 {40} **cm tall**, single or in small clumps; from slender, creeping underground stems (rhizomes).

Leaves: Stiff, **pale green**, 6–12, shorter than the stem; blades 5–30 cm long, 1–3 {5} mm wide, flat, usually curved and **spreading**; **dried leaves** of the previous year **conspicuous** at the stem base.

Flower clusters: Elongated, 5–20 cm long, with **2–5 widely spaced, stalked spikes**, with the lowest often borne near the base of the stem; **uppermost spike** with **male** flowers only, very narrow, {5} 10–20 mm long, on a rough, 1–7 cm long stalk; **lower spikes** with **female** flowers only, cylindrical, 10–20 {5–30} mm long, 5–6 mm wide, densely {5} 15–50-flowered, on short (1–3 cm long) stalks or stalkless; **bracts leaf-like with well-developed sheaths**, generally shorter than the flower cluster; **scales light reddish brown** with a pale green midrib and translucent edges, broadly egg-shaped, much shorter and narrower than the perigynia; 3 stigmas; {May} June–July.

Fruits: Dry, 3-sided seed-like fruits (achenes), **enclosed in slightly inflated perigynia**; perigynia **light green to tan, often speckled with tiny, reddish brown, resinous dots**, elliptic, very short-beaked, 2–3.5 {3.8} mm long, 1.2–2 mm wide, hairless, obscurely 15–25-veined.

Habitat: Calcareous meadows; elsewhere, in lime-rich wetlands, on lakeshores and in moist woods.

Notes: This species has also been called *Carex heterostachya* Torr. • Golden sedge (*Carex aurea* Nutt.) is a common sedge that resembles Crawe's sedge, but its styles have 2 stigmas, its perigynia are light green to orange-yellow or brownish, with rounded (not beaked) tips and its spikes are more loosely flowered but are not as widely spaced as those of Crawe's sedge. • Crawe's sedge was named for its discoverer, Ithamar Bingham Crawe (1792–1847).

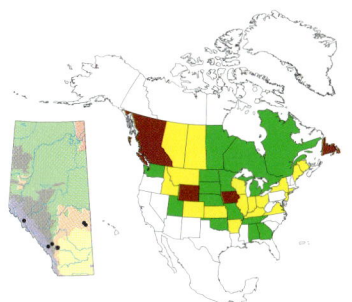

Grass-like Plants

SEDGE FAMILY

NODDING SEDGE

Carex misandra R. Br.
SEDGE FAMILY (Cyperaceae)

Plants: Tufted perennial herbs; **stems** slender, **erect but often nodding at the tip**, 10–35 cm tall, smooth; from short underground stems (rhizomes).

Leaves: Many, **basal**, thick, shorter than stems, 4–10 cm long, 1–3 mm wide, flat or channelled lengthwise, **often curved**, long-pointed; **dead leaves persisting for several years**.

Flower clusters: Loose, elongated, with 2–4 slender-stalked, nodding spikes, each 5–15 {20} mm long, 4–6 mm wide; **uppermost spike club-shaped, with male flowers at the base and female flowers at the tip** (gynaecandrous); **lower spikes** cylindrical, bearing **female flowers only**; lowermost bract inconspicuous, with a long sheath and short blade (sometimes almost bladeless); **scales brownish black** with translucent edges, narrowly egg-shaped, 2.5–3.5 mm long, wider and shorter than the perigynia; 3 stigmas (rarely 2); July {June–August}.

Fruits: Dry, 3-sided seed-like fruits (achenes) with 3 sharp angles, egg-shaped, enclosed in perigynia; **perigynia** purplish black at the tip, greenish to straw-coloured at the base, **narrowly lance-shaped**, 3.5–5 mm long, 1 mm wide, short-stalked, lacking flattened (winged) edges, **gradually tapered to a long, toothed, flattened beak on the upper half**.

Habitat: Dry alpine slopes.

Notes: This species has also been called *Carex fuliginosa* Kuk. var. *misandra* (R. Br.) Lang. • Stone sedge (*Carex petricosa* Dewey) resembles nodding sedge, in that it also has nodding spikes, 3-sided achenes with 3 stigmas, and beaked perigynia tipped with 2 teeth. However, it grows from spreading underground stems and therefore is not tufted like nodding sedge, and its uppermost spikes are tipped with male (rather than female) flowers. Stone sedge grows on dry to moist alpine slopes in southwestern Alberta, and flowers in July. • Nodding sedge and stone sedge are both arctic-alpine taxa with isolated (disjunct) populations in Alberta. • *Misandra* is derived from the Latin *miser* (wretched, unhappy) and *andro* (man), in reference to the inconspicuous male flowers on the uppermost spike.

C. PETRICOSA

SEDGE FAMILY   **Grass-like Plants**

PARRY'S SEDGE

Carex parryana Dewey var. *parryana*
SEDGE FAMILY (CYPERACEAE)

Plants: Loosely tufted perennial herbs; **stems** slender, **stiffly erect**, 15–40 {60} cm tall, bluntly 3-angled, **reddish-tinged at the base**; from long, creeping, scaly underground stems (rhizomes).

Leaves: Clustered near the base and usually much shorter than the stems, 5–12 per stem; blades 5–15 cm long, 2–4 mm wide, thin, stiff, **hairless**, usually flat with the edges rolled upward; dried **leaves of previous years persistent** and conspicuous.

Flower clusters: Elongated, 4–8 cm long, with **3–5 erect, stalkless or stiffly short-stalked spikes**; uppermost spike 1.5–3 cm long, **usually with female flowers at the tip** and male flowers at the base (gynaecandrous), but may be male throughout or may have female flowers scattered anywhere along its length; lower spikes 7–20 mm long, 2–3 mm wide, with 7–20 female flowers only, oblong to linear-oblong; **lowermost bract** short, inconspicuous, rarely as long as the flower cluster, reddish-tinged at the base, **essentially sheathless**; **scales of female flowers nearly round, 2–2.5 mm long, dark reddish brown**, with white, translucent edges and a **prominent midvein**, blunt or short-pointed, 1.5–2.5 mm long, **covering the perigynia**; 3 stigmas; {May–June} July.

Fruits: Dry, **3-sided** seed-like fruits (achenes), 1.4–1.8 mm long, enclosed in perigynia; **perigynia** straw-coloured (sometimes purplish near the tip), pressed against the main stalk, **hairless, 2-ribbed** (on edges), somewhat flattened, **2–2.3 {3} mm long**, 1–1.5 mm wide, egg-shaped, widest above the middle, **scarcely beaked** but sometimes tipped with 2 tiny (about 0.2 mm long) teeth or **slightly fringed** at the mouth (or both).

Habitat: Moist open meadows, swales and low ground near water, from the plains to moderate elevations in the mountains; elsewhere, on alkaline silt and marl flats.

Notes: Raynolds' sedge (*Carex raynoldsii* Dewey) is similar to Parry's sedge, but differs in that the uppermost spike usually has male flowers only, the scales of the female flower are larger (over 2.5 mm) and purplish black, and the perigynia are longer (3.5–4.5 mm). Raynolds' sedge

Rare *Vascular* Plants of Alberta

Grass-like Plants

SEDGE FAMILY

C. RAYNOLDSII

grows on moist open or wooded slopes and is reported to flower from June to August. • Both of these sedges resemble the common species Norway sedge (*Carex norvegica* Retz.). Norway sedge is most similar to Parry's sedge, but its weaker, often arched stems are sharply 3-angled, its scales are 1.5–2.5 mm long and purplish black and its perigynia are slightly larger (2–3.5 mm long) and never fringed at the mouth. • Parry's sedge was named for arctic explorer William Edward Parry (1790–1855).

Payson's sedge

Carex paysonis Clokey
SEDGE FAMILY (CYPERACEAE)

Plants: Loosely tufted perennial herbs; stems stiffly erect, 15–50 cm tall, 3-sided, with sharp edges, rough to the touch near the top; from tough, densely matted underground stems (rhizomes).

Leaves: Basal, **8–15 per stem**, 2–6 mm wide, flat with the edges rolled under, shorter than the stems, with **persistent fibrous remains of leaves from previous years**; lowermost leaves largest.

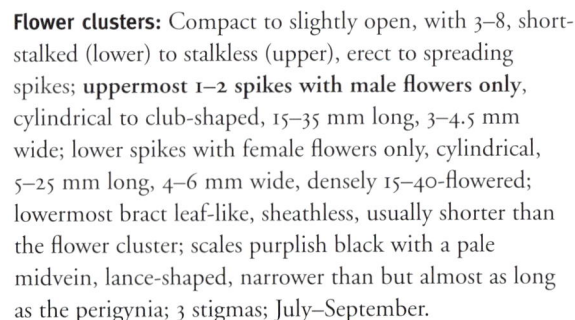

Flower clusters: Compact to slightly open, with 3–8, short-stalked (lower) to stalkless (upper), erect to spreading spikes; **uppermost 1–2 spikes with male flowers only**, cylindrical to club-shaped, 15–35 mm long, 3–4.5 mm wide; lower spikes with female flowers only, cylindrical, 5–25 mm long, 4–6 mm wide, densely 15–40-flowered; lowermost bract leaf-like, sheathless, usually shorter than the flower cluster; scales purplish black with a pale midvein, lance-shaped, narrower than but almost as long as the perigynia; 3 stigmas; July–September.

Fruits: Dry, light brown, 3-sided seed-like fruits (achenes) on short stalks, enclosed in perigynia; **perigynia purple-blotched**, pressed against the spike or pointed upward, 2–4 mm long, 1.5–2 mm wide, egg-shaped to nearly round, **strongly flattened**, abruptly narrowed to a short (0.2 mm) **purple beak**.

Habitat: Mountain meadows.

SEDGE FAMILY   **Grass-like Plants**

Notes: Alpine sedge (*Carex podocarpa* R. Br., also called *Carex montanensis* Bailey) and showy sedge (*Carex spectabilis* Dewey) are both very similar to Payson's sedge, but their stems have purplish bases and only 2–5 leaves, of which the lowermost are smallest. They also lack the dried remains of leaves from previous years. Alpine sedge is rare in Alberta, where it grows in alpine meadows, and flowers in June and July. It is distinguished from showy sedge by its nodding (rather than erect) lower spikes and its solid, brownish black scales, which lack the thick, whitish midribs found on the scales of showy sedge. • Payson's sedge is moderately to highly palatable to livestock, and it is an important forage plant when plentiful. It is regularly grazed by sheep, horses and cattle, and is resistant to damage from heavy grazing. In some regions, it is an important stabilizer of exposed soils. • This species was named in honour of Edwin Blake Payson (1893–1927), a professor of botany in Wyoming who studied the mustards (Brassicaceae) and borages (Boraginaceae).

C. PODOCARPA

Rare *Vascular* Plants of Alberta   297

Grass-like Plants

SEDGE FAMILY

OPEN SEDGE

Carex aperta Boott
SEDGE FAMILY (CYPERACEAE)

Plants: **Loosely clumped** perennial herbs; stems slender, stiff, **30–100 cm tall**, brownish or reddish at base, 3-sided with sharp edges, rough to touch near the top; from **stout, tough underground stems** (rhizomes).

Leaves: Erect, 3–5 per stem, **flat**, 2–5 mm wide, **much shorter than the stems**.

Flower clusters: Elongated, with 3 or 4 spikes, 15–20 cm long; **uppermost spike with male flowers only**, 2–3.5 cm long, 3–4 mm wide; **lower spikes** mostly with **female** flowers only, sometimes with male flowers at the tip, 25–75-flowered, **narrowly cylindrical**, 1–5 cm long, 5 mm wide, erect, stalkless or short-stalked; **lowermost bract leaf-like**, about as long as the flower cluster; upper bracts shorter; **scales** purplish black with a light midvein, **lance-shaped, longer and narrower than the perigynia**; 2 stigmas; {April–June} July–August.

Fruits: Dry, lens-shaped seed-like fruits (achenes) **enclosed in inflated perigynia**; perigynia **olive-green to straw-coloured**, somewhat flattened, pointing upward or spreading outward, 2.5–3.5 mm long, 1.5–2 mm wide, **egg-shaped to nearly round**, narrowed abruptly to a short beak.

Habitat: Open, wet ground; elsewhere, on floodplains and marshy lakeshores.

Notes: This species has also been called *Carex turgidula* Bailey. • Water sedge (*Carex aquatilis* Wahlenb. var. *aquatilis*) is a widespread species that is similar to open sedge, but its spikes are all erect (none nodding) on short (less than 3 cm) stalks, its lower leaves are generally larger than its upper leaves and its stems grow from extensive, spreading rhizomes and therefore are not tufted. • Holm's Rocky Mountain sedge (*Carex scopulorum* Holm) also resembles open sedge, but it is easily distinguished by its lowermost bracts, which are always shorter than the flower clusters. • The specific epithet *aperta* comes from the Latin word for 'open' or 'uncovered,' possibly referring to the fact that the spikes are borne well above the leaves, or perhaps to the exposed, spreading perigynia.

SEDGE FAMILY — Grass-like Plants

BLACKENED SEDGE

Carex heteroneura Boott var. *epapillosa* (Mack.) F.J. Hermann
SEDGE FAMILY (CYPERACEAE)

Plants: Densely tufted perennial herbs, with **brown, fibrous remains of old leaves**; stems 15–60 cm tall, **tinged purplish red at the base**; from short underground stems (rhizomes).

Leaves: Stiff, erect, 5–8 on the lower stem, 3–7 {8} mm wide, flat, shorter than the stems.

Flower clusters: Crowded, with 4 or 5 {3–6} spikes; spikes short, cylindrical, 10–25 mm long, 6–8 {10} mm wide, 30–60-flowered, on short, stiffly erect, 3–10 mm long stalks (lower spikes) to nearly stalkless (upper spikes); **uppermost spike with male flowers at the base** and female flowers at the tip (gynaecandrous); lower spikes with female flowers only; lowermost **bract leaf-like, dark red at the base**, shorter than the flower cluster; **scales reddish black with a lighter midvein**, lance-shaped, mostly pointed, **as long as the mature perygnia but much narrower**; 3 stigmas; {July–August}.

Fruits: Dry, 3-sided seed-like fruits (achenes), 1.5 mm long, enclosed in flattened perigynia; **perigynia yellowish green** (becoming brownish), pointing upward, widest at or above the middle, 3–4 {4.5} mm long, 1.5–2 {3.3} mm wide, rounded at base and tip, stalkless, dotted with tiny pits, **strongly flattened**, abruptly narrowing to a **reddish purple, 2-toothed beak** about 0.5 mm long.

Habitat: Moist to dry mountain meadows.

Notes: This species has also been called *Carex epapillosa* Mack., *Carex atrata* L. and *Carex hagiana* Kelso. • Purple sedge (*Carex mertensii* Prescott ssp. *mertensii*, also known as *Carex columbiana* Dewey) is also rare in Alberta. It grows in moist montane woods and along streambanks, forming dense, 30–100 cm tall tufts with short, stout rhizomes. Its stout, 3-sided stems are also purplish red at the base, but they have very sharp (winged), rough edges. The leaves of purple sedge are 4–7 mm wide, and their sheaths are often tinged cinnamon-brown. The uppermost spike has male flowers only (or sometimes a few female flowers at the tip), and the lower 4–9 spikes have female flowers only (or sometimes a few male flowers at the base). The numerous, large, clustered, dark-coloured spikes are cylindrical, 1–4 cm long and 7–9 mm wide, with slender,

Grass-like Plants   SEDGE FAMILY

C. MERTENSII

nodding stalks. The lower 2 or 3 bracts are leaf-like, sheathless, and longer than the flower cluster. The light green to pale brown (often purple-spotted) perigynia are broadly egg-shaped with tiny, purplish beaks. They measure 4–5 mm long and 2.5–3.5 mm wide, and their thin, flattened bodies are finely veined. The egg-shaped to lance-shaped scales are dark purplish brown with a light midvein and narrow, translucent edges; they are much narrower and shorter than the perigynia. Purple sedge is reported to flower from May to July and produces 3-sided achenes with 3 stigmas. • Dark-scaled sedge (*Carex atrosquama* Mack.) is a common species that resembles blackened sedge, but it has rougher perigynia that are scarcely flattened. Its scales are broader than those of blackened sedge and somewhat shorter than their perigynia. • Blackened sedge is eaten by all types of livestock, particularly horses, and is said to be of high value because of its abundant seed. • The specific epithet *epapillosa* means 'without small protuberances,' in reference to the smooth (not bumpy) perigynia.

SEDGE FAMILY Grass-like Plants

LENS-FRUITED SEDGE

Carex lenticularis Michx. var. *dolia* (M.E. Jones)
L.A. Standley
SEDGE FAMILY (CYPERACEAE)

Plants: Densely tufted perennial herbs; stems slender, erect, **10–60 cm tall**, 3-sided with sharp angles; from short underground stems (rhizomes).

Leaves: Erect, thin, 4–9 per stem, **as long as or longer than stems**, 20–65 cm long, 1–2 {3} mm wide, pleated near the base, flat toward the tip, slender-pointed, pale grey-green.

Flower clusters: Crowded, 5–12 cm long, with 4–6 erect, short-stalked (lower) to stalkless (upper) spikes; **uppermost spike** usually bearing **male** flowers only (sometimes tipped with a few female flowers), 1–3 cm long, 2.5 mm wide; **lowermost spikes** with **female** flowers only but **middle spikes with some male flowers at the base** (gynaecandrous), narrowly cylindrical, 15–40 {45} mm long, 3–4 mm wide; lowermost bract leaf-like with a short sheath, erect, longer than flower cluster; scales purplish or reddish brown with a broad, green, 3-nerved centre and translucent edges, blunt-tipped, smaller than the perigynia; 2 stigmas; {May–July}.

Fruits: Dry, brown, lens-shaped seed-like fruits (achenes) enclosed in perigynia; **perigynia bluish green**, pointed upward, **soon falling off**, dotted with a few yellow glands, flattened and sharply 2-edged, lightly 3–7-veined on both sides, short-stalked, **2–3 mm long**, 1–1.5 mm wide, egg-shaped, **abruptly narrowed to a short, toothless beak**.

Habitat: Moist lakeshores and marshes; elsewhere, on river flats and streambanks.

Notes: This species includes *Carex kelloggii* W. Boott in Wats., *Carex enanderi* Hult. and *Carex eurystachya* F.J. Herm. • The seeds of lens-fruited sedge are commonly eaten by small birds. • *Lenticularis* comes from the Latin words for 'like a lens' and refers to the shape of the perigynia.

Rare *Vascular* Plants of Alberta

Grass-like Plants

SEDGE FAMILY

NEBRASKA SEDGE

Carex nebrascensis Dewey
SEDGE FAMILY (CYPERACEAE)

Plants: Tufted perennial herbs, sometimes with single stems; stems 30–100 {20–120} cm tall, stout, 3-sided with sharp edges; from long, spreading, scaly underground stems (rhizomes).

Leaves: Thick, firm, blue-green, 8–15 per stem, 4–8 {3–12} mm wide, flat (sometimes channelled lengthwise), **usually with knobby cross-partitions** (like rungs on a ladder), shorter than or equal to the stems; **dead leaves conspicuous, persisting for several years** (marcescent).

Flower clusters: Open, elongated, with **3–7 erect**, stalked (lower) to stalkless (upper) **spikes**; upper 1 or 2 spikes with male flowers only, 1.5–4 cm long, 3–6 mm wide; **lower spikes** with female flowers only, cylindrical, **1–6 {7} cm long**, 5–9 mm wide, 30–150-flowered; lowermost bract leaf-like, sheathless, longer than the flower cluster; scales purplish or brownish black with light-coloured centres, lance-shaped, narrower than but usually equal to or longer than the perigynia; 2 stigmas; May–June {July}.

Fruits: Dry, lens-shaped seed-like fruits (achenes) enclosed in perigynia; **perigynia leathery, strongly 5–10-ribbed, straw-coloured with** tiny, **red dots**, oblong to egg-shaped, 3–3.5 {2.8–3.9} mm long, 2 mm wide, tapered to beaks 0.4–1 mm long tipped with 2 teeth.

Habitat: Marshy ground in the prairies; elsewhere, in seepage areas, often on alkaline ground.

Notes: The name of this species has been misspelled *Carex nebraskensis*. • Nebraska sedge is named after the state of Nebraska, from which it was first described.

SEDGE FAMILY — Grass-like Plants

SAND SEDGE

Carex houghtoniana Torr.
SEDGE FAMILY (CYPERACEAE)

Plants: Loosely tufted perennial herbs; **stems** rather stout, sharply triangular, **purplish at the base**, 20–60 {15–90} cm tall; from creeping, scaly underground stems (rhizomes).

Leaves: Alternate, 5–7 per stem, as long as the stems; **blades** flat, 10–20 cm long, 2–5 {10} mm wide, slender-pointed, **with rung-like cross-partitions** between the veins, very rough along the edges; sheaths loose and hairless.

Flower clusters: Elongated, 5–15 cm long, with **2–4 spikes**, with a single male spike at the tip (sometimes with a second, smaller one) and 1–3 widely spaced female spikes below; male spike 2–4 cm long; female spikes 1–4.5 cm long, 7–12 mm wide, erect, stalkless or short-stalked; **lowermost bract leaf-like**, 5–10 cm long, about as long as the flower cluster; sheathless or with a very short sheath; scales of female flowers reddish brown, shorter than the perigynia, broadly egg-shaped, pointed or tipped with a bristle (awn), translucent on the edges and green on the midvein; **3 stigmas**; {June–July}.

Fruits: 3-sided seed-like fruits (achenes), enclosed in perigynia; **perigynia** dull brownish green, **strongly many-nerved**, **short-hairy**, flattened, egg-shaped, **5–7 mm long**, 2–2.5 mm wide, with a {1} 2 mm long beak **tipped with 2 strong teeth**; August–September.

Habitat: Dry, acidic, sandy or gravelly places in the boreal forest, often in pine woods.

Notes: The name of this species has been misspelled *Carex houghtonii* Torr. • Sand sedge could be confused with the common species woolly sedge (*Carex pellita* Muhl. *ex* Willd., also incorrectly called *Carex lanuginosa* Michx.), but woolly sedge has narrower leaves, a more bunched growth form and smaller perigynia (less than 4 mm long) with fainter ribs that are hidden by the hairs. • Sand sedge responds vigorously to disturbance by growing more rhizomes. In natural systems, it would presumably require forest fires for survival.

Grass-like Plants

SEDGE FAMILY

LAKESHORE SEDGE

Carex lacustris Willd.
SEDGE FAMILY (Cyperaceae)

Plants: Tufted perennial herbs; stems stout, 50–120 {150} cm tall, 3-sided with sharp, rough edges, **conspicuously purplish-tinged at base**; from long underground stems (rhizomes).

Leaves: Usually longer than stems, up to 70 cm long, 5–15 mm wide, greyish blue to dark green with a thin, waxy coat, hairless, rough with **knobby cross-partitions** (like rungs on a ladder), bearing a **conspicuous** membranous collar (ligule) at the junction of sheath and blade; **lower sheaths bladeless, reduced to thread-like fibres with age**.

Flower clusters: Large, elongated, with about 4–6 spikes; upper 2 or 3 {1–5} spikes with male flowers only, very narrow, 1–8 cm long, 3–4 mm wide; **lower 2 or 3 {5} spikes** with **female** flowers only, thick, cylindrical, 2–10 cm long, 10–15 mm wide, **50–100-flowered**, well separated, short-stalked (upper sometimes stalkless), mostly erect (lower spikes sometimes nodding); lowermost bract leaf-like, with a short sheath, equal to or longer than the flower cluster; scales yellowish brown with a pale green midvein and translucent edges, **about ½ as long as the perigynia**, narrowly egg-shaped, pointed or tipped with a short bristle; 3 stigmas; {May–June} July–August.

Fruits: Dry, 3-sided seed-like fruits (achenes), enclosed in perigynia; **perigynia** olive-green, cylindrical to egg-shaped, 5.5–7 {5–8} mm long, **12–25-veined, leathery**, tapered to a **short**, thick **beak** tipped **with 2 straight or slightly spreading teeth** about 0.5 mm long.

Habitat: Marshes and swampy woods of the boreal forest; elsewhere, on lakeshores.

Notes: This species has also been called *Carex riparia* Curtis var. *lacustris* (Willd.) Kuk. • The scarcity of marshes with a constant water level limits the distribution of lakeshore sedge. • Awned sedge (*Carex atherodes* Spreng.) is a common species that resembles lakeshore sedge, but its leaf sheaths are hairy and the teeth of its perigynia are larger (1.5–3 mm long) and spreading. • Lakeshore sedge is the largest of the native sedges in Alberta. • The specific epithet *lacustris* is Latin for 'of lakes' and refers to the preferred habitat of this species.

SEDGE FAMILY　　　**Grass-like Plants**

PORCUPINE SEDGE

Carex hystericina Muhl. *ex* Willd.
SEDGE FAMILY (CYPERACEAE)

Plants: Densely tufted perennial herbs; stems slender, erect, **20–70 {100} cm tall**, sharply 3-sided, rough to the touch near the top, **reddish at the base**; from long, stout, horizontal underground stems (rhizomes).

Leaves: Thin, 3–7 per stem, 30–70 cm long, 3–8 {2–10} mm wide, flat, **often with knobby cross-partitions** (like rungs on a ladder); upper leaves **longer than stems**.

Flower clusters: Elongated, with 2–5 cylindrical spikes; uppermost spike with male flowers only, 1–5 cm long, 2.5–4 mm wide; lower **spikes** with female flowers only, 1–6 cm long, 1–1.5 cm wide, densely 100–200-flowered, **nodding slightly on long stalks** (lower spikes) to stalkless (upper spikes); lowermost bract leaf-like, sheathed, longer than the flower cluster; **scales** with small (1–2 mm), translucent bodies and **green midveins that extend past the tips as large** (2–6 mm), **broad, rough bristles**; 3 stigmas; {May–June}.

Fruits: Dry, 3-sided seed-like fruits (achenes) tipped with persistent, bony, bent or curved styles, enclosed in perigynia; **perigynia** light green, **inflated, pointing upward and slightly outward at maturity, shiny, strongly 15–20-nerved, 5–7 mm long**, 1.5–2 mm wide, narrowly egg-shaped, tapered to a **long, slender beak** tipped with 2 large (2–2.5 mm long), straight teeth; August–September.

Habitat: In heavy shade on mucky soils; elsewhere, sometimes in very open, sunny sites.

Notes: The name of this species has also been misspelled *Carex hystricina*. • Cyperus-like sedge (*Carex pseudocyperus* L., also misspelled *Carex pseudo-cyperus* L.) is a similar sedge that is also rare in swamps and marshes of Alberta. It differs from porcupine sedge most notably in having leathery, rigid perigynia that are only slightly inflated and that bend backward at maturity in 3–6 large (3–8 cm long, 8–20 mm wide) female spikes, some of which often hang on long stalks. It also has conspicuous ligules (much longer than wide) at the base of its leaf blades. Cyperus-like sedge forms large, dense clumps, 30–120 cm tall, from short, stout rhizomes. Its uppermost (male) spike is 2–6 cm long,

C. PSEUDOCYPERUS

Rare *Vascular* Plants of Alberta

3–4 mm wide and short-stalked. The 3-sided, yellowish brown perigynia are 3.5–5 {3–6} mm long, and they taper to a beak 1–2 mm long, tipped with 2 slender, slightly spreading to straight teeth. Cyperus-like sedge is reported to flower from June to July. It has a more northern distribution than porcupine sedge and is less dependent on heavy shade and mucky soils. To persist, it seems to require habitats that are relatively stable for lengthy periods.

FEW-FRUITED SEDGE

Carex oligosperma Michx.
SEDGE FAMILY (CYPERACEAE)

Plants: Perennial herbs; stems slender, stiffly erect, {20} 40–100 cm tall, 3-sided with sharp, rough edges; scattered along scaly underground stems (rhizomes).

Leaves: Stiff, thread-like, **with edges rolled inward**, 40–80 cm long, about as long as the stems.

Flower clusters: Open, elongated, 6–10 cm long, with 2–4 well-spaced, stalkless (upper) to short-stalked (lower) spikes; **uppermost spike** with **male** flowers only, 1–5 cm long, **very narrow** (about 1 mm wide), erect; **lower spikes** usually with **female** flowers only (upper ones sometimes tipped with male flowers), **round to short-cylindrical**, {7} 10–20 mm long, 5–8 mm wide, **3–15-flowered**; lowermost **bract slender, stiff, longer than the flower cluster**; scales chestnut-brown with a green midvein and translucent edges, broadly egg-shaped, pointed, ½–⅔ as long as the perigynia; 3 stigmas; {June} July.

Fruits: Dry, 3-sided, round to egg-shaped seed-like fruits (achenes) tipped with a persistent **bony style**, enclosed in perigynia; **perigynia 7–10-ribbed**, shiny, yellowish green, **inflated**, pointed upward, **4–7 mm long**, broadest above the middle, narrowed to a short (1–2 mm) beak tipped with 2 short teeth.

Habitat: Wet meadows and bogs.

Notes: This species has also been called *Carex depreauxii* Steud. • Few-fruited sedge is rarely grazed by wildlife, even when abundant. Birds occasionally eat its seeds.

SEDGE FAMILY **Grass-like Plants**

FEW-FLOWERED SEDGE

Carex pauciflora Lightf.
SEDGE FAMILY (CYPERACEAE)

Plants: Creeping perennial herbs; stems single or a few together, slender, stiff, curved at the base, rough above, 8–25 {40} cm tall; from long, slender underground stems (rhizomes).

Leaves: In clusters of 1–3 near the base of the stem, shorter than the stem; blades stiff, usually rolled inward, 3–10 {15} cm long, 0.7–2 mm wide, hairless.

Flower clusters: Narrow, solitary spikes, 7–10 mm long, with **male flowers at the tip** and 1–6 female flowers at the base (androgynous), bractless; **scales of female flowers** yellowish brown, with translucent edges and a greenish midvein, lance-shaped, about 5 mm long, shorter than the perigynia, **soon shed**; 3 stigmas; June–July.

Fruits: Oblong, **3-sided** seed-like fruits (achenes), enclosed in perygynia, tipped with a firm, continuous style; **perigynia straw-coloured**, faintly nerved, the lower 1–2 mm somewhat shrunken and spongy, **short-stalked, spreading or bent backward**, 6–7 mm long, 1–1.5 mm wide, **narrowly lance-shaped, gradually tapered to a long slender beak** not well distinguished from the body.

Habitat: Sphagnum bogs.

Notes: Few-flowered sedge resembles its much commoner relative, short-awned sedge (*Carex microglochin* Wahlenb.), but short-awned sedge has smaller perigynia (4–5 mm long) and more numerous stem leaves (4–8), and its achenes have a small thread-like appendage (rachilla) attached to the base. Short-awned sedge prefers lime-rich habitats, rather than acidic bogs. • Suitable habitat for this species is widespread in the boreal forest, but with so few perigynia per spike, it may just be a slow reproducer that has not yet succeeded in re-occupying its potential range following glacial withdrawal. • The specific epithet *pauciflora*, from the Latin *paucus* (few) and *flos* (flower), means 'few-flowered,' an appropriate name for this often-inconspicuous sedge.

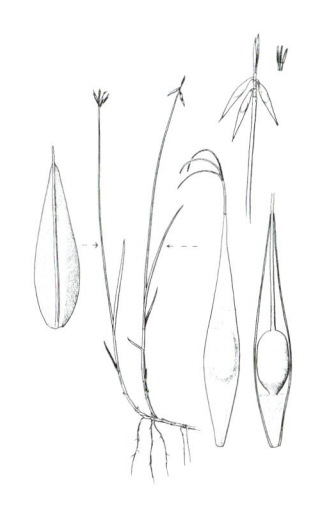

Rare Vascular Plants of Alberta

Grass-like Plants

SEDGE FAMILY

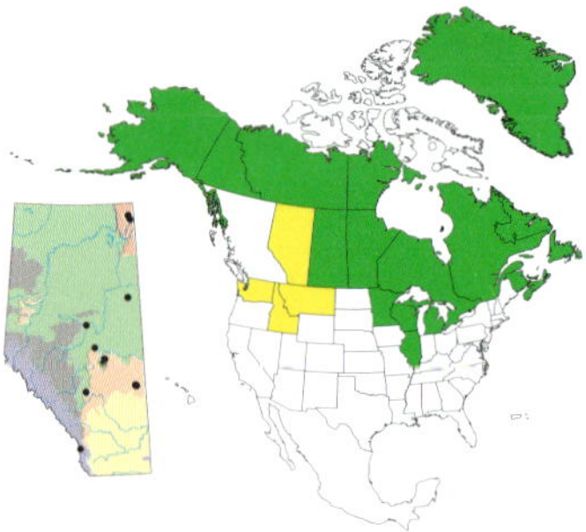

BEAKED SEDGE

Carex rostrata Stokes
SEDGE FAMILY (CYPERACEAE)

Plants: Stout, clumped perennial herbs; stems 50–100 cm tall, with spongy bases; from long, creeping underground stems (rhizomes).

Leaves: Narrow, 1.5–4 mm wide, **whitish or bluish with a fine, waxy powder on the upper surface** (bloom), dark green beneath; parallel **veins with thick cross-partitions** (like scattered rungs on a ladder) that appear as tiny warts when viewed under a microscope.

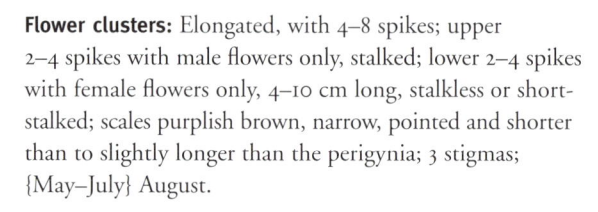

Flower clusters: Elongated, with 4–8 spikes; upper 2–4 spikes with male flowers only, stalked; lower 2–4 spikes with female flowers only, 4–10 cm long, stalkless or short-stalked; scales purplish brown, narrow, pointed and shorter than to slightly longer than the perigynia; 3 stigmas; {May–July} August.

Fruits: Yellowish, 3-sided seed-like fruits (achenes) tipped with bony, persistent, bent styles enclosed in the perigynia; perigynia shiny, pointing outward at maturity, veined, 4–8 mm long, egg-shaped, abruptly narrowed to a 1–2 mm long, 2-toothed beak.

Habitat: Floating fens at the edges of ponds and lakes.

Notes: Beaked sedge resembles bottle sedge (*Carex utriculata* Boott, formerly mistaken for *Carex rostrata* Stokes), but that species has broader (4–10 mm wide) leaves that lack the waxy powder found on beaked sedge plants. The cross-partitions between its veins are also finer and not wart-like. Bottle sedge is one of the most common sedges in Alberta. It grows in shallow water, often with water sedge (*Carex aquatilis* Wahlenb. var. *aquatilis*).

Rare *Vascular* Plants of Alberta

SEDGE FAMILY | **Grass-like Plants**

BLISTER SEDGE

Carex vesicaria L. var. *vesicaria*
SEDGE FAMILY (CYPERACEAE)

Plants: Tufted perennial herbs; **stems** erect, 30–100 cm tall, 3-sided with sharp edges, rough to the touch near the top, **often reddish at the base**; from short, stout underground stems (rhizomes).

Leaves: Flat, 4–10 per stem, shorter than or as long as stems, 20–80 cm long, 2–7 {8} mm wide, **with knobby cross-partitions** (like rungs on a ladder), rough to the touch and rolled under along the edges.

Flower clusters: Elongated, with 3–7 erect, essentially stalkless spikes; upper 2–4 spikes with **male** flowers only, very narrow, 2–7 cm long, 2–4 mm wide, held well above the female spikes; **lower spikes** with **female** flowers only, 1.5–8 cm long, 5–15 mm wide, 30–100-flowered; lowermost **bract leaf-like**, sheathless, usually **longer than the flower cluster**; scales yellowish to purplish brown with a lighter centre and translucent edges, lance-shaped to egg-shaped, pointed or tipped with a bristle, narrower and shorter than the perigynia; 3 stigmas; {June} July.

Fruits: Yellowish, 3-sided seed-like fruits (achenes) tipped with a **bony, persistent, bent style**, enclosed in perigynia; **perigynia** yellowish green to light brown, **inflated, shiny**, pointed upward at maturity, 7–20-ribbed, lance-shaped to egg-shaped, 4–8 {3–10} mm long, 3 mm wide, rounded at the base, gradually tapering to a **smooth, slender,** 2 {3} mm long **beak, tipped with 2 spreading teeth**.

Habitat: Swamps, marshes and shorelines.

Notes: This species has also been called *Carex inflata* Hudson, *Carex monile* Tuckerm. and *Carex raeana* Boott.
• Turned sedge (*Carex retrorsa* Schwein.) is also rare in Alberta, where it grows in swampy woods and wet meadows. It is very similar to blister sedge, but its lower bracts are several times longer than their flower clusters (rather than 1–several times longer), and the perigynia at the base of its spikes bend back (downward) at maturity. Turned sedge is reported to flower from May to September.
• Blister sedge has little nutritional value for livestock, but when it is abundant, it is grazed by both domestic and wild animals in the late summer and fall. It is a reasonable seed source for small birds.

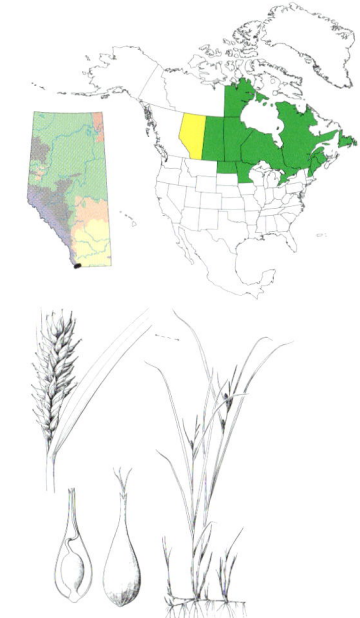

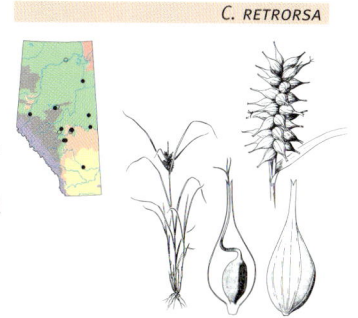

C. RETRORSA

Rare *Vascular* Plants of Alberta

Grass-like Plants

GRASS FAMILY

LITTLE BLUESTEM

Schizachyrium scoparium (Michx.) Nash ssp. *scoparium*
GRASS FAMILY (POACEAE [GRAMINEAE])

Plants: Densely tufted perennial herbs; stems wiry, erect or ascending, 20–60 {150} cm tall, often purplish, branched near the top; forming clumps from short, scaly underground stems (rhizomes).

Leaves: Alternate, light green to blue-green, usually with a whitish bloom; blades 2–5 {8} **mm wide**, flat or folded; ligules 1–1.5 mm long, finely hairy; **sheaths flattened, with a strong vertical ridge** (keel), usually **hairless**.

Flower clusters: Simple or branched (**panicles**), purplish, with **1–few, 2–5 {6} cm long, arching, spike-like racemes**, on stalks that are often enfolded by the upper leaf; central axis of each raceme (rachis) white-hairy, loosely zigzagged, **breaking apart** above a cupped structure at each joint (node); **spikelets 2-flowered**, borne **in pairs**, each with a **stalkless, {6} 7–9 mm long, fertile spikelet** and a **stalked, 4–5 mm long, sterile spikelet** with leathery, narrow, 5–10 mm long glumes and small, narrow and leathery lemmas; lemmas of fertile spikelets papery, tipped with a bent, 7–15 mm long bristle (awn) from between 2 small teeth; {July–August}.

Fruits: Grains; mature florets breaking free below the glumes, retaining a small, hairy segment that was part of the main stalk (rachis) of the flower cluster.

Habitat: Prairie grassland; elsewhere, in foothills; usually on calcareous soils.

Notes: This species has also been called *Schizachyrium scoparium* (Michx.) Nash var. *scoparium* and *Andropogon scoparius* Michx. • Little bluestem is an important prairie grass of the true prairie of central North America. In the west, it is usually restricted to areas with high water tables that can provide adequate moisture.

GRASS FAMILY **Grass-like Plants**

HOT-SPRINGS MILLET, THERMAL MILLET

Panicum acuminatum Swartz
GRASS FAMILY (POACEAE [GRAMINEAE])

Plants: Densely **tufted or matted, velvety-hairy**, greyish green perennial herbs, with 3 distinct seasonal phases; **winter phase a rosette** of short leaves; spring phase with unbranched stems producing sterile flower clusters; **autumn phase with widely spreading, repeatedly branching stems forming cushions**; stems 10–30 {100} cm tall, ascending or spreading, densely hairy at the joints (nodes); lacking underground stems (rhizomes).

Leaves: Basal and alternate; blades thick, flat, 5–12 mm wide; **ligules** reduced to a ring of **conspicuous hairs, 3 {5} mm long**.

Flower clusters: Open to contracted panicles, 3–6 {16} cm long; autumn panicles usually enfolded in upper leaf sheaths; **spikelets long-stalked, about 2 mm long, 2-flowered**, the **upper flower** with male and female parts (perfect) and the **lower one sterile or** with **male** parts only (staminate); lower glume tiny; **upper glume membranous, hairy**, about equal to the lower lemma; lemmas hairy; **upper lemma small, shiny, tightly rolled inward**; June {July–August}.

Fruits: Grains; mature florets breaking off below the glumes.

Habitat: Marshy places, around hot springs; elsewhere, on moist, sandy soil at woodland edges.

Notes: This species has also been called *Dichanthelium acuminatum* (Swartz) Gould & Clark var. *acuminatum* and *Panicum thermale* Boland (in the first edition of the *Flora of Alberta*).

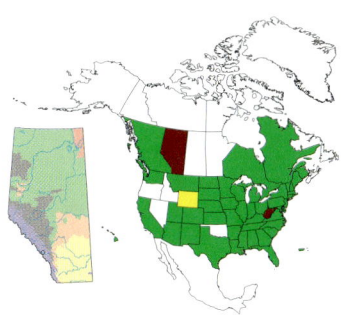

Grass-like Plants

GRASS FAMILY

LEIBERG'S MILLET

Panicum leibergii (Vasey) Scribn.
GRASS FAMILY (POACEAE [GRAMINEAE])

Plants: Tufted perennial herbs, with 3 distinct seasonal phases; **winter phase a rosette of short leaves**; spring phase with sterile flower clusters; **autumn phase with a few erect branches from middle and lower joints** (nodes); stems solitary or clustered, erect or ascending, abruptly bent near the base, 30–60 {70} cm tall, silky or rough-hairy.

Leaves: Basal and alternate, **covered with tiny, nipple-shaped bumps and coarse, spreading hairs** (sometimes almost hairless on the upper surface); blades flat, 7–11 cm long, 7–12 {6–15} mm wide; ligules reduced to a ring of hairs scarcely 0.5 mm long; sheaths separate (not overlapping).

Flower clusters: Contracted panicles, often somewhat enclosed in the upper leaf sheath, 5–10 {15} cm long, with ascending branches; **spikelets 2-flowered, 3–4 mm long**, with long, soft, spreading hairs, the **upper flower with male and female parts** (perfect) and the **lower one sterile or** with **male parts only** (staminate); lower glume 1.6–2.5 mm long; **upper glume membranous, hairy**, about equal to the lower lemma; lemmas hairy; **upper lemma small, shiny, tightly rolled inward**; {June–July}.

Fruits: Grains; mature florets breaking off below the glumes.

Habitat: Dry, sandy soil in grasslands and open woods.

Notes: This species has also been called *Dichanthelium leibergii* (Vasey) Freckman. • Leiberg's millet is also rare in much of Canada and the eastern United States. • Another rare species, sand millet (*Panicum wilcoxianum* Vasey, also called *Dichanthelium wilcoxianum* (Vasey) Freckman, *Panicum oligosanthes* Schultes and *Dichanthelium oligosanthes* (Schultes) Gould var. *wilcoxianum* (Vasey) Gould & Clark), differs from Leiberg's millet in having even smaller lower glumes (up to 1.5 mm long), shorter spikelets (2.7–3 mm long) and shorter flowering stalks (10–35 cm tall) with overlapping leaf sheaths and mostly hidden panicles. Its ligules are longer (1–1.5 mm), and its autumn phase is bushier, with many branches from the nodes and with tufts of overlapping reduced leaves. Sand millet grows in dry, open areas and is reported to produce fertile flowers in June and July.

P. WILCOXIANUM

RED THREE-AWN

Aristida purpurea Nutt. var. *longiseta* (Steud.) Vasey
GRASS FAMILY (POACEAE [GRAMINEAE])

Plants: Tufted perennial herbs, rough with short, stiff hairs; stems stiff, 20–30 {15–40} cm tall; often in large bunches, from fibrous roots.

Leaves: Alternate, mainly near the stem base, strongly rolled inward, 7–25 cm long, up to 1 {2} mm wide, often curved, very rough to touch; ligules membranous, fringed with short hairs, 0.5 mm long.

Flower clusters: Narrow panicles, 4–8 {10} cm long, few-flowered, **branches loosely ascending**, turning red {purple} when mature; **spikelets stalked, single-flowered**; glumes unequal, the lower one 8–10 {7–13} mm long, the upper one 16–20 {14–25} mm long; **lemmas hardened**, 10–15 mm long, cylindrical, slightly narrowed at the tip and with **3 similar, spreading bristles** (awns), each 4–8 cm long and rough at the tip; {July}.

Fruits: Grains, **tightly enfolded by hardened lemmas**; mature florets breaking off above the glumes; August–September.

Habitat: Dry, sandy plains.

Notes: This is the only *Aristida* species in Alberta. It has also been called *Aristida longiseta* Steud. • The generic name *Aristida* comes from the Latin *arista* (awn). The specific epithet *longiseta* was taken from the Latin *longus* (long) and *seta* (a stiff hair), in reference to the conspicuous, long bristles (awns). • The rough awns and the sharp points of the lemmas can injure the mouth and nostrils of grazing animals. • Red three-awn is intolerant of temperatures below 10°C at sunrise during the growing season, and therefore it is confined to warm, dry sites.

Grass-like Plants

GRASS FAMILY

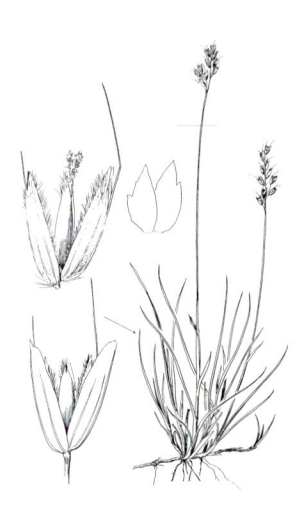

ALPINE SWEETGRASS

Anthoxanthum monticola (Bigelow) Y. Schouten & Veldkamp
GRASS FAMILY (POACEAE [GRAMINEAE])

Plants: Tufted, **sweet-smelling** perennial herbs, with leafy shoots at base; stems 10–40 cm tall, hairless; from short underground stems (rhizomes).

Leaves: Mostly basal, rolled inward, 1–2 mm wide, hairy on the upper (ventral) surface, hairless on the lower (dorsal) surface; stem leaves wider, less than 10 cm long; **ligules less than 1 mm long, ½ consisting of a fringe of hairs**; sheaths purplish, hairless.

Flower clusters: Narrow, bronze-coloured panicles 2–4 {1.5–4.5} cm long with short, ascending branches; spikelets tawny, {5} 6–8 mm long, **3-flowered**, with **2 male florets at the base** and a fertile **(male and female) floret at the tip**; glumes about as long as the florets, 5–7 {8} mm long, egg-shaped, thin and shiny; fertile lemmas hairy near the tip, pointed; **male lemmas** 5 mm long, hairy along the edges and **tipped with a bristle (awn) from between 2 teeth**; awn of the lowest floret straight and 2–4 mm long; awn of the second floret twisted and 5–8 mm long; June–August.

Fruits: Grains; mature florets breaking away above the glumes and between the florets; August.

Habitat: Dry alpine slopes; elsewhere, on dry, acidic peat and rocky tundra and outcrops.

Notes: This species has also been called *Hierochloe alpina* (Swartz *ex* Willd.) Roem. & Schult. • Alpine sweetgrass is similar to the more widespread common sweetgrass or holygrass (*Hierochloe odorata* (L.) Beauv.), but common sweetgrass usually grows at lower elevations, and it has longer ligules (1–2 mm long) which are almost entirely membranous, pyramid-shaped panicles (with spreading branches) and pointed (not awned) lemmas. • Sometimes germination of the seed occurs before the grains are dropped, and small plantlets develop in the panicle. • Common sweetgrass was widely used in ancient religious ceremonies, and it is still used in spiritual rituals by many native peoples in North America.

GRASS FAMILY · Grass-like Plants

LITTLE RICE GRASS

Oryzopsis exigua Thurb.
GRASS FAMILY (POACEAE [GRAMINEAE])

Plants: Densely tufted perennial herbs; stems hollow, stiffly erect, 10–30 {35} cm tall, rough to touch; from fibrous roots.

Leaves: Alternate; blades erect, **thread-like, rolled inward**, 5–10 cm long; ligules pointed, 3–4 mm long, with tiny hairs on the outer surface; sheaths smooth or minutely roughened.

Flower clusters: Slender panicles, **3–6 cm long**, with stiffly erect **branches pressed to the main stalk**; **spikelets single-flowered, plump**; glumes similar, about as long as the lemma, 4–6 mm long, broad, pointed (sometimes blunt-tipped), faintly 3–5-nerved; **lemmas firm, curled tightly around the developing grain** and over the palea, **4 mm long, covered with flat-lying hairs**, tipped with a **simple, twisted bristle** (awn) **4–6 mm long** from between 2 small lobes; anthers 1.5–2 mm long; {June–August}.

Fruits: Grains; mature florets breaking away above the glumes and between the florets.

Habitat: Dry subalpine slopes or open woods, usually on sandy and rocky soil.

Notes: Another rare Alberta species, Canadian rice grass (*Oryzopsis canadensis* (Poir.) Torr.), resembles little rice grass but has large (5–10 cm long), more open flower clusters and dark, hairy lemmas with long (6–12 mm) twisted bristles (awns). Canadian rice grass grows in open woods and on hillsides away from the mountains.
• Both of these species could be mistaken for their widespread relative, northern rice grass (*Oryzopsis pungens* (Torr.) A.S. Hitchc.), but northern rice grass is usually a taller plant (up to 50 cm tall) and its lemmas have tiny (0.5–2 mm long), inconspicuous bristles (awns) from blunt (not notched) tips. • The specific epithet *exigua* is Latin for 'small, short or meagre,' a reference to the small stature of this species.

O. CANADENSIS

Rare *Vascular* Plants of Alberta

LITTLE-SEED RICE GRASS

Oryzopsis micrantha (Trin. & Rupr.) Thurb.
GRASS FAMILY (POACEAE [GRAMINEAE])

Plants: Densely tufted perennial herbs; stems hollow, erect, 30–80 cm tall, smooth; from fibrous roots.

Leaves: Alternate; blades erect, **thread-like, usually rolled inward**, 1–2 mm wide; ligules squared or shorter in the middle, 0.5–1 mm long, fringed with tiny hairs; sheaths smooth or with fine, short hairs.

Flower clusters: Erect or nodding, rather **open panicles, 5–15 {20} cm long**, with spreading to erect branches; **spikelets single-flowered, plump**, borne near the branch tips; glumes similar, slightly longer than the lemma, 2–3 {4} mm long, pointed, broad, papery, 5-nerved, hairless or minutely roughened; **lemmas** eventually **curled tightly around the grain** and over the palea, **hairless** (or nearly so), **1.8–2.8 mm long**, tipped with a **simple, straight or weakly twisted bristle** (awn) **4–11 mm long**; anthers barely 1.5 mm long; {June–July}.

Fruits: Grains; mature florets breaking away above the glumes and between the florets.

Habitat: Dry open slopes, often on rocky ground; elsewhere, in open woods.

Notes: Little-seed rice grass is distinguished from similar species by its hairless lemmas. • The generic name *Oryzopsis*, from the Latin *oryza* (rice) and *opsis* (like), and the common name 'rice grass' both refer to the similarity between these grasses and rice (*Oryza sativa*). The specific epithet *micrantha*, from the Greek *mikros* (small) and *anthos* (flower), refers to the small spikelets of this species.

MARSH MUHLY

Muhlenbergia racemosa (Michx.) BSP.
GRASS FAMILY (POACEAE [GRAMINEAE])

Plants: Perennial herbs; **stems** 30–60 cm tall, slightly flattened, hollow, **hairy at the joints** (nodes) and **smooth and shiny in between**, often branching above; from creeping, scaly, branched underground stems (rhizomes).

Leaves: Alternate on the stem; blades flat, 2–7 mm wide, erect to ascending; **ligules 0.6–1.5 {3} mm long**, blunt, membranous; **sheaths with a lengthwise ridge on the back** (keel), loose.

Flower clusters: Dense, spike-like panicles, 3–7 {2.5–14} cm long, 5–15 mm wide, with branches tightly pressed against the stem; **spikelets short-stalked, single-flowered**, 5–6 mm long, green (sometimes purplish); **glumes narrow, tipped with long bristles** (awns) equal to or longer than the body, {4} 5–6.5 mm long (including awns); **lemmas hairy on the lower half**, 2.5–3.5 mm long, 3-nerved, gradually **tapered to slender bristles** (**awns**); anthers 0.4–0.8 mm long; {late July–August}.

Fruits: Grains; mature florets breaking away above the glumes and between the florets; August–September.

Habitat: Dry sand hills, slopes and eroded banks; elsewhere, in a wide variety of habitats including prairies, meadows, streambanks, edges of woodland, dry rocky slopes and waste ground.

Notes: Marsh muhly is easily confused with the more common bog muhly (*Muhlenbergia glomerata* (Willd.) Trin.), which is usually found in wet sites such as fens and marly shores, but can also grow near ephemeral springs that become dry in summer. However, the stems of bog muhly are hairy and dull between the joints (nodes) and slightly more slender and the leaf sheaths are only slightly ridged. Bog muhly has shorter ligules (0.2–0.6 mm) and longer anthers (0.8–1.5 mm), and its lemmas are hairy almost to the tip. • Another rare species, scratch grass (*Muhlenbergia asperifolia* (Nees & Meyen *ex* Trin.) Parodi), is the only species of the genus in Alberta with open, diffuse panicles (about as long as wide) that break away at maturity. Scratch grass is the main host of the smut fungus *Tilletia asperifolii* Ell. & Ev. When this fungus infects the grass, its spore clusters replace the ovary, eventually releasing a

Grass-like Plants — GRASS FAMILY

M. ASPERIFOLIA

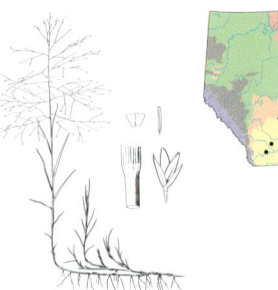

dark-coloured mass of spores. • The muhlys are considered warm-season grasses, growing most rapidly during the warmest part of the year, when many other grasses become senescent.

ALPINE FOXTAIL

Alopecurus alpinus J.E. Smith
GRASS FAMILY (POACEAE [GRAMINEAE])

Plants: Coarse perennial herbs; **stems hairless**, 10–80 cm tall, rather **stiff and erect** (sometimes with bases spreading on the ground); from **slender, creeping underground stems** (rhizomes).

Leaves: Alternate, few; blades flat, 3–5 {6} mm wide, rough to touch; ligules squared, 1–3 mm long, irregular or finely torn along the upper edge; **sheaths** often **inflated**, hairless.

Flower clusters: Woolly, often purplish, spike-like panicles, oblong to egg-shaped, **1–4 cm long**, about 1 cm wide, with dense, short-stalked spikelets; **spikelets single-flowered**, flattened; glumes silky to woolly, 3–4 mm long, slightly longer than the lemmas; lemmas 5-nerved, with a **bristle (awn) attached on the back near the base and sticking out** about 2–3 mm past the glumes; anthers 0.3–0.4 mm long; June–August.

Fruits: Grains; mature florets breaking free below the glumes.

Habitat: Shores and open woodland; elsewhere, in moist to wet montane, subalpine and alpine areas.

Notes: This species has also been called *Alopecurus occidentalis* Scribn. & Tweedy and *Alopecurus borealis* Trin. • The fuzzy spikelets catch on the fur of passing animals, and in the arctic, plants are often found growing around the burrows of ground squirrels and the dens of foxes and wolves. • The generic name *Alopecurus* was derived from the Greek *alopex* (fox) and *oura* (tail), because the fuzzy flower clusters of some species were likened to the tails of foxes.

GRASS FAMILY | Grass-like Plants

POLAR GRASS

Arctagrostis arundinacea (Trin.) Beal
GRASS FAMILY (POACEAE [GRAMINEAE])

Plants: Coarse perennial herbs; stems 40–100 {25–150} cm tall, single or tufted; from trailing stems (stolons).

Leaves: Alternate; blades 5–20 cm long, 4–10 {2–15} mm wide, flat or rolled inward; ligules prominent.

Flower clusters: Open, pyramidal panicles, 10–20 {25} cm long; **spikelets** purplish (sometimes green to brownish), 3–4 mm long, **single-flowered, stalked**; glumes unequal, shorter than the lemma, pointed; **lemmas awnless**, 3-nerved, densely covered with tiny, bristly hairs; paleas resembling lemmas; anthers 1.3–4 mm long.

Fruits: Grains; mature florets **breaking off below the glumes**.

Habitat: Marshy ground and moist meadows; elsewhere, in damp turfy tundra, heathland and open woodland.

Notes: This species has also been called *Arctagrostis latifolia* (R. Br.) Griseb. ssp. *arundinacea* (Trin.) Griseb.
• Because this is a highly variable species, as many as 5 subspecies have been proposed, but comparisons of morphology and chromosome number do not support these separations.

Rare *Vascular* Plants of Alberta

Grass-like Plants

GRASS FAMILY

LOW BENT GRASS

Agrostis humilis Vasey
GRASS FAMILY (POACEAE [GRAMINEAE])

Plants: Tufted perennial herbs; **stems 5–15 cm tall; without spreading underground stems** (rhizomes).

Leaves: Crowded at the base, **thread-like**; blades less than 1 mm wide, hairless; **ligules membranous, 0.5–1 mm long**, blunt.

Flower clusters: Narrow panicles, 1–3 cm long and less than 5 mm wide; branches flattened, strongly ascending; **spikelets** deep purple, **stalked, single-flowered, 2–3 mm long; glumes and lemma about equal** in length; **lemmas awnless; palea clearly visible,** ⅔–¾ as long as the lemma; **axis of the spikelet** (rachilla) **not prolonged past the palea or very tiny;** callus hairless; anthers 0.6–0.7 mm long; {July} August–September.

Fruits: Grains; mature florets breaking off above the glumes.

Habitat: Moist meadows and streambanks in alpine areas.

Notes: This species has also been called *Podagrostis humilis* (Vasey) Björkm. • Another rare Alberta species of moist alpine sites, Thurber's bent grass (*Agrostis thurberiana* A.S. Hitch.), differs from low bent grass in that the axis of each spikelet (rachilla) projects from the base of the palea as a tiny (0.1–0.3 mm) bristle. Thurber's bent grass is generally taller (10–40 cm), with longer (5–7 {3–10} cm), looser (often nodding) flower clusters and tufts of flat, relatively wide (about 2 mm) leaves, from short underground stems (rhizomes). It grows in moist alpine areas and flowers in July and August. These species sometimes hybridize. • A small (2–15 cm tall) arctic-alpine grass, frigid phippsia (*Phippsia algida* (Sol.) R. Br.) has also been reported to occur in Alberta. Like low bent grass, frigid phippsia has single-flowered spikelets in small, narrow clusters, but their glumes are tiny and much shorter than the lemmas. Frigid phippsia grows on wet alpine slopes and river flats.

A. THURBERIANA

PHIPPSIA ALGIDA

Rare *Vascular* Plants of Alberta

GRASS FAMILY | Grass-like Plants

NORTHERN BENT GRASS

Agrostis mertensii Trin.
GRASS FAMILY (POACEAE [GRAMINEAE])

Plants: Densely tufted, dwarf perennial herbs; stems 10–20 {4–40} cm tall; **without underground stems** (rhizomes).

Leaves: Mostly basal; dark green, smooth, erect; blades 0.2–2 {3} mm wide, **usually rolled inward**; ligules 1–2 mm long, blunt, with an irregular edge.

Flower clusters: Open, pyramid-shaped panicles, 1–14 cm long; **spikelets** purplish, 2–3 {4} mm long, single-flowered, **at tips of long, forked, smooth branches** (none at the base); glumes with a rough ridge (keel) on the back, almost equal, slightly longer than the lemma; **lemmas** rough-hairy, **with a bent** or sometimes straight bristle (awn), {1} 3–6 mm long near the middle of the back and sticking out from the spikelet; **paleas inconspicuous or absent**; {July–August}.

Fruits: Grains; mature florets breaking off above the glumes; August.

Habitat: Moist alpine slopes; usually in areas that hold snow late in the growing season.

Notes: This species has also been called *Agrostis borealis* Hartm. • Another rare species, spike redtop (*Agrostis exarata* Trin.), may be found in habitats similar to those of northern bent grass, but spike redtop is a larger grass (20–100 cm tall) with wider (2–6 mm), flat leaves, longer (2–6 mm) ligules and narrower panicles in which some branches bear spikelets to near their base. It is reported to flower from late June to August.

A. EXARATA

Rare *Vascular* Plants of Alberta

Grass-like Plants

GRASS FAMILY

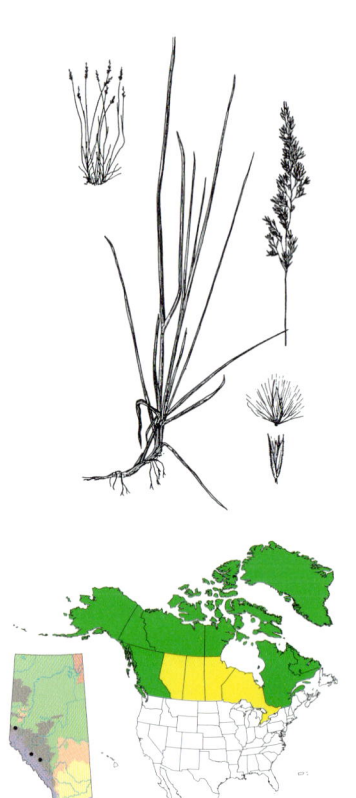

LAPLAND REED GRASS

Calamagrostis lapponica (Wahl.) Hartm.
GRASS FAMILY (POACEAE [GRAMINEAE])

Plants: Loosely tufted perennial herbs; **stems** slender, erect, 30–60 {80} cm tall, **smooth** (sometimes slightly rough just below the panicle), with **2–3 joints** (nodes); from short, thin underground stems (rhizomes).

Leaves: Alternate; blades short, flat or strongly rolled inward, 2–3 {1–4} mm wide, rough to the touch toward the tip; **ligules membranous, 0.5–3 {5} mm long**; sheaths smooth.

Flower clusters: Narrow, soft panicles, often **purplish**, {4} 5–10 (rarely to 15) cm long, 1–2 cm wide, with ascending branches; **spikelets stalked, single-flowered**; glumes purplish, turning bronze, slightly shiny, 4–5.5 mm long; **lemma slightly shorter than the glumes**, 3.5–5 mm long, **with a short, straight bristle** (awn) **from the middle and a conspicuous tuft of white hairs** (callus hairs) **from the base**; callus hairs abundant, of various lengths, the longest **equalling or surpassing the lemma**; August.

Fruits: Grains; mature florets breaking off above the glumes.

Habitat: Moist to dry, gravelly alpine slopes and ridges; elsewhere, on stable dunes and on peaty soil in open heath or muskeg.

Notes: Populations of this species in Alberta are widely separated (disjunct) from more widespread northern populations. • The spikelets of Lapland reed grass resemble those of bluejoint (*Calamagrostis canadensis* (Michx.) Beauv.), but the flower clusters of bluejoint are larger and more open (10–20 cm long and at least 2 cm wide), the callus hairs are fairly equal in length, and the stems are usually taller (60–120 cm) with 5 or 6 joints. • Northern reed grass (*Calamagrostis inexpansa* A. Gray, also called *Calamagrostis stricta* (Timm) Koeler ssp. *inexpansa* (A. Gray) C.W. Greene) and narrow reed grass (*Calamagrostis stricta* (Timm) Koeler) are also very similar, but northern reed grass has very rough stems (sandpapery with stiff hairs) and longer (4–7 mm) ligules, and narrow reed grass has smaller (2–3 mm long) glumes and relatively short callus hairs. • Lapland reed grass is also considered rare in Saskatchewan and Ontario. • The generic name *Calamagrostis* was taken from the Greek *calamus* (reed) and *agrostis* (grass).

GRASS FAMILY   **Grass-like Plants**

> **SLENDER HAIR GRASS**

Deschampsia elongata (Hook.) Munro *ex* Benth.
GRASS FAMILY (POACEAE [GRAMINEAE])

Plants: Densely tufted perennial herbs; stems slender, erect, {10} 30–100 cm tall; from fibrous roots.

Leaves: Mainly basal, a few alternate, soft; blades flat or folded, **1–1.5 mm wide** on the stem, **thread-like** at the base and usually only 2–4 cm long; ligules pointed, 3–9 mm long, finely hairy; sheaths usually hairless, sometimes slightly rough to touch.

Flower clusters: Pale green or purplish, shiny, narrow panicles, {5} 10–30 cm long, with slender, ascending branches; **spikelets 2-flowered, stalked**, pressed to the main branch; **glumes** similar, equal to or **longer than the lemmas**, {3} 4–6 mm long, **thin**, 3-nerved, ridged down the back (keeled), slender-pointed; **lemmas** thin, shiny, squared and torn at the tip, 2–3 mm long **with a straight, 3–5 mm long bristle** (awn) **attached at or below the middle and a tuft of stiff hairs at the base**; {June–July}.

Fruits: Grains; mature florets breaking free above the glumes and between the florets.

Habitat: Meadows and open slopes; elsewhere, on sandy or gravelly slopes by streams and lakes and occasionally in woods.

Notes: This genus was named in honour of L.A. Deschamps, a French botanist who lived 1774–1849. The specific epithet *elongata* refers to the narrow, elongated flower clusters.

Grass-like Plants

GRASS FAMILY

NODDING TRISETUM

Trisetum cernuum Trin.
GRASS FAMILY (POACEAE [GRAMINEAE])

Plants: Loosely tufted perennial herbs; stems 50–100 {120} cm tall, hairless, somewhat lax; from fibrous roots.

Leaves: Mostly basal; blades flat, lax, 5–10 {7–12} mm wide, with thin, prominent tips, rough to the touch to soft-hairy; ligules 1.5–3 {4} mm long, irregularly cut; **sheaths split down 1 side** (open), hairy.

Flower clusters: Open, loose panicles, often nodding, 15–25 {10–30} cm long, with slender branches bearing spikelets mainly near the ends; **spikelets 3 {2–5}-flowered**, usually with 3 fertile florets (with both male and female parts) and sometimes with 1 or 2 small, sterile florets above these; **glumes shorter than the lowest floret**, unequal, with a narrow, 1–2 {4} mm long lower glume, and a broader, 3.5–4 {6} mm long upper glume; **lemmas 5–6 {7} mm long**, faintly 5-nerved, tipped with 2 slender teeth, **bearing a curved or bent, 5–10 {14} mm long bristle (awn) on the upper back** and a tuft of short, **stiff hairs at the base** (bearded) and on the stalk; {May–July}.

Fruits: Grains; mature florets breaking off above the glumes and between the florets.

Habitat: Moist woods; elsewhere, also on banks of lakes and streams.

Notes: This species includes 2 varieties (var. *canescens* (Buckl.) Beal and var. *cernuum*), and it has also been called *Trisetum canescens* Buckl. (in part). • A similar grass, mountain trisetum (*Trisetum montanum* Vasey, sometimes considered a variety of *Trisetum cernuum*), is distinguished by its hairless grains, and its denser flower clusters, with branches bearing spikelets almost to their bases (rather than just near the tips). This rare Alberta species grows in moist, shady sites and flowers from July to August.

T. MONTANUM

GRASS FAMILY **Grass-like Plants**

WOLF'S TRISETUM

Trisetum wolfii Vasey
GRASS FAMILY (POACEAE [GRAMINEAE])

Plants: Loosely tufted perennial herbs; stems {40} 50–100 cm tall; from fibrous roots, sometimes with short underground stems (rhizomes) also.

Leaves: Mostly basal; blades flat, 2–4 {6} **mm wide**, rough to touch (rarely hairy); ligules 2.5–4 mm long, blunt, irregularly cut, fringed with hairs; sheaths split down 1 side (open), short-hairy to long-hairy.

Flower clusters: Narrow, erect panicles, 8–15 cm long, fairly dense with short, ascending branches; **spikelets 2 {3}-flowered, green or purplish**; glumes usually equalling or longer than the upper florets, with the lower glume 4–6 {6.5} mm long and the **upper glume 5.5–7 mm long**; lemmas 4–5 {6} **mm long**, ridged on the back (keeled), **with a few stiff hairs at the base** and many long hairs on the floret stalk (rachilla), blunt-tipped, **awnless** or with a very small awn (less than 2 mm long) on the upper back hidden by the glumes; anthers 1.5 mm long; {June} July–August.

Fruits: Hairy grains; mature florets breaking away above the glumes.

Habitat: Moist meadows and woodlands in the southern Rocky Mountains; elsewhere, in wet meadows and along streams in the mountains.

Notes: Wolf's trisetum might be mistaken for a depauperate June grass (*Koeleria macrantha* (Ledeb.) J.E. Schultes), but June grass has smaller ligules (usually about 0.5–2 mm long), its florets do not have conspicuous, stiff hairs at their bases and on their stalks, and its glumes reach only to the tip of the lowest lemma. • The generic name *Trisetum* is from the Latin *tres* (three) and *seta* (bristle), in reference to the 3 bristles characteristic of the lemmas of Old World species of this genus.

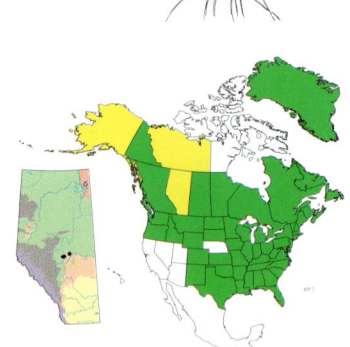

Grass-like Plants

GRASS FAMILY

POVERTY OAT GRASS

Danthonia spicata (L.) Beauv. *ex* R. & S.
GRASS FAMILY (POACEAE [GRAMINEAE])

Plants: Tufted perennial herbs; stems slender, {10} 20–70 cm tall; without underground stems (rhizomes).

Leaves: Mostly basal; blades up to 18 cm long, straight to curled, usually rolled inward, thread-like, hairless to sparsely hairy; **ligules reduced to a ring of hairs**, 0.5 mm long; sheaths with a tuft of long hairs at the throat.

Flower clusters: Panicles 2–5 cm long; branches erect, with 1 or 2 spikelets; **spikelets stalked, 3–8-flowered; glumes 9–12 mm long**, broad, papery, longer than the lower lemmas; **lemmas 3.5–5 mm long, sparsely hairy across the back**, tipped with a **bristle (awn) from between 2 slender, pointed teeth**; awns 5–8 {9} mm long, tightly twisted at the base (especially when dry); {June} July.

Fruits: Grains; mature spikelets breaking off above the glumes and between the florets; late July–September.

Habitat: Sandy and rocky sites, mostly in dry woods but sometimes in moist meadows; elsewhere, on rock outcrops and dry prairie.

Notes: The name 'poverty oat grass' reflects the ability of this species to grow on nutrient-poor sites and also its resemblance to oats (*Avena* spp.). • Another rare species, one-spike oat grass (*Danthonia unispicata* (Thurb.) Munro *ex* Macoun), differs from poverty oat grass in having a panicle usually consisting of only 1 spikelet and lemmas that are hairless (except on their lower edges). One-spike oat grass grows on open rocky or sandy ground and flowers in June {July}. • California oat grass (*Danthonia californica* Bol.) is also rare in Alberta. It is recognized by its largely hairless lemmas (silky-hairy along the edges and at the base only) and its small, open flower clusters, which usually have 1–4 spikelets on spreading branches. A widespread relative, timber oat grass (*Danthonia intermedia* Vasey), has narrower clusters of 5–10 spikelets, and its leaf sheaths are usually hairless (rather than soft-hairy). California oat grass grows in open, grassy meadows and on rocky ridges. It

D. UNISPICATA

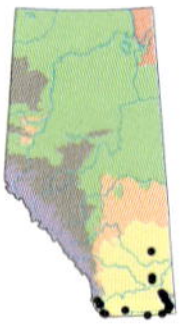

flowers in June and July. • Many oat grasses (*Danthonia* spp.) are often cleistogamous, which means that the florets remain closed at the time of fertilization. Their tiny anthers (often only a fraction of a millimetre long) produce very few pollen grains (perhaps 10–12), which grow pollen tubes through the anther wall and eventually fertilize the ovary. These self-fertilized flowers are usually found on spikelets within the lower leaf sheaths. • The long, bent bristles (awns) of these grasses twist and untwist with changes in humidity. This action may help to drill the seeds into the ground.

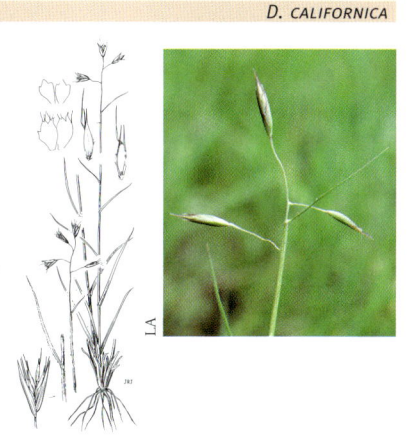

PRAIRIE CORD GRASS

Spartina pectinata Link
GRASS FAMILY (POACEAE [GRAMINEAE])

Plants: Coarse perennial herbs; stems erect, **50–200 cm tall**, hairless; from stout, elongated underground stems (rhizomes).

Leaves: Alternate on the stem; blades fibrous, **5–15 mm wide**, flat or rolled inward, rough with **small, sharp teeth** along the edges; **ligules a fringe of hairs** {1.5} **2–3 mm long**; sheaths distinctly veined, hairy at the collar only.

Flower clusters: Linear, 10–30 cm long, with **10–20 erect, stalkless spikes**, each 4–8 {2–10} cm long; **spikelets single-flowered, stalkless, strongly flattened**, crowded along 1 side of the spike branch **like teeth on a comb; glumes rough to touch**; lower glume 5–6 mm long, bristle-tipped; **upper glume 8–12 mm long, longer than the lemma, tipped with a 4–10 mm long bristle** (awn); lemmas 7–9 mm long, rough along the ridge on the back (keel); **paleas conspicuous**, pointed, longer than the lemmas; {late June–July}.

Fruits: Grains; mature florets breaking free below the glumes; August–September.

Grass-like Plants

GRASS FAMILY

Habitat: Saline shores and marshes; elsewhere, on moist prairie, riverbanks and shores, and in ditches and marshes.

Notes: Prairie cord grass is similar to the more common alkali cord grass (*Spartina gracilis* Trin.). The glumes of alkali cord grass are relatively smooth to the touch and the upper glume does not have a bristle. Alkali cord grass has shorter, less spreading panicle branches, shorter spikes (2–4 cm), shorter ligules (about 1 mm) and smaller, less robust plants, and it appears to be more salt tolerant.
• Prairie cord grass is a warm-season grass, meaning that most vegetative growth occurs during the hottest part of the year. • The Omaha Indians used prairie cord grass for thatch on permanent lodges before covering them with earth. Early pioneers used it in a similar manner, as thatching on buildings, haystacks and unroofed granaries. In the southern US, the stems were used for brooms and brushes. • This species provides poor to fair cattle forage when young and soon becomes unpalatable. It does not do well under grazing pressure due to the effects of trampling on the brittle mature stems. • Prairie cord grass is an indicator of native tall-grass prairie. It is unusual to find it as far west as Alberta. Most plants here probably do not set seed. Instead, they increase vegetatively, producing local, concentrated populations. • 'Ripgut' is another common name for prairie cord grass, referring to the sharp teeth along the leaf edges, which **can cut flesh**. • The generic name *Spartina* was derived from the Greek *spartine* (a cord), probably because of its wiry leaves and extensive tough rhizomes. The specific epithet *pectinata* refers to the pectinate (arranged like the teeth on a comb) spikelets.

FALSE BUFFALO GRASS

Munroa squarrosa (Nutt.) Torr.
GRASS FAMILY (POACEAE [GRAMINEAE])

Plants: Mat-forming annual herbs, 3–15 cm tall; **stems spreading**, often rooting at the joints (nodes), **branched**, hollow, hairy, rough to touch; from fibrous roots.

Leaves: In **clusters** at stem joints and stem tips; **blades spine-tipped**, stiff, 1–3 cm long, 1–3 mm wide, **with rough, white edges**; **ligules** consisting of **a ring of hairs**; sheaths crowded, split down 1 side (open), hairy on the edges at the collar.

Flower clusters: Short panicles nestled among the leaves; spikelets crowded in clusters of 2–3, 6–9 mm long; lower spikelets smaller, 3–4-flowered, with equal, narrow, sharp-pointed, 1-nerved glumes; **upper spikelets larger**, 2–3 flowered, with unequal glumes, the lower glume short or absent; lemmas short-stalked, 3-nerved, tipped with a short bristle (awn) from between 2 lobes, hairy on the edges near mid-length, membranous in upper spikelets, leathery in lower spikelets; anthers 1.5 mm long; {June–August}.

Fruits: Grains; mature florets breaking away above the glumes and between the florets.

Habitat: Dry plains and slopes, often on disturbed ground; elsewhere, on sandy and gravelly soil, on open ground and in old fields.

Notes: The name of this genus has also been misspelled *Monroa*. • This is a very distinctive grass, the only species of its genus in North America, not easily confused with any other grass. It bears superficial resemblance to buffalo grass (*Buchloe dactyloides* (Nutt.) Engelm.) and hence the common name 'false buffalo grass.' • False buffalo grass is a warm-season species, meaning it grows most rapidly during the hottest part of the year. • It is unpalatable to livestock. • The specific epithet *squarrosa* is Latin for 'rough with outward-projecting tips' and describes the spreading leaves and stems of these plants.

Grass-like Plants

GRASS FAMILY

SMITH'S MELIC, MELIC GRASS

Melica smithii (Porter *ex* A. Gray) Vasey
GRASS FAMILY (POACEAE [GRAMINEAE])

Plants: Tufted perennial herbs; stems 50–100 {30–130} cm tall, **usually lacking the bulb-like base** that is typical of this genus; from underground stems (rhizomes).

Leaves: Alternate on the stem; blades flat, {5} 6–12 mm wide, lax, rough to touch, with prominent veins; ligules 3–9 mm long, blunt, irregularly cut; **sheaths** with stiff, downward-pointing hairs, **closed to the top**.

Flower clusters: Loose panicles, 10–25 {30} cm long, with solitary, widely spaced branches eventually spreading or nodding; **spikelets** {3} **4–6-flowered**, {15} 16–20 mm long, often purplish; **lower glume** {3} 4–6 mm long, **faintly 3-nerved**; **upper glume** {4} 5–8 mm long, **shorter than the lowest lemma**, distinctly {3} 5-nerved; **lemmas about 10 mm long, strongly 7-nerved**, tipped with a 3–5 mm long bristle (awn) from between 2 teeth; July {June–August}.

Fruits: Grains; mature florets breaking away above the glumes and between the florets.

Habitat: Moist subalpine woodlands; elsewhere, in moist woods.

Notes: This is the only native onion grass (*Melica* spp.) in Alberta that has awned lemmas. • Another rare Alberta species, showy onion grass (*Melica spectabilis* Scribn.), also known as purple onion grass, has fleshy, bulb-like swellings at the base of its flowering stalks. It is distinguished from similar species by its broader, blunt or abruptly pointed, hairless, purplish lemmas, its broad, papery glumes and its solitary stems, which arise from spreading rhizomes. Showy onion grass grows in wet to moderately dry, fairly open sites and flowers in {May–July} August. • The 'bulbs' of onion grasses are edible, with a pleasant, nutty flavour, but they do not seem to have been used by aboriginal peoples.

M. SPECTABILIS

GRASS FAMILY | **Grass-like Plants**

PRAIRIE WEDGE GRASS

Sphenopholis obtusata (Michx.) Scribn.
GRASS FAMILY (POACEAE [GRAMINEAE])

Plants: Slender, densely tufted perennial herbs; stems 20–80 {100} cm tall; from fibrous roots.

Leaves: Alternate; blades flat, **2–5 {6} mm wide**, **rough** to touch on both sides; ligules 1.5–2 {1–4} mm long, irregularly torn, toothed and fringed; sheaths hairless or rough to touch, distinctly veined.

Flower clusters: Erect, lustrous, spike-like panicles 3–10 {2–15} cm long and 5–15 mm wide, with erect branches; **spikelets stalked, 2-flowered** (sometimes with a small, third flower), 2.5–3 mm long; **glumes** very **unequal**, the lower slender, the **upper broadly egg-shaped to almost round**, tapered to a wedge-shaped base and **slightly hooded at the tip**; **lemmas firm, rough**, covered with tiny bumps (papillose), faintly nerved, 1.5–2.6 mm long, longer than the glumes; {June–July}.

Fruits: Grains; mature florets breaking free below the glumes.

Habitat: Moist sites in meadows and open woods and on shores; elsewhere, on moist to wet prairie.

Notes: A more common species, slender wedge grass (*Sphenopholis intermedia* (Rydb.) Rydb.), can be distinguished by its open, lax panicles and its second glume that is 3 times as long as wide and pointed to rounded (not hooded). • The generic name *Sphenopholis* was derived from the Greek *sphen* (wedge) and *pholis* (scale), in reference to the wedge-shaped upper glumes. The specific epithet *obtusata*, from the Latin *obtusus* (blunt), refers to the broad, rounded tips of the upper glumes.

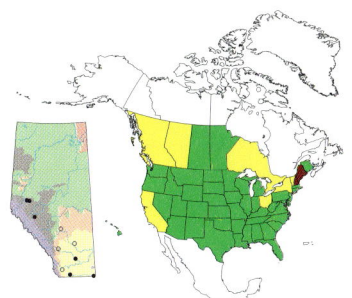

Rare *Vascular* Plants of Alberta

Grass-like Plants
GRASS FAMILY

WHEELER'S BLUEGRASS

Poa nervosa (Hook.) Vasey var. *wheeleri* (Vasey) C.L. Hitchc.
GRASS FAMILY (POACEAE [GRAMINEAE])

Plants: Slender perennial herbs; stems rounded, ascending or erect, 30–60 {70} cm tall, hairless; from **long, spreading underground stems** (rhizomes).

Leaves: Mostly basal; blades flat or folded with boat-shaped (keeled) **tips**, 2–4 {5} mm wide; **ligules 1–2 mm long, thickish, short-hairy and fringed with hairs**; sheaths with downward-pointing hairs (sometimes hairless), often purplish near the base.

Flower clusters: Open panicles, 5–10 cm long, with slender, nodding or spreading branches mostly in 2s or 3s, **usually entirely female**, with a few small, non-functioning anthers; **spikelets stalked near branch tips**, {2} 4–7-flowered, 4–6 {8} mm long, slightly flattened; **glumes 2–3.5 mm long, shorter than the lowest lemma; lemmas 4–5 {3–6} mm long**, clearly 5-nerved (with nerves converging at the tip), ridged down the back (keeled), sometimes hairy on the lower nerves but never with a tuft of cobwebby hairs at the base; {April–July} August.

Fruits: Grains; mature florets breaking away above the glumes and between the florets.

Habitat: Open woods; elsewhere, on exposed ridges and talus slopes in montane to alpine zones.

Notes: Perhaps the most distinctive characteristic of this species is its peculiarly thickened, hairy and fringed ligules, which are squared at their tips (at least on lower leaves).
• Wheeler's bluegrass resembles early bluegrass (*Poa cusickii* Vasey), but that species lacks rhizomes and has much narrower flower clusters. Its ligules are also thin, sparsely hairy and 2.5 mm long, and it grows in dry, open sites (grasslands, sandhills) at low elevations and on the prairie.

Letterman's bluegrass

Poa lettermanii Vasey
GRASS FAMILY (POACEAE [GRAMINEAE])

Plants: Small, densely tufted perennial herbs; stems {2} 3–12 cm tall; creeping underground stems (rhizomes) lacking.

Leaves: Basal and alternate on the stem; blades lax, mostly flat, 1–2 mm wide, with boat-shaped (keeled) tips; ligules 0.5–2 {3} mm long, blunt, irregularly cut; sheaths split down 1 side (open) for most of their length.

Flower clusters: Narrow panicles, 1–2 {3} cm long; spikelets 2–5-flowered, flattened (keeled), purple-tinged, 2.5–4 {4.5} mm long; glumes shorter than the lowest lemma, 2–4 mm long, 3-nerved, rough along the ridged back (keel); lemmas hairless and smooth, 2–3 {3.5} mm long, with 3–5 faint nerves converging toward the rather blunt tip; anthers about 0.5 mm long; {June–August}.

Fruits: Grains; mature florets breaking away above the glumes and between the florets; July–August.

Habitat: Exposed alpine ridges, in dry, rocky fellfields.

Notes: The range of Letterman's bluegrass overlaps with that of another rare species, bog bluegrass (*Poa leptocoma* Trin.), but bog bluegrass grows along streams and in seepage areas and wet meadows. Bog bluegrass also differs from Letterman's bluegrass in having a tuft of cobwebby hairs at the base of its lemmas, nodding panicles with spreading branches, and longer (3–4 mm) ligules. It is reported to flower from late June to August. • Letterman's bluegrass provides nutritious, palatable forage for mountain goats and bighorn sheep. • This species was named in honour of G.W. Letterman, a botanist who collected the type specimen at Grays Peak, Colorado.

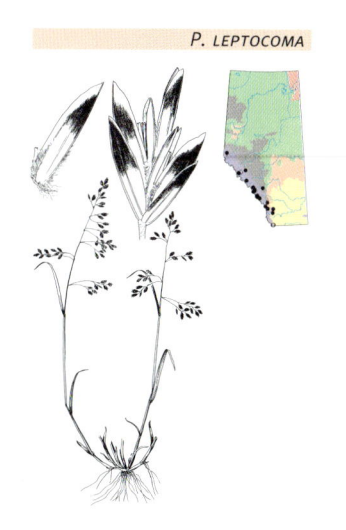

P. LEPTOCOMA

Grass-like Plants

GRASS FAMILY

NARROW-LEAVED BLUEGRASS

Poa stenantha Trin.
GRASS FAMILY (POACEAE [GRAMINEAE])

Plants: Tufted perennial herbs; stems erect, 30–50 {25–60} cm tall; **without creeping underground stems** (rhizomes).

Leaves: Mostly basal, but also a few alternate on the stem; **blades with a boat-shaped tip**, flat or loosely rolled inward, rather soft and lax, 1–2 {2.5} mm wide; **ligules 1.5–3.5 {1–5} mm long**, finely hairy and rough to touch, pointed and torn at the tip; sheaths hairless, fused into a tube (closed) for about ¼ of their length.

Flower clusters: Nodding, **open panicles**, 5–15 cm long, with spreading to drooping lower branches in 2s and 3s; **spikelets flattened, 3–5-flowered, 6–8 {10} mm long**; glumes 3-nerved, unequal, 2.7–5.7 mm long, rough to the touch; **lemmas 4–6 mm long, with a clearly defined ridge** (keel) and nerves, hairy on the lower keel and nerves and sparsely hairy between; anthers 1.2–2 mm long; {July} August.

Fruits: Grains; mature florets breaking away above the glumes and between the florets.

Habitat: Open woods at montane elevations; elsewhere, in subalpine meadows and on talus slopes.

Notes: Narrow-leaved bluegrass can be very similar to Sandberg bluegrass (*Poa secunda* J. Presl, which includes *Poa sandbergii* Vasey), a much more common species that grows in similar habitats. However, the lemmas of Sandberg bluegrass are rounded on the back (rather than keeled), and the spikelets are only slightly flattened.
• Wavy bluegrass (*Poa laxa* Haenke ssp. *banffiana* Soreng) is a rare perennial species with soft, slender leaves, similar to those of narrow-leaved bluegrass. It is recognized by its smaller (rarely over 6 cm long), erect panicles with spikelets near the tips of smooth, curved, single or paired branches. Its plants have few, solitary or loosely tufted, thread-like stems. Wavy bluegrass grows on alpine slopes.

P. LAXA

GRASS FAMILY **Grass-like Plants**

TUFTED TALL MANNA GRASS

Glyceria elata (Nash) A.S. Hitchc.
GRASS FAMILY (POACEAE [GRAMINEAE])

Plants: Loosely tufted perennial herbs; stems 60–150 cm tall, fleshy; often in large clumps, from creeping underground stems (rhizomes).

Leaves: Alternate on the stem, dark green; blades lax, flat, 20–40 cm long, 4–10 {12} mm wide; **ligules 3–6 mm long, blunt, with short hairs; sheaths fused into a tube** (closed) to near the top, rough to the touch.

Flower clusters: Loose panicles 15–30 cm long, with **spreading branches**, the lower branches often bent downward (reflexed); **spikelets 4–8-flowered**, 3–5 mm long; **glumes 1–2 mm long**, shorter than the lowest lemma, with irregular, fringed edges; **lemmas firm, 2–2.5 mm long, with 7 prominent, parallel veins** and a **blunt, translucent tip**; {May–July}.

Fruits: Grains; mature florets breaking away above the glumes and between the florets.

Habitat: Stream edges and wet meadows in the southern Rocky Mountains.

Notes: The range of tufted tall manna grass overlaps with the ranges of other, more common manna grasses, which it resembles, including common tall manna grass (*Glyceria grandis* S. Wats.) and fowl manna grass (*Glyceria striata* (Lam.) A.S. Hitchc.). Common tall manna grass differs in having hairless ligules, a longer lowest (lower) glume (1.3–2 mm) and lemmas that are purplish rather than greenish. Fowl manna grass has narrower leaves (2–5 mm wide), shorter lower ligules (1.5–3 mm) and shorter lemmas (about 2 mm). • Manna grass is tender and readily eaten by livestock and wildlife when available. • The generic name *Glyceria* comes from the Greek *glykeros* (sweet), presumably in reference to the grains.

Map not available

Rare *Vascular* Plants of Alberta   335

Grass-like Plants

GRASS FAMILY

FEW-FLOWERED SALT-MEADOW GRASS

Torreyochloa pallida (Torr.) Church var. *pauciflora* (J. Presl) J.I. Davis
GRASS FAMILY (POACEAE [GRAMINEAE])

Plants: Perennial herbs; stems loosely tufted or solitary, 40–100 {15–140} cm tall, often somewhat spreading; from creeping underground stems (rhizomes).

Leaves: Alternate on the stem; blades lax, thin, flat, 8–15 cm long, 4–12 {3–15} mm wide, slightly rough; **ligules 5–6 {3–9} mm long**, sharp-pointed, usually irregularly cut; **sheaths split down 1 side** (open), sometimes inflated.

Flower clusters: Rather **loose, nodding panicles**, {5} 10–20 cm long, **often purplish**, mostly with 2–4 curved branches per joint (node); **spikelets 4–5 {3–7}-flowered**, 4–5 mm long, crowded into the upper half of the panicle; **glumes broadly egg-shaped**, {0.8} **1–2 mm long**, irregularly cut along the edges, the lower glume 1-nerved and the **upper glume 3-nerved**; **lemmas 2–2.5 {1.9–2.9} mm long**, with **5 conspicuous parallel nerves** (not converging toward the tip) and usually with 2 faint nerves at the edges, short-hairy on the nerves, blunt and broadly translucent at the tip; anthers 0.5–0.7 mm long; {June–August}.

Fruits: Grains; mature florets breaking away above the glumes and between the florets; August.

Habitat: Wet places; elsewhere, in shallow water, wet meadows, woods and thickets.

Notes: This species has also been called *Puccinellia pauciflora* (J. Presl) Munz var. *holmii* (Beal) C.L. Hitchc., *Torreyochloa pauciflora* (J. Presl) Church and *Glyceria pauciflora* Presl (in the first edition of the *Flora of Alberta*). • Another variety of this species, *Torreyochloa pallida* (Torr.) Church var. *fernaldii* (Hitchc.) Dore, has also been reported to occur in Alberta. It is distinguished by its larger (to 3 mm long), distinctly 7-nerved lemmas. • Conspicuously nerved lemmas are also common in the manna grasses (*Glyceria* spp.), and small-flowered salt-meadow grass was once classified as a species of that genus. Manna grasses are distinguished by their closed sheaths (fused in a tube almost to the top) and their 1-nerved glumes. • The generic name *Torreyochloa* commemorates John Torrey, an American botanist who lived 1796–1873.

T. PALLIDA var. FERNALDII

TINY-FLOWERED FESCUE

Festuca minutiflora Rydberg
GRASS FAMILY (POACEAE [GRAMINEAE])

Plants: Delicate, tufted perennial herbs, slightly bluish green; stems 7.5–20 {4–30} cm tall, with the joints (nodes) rarely exposed; old leaf bases not evident.

Leaves: Mostly basal; **blades soft, thread-like**, rolled inward, tipped with a slender bristle (awn); **stem leaves** 0.7–3 cm long; erect, with a **swelling at the base of the blade**; ligules 0.1–0.3 {0.75} mm long, irregularly cut; sheaths split down one side (open), hairless, with a prominent midvein.

Flower clusters: Narrow, somewhat lax panicles, 1–4 {5} cm long, with short branches; **spikelets 3–4 {2–5}-flowered, 2.5–5 mm long; glumes** unequal, shorter than the lowest lemma, 1.3–3.5 mm long, short-hairy near the tips; **lemmas 5-nerved**, rounded on the back, 2.2–3.4 {4} mm long, abruptly narrowed at the tip to a 0.7–1.5 {0.5–1.7} mm long awn; August.

Fruits: Grains; mature florets breaking away above the glumes and between the florets.

Habitat: Alpine tundra and meadows and subalpine openings in the mountains.

Notes: This species has also been included in *Festuca brevifolia* R. Br. var. *endotera* Saint-Yves, *F. brevifolia* var. *utahensis* Saint-Yves and *F. ovina* L. var. *minutiflora* (Rydb.) Howell. • Tiny-flowered fescue is similar to the more common alpine fescue (*Festuca brachyphylla* Schultes & Schultes f.), but alpine fescue has larger spikelets, broadly lance-shaped lemmas, longer awns (1–3 mm) and stiffer, more erect leaves and flower clusters. • *Festuca* is the ancient Latin name for a grass.

Grass-like Plants

GRASS FAMILY

VIVIPAROUS FESCUE

Festuca viviparoidea Krajina *ex* Pavlick ssp. *krajinae* Pavlick
GRASS FAMILY (POACEAE [GRAMINEAE])

Plants: **Densely tufted perennial** herbs; stems erect, **6–30 cm tall**, rough to touch on the upper part; from fibrous roots.

Leaves: Basal, **½–⅓ as long as the stem**, slender; blades 0.5–1 mm wide, **rolled inward**, hairless, sometimes rough to touch near the tips; sheaths parchment-like, persistent.

Flower clusters: Dense, narrow, almost **spike-like panicles, 1–6 cm long**, with rough, ascending branches; **spikelets few-flowered, upper spikelets with small leaves** projecting from their tips (viviparous); lower spikelets sometimes producing seed; glumes lance-shaped, 2–3 mm long; lemmas purplish, broadly lance-shaped, 2–4 mm long, tapered to a short, firm bristle (awn); anthers 0.3–1 mm long; {July–August}.

Fruits: Grains; mature florets breaking free above the glumes and between the florets.

Habitat: Rocky alpine slopes.

Notes: This species was previously called *Festuca vivipara* (L.) Sm. ssp. *glabra* Frederiksen. It is sometimes considered a viviparous form of alpine fescue (*Festuca brachyphylla* Schultes & Schultes f.) or arctic fescue (*Festuca baffinensis* Polunin). • The flowering spikes of this northern grass are viviparous—that is, small plants develop in the flower clusters, producing leaves and rootlets before falling to the ground. This type of reproduction is often seen in arctic plants, where growing seasons are short, and environmental conditions are harsh. Young plants get a headstart by growing while still protected in the spikelets of the parent plant, and when they fall to the ground, they have a better chance of becoming established and maturing.

WESTERN FESCUE

Festuca occidentalis Hook.
GRASS FAMILY (POACEAE [GRAMINEAE])

Plants: Densely **tufted perennial herbs**; bright green to blue-green; stems slender, 40–80 {25–110} cm tall, with 2 exposed joints (nodes); underground stems (rhizomes) absent.

Leaves: Mostly basal; **blades thread-like**, rolled inward, **soft, hairless**, with an erect swelling at the base; ligules 0.1–0.4 mm long, irregularly cut; **sheaths** hairless, **split** down 1 side to near the base (open), with a prominent midvein.

Flower clusters: Nodding panicles, 5–20 cm long, with the **lower branches** unequal and **strongly bent downward** (reflexed); **spikelets 3–5 {7}-flowered**, 6–10 mm long, with the florets on slender stalks; **glumes unequal**, shorter than the lowest lemma, 3–4 {2–4.5} **mm long**, usually sharp-pointed, hairless or short-hairy at the tip; **lemmas** oblong to lance-shaped, **5-nerved**, {4} **5–6.5 mm long**, thin and translucent, **with a long, 4–7 {12} mm bristle** (awn); {late May–July}.

Fruits: Grains; mature florets breaking away above the glumes and between the florets.

Habitat: Dry, rocky, wooded slopes; associated with lodgepole pine and aspen, in the southern Rocky Mountains; elsewhere, in moist woods and along streambanks and lake shores.

Notes: Immature plants of western fescue may be difficult to distinguish from bluebunch fescue (*Festuca idahoensis* Elmer). • Western fescue's fine leaves and sparse occurrence in shade can make it difficult to detect. • Most native fescues are valuable forage grasses.

Grass-like Plants

GRASS FAMILY

BEARDED FESCUE

Festuca subulata Trin.
GRASS FAMILY (POACEAE [GRAMINEAE])

Plants: Loosely tufted perennial herbs; stems 50–100 {130} cm tall, with 2–4 exposed joints (nodes).

Leaves: Alternate on the stem; soft, **flat, 3–10 mm wide, hairless or short-hairy**, deep green; **ligules** 0.2–1 mm long, **fringed with hairs** (ciliate) along the **irregular edge**; sheaths hairless, fused into a tube (closed) for at least ½ their length.

Flower clusters: Open, nodding panicles, {10} 15–40 cm long, with slender branches spreading or bent downward (reflexed); **spikelets 3–5 {2–6}-flowered**, 7–10 {6–12} mm long, with stalkless florets; glumes unequal, 3.5–5.5 {2–6} mm long, shorter than the lowest lemmas; **lemmas** {4} **5–7 mm long**, usually with a lengthwise ridge on the back (keel), sparsely hairy, with a straight or sometimes kinked **bristle** (awn) **5–10** {20} **mm long**; {May–June} July.

Fruits: Grains; mature florets breaking away above the glumes and between the florets.

Habitat: Moist woods, thickets and shaded banks up to 1,500 m in the southern Rocky Mountains.

Notes: The bristles (awns) of bearded fescue curve or kink on drying. • This species has also been called *Festuca jonesii* Vasey. • Northern rough fescue (*Festuca altaica* Trin. *ex* Ledeb., treated under *Festuca scabrella* Torr. in the *Flora of Alberta*) is a rare grass of boreal and alpine grasslands. Like bearded fescue, it is up to 120 cm tall and has an open, nodding panicle, but it differs in having yellow-green to dark green leaves and purplish spikelets with transparent edges that give a lustrous sheen to the panicle. The leaves of northern rough fescue are less than 2.5 mm wide, the lemmas are rounded (rather than ridged), and the lemma awns are less than 1 mm long. The dead leaf blades of northern rough fescue break off at the collars, leaving entire sheaths that persist for many years. Northern rough fescue flowers in June {July}.

F. ALTAICA

GRASS FAMILY   **Grass-like Plants**

CANADA BROME, HAIRY WOODBROME

Bromus latiglumis (Shear) A.S. Hitchc.
GRASS FAMILY (POACEAE [GRAMINEAE])

Plants: Loosely tufted perennial herbs; **stems** 80–150 cm tall, essentially hairless, **with {8} 10–20 hairless joints** (nodes); without underground stems (rhizomes).

Leaves: Alternate; blades flat, mostly hairless, **5–15 mm wide**, 20–30 cm long, **with broad lobes at the base extending into small projections** (auricles); ligules blunt, to 1.4 mm long, irregularly cut, fringed with hairs (ciliate); sheaths overlapping, hairless to densely soft-hairy, with woolly collars.

Flower clusters: Spreading panicles, **10–22 cm long**; spikelets stalked, 4–9-flowered, 1.5–3 cm long; glumes hairy, shorter than the lowest lemma, the **lower usually 1-nerved** (rarely 3-nerved) and 4–7 {7.5} mm long, the upper 3-nerved and 6–9 mm long; **lemmas evenly hairy across the back**, 7-nerved, {8} 9–14 mm long, **tipped with a 3–4.5 {7} mm long bristle** (awn); anthers 2–3 mm long; {late June–August}.

Fruits: Grains; mature florets breaking off above the glumes and between the florets; August–September.

Habitat: Moist streambanks; elsewhere, in meadows and riparian thickets and forests.

Notes: This species has also been called *Bromus altissimus* Pursh, *Bromus incanus* (Shear) A.S. Hitchc., *Bromus purgans* L. var. *incanus* Shear and *Bromus ciliatus* L. var. *latiglumis* Scribn. *ex* Shear. • The generic name *Bromus* comes from the ancient Greek name for the oat, *bromos*. The specific epithet *latiglumis* means 'broad-glumed.'

Rare *Vascular* Plants of Alberta   341

Grass-like Plants

GRASS FAMILY

SPREADING WHEAT GRASS, SCRIBNER'S WHEAT GRASS

Elymus scribneri (Vasey) M.E. Jones
GRASS FAMILY (POACEAE [GRAMINEAE])

Plants: Perennial herbs with **a dense, leafy basal tussock**; **stems spreading, curved to ascending**, 10–40 {60} cm tall; **without underground stems** (rhizomes).

Leaves: Mostly basal, finely hairy; blades flat, 1–3 {4} mm wide, with edges slightly rolled inward; stem leaves less than 5 cm long; ligules barely 0.5 mm long; auricles usually present.

Flower clusters: Dense, **nodding spikes**, 3–7 {12} cm long, breaking into pieces and **separating** from the rest of the stem **at maturity**; **spikelets stalkless**, **single** at each joint (node), with 3–5 flowers; glumes 6–8 mm long, tapered to a bristle-tip (awn); **lemmas** {7} **8–10 mm long, essentially hairless**, faintly 5-nerved to the tip, tipped **with a 15–30 mm long, curved, spreading awn**; anthers 1.5–2.5 mm long; {late June–August}.

Fruits: Grains; mature florets breaking off above the glumes and between the florets; August.

Habitat: Dry alpine scree slopes; elsewhere, on montane slopes and in grassland.

Notes: This species has also been called *Agropyron scribneri* Vasey. • Similar species with curving or spreading bristles (awns) in Alberta, awned northern wheat grass (*Agropyron albicans* Scribn. & Smith) and blue-bunch wheat grass (*Elymus spicatus* (Pursh) Gould, also called *Agropyron spicatum* (Pursh) Scribn. & Smith), are normally found in the grassland region or at lower elevations. • Spreading wheat grass hybridizes with a number of species. • The common name 'spreading wheat grass' refers to the curved culms, which 'spread out' from the basal tussock.

Grass-like Plants

GRASS FAMILY

SQUIRRELTAIL
Elymus elymoides (Raf.) Swezey
GRASS FAMILY (POACEAE [GRAMINEAE])

Plants: Tufted perennial herbs; stems stiff, erect or spreading, 10–50 {60} cm tall; from fibrous roots.

Leaves: Alternate, hairless or short-hairy; blades flat, folded or rolled inward, 1–4 mm wide, usually with small (less than 1 mm long) auricles at the base; ligules less than 1 mm long, membranous, fringed with fine hairs (ciliate); **upper sheaths often inflated**.

Flower clusters: Green or purplish spikes, 2–10 {15} cm **long and wide** (including awns), often partly enveloped by the upper leaf sheath, **breaking apart** at the joints (nodes) when mature; **spikelets stalkless, usually in 2s** (sometimes single or in 3s), **2–6 {1–8}-flowered**, often with lower lemmas sterile and glume-like; **glumes** more or less awl shaped, 1–2-nerved, **tipped with 1 or 2 spreading** (eventually), **5–9 cm long bristles** (awns), sometimes tipped with 2 teeth or with an awn on one side; **lemmas firm**, rounded on the back, 5-nerved, rough to touch or short-hairy, **tipped with 1–3 widely spreading, 2–10 cm long bristles** (awns); {May–June} July.

Fruits: Grains; mature florets breaking free with the glumes and part of the central stalk (rachis).

Habitat: Dry plains and open woods; elsewhere, in dry to moist, often rocky sites from sea level to above treeline.

Notes: This species has also been called *Sitanion hystrix* (Nutt.) J.G. Smith. • Squirreltail is a highly variable species, and several varieties have been identified within it, based mainly on differences in hairiness, number of spikelets, number of fertile lemmas and the presence of glume-like florets. It hybridizes readily with smooth wild rye (*Elymus glaucus* Buckley), producing plants that are intermediate between the 2 species, with spikes that break apart at their joints (nodes). Several of these hybrids have been described as species in the past.

Grass-like Plants

GRASS FAMILY

LITTLE BARLEY

Hordeum pusillum Nutt.
GRASS FAMILY (POACEAE [GRAMINEAE])

Plants: Tufted, greyish green **annual** herbs; stems 10–40 {60} cm tall, usually abruptly bent at the base, with dark, hairless joints (nodes); from fibrous roots.

Leaves: Alternate on the stem; blades flat, 1–12 cm long, 2–4 {5} mm wide, **without small projections** (auricles) **at the base**; ligules 0.4–0.7 mm long, blunt, irregularly cut, short-hairy; sheaths soft-hairy.

Flower clusters: Erect, **dense spikes**, {2} 3–7 cm long, 10–14 mm wide; **spikelets 1-flowered, 3 per joint** (node), with **2 short-stalked, sterile side spikelets and 1 stalkless, fertile central spikelet**; **glumes** rough to touch, mostly narrowly lance-shaped, {0.5} **0.8–1.8 mm wide** and **tipped with an 8–15 mm long bristle** (awn), but the **lower glumes of the side spikelets bristle-like** (resembling awns); fertile lemma (central spikelet) tapered to a long awn, 10–15 mm long (including awn); sterile lemmas (side spikelets) awn-tipped; {May–June}.

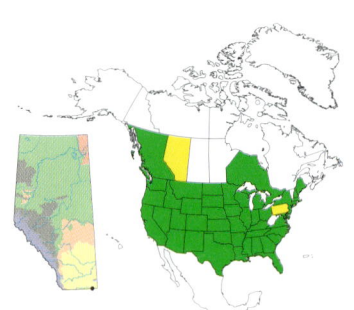

Fruits: Grains; mature florets breaking away above the glumes.

Habitat: Saline prairie, eroded areas, grassland and waste land; elsewhere, alkali flats, roadsides and wetland borders.

Notes: This annual species may be mistaken for the common perennial foxtail barley (*Hordeum jubatum* L.), if foxtail barley is flowering in the first year. However, foxtail barley has slightly nodding spikes, and its glumes are hair-like throughout. • Grains of little barley are common in archaeological digs, suggesting domestication by early peoples prior to the widespread use of corn. • The sharp, rough awns can cause injury to the nose and mouth of grazing animals and can even penetrate hide and flesh.

GRASS FAMILY   **Grass-like Plants**

AMERICAN DUNE GRASS, SEA LYME GRASS

Leymus mollis (Trin.) Pilger ssp. *mollis*
GRASS FAMILY (POACEAE [GRAMINEAE])

Plants: Coarse perennial herbs; stems erect, 60–120 {15–150} **cm tall**, with dense, fine hairs; old withered leaves persisting at the stem base; mat-forming, **with long, thick underground stems** (rhizomes).

Leaves: Basal and alternate on the stem; blades stiff, flat (sometimes rolled), 6–12 {3–15} mm wide, prominently nerved, hairless beneath, rough on the upper surface, **with small, projecting lobes** (auricles) at the base; ligules up to 1 mm long, finely ciliate; sheaths crowded and overlapping.

Flower clusters: Erect spikes, 10–20 {8–30} cm long, 1–2 cm thick; **spikelets** dense, **paired, stalkless** (sometimes short-stalked), on opposite sides of the main stalk (rachis), **4–6-flowered**; glumes lance-shaped, flat, 3–5 {6}-nerved, **hairy**, 12–25 mm long; **lemmas sharp-pointed**, with thin, translucent edges, **soft-hairy**, 10–20 mm long; anthers 5–9 mm long; {June–August}.

Fruits: Grains; mature florets **breaking away above the glumes** and between the florets.

Habitat: Sand dunes; elsewhere, primarily on coastal sand dunes.

Notes: This species has also been called *Elymus mollis* Trin. and *Elymus arenarius* L. ssp. *mollis* (Trin.) Hult. • In some parts of the arctic, leaves of American dune grass were used for basket weaving and for insulation in boots. • This species forms associations with mycorrhizal fungi. • Virginia wild rye (*Elymus virginicus* L.) is a similar rare grass that grows in open woods and thickets farther south. It is also stout and erect, but it lacks underground stems. Unlike American dune grass, Virginia wild rye has hairless or sparsely hairy lemmas tipped with a bristle (awn), and its glumes curve outward at the base.

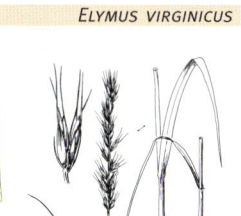

ELYMUS VIRGINICUS

Rare *Vascular* Plants of Alberta

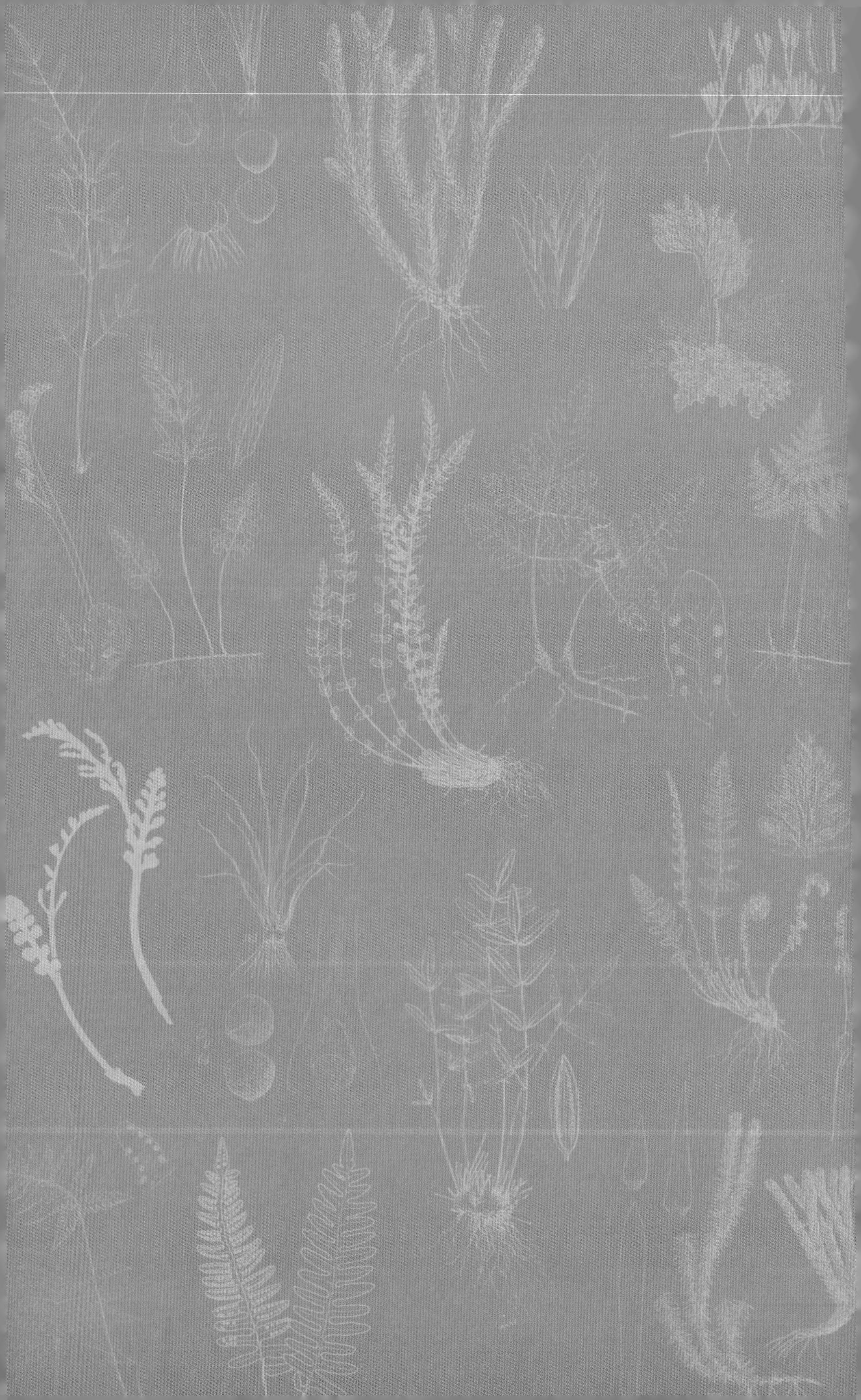

Ferns and Fern Allies

CLUB-MOSS FAMILY

Ferns and Fern Allies

BOG CLUB-MOSS

Lycopodiella inundata (L.) Holub
CLUB-MOSS FAMILY (LYCOPODIACEAE)

Plants: Dwarf perennial herbs, with prostrate, short-creeping stems 5–10 {3–20} cm long, producing unbranched, ascending to erect fertile branches 3–6 {10} **cm tall**; anchored by many irregular roots on underside of stems.

Leaves: Crowded, in 8–10 vertical rows, **awl-shaped, 4–6 {8} mm long**, less than 1 mm wide, curved toward the upper side of the stem, broadest near base, tapered to a soft, slender point.

Spore clusters: Round dots (sporangia) about 1 mm wide, in the axils of broad-based, specialized leaves (sporophylls), in stalkless **cones** (strobili) **1–2 cm long** at the tips of fertile branches; spores sulphur-yellow, shed as fine powder through a horizontal opening in the sporangia.

Habitat: Sphagnum bogs; elsewhere, on sandy shores and in marshes and other wet sites.

Notes: This species has also been called *Lycopodium inundatum* L. It is also found in scattered boreal regions around the world, including western Europe and Japan. These inconspicuous little plants are easily overlooked.
• Another rare Alberta club-moss, ground-fir (*Diphasiastrum sitchense* (Rupr.) Holub, also called *Lycopodium sitchense* Rupr. and *Lycopodium sabinifolium* Willd. var. *sitchense* (Rupr.) Fern., also misspelled *Lycopodium sabinaefolium*), is a low (5–15 {17.5} cm tall) species with small (3–4 {5–6} mm long), broadly lance-shaped leaves in 5 vertical spiralling rows on repeatedly branched stems from creeping underground stems (rhizomes). Its spore clusters are borne in the axils of specialized leaves that are very different from the vegetative leaves, and these are crowded in stalkless or short-stalked strobili at the tips of fertile branches, taller than the sterile branches. Ground-fir grows in open woods and alpine tundra. • The Cree used club-moss spores to determine the fate of sick people. The spores were placed in a container of water, and if they radiated toward the sun, the patient would survive his illness. • Club-mosses produce 2 very different forms of plants: the green, leafy plants (sporophytes) that we call club-mosses and tiny, scale-like plants (gametophytes) that are rarely seen.

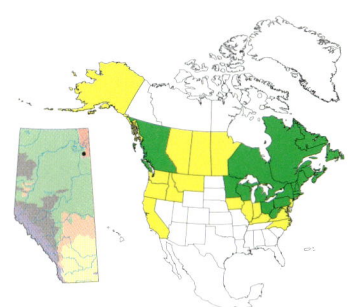

DIPHASIASTRUM SITCHENSE

Rare *Vascular* Plants of Alberta

Ferns and Fern Allies

CLUB-MOSS FAMILY

NORTHERN FIR-MOSS, MOUNTAIN CLUB-MOSS

Huperzia selago (L.) Bernh. *ex* Schrank & Mart.
CLUB-MOSS FAMILY (LYCOPODIACEAE)

Plants: Small, glossy, yellow-green perennial herbs, with erect, forked, bottle-brush-like stems **5–12 {20} cm tall**, with **annual rings** as a result of alternating fertile (lower) and sterile (upper) zones which are produced each year, often also producing **small bud-like growths** (gemmae) in the upper leaf axils; from short (mostly 2–5 cm long) creeping stems with roots from among persistent leaves.

Leaves: Densely crowded in 8–10 vertical rows, **broadly awl-shaped, 3–8 mm long**, usually hollow at the base.

Spore clusters: Small round dots (sporangia) about 1 mm wide, **scattered along the stem in the axils of ordinary** (not specialized) **leaves**; spores sulphur-yellow, shed as fine powder through a horizontal opening in the sporangia.

Habitat: Bogs and cold woods in northern Alberta; elsewhere, in moist tundra.

Notes: Another rare species, alpine fir-moss (*Huperzia haleakalae* (Brack.) Holub) grows on damp, mossy ledges in alpine and subalpine zones in southern Alberta. It is distinguished from northern fir-moss by the lack of weak annual constrictions on its bulblet-producing (gemmiferous) branches and by the presence of 1–3 rings (pseudowhorls) of gemmae at the tips of the current year's growth or scattered throughout the mature shoots (rather than in a single ring at the end). • These 2 species are distinguished from other club-mosses in this province by their spore clusters, which are scattered along the stems in the axils of normal leaves, rather than clustered near the stem tips in cone-like clusters or on specialized leaves. Both have been called *Lycopodium selago* L.

H. HALEAKALAE

SPIKE-MOSS FAMILY **Ferns and Fern Allies**

WALLACE'S LITTLE CLUB-MOSS, WALLACE'S SPIKE-MOSS

Selaginella wallacei Hieron.
SPIKE-MOSS FAMILY (SELAGINELLACEAE)

Plants: Pale green to greyish, **moss-like**, **matted** herbs; main stems slender, trailing, irregularly branched, up to 10–20 cm long; branches ascending, 1–2-forked; producing loose or compact mats with fibrous roots along the stems.

Leaves: Densely **overlapping**, spirally arranged, thick, firm, **needle-like**, 1-nerved, **2–3 {1.5–3.5} mm long**, **attached squarely** to the stem, **leaving no conspicuous remnants** when detached, usually hairy and edged with 5–12 stiff hairs, **tipped with a white, 0.2–0.5 {0.9} mm long bristle**.

Spore clusters: In the axils of triangular to egg-shaped, specialized leaves (sporophylls) that form sharply **4-angled, 1–3 {9} cm long cones** (strobili) at the stem tips (often paired); spores bright to pale orange, of 2 kinds (megaspores and microspores) which are borne in separate spore cases (sporangia) in the same cone (usually).

Habitat: Dry, rocky slopes, cliffs and ledges; elsewhere, also on moist, shaded, rocky banks and in meadows.

Notes: This small plant resembles and often grows intermingled with prairie spike-moss (*Selaginella densa* Rydb.), also known as prairie selaginella, but that species has smaller, denser mats with shorter, tufted branches. The leaves of prairie selaginella extend down the sides of the stem at their bases, and the lobes remain attached to the stems when the leaves are removed. • The form of Wallace's little club-moss can vary greatly with habitat. In dry, exposed conditions it forms small, compact mats of short stems bearing tightly appressed leaves that narrow abruptly to a bristle at their tip. Plants from moister habitats form loosely creeping mats of longer stems with looser, fleshier leaves that extend down the stem at their bases and taper gradually to a bristle at their tips. • The spores of the spike-mosses produce microscopic plants called prothalli. These are seldom observed, as they usually develop within the walls of the spores. The prothalli are gametophytes, which produce male and female sexual organs. When an egg cell is fertilized, it grows into a green plant that we recognize as a spike-moss. These plants are sporophytes, which produce spores, and thus the cycle begins again.

Rare *Vascular* Plants of Alberta

Ferns and Fern Allies

QUILLWORT FAMILY

QUILLWORTS

Isoetes L. spp.
QUILLWORT FAMILY (Isoetaceae)

(SEE KEY, APPENDIX 1, PP. 391)

Plants: Aquatic perennial **herbs**; stems short, fleshy, **corm-like, 2–3-lobed, nearly round, at the centre of a dense tuft of onion-like leaves**; from fibrous roots.

Leaves: Basal, several to many, linear with broad spoon-shaped bases, **grass-like**, 2–20 cm long, erect to spreading, straight or curved downward, with **an inconspicuous membrane** (ligule) **up to 2.5 mm long on the inner side** just above the spore cluster.

I. BOLANDERI

I. MARITIMA

Spore clusters: Single, 4–10 mm long, 2.5–3 mm wide, in a depression **on the inner side of broad, spoon-shaped leaf bases**, covered by a thin membrane (velum); spores of two forms, female (megaspores, 350–700 microns in diameter) and male (microspores, 23–43 microns in diameter), which are borne in separate spore cases (sporangia); variously **ornamented**; late August.

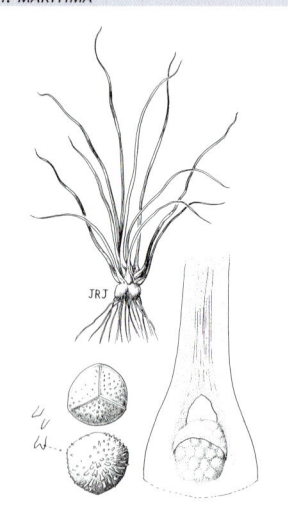

Habitat: The Alberta quillworts are all submerged aquatic species growing in permanent lakes and ponds. Northern quillwort (*Isoetes echinospora* Durieu) and coastal quillwort (*Isoetes maritima* L. Underwood) grow in clear, nutrient-poor, non-calcareous lakes and ponds, in shallow water up to about 1 m deep. Western quillwort (*Isoetes occidentalis* L.F. Henderson) occurs in deep lake water, and Bolander's quillwort (*Isoetes bolanderi* Engelm.) grows in deep water of nutrient-poor alpine or subalpine lakes and ponds in unglaciated areas.

Notes: Western quillwort and coastal quillwort are known only from Jasper National Park. Northern quillwort is known from northern and northeastern Alberta. Bolander's quillwort is known in Canada only from Waterton Lakes National Park. • A fifth species, Howell's quillwort (*Isoetes*

QUILLWORT FAMILY **Ferns and Fern Allies**

I. ECHINOSPERMA

I. OCCIDENTALIS

I. HOWELLI

I. x *TRUNCATA*

ELEOCHARIS ENGELMANNII

howellii Engelm.), has not been reported for Alberta but is found within a few hundred metres of the Continental Divide in BC at Akamina Pass. Howell's quillwort is very similar to *I. bolanderi* but is distinguished by its larger (often more than 15 cm) leaves with well-developed peripheral strands and by the larger (over 1 cm long) wing margin above each sporangium. • Quillworts could be mistaken for spike-rushes (*Eleocharis* spp.) at first glance, but quillworts are easily identified by their swollen leaf bases and short, thickened stems (corms). • Hybrids between species are frequent when 2 species grow in the same habitat. Hybrids can be distinguished by their malformed spores. *Isoetes* x *truncata* (A.A. Eaton) Clute, a hybrid between *I. maritima* and *I. occidentalis*, is the only known hybrid in Alberta. • Quillworts are very sensitive to changes in water quality, and the length, colour, shape and rigidity of leaves varies greatly with changes in habitat. • The generic name *Isoetes* was taken from the Greek *isos* (equal) and *etos* (year), in reference to the evergreen leaves of these plants. Two of the specific epithets refer to distribution. *Occidentalis* means 'western' and *maritima* means 'growing by the sea.' *Echinospora*, from the Greek *echinos* (sea urchin) and *spora* (seed), refers to the spiny megaspores, and *bolanderi* commemorates Henry Nicholas Bolander (1831–1897), a botanical collector who worked in the Sierra Nevada with the California State Geological Survey.

Ferns and Fern Allies

ADDER'S-TONGUE FAMILY

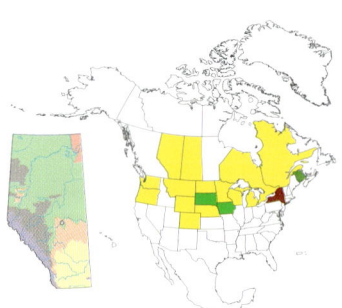

FIELD GRAPE FERN, PRAIRIE MOONWORT

Botrychium campestre W.H. Wagner & D.R. Farrar
ADDER'S-TONGUE FAMILY (OPHIOGLOSSACEAE)

(SEE KEY, APPENDIX 1, PP. 389–90)

Plants: Small, **fleshy** perennial herbs **6–12 cm tall**; common basal stalk usually 5–10 cm tall; from **underground stems with tiny, round bodies** (gemmae) 0.4–0.8 mm in diameter, in grape-like clusters; **appear in early spring** and **usually die by late spring** to early summer before plants of associated moonwort species (occasionally persisting until late summer).

Leaves: Of 2 types, a single short-stalked, sterile blade below a single fertile blade (see description of spore clusters); **sterile blades** fleshy, dull, whitish green, oblong to linear-oblong in outline, commonly widest above the middle, up to 4 cm long and 1.3 cm wide, **folded lengthwise, once-divided** into as many as 5 {9} pairs of segments; segments spreading, usually not overlapping, separated by 1–3 times the segment width and with the basal pair about equal in size, mostly long and narrow to narrowly spatula-shaped, **edged with shallow, rounded or sharp teeth**, usually notched at the tip; veins in a fan-shaped pattern; midribs absent.

Spore clusters: Small, spherical, yellow, stalkless spore sacs (sporangia) borne on specialized, fertile leaf blades; fertile blades **single**, once-divided (rarely twice-divided), **usually stalkless** but sometimes broadly tapered to stalk that is 1–1½ times as long as the sterile blade.

Habitat: Grassy fields and ditches.

Notes: Another rare species, spatulate grape fern (*Botrychium spathulatum* W.H. Wagner), also called spatulate moonwort, has a nearly stalkless, leathery sterile blade that is shiny yellow-green, triangular in outline and divided into as many as 8 pairs of ascending, widely spaced segments. Its fertile blade is once- or twice-divided, and its leaves appear in late spring through summer. Spatulate grape fern grows in fields and grassy openings in the mountains. • Mingan grape fern (*Botrychium minganense* Victorin), also known as Mingan moonwort and *Botrychium lunaria* (L.) Swartz var. *minganense* (Victorin) Dole, is a larger plant, up to 30 cm tall. Its dull yellow-green, linear sterile blade is once-divided into as many as 10 pairs of horizontal to

B. SPATHULATUM

Rare *Vascular* Plants of Alberta

ADDER'S-TONGUE FAMILY   **Ferns and Fern Allies**

spreading segments. The fertile blade is also once-divided (occasionally twice-divided), and the leaves appear in spring through summer. Mingan grape fern is a rare species that grows in a variety of habitats at elevations up to 3,700 m. Many specimens of *B. minganense* have been misidentified as *Botrychium dusenii* (H. Christ) Alston. • Scalloped grape fern (*Botrychium crenulatum* W.H. Wagner), also called dainty moonwort, has a thin, yellow-green sterile blade that is once-divided into as many as 5 pairs of spreading, well-separated segments. Its fertile blade is once- or twice-divided, and its leaves appear in mid to late spring and die in late summer, although they may not appear every year. Scalloped grape fern grows in wet areas and in Alberta is known only from Banff National Park.
• Ascending grape fern (*Botrychium ascendens* W.H. Wagner), also called upswept moonwort, has a thin but firm, bright yellow-green sterile blade that is longer than wide (often broadly lance-shaped), up to 6 cm long and 1.5 cm wide and once-divided into as many as 5 pairs of rounded, strongly ascending, well-separated, sharply toothed or shallowly lobed segments. The segments lack midribs, and their veins form a fan-shaped pattern. Some of the basal segments often develop spore clusters (sporangia). The twice-divided, spore-bearing (fertile) blade is ⅓–½ again as long as the sterile blade and 1⅓–2 times as long as the common basal stalk. Ascending grape fern appears in late spring to midsummer in grassy openings in mountain forests. It grows with *Botrychium lunaria*, *B. minganense* and *B. crenulatum*, and hybrids with *B. crenulatum* have been reported. • Common moonwort (*Botrychium lunaria* (L.) Swartz) is a more widespread species of open fields. It has a dark green, once-divided, thick, fleshy sterile blade, with up to 9 pairs of tooth-less, spreading and mostly overlapping segments, and its fertile blade is once- or twice-divided. Common moonwort appears in spring and dies in the latter half of summer.

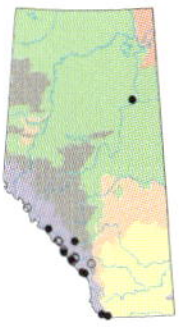

Rare *Vascular* Plants of Alberta

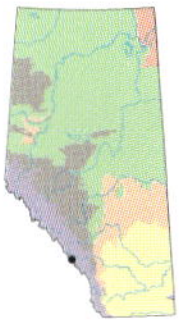

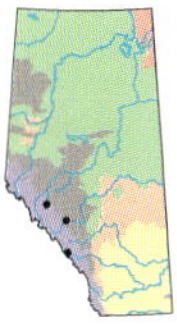

Ferns and Fern Allies

ADDER'S-TONGUE FAMILY

WESTERN GRAPE FERN, WESTERN MOONWORT

Botrychium hesperium (Maxon & R.T. Clausen)
W.H. Wagner & Lellinger
ADDER'S-TONGUE FAMILY (OPHIOGLOSSACEAE)

(SEE KEY, APPENDIX 1, PP. 389–90)

Plants: Small perennial herbs {5} 10–20 cm tall; common basal stalk usually 5–10 cm tall; **appear in early spring and die in early fall.**

Leaves: Of 2 types: a single short-stalked, sterile blade below a single fertile blade (see description of spore clusters); **sterile blades oblong-linear to triangular in outline**, up to 6 cm long and 5 cm wide, dull grey-green, **once- or twice-divided**, with up to 6 pairs of ascending, **usually crowded to overlapping segments**; lowest pair of segments commonly much larger and more divided than the others and lobed along the edges; upper segments rounded, smooth-edged to shallowly lobed.

Spore clusters: Small, spherical, yellow, stalkless spore sacs (sporangia) borne on erect, fertile leaves; **fertile blades single**, 1–3 times divided, 2–10 cm long, borne on slender stalks 1–8 cm long.

Habitat: Wooded areas, often with other moonworts, including *Botrychium lanceolatum*, *B. lunaria*, *B. pinnatum*, *B. minganense*, *B. simplex* and *B. paradoxum*.

Notes: This species is also known as *Botrychium matricariifolium* (Döll) W.D.J. Koch ssp. *hesperium* Maxon & R.T. Clausen. • Another rare species, lance-leaved grape fern (*Botrychium lanceolatum* (Gmelin) Angström ssp. *lanceolatum* and ssp. *angustisegmentum* (Pease & A.H. Moore) R.T. Clausen), also called triangle moonwort, grows up to 20 cm in height. It has a green to pale yellow-green, somewhat shiny, triangular sterile blade up to 6 cm long and 10 cm wide, that is once- or twice-divided into as many as 5 pairs of ascending, well-spaced (not overlapping) segments. The leaves appear in late spring to early summer and wither in midsummer. The fertile blade is divided 1–3 times. In Alberta, this rare species has been found only on mountain slopes, but outside of Alberta it grows mainly in open fields and also on peaty slopes, in mountain meadows and in open woods at elevations up to 3,700 m. • Paradoxical grape fern (*Botrychium paradoxum* W.H. Wagner), also called paradox

356   Rare *Vascular* Plants of Alberta

ADDER'S-TONGUE FAMILY **Ferns and Fern Allies**

moonwort, does not have sterile blades. Instead, it produces 2 fertile blades (a large blade 7–15 cm long below a smaller blade about ⅔ as long) on a once-divided, 0.5–2 mm wide stalk that reaches up to ½ the total height. Its leaves appear in early summer and disappear in late summer. In Alberta, this rare species grows in grassy openings in mountain forests, but in other areas it is also found in fields and meadows at elevations from 1,500–3,000 m. Paradoxical grape fern does not have a green sterile blade that could provide food through photosynthesis. Instead, it probably acquires nutrients through the fungi associated with its roots (mycorrhizae). The specific epithet *paradoxum* refers to the unique, paradoxical condition of having 2 spore-bearing blades and no sterile leaves. • As its name suggests, stalked grape fern (*Botrychium pedunculosum* W.H. Wagner), also called stalked moonwort, is distinguished by its reddish brown common basal stalk and its long leaf stalks, which approximately equal their blades. The lowest segments of the sterile leaves are relatively wide (egg-shaped to diamond-shaped), and they often develop spore clusters (sporangia). Stalked grape fern grows in brushy, disturbed sites along streams and roads at 300–1,000 m. • Western grape fern apparently hybridizes with paradoxical grape fern in Wateron Lakes National Park to produce the very rare *Botrychium* x *watertonense* W.H. Wagner. This hybrid is distinguished by its broad, sterile blades, which have spore clusters along their edges, rather than on separate fruiting stalks. It grows in grassy openings in coniferous forests in the mountains and is not known from outside of Alberta. The leaves appear in early summer.

B. PEDUNCULOSUM

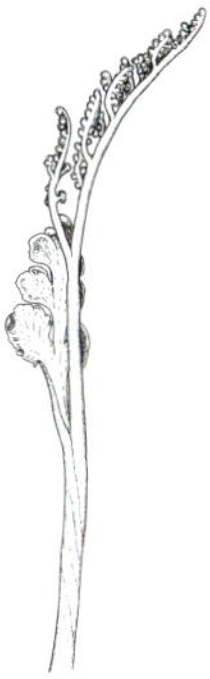

B. PARADOXUM

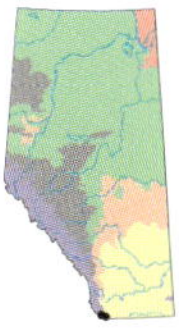

B. X WATERTONENSE

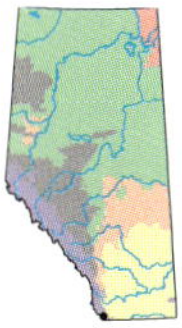

B. LANCEOLATUM

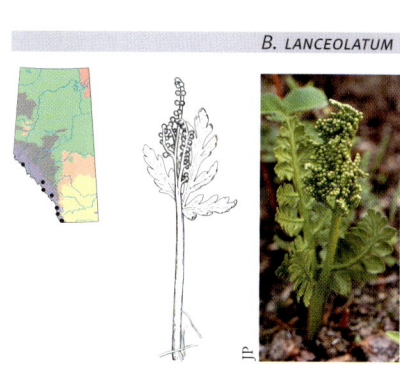

Rare *Vascular* Plants of Alberta

Ferns and Fern Allies

ADDER'S-TONGUE FAMILY

DWARF GRAPE FERN, LEAST MOONWORT

Botrychium simplex E. Hitchc.
ADDER'S-TONGUE FAMILY (Ophioglossaceae)

(SEE KEY, APPENDIX 1, PP. 389–90)

Plants: Small perennial herbs, 3–15 {25} cm tall; common basal stalk small or absent; appearing from mid spring to early fall.

Leaves: Of 2 types, a single short-stalked, sterile blade below a single fertile blade (see description of spore clusters); **sterile blades thin and herbaceous**, 0.5–4 {6} cm long, dull to bright green or whitish green, linear to triangular in outline, usually **divided** into **3 similar, pinnately divided segments** (pinnae); pinnae with 2–4 {5} pairs of well-developed segments (pinnules), the lowest pair of pinnules usually larger than the others and again divided; stalks of sterile blades 0–3 cm long.

Spore clusters: Small, spherical, yellow, stalkless spore sacs (sporangia) borne on erect, fertile leaf blades; fertile blades single, 3–8 times as long as the sterile blade, mostly once-divided, borne on a stout stalk more than twice as long as the blade.

Habitat: In moist meadows and along the edges of wetlands; elsewhere, also in dry fields and roadside ditches up to 2,200 m.

Notes: This species has also been called *Botrychium tenebrosum* A.A. Eaton. • Another rare species, leather grape fern (*Botrychium multifidum* (S.G. Gmelin) Rupr.), also has the sterile blade attached near the base of the stalk. It is 2–15 {20} cm tall with a shiny, green sterile blade on a stalk 2–15 cm long. The sterile blade can be up to 25 cm long and 35 cm wide, and it is divided 2–3 times, with up to 10 pairs of horizontal or ascending segments. These rather leathery leaves remain green through winter, reappearing in spring. The fertile blade is also 2–3 times divided. In Alberta, leather grape fern has been found in moist sandy areas, but in other areas, it grows mainly in fields.• The generic name *Botrychium*, from the Greek *botryos* (a bunch of grapes), refers to the grape-like clusters of spore sacs on the fertile leaves.

B. MULTIFIDUM

Rare *Vascular* **Plants** of Alberta

MAIDENHAIR FERN FAMILY Ferns and Fern Allies

WESTERN MAIDENHAIR FERN

Adiantum aleuticum (Rupr.) Paris
MAIDENHAIR FERN FAMILY (PTERIDIACEAE [POLYPODIACEAE])

Plants: **Delicate** perennial herbs, with a single to a few **annual leaves** on **10–40 {60} cm tall stalks**; from stout, short-creeping or slanted underground stems (rhizomes) 2–5 mm thick, with small (3–6 mm long, {0.5} 1–2 mm wide) scales.

Leaves: With spreading, horizontal blades on erect, wiry, lustrous, reddish brown to purplish black stalks; blades round to kidney-shaped in outline, 10–40 cm long, **twice-divided, with 6–10 radiating, finger-like branches** (pinnae), each with 15–35 pinnately arranged leaflets (pinnules); **pinnules very thin, greyish green**, with a thin, waxy coating (glaucous), **oblong**, mostly 12–22 mm long, 5–9 mm wide, **spreading** at right angles to the central stalk (rachis), usually deeply and irregularly **cut and rolled under along the upper edge**.

Spore clusters: Oblong to moon-shaped clusters of spore cases (sori) along the upper edge of the pinnules, **covered by the down-rolled leaf edge**.

Habitat: Cliffs and boulder-strewn slopes in subalpine areas; elsewhere, in moist woods and seepage areas and on protected creek sides.

Notes: This species has also been called *Adiantum pedatum* L. • This delicate fern requires protection from early or late frost, and is adapted to the wet zones of the western cordillera. In Alberta it is limited climatically by its need for abundant precipitation in conjunction with sheltered rocky habitats. • The generic name *Adiantum* means 'not wetted,' referring to the fact that the leaves shed water.
• In the 19th century, maidenhair fern was highly valued as a medicine. It was used in teas and syrups for treating colds, sore throats, flu and asthma. The stems were also used in hair rinses to give hair shine.

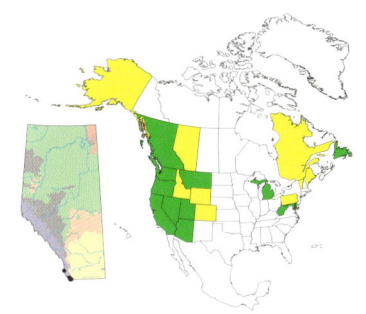

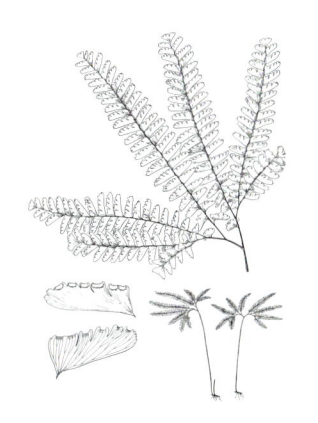

Ferns and Fern Allies

MAIDENHAIR FERN FAMILY

STELLER'S ROCK BRAKE

Cryptogramma stelleri (S.G. Gmel.) Prantl
MAIDENHAIR FERN FAMILY (PTERIDIACEAE [POLYPODIACEAE])

Plants: Small, **delicate herbs** with 2 types of leaves (fronds); from **slender** (1–1.5 mm thick), **creeping, few-branched** stems that are succulent and brittle, and that shrivel in the second year.

Leaves: **Scattered along stems**, delicate, soon shed, of 2 types; sterile leaves, green, 3–15 cm tall, with **blades 5 {3–8} cm long**, broadly lance-shaped in outline, 1–2 times pinnately divided into 9–13 wide, **flat, round-toothed segments**; stalks yellowish green above, reddish brown to purplish and scaly at the base; fertile leaves described under 'spore clusters.'

Spore clusters: Borne on the lower surface of **specialized, fertile leaves**, forming an almost continuous band, covered by **down-rolled leaf edges**, lacking membranous coverings (indusia); fertile leaves stiffly erect, 5–20 cm tall, longer than sterile leaves, with blades shorter than their stalks, 2–3 times pinnately divided into **slender, toothless segments** that are up to 2 cm long.

Habitat: Cool, shaded, **calcareous sites**, on rock or in springs.

Notes: Steller's rock brake is a more or less circumpolar species. • The generic name *Cryptogramma* was derived from the Greek *cryptos* (hidden) and *gramme* (line), referring to the hidden lines of spores clusters under the down-rolled leaf edges.

MAIDENHAIR FERN FAMILY • Ferns and Fern Allies

LACE FERN

Cheilanthes gracillima D.C. Eat.
MAIDENHAIR FERN FAMILY (PTERIDIACEAE [POLYPODIACEAE])

Plants: Delicate, densely tufted perennial herbs with evergreen leaves, 8–25 cm tall; from branched crowns or short, scaly underground stems (rhizomes).

Leaves: Erect, linear-oblong, **slender, 4–11 cm long, 1–2 cm wide, 2–3 times pinnately divided** into leaflets (bipinnate to tripinnate), hairless or nearly so on the upper surface, **rusty-woolly beneath and with conspicuous scales on the midribs, central axes** (rachises) **and stalks,** rolled under along the edges; 19–41 leaflets (pinnae), crowded but not overlapping; stalks wiry, 3–12 cm long, chestnut-coloured.

Spore clusters: Protected by and developing **under down-rolled leaf edges,** often appearing to cover the entire lower surface of the leaflet when mature.

Habitat: Ledges and crevices of igneous rocks and shale; elsewhere, in exposed, sunny sites from lower slopes to timberline in mountainous regions.

Notes: Slender lip fern (*Cheilanthes feei* T. Moore) resembles lace fern, but it has wider (1–3.5 cm) leaves without scales, and it grows on calcareous rock. • The generic name *Cheilanthes* was derived from the Greek *cheilos* (lip) and *anthos* (flower), in reference to the down-rolled leaf edges (lips), which overlap the spore clusters. The specific epithet *gracillima* means 'very slender.'

Ferns and Fern Allies

MAIDENHAIR FERN FAMILY

Gaston's cliff brake

Pellaea gastonyi Windham
MAIDENHAIR FERN FAMILY (PTERIDIACEAE [POLYPODIACEAE])

Plants: Delicate, tufted perennial herbs, **up to 25 cm tall**; from compact, ascending underground stems (rhizomes), 5–10 mm in diameter, covered with thin, reddish brown scales, 0.1–0.3 mm wide.

Leaves: Clumped, 8–25 cm long, of **2 slightly different forms** (dimorphic) with sterile leaves shorter than the fertile leaves; **blades leathery**, elongate-triangular to lance-shaped, 3–6 cm wide, **twice pinnately divided** into widely spreading leaflets (pinnae) with abruptly narrowed **bases that do not extend down the stalk**; leaflets usually divided into 3–7 oblong–lance-shaped, 7–30 **mm long leaflets** (pinnules) with blunt or abruptly short-pointed tips; **edges rolled under**; stalks and **main leaf axes** (rachises) reddish **purple to dark brown, lustrous**, rounded on the upper (inner) side, sparsely hairy; young, coiled leaves soft-hairy.

Spore clusters: Long-stalked clusters (sporangia) containing 32 spores, borne on the underside of fertile leaflets in a **band along the edges, protected by down-rolled leaf edges** which cover less than ½ the lower surface; summer and fall.

Habitat: Calcareous cliffs and ledges.

Notes: Gaston's cliff brake originated through repeated hybridization between purple cliff brake (*Pellaea atropurpurea* (L.) Link) and smooth cliff brake (*Pellaea glabella* Mett. *ex* Kuhn), and it always reproduces asexually. It is an apogamous tetraploid. • Gaston's cliff brake is most commonly mistaken for purple cliff brake, but purple cliff brake has densely soft-hairy leaf axes (rachises), large leaflets (pinnules) and smaller spores (averaging less than 62 microns in diameter). Purple cliff brake is an eastern species that is not found in Alberta. • Smooth cliff brake is more common in Alberta. It is distinguished by the hairless lower surface of its leaflets and its nearly hairless stalks (rachises). It also generally has narrower, linear-oblong to lance-shaped leaflets with bases that extend slightly down the main stalk (slightly decurrent).

P. GLABELLA

WOODFERN FAMILY　　Ferns and Fern Allies

ALPINE SPLEENWORT, AMERICAN ALPINE LADY FERN

Athyrium alpestre (Hoppe) Clairville var. *americanum* Butters
WOODFERN FAMILY (DRYOPTERIDACEAE [POLYPODIACEAE])

Plants: Tufted perennial herbs, **20–80 cm tall**, with **annual leaves** (fronds); from short, stout, densely scaly rhizomes covered with the bases of old leaf stalks.

Leaves: Basal, clustered, **feathery**; blades thin, delicate, **obscurely veined**, narrowly elliptic to lance-shaped, **2–3 {4} times pinnately divided** into slender leaflets; main leaflets (pinnae) in about 20–25 offset pairs, well-spaced, short-stalked, the larger ones 2–8 cm long and with 8–15 offset pairs of deeply toothed or lobed sub-leaflets (pinnules); stalks straw-coloured or red-brown with brown to dark brown, lance-shaped scales at the base, shorter than the blades.

Spore clusters: Round dots (sori) less than 1 mm across, on the lower side of the leaflets, **set in from the edge**; **indusia completely absent** (sometimes present as tiny scales; independent of season.

Habitat: Rocky alpine slopes and alpine meadows; elsewhere, also along streams, often near timberline.

Notes: This species has also been called *Athyrium americanum* (Butters) Maxon and *Athyrium distentifolium* Tausch *ex* Opiz ssp. *americanum* (Butters) Hult. It is limited in Alberta by its requirement for abundant precipitation in conjunction with rocky habitats. • Alpine spleenwort is very similar to a more widespread species, lady fern (*Athyrium filix-femina* (L.) Mertens), but that fern has longer (40–200 cm), wider leaves, with larger (up to 15 cm long) leaflets. Its crescent-shaped spore clusters have delicate indusia attached along 1 side. Lady fern grows on organic-rich soils in moist woods and thickets. • The fiddleheads of lady fern are sometimes collected for food, but most people consider them 'survival food' only because of their bitterness. • Oil from lady fern rootstocks was used for hundreds of years to expel worms in humans and animals. A single dose was often enough, but too much could cause **muscular weakness, coma and, most frequently, blindness.** • The generic name *Athyrium* was derived from the Greek *athyros* (without a door), in reference to the lack of a protective membrane (indusium) over the spore clusters of some species.

Ferns and Fern Allies

WOODFERN FAMILY

NORTHERN OAK FERN, NAHANNI OAK FERN

Gymnocarpium jessoense (Koidz.) Koidz.
WOODFERN FAMILY (DRYOPTERIDACEAE [POLYPODIACEAE])

Plants: Delicate herbs 5–50 cm tall, with **annual leaves** (fronds) that are **glandular-hairy** on both surfaces; from dark, slender creeping underground stems (rhizomes), 0.5–1.5 mm in diameter.

Leaves: Single, along rhizomes, with blades **triangular** in outline and on long, slender **stalks** (clearly **longer than the blades**); blades 5–14 {3–18} cm long, longer than wide, **firm** and rather stiff or lax and delicate, 2–3 times pinnately divided, with **3 main parts** but the **lower 2 half as long as the upper** one, the smallest leaflets blunt-tipped; **stalks** straw-coloured, dark at the base, **dull and glandular**, 5–25 cm long, with 2–6 mm long scales.

Spore clusters: Round dots (sori) **without protective membranes** (indusia), set in slightly from the leaf edge.

Habitat: Acidic to neutral rock crevices and slopes; elsewhere, on cliffs and in moist woods.

Notes: Northern oak fern is distinguished from its more common relative, common oak fern (*Gymnocarpium dryopteris* (L.) Newm.) by its narrower, firmer leaves. The 3 main leaflets of common oak fern are almost equal in size, and they are also thinner and hairless (not glandular-hairy). • Another species reported for western Alberta, western oak fern (*Gymnocarpium disjunctum* (Ruprecht) Ching), differs from common oak fern in having leaf blades that are about twice as large (8–24 cm vs. 3–14 cm), with the smallest leaflets pointed rather than blunt-tipped. • Northern oak fern could also be confused with another rare species, mountain bladder fern (*Cystopteris montana* (Lam.) Bernh. *ex* Desv., p. 365), but mountain bladder fern has delicate membranes (indusia) cupped over its young spore clusters (these membranes disintegrate with age), and its leaves are usually more finely divided and feathery in appearance. • The generic name *Gymnocarpium* was taken from the Greek *gymnos* (naked) and *karpos* (fruit), in reference to the bare, unprotected spore clusters, which never have indusia.

WOODFERN FAMILY — Ferns and Fern Allies

MOUNTAIN BLADDER FERN

Cystopteris montana (Lam.) Bernh. *ex* Desv.
WOODFERN FAMILY (DRYOPTERIDACEAE [POLYPODIACEAE])

Plants: Delicate, hairless herbs, 4–35 {45} cm tall, with **annual leaves** (fronds); from dark, cord-like, branched underground stems (rhizomes).

Leaves: Single, along rhizomes; blades triangular to egg-shaped in outline, 4–15 cm long and **about as wide as long**, 3 {2–4} **times pinnately divided** into irregularly toothed leaflets, appearing to have 3 main leaflets but the lower pair very asymmetric and clearly smaller than the upper one; **stalks** 6–20 {30} cm long, **green or straw-coloured** on upper part, dark brown to black at the base, **longer than the blade**, with scattered glands and scales.

Spore clusters: Small, round **dots** (sori) **on veins** on lower side of leaf, partly covered by an inconspicuous, whitish, **hood-like membrane** (indusium) which disintegrates as the spore clusters mature; summer–autumn.

Habitat: Damp, calcareous sites, often by springs or along streams in mixed or coniferous forest.

Notes: Mountain bladder fern might be confused with a much more common species, common oak fern (*Gymnocarpium dryopteris* (L.) Newm.), but oak fern has no indusia on its spore clusters and its leaves are usually less finely divided (less feathery in appearance) with 3 almost equal main leaflets. • The generic name *Cystopteris* was derived from the Greek *kystos* (bladder) and *pteris* (fern), in reference to the hood-like membranes (indusia) that cover the spore clusters. The specific epithet *montana* means 'of the mountains,' although this is not always the case in Alberta.

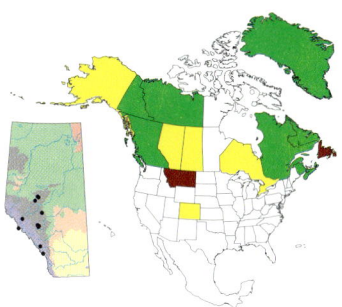

Rare *Vascular* Plants of Alberta

Ferns and Fern Allies   WOODFERN FAMILY

SMOOTH WOODSIA

Woodsia glabella R. Br. *ex* Richardson
WOODFERN FAMILY (DRYOPTERIDACEAE [POLYPODIACEAE])

Plants: Delicate perennial herbs 3–10 cm tall, with annual leaves (fronds); in dense clumps from compact stems covered in brown, finely toothed scales.

Leaves: Pale green, **hairless**; blades linear to linear–lance-shaped, {3.5} **4–16 cm long**, 6–15 mm wide, pinnately divided into **8–15 pairs of egg-shaped to almost round, 3–7-lobed leaflets**, sometimes edged with broad, rounded teeth; **stalks** delicate, hairless, **straw-coloured or greenish** throughout, breaking at a **dark joint near the base** and leaving **persistent brown-scaly bases**.

Spore clusters: Round, borne on the veins; **protective membrane (indusium) an inconspicuous disc with 5–8 thread-like segments**, under the spore cluster; summer–early autumn.

Habitat: Moist, shaded, usually calcareous sites, among boulders, on cliff ledges, and in crevices.

Notes: Rusty woodsia (*Woodsia ilvensis* (L.) R. Br.) is also rare in Alberta. It also has jointed leaf stalks, but the stalks are brown, firm, scaly and hairy along their length. The blades are wider than those of smooth woodsia (10–35 mm across), and they are dark green, hairy and scaly. The indusia are more conspicuous, and they have numerous (10–20) dark, thread-like segments. • The genus *Woodsia* was named in honour of Joseph Woods, the English botanist and architect who studied roses (*Rosa* spp.). *Glabella* is from the Latin word for 'smooth,' possibly referring to the relative lack of hairs and scales.

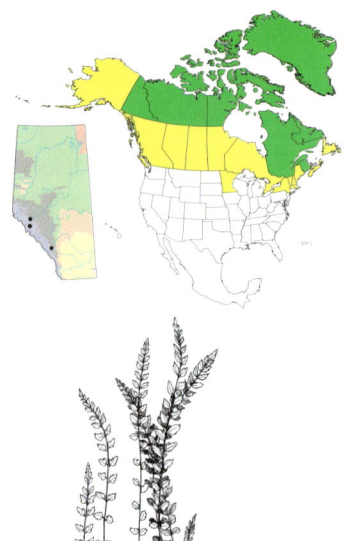

W. ILVENSIS

Rare *Vascular* Plants of Alberta

WOODFERN FAMILY **Ferns and Fern Allies**

CRESTED SHIELD FERN

Dryopteris cristata (L.) A. Gray
WOODFERN FAMILY (DRYOPTERIDACEAE [POLYPODIACEAE])

Plants: Loosely tufted, **glossy, somewhat leathery herbs, with some leaves persisting** through the winter, 30–80 cm tall; from short, stout, creeping underground stems (rhizomes) covered with scales and the bases of old leaf stalks.

Leaves: In clusters with **deciduous fertile leaves** and **evergreen sterile leaves**; blades firm, lance-shaped in outline, **twice pinnately divided** into 10–20 pairs of leaflets; leaflets (pinnae) oblong–lance-shaped to broadly egg-shaped (lower leaflets clearly wider than the upper ones), up to 8 {10} cm long, lobed about ⅔–¾ of the way to the middle and edged with spiny teeth, twisted to lie perpendicular to the plane of the blade; sterile leaves spreading and shorter (15–30 cm long) than the erect fertile leaves; **stalks stout, 10–40 cm long, ¼–⅓ as long as the blades**, with scattered tan scales.

Spore clusters: Round dots (sori) on veins on the lower surface of fertile leaves **halfway between the midvein and the edge**, partly covered by a round to **kidney-shaped membrane** (indusium) **0.7–2 mm wide** and with a deep notch on one side.

Habitat: Marshes, swamps and moist woods and thickets.

Notes: This is the only shield fern (*Dryopteris* spp.) in North America in which the fertile and sterile leaves are clearly different. Because it has evergreen leaves, it can take advantage of high light levels in the fall and spring, when most other plants do not have leaves. The factors controlling its survival in Alberta are not understood, but it is probably important that it have good snow cover to protect its evergreen leaves in cold winters. • Crested shield fern sometimes hybridizes with narrow spinulose shield fern (*Dryopteris carthusiana* (Vill.) H.P. Fuchs), producing ferns with characteristics of both parents.
• Another rare Alberta fern, male fern (*Dryopteris filix-mas* (L.) Schott), is very similar to crested shield fern, but its leaflets are narrower (lance-shaped to almost linear), longer (up to 18 cm), more uniform, and more deeply cut

D. FILIX-MAS

Ferns and Fern Allies

MARSH FERN FAMILY

(almost to the midrib). Male fern grows on moist, wooded slopes. It is evidently adapted to a moister climate than Alberta's and is probably limited here by hot, dry conditions in the summer and by lack of protective snow cover in winter. • Both of these rare ferns are sometimes grown in gardens, but neither appears to have escaped from cultivation in Alberta. • The leaves and stalks of male fern contain a chemical that paralyzes the voluntary muscles of the intestinal wall and similar tissues in the tapeworm. This does not kill the worms, but they are easily flushed out of the gut by an active purge.

NORTHERN BEECH FERN, LONG BEECH FERN

Phegopteris connectilis (Michx.) Watt
MARSH FERN FAMILY (THELYPTERIDACEAE [POLYPODIACEAE])

Plants: Small but rather coarse herbs, 10–50 cm tall; from long, slender, creeping underground stems (rhizomes), 1–2 {3} mm in diameter.

Leaves: Single, from rhizomes, erect; blades narrowly triangular in outline, 5–15 {4–25} cm long, 5–15 {4–20} cm wide, **pinnately divided into 10–25 pairs of leaflets** (pinnae), **fringed with hairs** and **also** with conspicuous, **tiny, needle-like hairs** (especially on the lower side of the main axis and veins); leaflets shallowly to deeply cut into rounded teeth or lobes, spreading from the main axis (rachis), but the **lowest pair bent sharply downward**; **stalks hairy**, straw-coloured above, dark and slightly scaly near the base, 5–30 cm long, **1–2 times as long as the blade**.

Spore clusters: Round dots (sori) **near the edges** on the underside of the lower leaflets, **without protective membranes** (indusia).

Habitat: Moist woodlands, on moderately to strongly acidic soil; elsewhere, in moist, shaded rock crevices and on streambanks.

Notes: This species has also been called *Thelypteris phegopteris* (L.) Slosson and *Dryopteris phegopteris* (L.) C. Chr. • The generic name *Phegopteris*, from *fagus* (beech) and *pteris* (fern), means 'beech fern,' perhaps in reference to the habitats of some species.

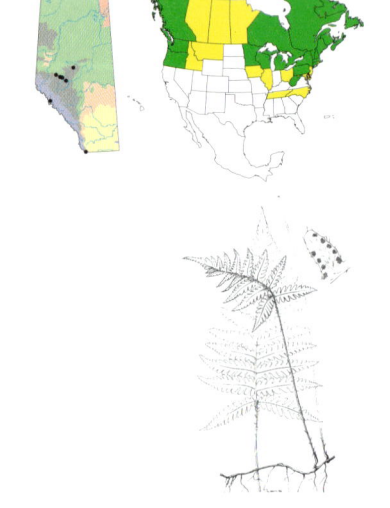

POLYPODY FAMILY　　　**Ferns and Fern Allies**

WESTERN POLYPODY

Polypodium hesperium Maxon
POLYPODY FAMILY (Polypodiaceae)

Plants: Evergreen herbs, erect or spreading, usually 5–20 cm tall, **scattered** along thick, creeping, reddish brown, scaly underground stems (rhizomes) up to 6 mm in diameter.

Leaves: Herbaceous to somewhat leathery, oblong to broadly lance-shaped in outline, **4–20 {35} cm long**, 1–5 {7} cm wide, **once pinnately divided** into 5–18 opposite or slightly offset pairs of **blunt** (sometimes pointed), oblong to linear–lance-shaped **leaflets less than 12 mm wide**; stalks 1–10 {15} cm long, hairless, straw-coloured, shorter than the blades; scales on the lower side linear to lance-shaped, less than 6 cells wide, not toothed.

Spore clusters: Oval (when immature) **dots** (sori) less than 3 mm in diameter on underside of leaflets, **midway between the leaf edge and central veins**; lacking a protective covering (indusium); summer–autumn.

Habitat: Moist, rocky, non-calcareous montane slopes and crevices; rarely on limestone.

Notes: This species has also been called *Polypodium vulgare* L. var. *columbianum* Gilbert. • Rock polypody (*Polypodium virginianum* L., p. 370) is very similar to western polypody and grows in similar habitats. It is distinguished by its round spore dots (sori), which are located along the leaf edges, and by the scales on its rhizomes, which have dark centres and lighter edges. • The acrid- to sweet-tasting, licorice-flavoured rhizomes of western polypody contain sucrose and fructose, as well as osladin, a compound that is said to be 300 times sweeter than sugar. Although polypody rhizomes have seldom been used as food, they were occasionally used for flavouring, and they were often chewed as an appetizer. However, **all ferns should be used with caution**, as the toxicity of most species is not known. • The generic name *Polypodium* and the common name 'polypody' were derived from the Greek *polys* (many) and *podos* (foot), in reference to the many-branched rhizomes of these ferns.

Ferns and Fern Allies

POLYPODY FAMILY

ROCK POLYPODY

Polypodium virginianum L.
POLYPODY FAMILY (POLYPODIACEAE)

Plants: Evergreen herbs, erect or spreading, usually 10–30 cm tall, **scattered along** thick (up to 6 mm in diameter), creeping, **reddish brown, scaly underground stems** (rhizomes); scales weakly 2-coloured or with a dark central stripe, coarsely toothed along the edges and twisted at the tips.

Leaves: Somewhat leathery, oblong to narrowly lance-shaped in outline, 5–25 {40} **cm long**, 2–7 cm wide, **once pinnately divided** into 10–20 opposite or slightly offset pairs of **blunt** (sometimes broadly pointed), oblong leaflets; leaf axis (rachis) with broadly lance-shaped scales more than 6 cells wide; stalks 5–15 cm long, hairless, shorter than the blade.

Spore clusters: Round dots (sori) less than 3 mm in diameter on underside of leaflets, from halfway between the midvein and the leaf edge to along the leaf edge; lacking a protective covering (indusium) but with glandular-hairy modified spore cases (sporangiasters); **spores more than 52 microns in diameter**; summer–autumn.

Habitat: Moist cliffs and rocky sites in northern Alberta.

Notes: This species was called *Polypodium vulgare* L. var. *virginianum* (L.) Eat. in the first edition of the *Flora of Alberta*. • Western polypody (*Polypodium hesperium* Maxon, p. 369) is distinguished from rock polypody by its oval spore dots, by the uniformly coloured, usually smooth-edged scales on its stems and by the slender (less than 6 cells wide) scales on the underside of its leaf axis (rachis). • Siberian polypody (*Polypodium sibiricum* Siplivinsky, sometimes included in *Polypodium virginianum* L.) is also rare in northern Alberta. It is distinguished from rock polypody by the general lack of gland-tipped hairs on its sporangiasters, by the uniformly dark brown scales on its stems (though these are sometimes obscurely 2-coloured, with lighter edges) and by its smaller (less than 52 microns) spores. Siberian polypody grows on a variety of rocky substrates, including granite and limestone.

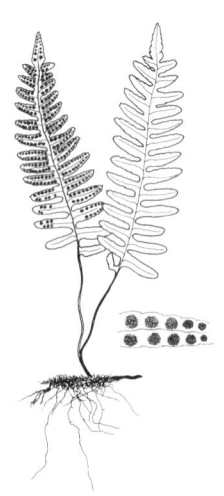

P. SIBIRICUM

MARSILEA FAMILY

Ferns and Fern Allies

HAIRY PEPPERWORT, WATER FERN, HAIRY WATER-CLOVER

Marsilea vestita Hook. & Grev.
MARSILEA FAMILY (Marsileaceae)

Plants: Delicate perennial herbs with aerial or floating leaves, **forming colonies** from thin (0.4–0.6 mm thick), creeping underground stems (rhizomes).

Leaves: Basal, resembling a 4-leaf clover, palmately divided into 4 egg-shaped to spatula-shaped leaflets 5–15 {3–17} **mm long**, with wedge-shaped bases and smooth to wavy tips, somewhat hairy (at least on the lower surface) to hairless, folded upward or spreading on slender stalks 2–10 {20} cm long.

Spore clusters: Bony, bean-shaped cases (sporocarps) **with 2 teeth** near the base (usually), 4–6 {8} mm long and **covered with coarse, flat-lying stiff hairs**, greenish when young, brown when mature, containing about 10–20 spore clusters (sori), each with a cluster of 5–20 large female spores (megaspores) and 2 side clusters of smaller male spores (microspores), splitting in 2 to release spores; few, borne singly on erect, short stalks from leaf axils; late May–August {April–October}.

Habitat: Shallow water of ponds, ditches and depressions, often where ground has dried late in the season; elsewhere, on shores of lakes and streams, often tolerant of alkali.

Notes: This species has also been called *Marsilea mucronata* A. Braun. • Hairy pepperwort is the only species of this genus found in Alberta. • The hard, bony covering of the spore cases helps the spores to survive dry periods. • This genus was named in honour of Count Luigi Marsigli, an Italian mycologist who lived 1656–1730. The specific epithet *vestita* was taken from the Latin *vestitus* (clothed), possibly in reference to the shaggy covering of hairs on the sporocarps.

Rare *Vascular* Plants of Alberta

Addendum

Addendum

Species New to The Alberta Natural Heritage Information Centre (ANHIC) Tracking List Since 1998

Two field seasons have passed since the original text for this guide was written, and in that time there have been many changes to the Alberta Natural Heritage Information Centre (ANHIC) tracking list. The status of many species in the original manuscript has changed (see Appendix Three). Several other species have been reclassified as rare or have been discovered for the first time in Alberta. This addendum includes descriptions of taxa that have been added to the ANHIC tracking list since 1998.

Two species from the February 2000 tracking list have been excluded. Many taxonomists now include Macoun's cryptanthe (*Cryptantha macounii* (Eastw.) Payson) in the more widespread species *Cryptantha celosioides* (Eastw.) Payson or *Cryptantha interrupta* (Greene) Payson, so this taxon is not described. Similarly, many taxonomists now include northern wormwood (*Artemisia borealis* Pall.) in the more widespread species *Artemisia campestris* L. as ssp. *borealis* (Pall.) H.M. Hall & Clem. var. *borealis*, so it has also been excluded from this addendum.

Addendum

ADDER'S-TONGUE FAMILY

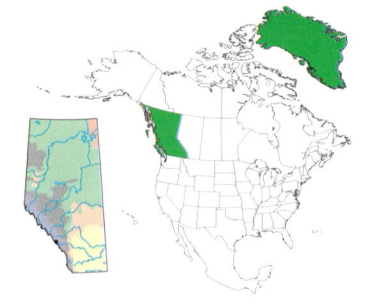

NORTHERN GRAPE FERN

Botrychium boreale J. Milde
ADDER'S-TONGUE FAMILY (OPHIOGLOSSACEAE)

Plants: Small, fleshy perennial herbs **less than 15 cm tall**; appear in July–August.

Leaves: Of 2 types: a single sterile blade below a single fertile blade (see spore clusters below); **sterile blades** fleshy, **shiny** green, **stalkless** or short-stalked, **subdeltoid to deltoid** in outline, up to 6 cm long and about as wide, **twice-divided** into as many as 6 pairs of segments on a single main axis; **segments** ascending, mostly overlapping, with the basal pair usually considerably larger than the adjacent pair, edged with shallow (rarely deeply cut) lobes, **pointed at the tip**; **veins** pinnate near the leaf base, otherwise **in a fan-like pattern**.

Spore clusters: Small, spherical, yellow, stalkless spore sacs (sporangia) borne on specialized, fertile leaf blades; fertile blades single, once- or twice-divided, about 1–1½ times as long as the sterile blade.

Habitat: Dry meadows on south-facing slopes from Greenland to Eurasia; not well understood in North America.

Notes: This species has also been called boreal moonwort and northern moonwort. • The *Flora of North America* describes northern grape fern as a well-marked species of northern Eurasia (especially Scandinavia) and Greenland. However, this small grape fern has recently been identified in BC and Alberta.

NORTHWESTERN GRAPE FERN

Botrychium pinnatum H. St. John
ADDER'S-TONGUE FAMILY (OPHIOGLOSSACEAE)

Plants: Small, **fleshy** perennial herbs {3.5} **8–15 cm tall**; appear in June–August.

Leaves: Of 2 types: a single sterile blade below a single fertile blade (see spore clusters below); **sterile blades** fleshy, **shiny** bright green, **stalkless** or with a 1–2 mm stalk, **oblong to subdeltoid** in outline, up to 8 cm long and 5 cm wide, **divided 1–2 times** into as many as 7 pairs of segments on a single main axis; **segments at right angles** to the main axis or only slightly ascending, close to overlapping, with the **lowermost pair symmetrical**, edged with deep, regular lobes, squared to somewhat pointed at the tip; **veins pinnately branched**.

Spore clusters: Small, spherical, yellow, stalkless spore sacs (sporangia) borne on specialized, fertile leaf blades; fertile blades single, twice-divided, about 1–2 times as long as the sterile blade.

Habitat: Open, moist to mesic sites in montane, subalpine and alpine zones.

Notes: This species has also been called *Botrychium boreale* J. Milde ssp. *obtusilobum* (Ruprecht) R.T. Clausen and northwestern moonwort. • The range of northwestern grape fern extends from Alaska to the western NWT, and south to California, Nevada, Utah and Colorado.

Addendum

GRASS FAMILY

WOODLAND BROME

Bromus vulgaris (Hook.) Shear
GRASS FAMILY (POACEAE [GRAMINEAE])

Plants: Slender perennial herbs; stems 80–100 {60–120} cm tall, with hairy (sometimes hairless) joints (nodes); **lacking elongated underground stems** (rhizomes).

Leaves: Alternate; blades flat, usually soft-hairy, 5–8 {15} mm wide; **ligules prominent,** {2} **3–5 mm long,** lacking small projections (auricles); sheaths usually soft-hairy.

Flower clusters: Slender panicles, {10–20} cm long; spikelets few, **drooping,** narrow, usually 5–7-flowered and {20–28} cm long; glumes shorter than the lowest lemma, the lower single-nerved, the upper 3-nerved; **lemmas** sparsely hairy to nearly hairless across the back, **hairier at the edges,** narrow; **lower lemmas** 8–10 {13} mm long and **about 2 mm wide,** tipped with a {3} **6–8 mm long bristle** (awn); anthers 3–5 mm long; {late June–August}.

Fruits: Grains; mature florets breaking off above the glumes and between the florets.

Habitat: Open woods; elsewhere, in montane meadows and on rocky slopes.

Notes: This species has also been called Colombian brome.
• Woodland brome is found from BC and southwestern Alberta south to California and Wyoming.

Pacific bluegrass

Poa gracillima Vasey
GRASS FAMILY (POACEAE [GRAMINEAE])

General: Loosely tufted perennial herbs; stems {10} 30–60 cm tall, frequently spreading at the base; **without creeping underground stems** (rhizomes).

Leaves: Mainly basal; sheaths hairless, closed less than half their length; **blades with a boat-shaped tip** (keeled), lax; **basal leaves hair-like**; stem leaves flat (rarely folded), 0.5–1.5 mm wide; ligules mostly pointed, {0.5} 2–5 mm long.

Flower clusters: Open, pyramid-shaped panicles, 4–10 {15} cm long, **longer than wide**; spikelets **rounded, not compressed**, 4–6 mm long, {2} 3–5-flowered; glumes 2.5–5 mm long, the first usually shorter than the second; **lemmas rounded, with a slight, hairy ridge (keel) on the back, with crisp (not cobwebby) hairs at the base**, about 4 mm long; {May} June–July.

Fruits: Grains; breaking away above the glumes and between the florets; August {September}.

Habitat: Moist montane woods and meadows; elsewhere, on rocky slopes, shaded cliffs, riverbanks and lake shores.

Notes: Pacific bluegrass is found from BC and south-western Alberta south to California, Utah and Colorado. • Two relatively common species, Canby bluegrass (*Poa canbyi* (Scribn.) Piper) and Sandberg bluegrass (*Poa sandbergii* Vasey), also have non-flattened spikelets and rounded, obscurely ridged, hairy lemmas. However, Canby bluegrass has a narrow, dense panicle and grows in dry to moist grassland. Sandberg bluegrass grows in dry grassland and has crowded, soft, curled basal leaves, very prominent ligules and short panicle branches that are often flattened against the stem.

Addendum

PINK FAMILY

SHORT-STALK MOUSE-EAR CHICKWEED

Cerastium brachypodium (Engelm. *ex* A. Gray) B.L. Robins.
PINK FAMILY (CARYOPHYLLACEAE)

Plants: Delicate annual herbs, 5–35 cm tall, with **short, glandular hairs**; stems slender, **erect to ascending**, single or branched from the base; from **shallow taproots**.

Leaves: Opposite, stalkless; blades narrowly elliptic to lance-shaped, sometimes widest above mid-leaf, 5–30 mm long, 2–8 mm wide.

Flowers: Green to whitish, highly variable, usually with 5 petals and 5 sepals; **petals absent or shorter than to twice as long as the sepals**; sepals 3–4.5 mm long, **with short, glandular hairs** (sometimes hairless), entirely green or with narrow, translucent edges; styles commonly 5; flowers few to many (rarely single), on **stalks ½–1¼ times as long as the sepals, never bent sharply downwards**, forming open, spreading clusters (cymes) above **green bracts**; {April–July}.

Fruits: Cylindrical capsules, 6–12 mm long, on straight or slightly up-curved **stalks up to 3 times as long as the calyx**, opening at the tip via 10 small teeth to release rough, grooved, golden-brown seeds.

Habitat: Wet or dry, open sites in grasslands, meadows, open woods and waste places, often on rocky or sandy soil.

Notes: This species has also been called *Cerastium nutans* Raf. var. *brachypodium* Engelm. • Short-stalk mouse-ear chickweed is found across the US, from Washington and Oregon to Virginia and Georgia, and south through Mexico to Central America. In Canada, it is found only in southern Alberta.

POPPY FAMILY

Addendum

ALPINE POPPY

Papaver radicatum Rottbøll ssp. *kluanensis* (D. Löve)
D.F. Murray
POPPY FAMILY (PAPAVERACEAE)

General: **Hairy perennial** herbs with milky sap, loosely tufted, with dull brown, soft, persistent leaf bases; **flowering stems leafless**, stiffly hairy, **6–9 {18} cm tall**; from underground stems (rhizomes).

Leaves: **Basal, grey-green to blue-green** and **hairy** above and below, lance-shaped, 3–7 cm long, pinnately cut into **5 {9} lance-shaped lobes**; stalks to ⅔ of the leaf length.

Flowers: **Sulphur-yellow** (drying pale green), rarely pink-tinged to brick-red, **saucer-shaped, up to 2 cm across**, with **4 large, thin petals**; **buds** oval, **erect**, with 2 **dark-hairy** sepals, 8–14 mm long; solitary; June–August.

Fruits: Erect, **stiffly brown-hairy capsules**, usually oblong and broad-tipped (sometimes almost round), 8–12 mm long, tipped with a disc-like stigma with 4–7 rays.

Habitat: Rocky alpine slopes, on shale in Alberta.

Notes: This taxon includes *Papaver kluanensis* D. Löve and *Papaver freedmanianum* D. Löve. • Alpine poppy is a circumpolar species. In North America, it is found from Alaska to the NWT, and its range extends south through the cordillera to Colorado and New Mexico.

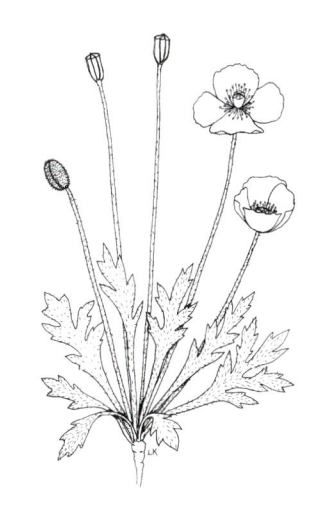

Addendum

MUSTARD FAMILY

PORSILD'S WHITLOW-GRASS

Draba porsildii G.A. Mulligan
MUSTARD FAMILY (BRASSICACEAE [CRUCIFERAE])

Plants: Small, **tufted** perennial herbs; flowering stems erect, **2–6.5 cm tall**, hairless or sparsely hairy; from taproots.

Leaves: In basal rosettes, lance-shaped and widest towards the tip, **3–12 mm long**, 1.5–2 mm wide, not toothed, **with simple, 2-branched and 4–8-rayed hairs on both surfaces; stem leaves absent or single.**

Flowers: White, about **5 mm across**, with 4 sepals and 4 petals; petals 2–3 mm long; sepals 1.5–2 mm long; styles 0.25 mm long; 2–5 {6} flowers in compact, branched clusters (racemes).

Fruits: Narrowly egg-shaped pods (siliques), widest towards the tip, 4–8 mm long, 2–3 mm wide, **hairless**, splitting in 2 to release 5–9 seeds 1 mm long; **stalks shorter than pods.**

Habitat: Moist, turfy alpine sites; elsewhere, in mesic to dry, rocky subalpine to alpine sites.

Notes: Porsild's whitlow-grass is found in southeastern Alaska, the southern Yukon and western NWT, northeastern and southeastern BC, and southwestern Alberta.

WOOLLEN-BREECHES

Hydrophyllum capitatum Douglas ex Benth.
WATERLEAF FAMILY (Hydrophyllaceae)

Plants: Small perennial herbs; stems single to few, 10–20 {40} cm tall, hairy; from short rootstocks with slender, fleshy roots.

Leaves: Alternate, mostly **basal, long-stalked**; blades egg-shaped to oval, {10–15} cm long, hairy, **pinnately divided into 5–7 {11} toothed leaflets**.

Flowers: Purplish blue to white, **broadly funnel-shaped**, 5–9 mm long, 5-lobed; lobes with slender appendages on their inner surfaces; calyx deeply 5-lobed, covered with long, stiff hairs; styles 2-pronged; **5 stamens, longer than the petals**, hairy at the middle; flowers in **compact, head-like clusters** (cymes) on 1–5 cm stalks, **shorter than the leaves**; {March–July}.

Fruits: Round capsules containing 1–4 seeds.

Habitat: Moist, open or thinly wooded montane slopes; elsewhere, from foothills to well up in the mountains.

Notes: This species has also been called cat's-breeches.
• The range of woollen-breeches extends from southern BC and Alberta to California and Colorado.

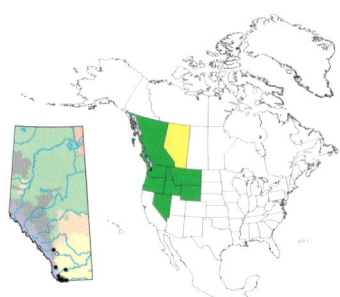

Addendum

FIGWORT FAMILY

YELLOW PAINTBRUSH

Castilleja cusickii Greenm.
FIGWORT FAMILY (SCROPHULARIACEAE)

Plants: Loosely clumped, **sticky-hairy** perennial herbs; stems simple, erect or ascending, 10–60 cm tall; from root crowns.

Leaves: Alternate, numerous, 2–4 cm long, covered with sticky hairs; lower leaves linear and undivided; upper leaves broader, **tipped with 3–7 slender, ascending lobes**.

Flowers: Sulphur-yellow, densely **glandular-hairy**; corolla shorter than to longer than the calyx, with a **short upper lip** (galea) less than ½ as long as the corolla tube and a prominent lower lip ⅓ as long to almost equal to the galea; calyx tubular, 2–3 cm long, tipped with 4 rounded lobes; **bracts yellow**, oblong, wider than the leaves, with 3–5 slender, ascending lobes, **equalling or longer than the flowers**; 4 stamens, **enclosed in the upper lip**; styles slender, usually protruding from the tip of the upper lip; numerous, in the axils of bracts in dense spikes that elongate in fruit; {April–August}.

Fruits: Capsules, splitting open lengthwise, containing many seeds with netted veins; May–July {April–August}.

Habitat: Grasslands; elsewhere, in meadows at intermediate elevations.

Notes: This species has also been called *Castilleja lutea* Heller and Cusick's Indian-paintbrush. It was included in *Castilleja septentrionalis* Lindl. in the first edition of the *Flora of Alberta*. • Yellow paintbrush is found from BC and Alberta to Nevada and Wyoming.

CLAMMY HEDGE-HYSSOP

Gratiola neglecta Torr.
FIGWORT FAMILY (SCROPHULARIACEAE)

Plants: Slender annual herbs, **clammy with fine glandular hairs** on upper parts; stems more or less erect, 10–30 cm tall, usually diffusely branched; from **fibrous roots**.

Leaves: Opposite, linear to lance-shaped and widest above the middle, usually 2–4 {5} cm long, toothless or wavy-toothed, tapered to a narrow **stalkless** base.

Flowers: Pale yellow to whitish (lips sometimes purplish), {6} **8–10 mm long, tubular, 2-lipped**, hairy in the throat; calyxes 4–6 {3–7} mm long, with 4 deeply cut lobes immediately above 2 similar, slender bracts; 2 fertile stamens, with a large, flattened disc (connective) bearing 2 parallel pollen sacs; sterile stamens tiny or none; flowers **borne singly on slender, spreading, 1–2 {3} cm long stalks** from upper leaf axils; {June–August}.

Fruits: Broadly egg-shaped, pointed capsules, 3–5 {7} mm long, splitting in 4 to release short-cylindrical seeds about 0.5 mm long.

Habitat: Wet, muddy sites, often in shallow water.

Notes: The range of clammy hedge-hyssop extends east from southern BC to Quebec and Nova Scotia, and south through the US to California, Texas and Georgia.

Addendum

FIGWORT FAMILY

FLAME-COLOURED LOUSEWORT

Pedicularis flammea L.
FIGWORT FAMILY (SCROPHULARIACEAE)

Plants: Small perennial herbs; stems single to several, reddish purple, **6–10 cm tall**, **usually hairless, unbranched**, with **1 or 2 leaves**; from spindly roots.

Leaves: Alternate, mainly basal, lance-shaped to linear, 2–6 cm long, edged with **numerous oblong to oval, 3–5 mm, round-toothed lobes**.

Flowers: Yellow, {6–7} 10–20 mm long and about 2 mm wide, 2-lipped, with a **small, yellow, 3-lobed lower lip** and a **deep crimson to purple** (usually), arching upper lip (galea), **style not protruding**: calyxes equal to or slightly longer than the corolla tube, 5-toothed; flowers borne in the axils of short, linear to lance-shaped bracts, forming hairless to woolly, 2–5 cm long, **spike-like clusters** (racemes).

Fruits: Curved, lance-shaped capsules, 2–3 times longer than the calyx.

Habitat: Calcareous alpine meadows; elsewhere, also in moist sites such as snow beds and lake shores.

Notes: This species includes *Pedicularis albertae* Hult. and *Pedicularis oederi* M. Vahl *ex* Hornem. var. *albertae* (Hult.) B. Boivin. It has also been called red-rattle. • Flame-coloured lousewort occurs in Greenland, Iceland and Lapland. In North America it is found in Alberta, scattered across the NWT and Nunavut, around Hudson Bay and James Bay (Manitoba, Ontario and Quebec) and in Labrador and Newfoundland. • Oeder's lousewort (*Pedicularis oederi* M. Vahl) was recently collected from an alpine site in the Willmore Wilderness Park. This is the first verified record of Oeder's lousewort from Alberta. This species is very similar to flame-coloured lousewort, but Oeder's lousewort is distinguished by its larger (over 20 mm long) flowers, with styles protruding from the tips of galeas that are not strongly reddish or purple-coloured. Oeder's lousewort plants are also larger (10–20 cm tall) and usually somewhat hairy, at least in the flower clusters. They may occasionally be hairless except for a fringe of hairs on the floral bracts and sepals. Oeder's lousewort grows in alpine meadows, often on rocky slopes.

ASTER FAMILY

Addendum

PATHFINDER

Adenocaulon bicolor Hook.
ASTER FAMILY (ASTERACEAE [COMPOSITAE])

Plants: Erect perennial herbs; stems essentially hairless to white-woolly, 30–90 {100} cm tall; from short rootstocks with fibrous roots.

Leaves: Alternate, mainly basal; thin, green and essentially hairless above, **densely white-woolly beneath, heart-shaped to triangular** with somewhat notched bases, 3–15 cm wide, usually wavy-toothed or shallow-lobed; stalks long, winged.

Flowerheads: Whitish, about 3 mm long, with **disc florets only**; inner florets functionally male; outer 4–6 {3–7} florets female; **involucres with 4 or 5 egg-shaped, pointed bracts, 2 mm long,** that spread or bend backwards at maturity; flowerheads borne on **sticky, glandular-hairy branches** in open clusters (panicles); {June–September}.

Fruits: Dry seed-like fruits (achenes), **club-shaped, 5–8 mm long, lacking bristles** (pappus) but **tipped with stalked glands**.

Habitat: Moist, shady montane woods.

Notes: This species has also been called American trail-plant. • Pathfinder is found from southern Alberta and BC south to California and Wyoming, and also in North and South Dakota, Michigan, Minnesota and Ontario.

Addendum

ASTER FAMILY

NODDING SCORZONELLA

Microseris nutans (Hook.) Sch.-Bip.
ASTER FAMILY (ASTERACEAE [COMPOSITAE])

Plants: Perennial herbs with **milky juice**, highly variable; **stems usually with a least 1 leaf**, erect or somewhat reclining at the base, 10–50 {60} cm tall, hairless or with minute hairs; **main stem usually with a few branches**; from thick, **fleshy taproots**.

Leaves: Alternate, mainly basal, **linear**, tapered to a slender point, 10–30 cm long; toothless or with slender teeth to spreading lobes.

Flowerheads: Yellow, with **strap-like (ligulate) florets only**, nodding in bud; florets 3–8 mm long, with both male and female parts; involucres 10–20 {8–22} mm high, usually granular, hairless or minutely hairy, with almost **equal, lance-shaped, long-tapered bracts** above a few short bractlets; flowerheads solitary to several, on **long, slender stalks**; {April–July}.

Fruits: Dry seed-like fruits (achenes), {5–8} mm long, **beakless**, tapered to both ends, hairless or finely hairy, tipped with a tuft of 15–20 **feathery-hairy bristles** (pappus) **with scale-like bases**.

Habitat: Grassy slopes and open montane woods.

Notes: This species has also been called nodding silverpuffs. • Nodding scorzonella is found from southern Alberta and BC south to California, Utah and Colorado.

Appendix 1

APPENDIX 1
KEYS TO RARE *BOTRYCHIUM* AND *ISOETES* SPECIES FOUND IN ALBERTA

The moonworts and quillworts, as they are commonly known, are complex genera whose species are distinguished by minute diagnostic features. These keys present only species known to occur in Alberta.

A. Key to rare *Botrychium* species
Key adapted from Flora of North America *Vol. 2, pp. 87–90.*

1. Sterile leaf blades triangular, mostly 5–25 cm long; plants mostly over 12 cm tall, commonly sterile, fertile blades absent or misshapen; leaf sheaths open or closed — *Botrychium multifidum*

1. Sterile leaf blades mainly oblong to linear, mostly 2–4 cm long; plants mostly less than 15 cm tall, fertile blades always present; leaf sheaths closed — 2

   2. Sterile leaf blades linear to linear-oblong, simple to lobed, lobes rounded to square and angular; stalks usually ⅓–⅔ the length of the sterile blade; plants usually in shaded sites — *Botrychium simplex*

   2. Sterile leaf blades oblong–lance-shaped to oblong, pinnate, rarely simple, lobes, if present, of various shapes; stalks usually less than ¼ the length of the sterile blade; plants usually in exposed sites — 3

      3. Sterile blades present; basal pinnae fan-shaped to spoon-shaped, with veins like the ribs of a fan, midrib absent — 4

      4. Basal pinnae broadly fan-shaped — 5

         5. Plants herbaceous; sterile blades mostly less than 4 cm long and 1.5 cm wide; pinnae 2–5 pairs, well separated, margins commonly rounded to toothed; fertile blades 1⅓–3 times the length of the sterile blade; in damp sites — *Botrychium crenulatum*

         5. Plants fleshy; sterile blades mostly more than 5 cm long and 2 cm wide; pinnae 4–9 pairs, approximate to overlapping; margins usually entire, rarely toothed; fertile blades ⅘–2 times the length of the sterile blade; in dry sites — *Botrychium lunaria*

      4. Basal pinnae narrowly fan-shaped or wedge-shaped to lance-shaped or linear — 6

         6. Pinnae strongly ascending; margins conspicuously sharply toothed — *Botrychium ascendens*

         6. Pinnae spreading or only moderately ascending; outer margins entire to rounded, rarely toothed — 7

            7. Sterile blades folded lengthwise, usually 4 cm long and 1 cm wide; pinnae up to 5 pairs; basal pinnae 2-lobed — *Botrychium campestre*

Appendix 1

      7. Sterile blades flat or folded only at the base, usually 10 cm long and 2.5 cm wide; pinnae up to 10 pairs; basal pinnae unlobed, or if lobed, not usually 2-cleft — 8

      8. Sterile blades narrowly oblong, herbaceous; pinnae nearly round to fan-shaped; margins shallowly lobed; lowest branches of fertile blades 1-pinnate — *Botrychium minganense*

      8. Sterile blades narrowly triangular, leathery; pinnae spoon-shaped to narrowly spoon-shaped; margins entire to coarsely and irregularly toothed; lowest branches of fertile blades 2-pinnate — *Botrychium spathulatum*

3. Sterile blades present or replaced by fertile blades; if present, basal pinnae linear to egg-shaped, with pinnate venation, midrib present — 9

    9. Sterile blade replaced by a fertile blade, yielding 2 fertile blades — *Botrychium paradoxum*

    9. Sterile blade present, distinct from the fertile blade — 10

      10. Sterile blades triangular; fertile blades divided at base into several equally long branches — *Botrychium lanceolatum*

      10. Sterile blades egg-shaped to oblong; fertile blades with single midrib or 1 dominant midrib and 2 smaller ribs — 11

        11. Stalk of sterile blade as long as the blade; blades oblong–egg-shaped to triangular-oblong; basal pinnae rhombic — *Botrychium pedunculosum*

        11. Stalk of sterile blade short to nearly absent, up to ¼ the length of the blade; blades oblong–lance-shaped to nearly triangular — 12

          12. Sterile blades nearly triangular, basal pinna pair elongate; segments and lobes rounded at tip — *Botrychium hesperium* (This species has been reported falsely from Alberta; the Alberta material is most likely *Botrychium michiganense*.)

          12. Sterile blades oblong-triangular to oblong–egg-shaped; basal pinna pair not elongate; segments and lobes with blunt or slightly pointed tips — 13

            13. Pinnae of mature sterile blades nearly as wide as long, with slightly pointed tips, veins like ribs of a fan; basal pinnae with shallow, narrow sinuses and 1–3 lobes — *Botrychium boreale*

            13. Pinnae of mature sterile blades much longer than wide, with blunt tips, veins mostly pinnate; basal pinnae with deep, wide sinuses and 3–8 lobes — *Botrychium pinnatum*

Appendix 1

B. Key to *Isoetes* species

Key adapted from Flora of North America *Vol. 2, pp. 66–68. Characters based on geography, habitat, megaspore ornamentation, and size and shape of the ligule and velum are useful in distinguishing these species, but microscopic examination of the mature megaspores is necessary for positive identification.*

1. Megaspores with spines — 2

    2. Spines long and pointed, not smaller around the middle of the megaspore; microspores spineless or with fine, thread-like spines — *Isoetes echinospora*

    2. Spines blunt-tipped, sometimes joined together to form ridges, notably smaller around the middle of the megaspore; microspores covered with coarse spines — *Isoetes maritima*

1. Megaspores with ridges, not spiny — 3

    3. Megaspores with distinct, roughly or sharply crested, branching ridges; ligules heart-shaped — *Isoetes occidentalis*

    3. Megaspores with low, scattered ridges that often merge together; ligules various — 4

        4. Plants in or out of water; broad, translucent leaf edges (wing margins) extending 1–5 cm above the spore cluster; ligules elongated-triangular — *Isoetes howellii*

        4. Plants underwater; broad, translucent leaf edges (wing margins) not extending more than 1 cm above the spore cluster; ligules heart-shaped — *Isoetes bolanderi*

APPENDIX 2
RARE VASCULAR PLANTS OF ALBERTA BY NATURAL REGION

Note that not all species in this list are included in the text. See also Addendum, p. 375–88.

SCIENTIFIC NAME	CANADIAN SHIELD	BOREAL FOREST	ROCKY MTN.	FOOT- HILLS	PARK- LAND	GRASS- LANDS	S RANK 1998	S RANK 2000
Adenocaulon bicolor Hook.			•				S3	S2S3
Adiantum aleuticum (Rupr.) Paris			•				S1S2	S2
Agoseris lackschewitzii D. Henderson & R. Moseley			•				S2	S2
Agrostis exarata Trin.		•	•	•		•	S2	S2
Agrostis humilis Vasey			•				S1S2	S1
Agrostis mertensii Trin.			•				S2	S2
Agrostis thurberiana A.S. Hitchc.			•				S2	S2
Allium geyeri S. Wats.			•			•	S2	S2
Alopecurus alpinus J.E. Smith			•		•	•	S2	S2
Amaranthus californicus (Moq.) S. Wats.						•	S1	S1
Ambrosia acanthicarpa Hook.						•	S2	S2
Anagallis minima (L.) Krause						•	S1	S1S2
Anemone quinquefolia L. var. *quinquefolia*				•			S1	S1
Antennaria aromatica Evert			•				S2	S2
Antennaria corymbosa E. Nels.			•			•	S1	S1
Antennaria luzuloides T. & G.			•				S1	S1
Antennaria monocephala D.C. Eaton			•				S2	S2
Anthoxanthum monticola (Bigelow) Y. Schouten & Veldkamp			•				S2	S2
Aquilegia formosa Fisch. *ex* DC. var. *formosa*			•	•			S2	S2
Aquilegia jonesii Parry			•				S2	S2
Arabidopsis salsuginea (Pallas) N. Busch		•					S1	S1
Arabis lemmonii S. Wats.			•				S2	S2
Arctagrostis arundinacea (Trin.) Beal		•	•				S1	S1
Arenaria longipedunculata Hult.			•				S1	S1
Aristida purpurea Nutt. var. *longiseta* (Steud.) Vasey						•	S1	S1
Arnica amplexicaulis Nutt.			•				S2	S2
Arnica longifolia D.C. Eat.			•				S2	S2
Arnica parryi A. Gray			•				S2	S2
Artemisia borealis Pall.			•				S2?	S2
Artemisia furcata Bieb. var. *furcata*			•				S1	S1
Artemisia tilesii Ledeb. ssp. *elatior* (T. & G.) Hult.		•		•			S2	S2

Appendix 2

Scientific Name	Canadian Shield	Boreal Forest	Rocky Mtn.	Foot- Hills	Park- Land	Grass- Lands	S Rank 1998	S Rank 2000
Artemisia tridentata Nutt.			•				S2	S2
Asclepias ovalifolia Dcne.		•			•	•	S3	S3
Asclepias viridiflora Raf.						•	S1	S1
Aster campestris Nutt.			•			•	S2	S2
Aster eatonii (A. Gray) J.T. Howell			•		•	•	S2	S2
Aster pauciflorus Nutt.					•		S1	S1
Aster umbellatus Mill.		•			•		S2	S2
Astragalus bodinii Sheldon		•	•				S1	S1
Astragalus kentrophyta A. Gray var. *kentrophyta*						•	S1	S1S2
Astragalus lotiflorus Hook.						•	S2	S2
Astragalus purshii Dougl. *ex* Hook. var. *purshii*						•	S2	S2
Athyrium alpestre (Hoppe) Clairville var. *americanum* Butters			•				S1	S1
Atriplex canescens (Pursh) Nutt.						•	SU	SU
Atriplex powellii S. Wats.						•	S1	S1
Atriplex truncata (Torr.) A. Gray						•	S1	S1
Bacopa rotundifolia (Michx.) Wettst.						•	S1	S1
Barbarea orthoceras Ledeb.	•	•	•	•	•		S2	S2
Bidens frondosa L.						•	S1S2	S1
Blysmus rufus (Hudson) Link		•					S1	S1
Boisduvalia glabella (Nutt.) Walp.						•	S2	S2
Bolboschoenus fluviatilis (Torr.) J. Sojak.			•			•	S1	S1
Boschniakia rossica (Cham. & Schlecht.) Fedtsch.		•					S1	S1
Botrychium ascendens W.H. Wagner			•					S1
Botrychium boreale (Fries) Milde			•				SRF	S1
Botrychium campestre W.H. Wagner & D.R. Farrar					•		S1	S1
Botrychium crenulatum W.H. Wagner			•				S1	S1
Botrychium hesperium (Maxon & R.T. Clausen) W.H. Wagner & Lellinger			•				S1	SRF
Botrychium lanceolatum (Gmelin) Angström ssp. *angustisegmentum* (Pease & A.H. Moore) R.T. Clausen			•				S2	S2

Appendix 2

Scientific Name	Canadian Shield	Boreal Forest	Rocky Mtn.	Foot-hills	Park-land	Grass-lands	S Rank 1998	S Rank 2000
Botrychium lanceolatum (Gmelin) Angström ssp. *lanceolatum* (Pease & A.H. Moore) R.T. Clausen			•				S2	S2
Botrychium michiganense			•					S1
Botrychium minganense Victorin		•	•	•			S2	S2S3
Botrychium multifidum (S.G. Gmelin) Rupr.	•	•	•	•	•		S2	S2
Botrychium paradoxum W.H. Wagner			•				S1	S1
Botrychium pedunculosum W.H. Wagner			•				S1	S1
Botrychium pinnatum H. St. John			•					S1
Botrychium simplex E. Hitch.			•				S1S2	S2
Botrychium spathulatum W.H. Wagner			•	•	•		S1S2	S2
Botrychium x *watertonense* W.H. Wagner			•				S1	S1
Boykinia heucheriformis (Rydb.) Rosendahl			•				S2	S2
Brasenia schreberi J.F. Gmelin		•						S1
Braya purpurascens (R. Br.) Bunge *ex* Ledeb.			•				S1	S1
Brickellia grandiflora (Hook.) Nutt.			•				S2	S2
Bromus latiglumis (Shear) A.S. Hitchc.			•		•		S1	S1
Bromus vulgaris (Hook.) Shear			•				S3	S2S3
Calamagrostis lapponica (Wahl.) Hartm.			•	•			S1	S1
Calylophus serrulatus (Nutt.) Raven					•	•	S2	S2
Camassia quamash (Pursh) Greene var. *quamash*			•		•	•	S2	S2
Camissonia andina (Nutt.) Raven						•	S1	S1
Camissonia breviflora (Torr. & Gray) Raven						•	S1	S1
Campanula uniflora L.			•				S2	S2
Cardamine bellidifolia L.			•				S2	S2
Cardamine parviflora L.		•					S1	S1
Cardamine pratensis L.	•	•		•			S1	S1S2
Cardamine umbellata Greene			•	•			S2	S2
Carex adusta Boott		•	•	•	•		S1	S1
Carex aperta Boott			•		•		S2	S1

Rare *Vascular* Plants of Alberta   395

Appendix 2

Scientific Name	Canadian Shield	Boreal Forest	Rocky Mtn.	Foot- hills	Park- land	Grass- lands	S Rank 1998	S Rank 2000
Carex arcta Boott		•	•	•			S1	S1
Carex backii Boott		•	•		•		SU	SU
Carex capitata L.	•	•	•	•			S2	S2
Carex crawei Dewey		•		•		•	S2	S2
Carex franklinii Boott		•					S2	S3
Carex glacialis Mack.		•					S2	S2
Carex haydeniana Olney		•			•		S2	S3
Carex heleonastes Ehrh. *ex* L. f.		•	•	•			S2	S2
Carex heteroneura Boott var. *epapillosa* (Mack.) F.J. Hermann		•					S1	S1
Carex hookerana Dewey		•	•		•		S2	S2
Carex houghtoniana Torr.	•	•		•			S2	S2
Carex hystericina Muhl. *ex* Willd.		•					S1	S1
Carex illota Bailey			•				S1	S1
Carex incurviformis Mack. var. *incurviformis*			•	•			S2	S2
Carex lachenalii Schk.			•				S2	S2
Carex lacustris Willd.		•		•			S2	S2
Carex lenticularis Michx. var. *dolia* (M.E. Jones) L.A. Standley	•		•				S2	S1
Carex leptopoda Mack.			•					S1
Carex loliacea L.		•	•	•			S2	S3
Carex mertensii Prescott ssp. *mertensii*			•	•			S1	S1
Carex misandra R. Br.			•				S1S2	S1S2
Carex nebrascensis Dewey					•		S2	S2
Carex oligosperma Michx.	•	•					S1	S1
Carex parryana Dewey var. *parryana*			•		•	•	S1S2	S1S2
Carex pauciflora Lightf.		•	•	•	•		S3	S3
Carex paysonis Clokey			•	•			S2	S1S2
Carex pedunculata Muhl.		•					S1	S1
Carex petasata Dewey			•				S1S2	S1S2
Carex petricosa Dewey			•				S2	S2
Carex platylepis Mack.			•				S1S2	S1S2
Carex podocarpa R. Br.			•				SU	S2
Carex preslii Steud.			•	•			S2	S2
Carex pseudocyperus L.	•	•			•		S2	S2
Carex raynoldsii Dewey			•	•			S3	S3
Carex retrorsa Schwein.		•			•	•	S2	S2
Carex rostrata Stokes	•	•	•	•	•		S2	S2
Carex saximontana Mack.			•				SU	SU
Carex scoparia Schk. *ex* Willd.			•				S1	S1

Appendix 2

Scientific Name	Canadian Shield	Boreal Forest	Rocky Mtn.	Foot- hills	Park- land	Grass- lands	S Rank 1998	S Rank 2000
Carex supina Willd. ssp. *spaniocarpa* (Steud.) Hult.			•				S1	S1
Carex tincta (Fern.) Fern.		•	•				S1	S1
Carex tonsa (Fern.) Bickn. var. *tonsa*	•	•	•	•	•	•	S3	S3
Carex trisperma Dewey		•		•			S3	S3
Carex umbellata Schkuhr *ex* Willd.		•						S1
Carex vesicaria L. var. *vesicaria*			•		•	•	SU	S1
Carex vulpinoidea Michx.		•					S2	S2
Castilleja cusickii Greenm.					•		S3	S2S3
Castilleja lutescens (Greenm.) Rydb.				•	•		S3	S2S3
Castilleja pallida (L.) Spreng. ssp. *caudata* Pennell			•				SU	SU
Castilleja sessiliflora Pursh					•		S1	S1
Cerastium brachypodum (Engelm. *ex* A. Gray) B.L. Rob.						•		S2
Cheilanthes gracillima D.C. Eat.			•				S1	S1
Chenopodium atrovirens Rydb.					•		S1	S1
Chenopodium desiccatum A. Nels.						•	S1S2	S1S2
Chenopodium incanum (S. Wats.) Heller			•				S1	S1
Chenopodium leptophyllum (Nutt. *ex* Moq.) Nutt. *ex* S. Wats.		•			•	•	SU	SU
Chenopodium subglabrum (S. Wats.) A. Nels.						•	S1	S1
Chenopodium watsonii A. Nels.						•	S1	S1
Cirsium scariosum Nutt.			•				SU	SU
Conimitella williamsii (D.C. Eaton) Rydb.			•		•		S2	S2
Coptis trifolia (L.) Salisb.		•	•				S3	S3
Coreopsis tinctoria Nutt.					•		S1S2	S2
Crepis atribarba Heller			•			•	S2	S2
Crepis intermedia A. Gray			•		•		S2	S2
Crepis occidentalis Nutt.			•			•	S2	S2
Cryptantha macounii (Eastw.) Payson					•		S3	S2
Cryptantha minima Rydb.						•	S1	S1
Cryptogramma stelleri (S.G. Gmel.) Prantl			•				S2	S2
Cuscuta gronovii Willd.						•	S1	S1
Cynoglossum virginianum L. var. *boreale* (Fern.) Cooperrider					•		S1	S1
Cyperus schweinitzii Torr.					•	•	S2	S2

Rare *Vascular* Plants of Alberta

Appendix 2

Scientific Name	Canadian Shield	Boreal Forest	Rocky Mtn.	Foothills	Parkland	Grasslands	S Rank 1998	S Rank 2000
Cyperus squarrosus L.						•	S1	S1
Cypripedium acaule Ait.	•	•					S2	S2
Cypripedium montanum Dougl. *ex* Lindl.			•		•		S2	S2
Cystopteris montana (Lam.) Desv.		•	•	•			S2	S2
Danthonia californica Bol.		•	•	•	•	•	S3	S3
Danthonia spicata (L.) Beauv. *ex* R. & S.	•	•	•		•		S2	S1S2
Danthonia unispicata (Thurb.) Munro *ex* Macoun			•			•	S2	S3
Deschampsia elongata (Hook.) Munro *ex* Benth.			•	•			S1S2	S1
Diphasiastrum sitchense (Rupr.) Holub		•	•	•			S2	S2
Douglasia montana A. Gray			•				S1	S1
Downingia laeta Greene						•	S1S2	S1S2
Draba densifolia Nutt.			•				S1S2	S1S2
Draba fladnizensis Wulfen			•				S1	S1
Draba glabella Pursh			•				S1	S1
Draba kananaskis Mulligan			•				S1	S1
Draba longipes Raup			•				S1S2	S1S2
Draba macounii O.E. Schulz			•				S2	S2
Draba porsildii G.A. Mulligan			•				S2	S2
Draba reptans (Lam.) Fern.						•	S1S2	S1
Draba ventosa A. Gray			•				S2	S2
Drosera anglica Huds.	•	•	•	•	•		S2	S3
Drosera linearis Goldie		•	•	•	•		S2	S2
Dryopteris cristata (L.) A. Gray		•					S1	S1
Dryopteris filix-mas (L.) Schott		•					S1	S1
Elatine triandra Schk.		•			•		S1	S1
Eleocharis elliptica Kunth		•			•		SU	SU
Eleocharis engelmannii Steud.		•					S1?	S1?
Ellisia nyctelea L.					•	•	S2	S2
Elodea bifoliata St. John		•				•	S1S2	S1
Elymus elymoides (Raf.) Swezey			•			•	S3	S3
Elymus scribneri (Vasey) M.E. Jones			•				S2	S2
Elymus virginicus L.			•			•	S1	S1
Epilobium clavatum Trel.			•	•			S2	S2
Epilobium glaberrimum Barbey ssp. *fastigiatum* (Nutt.) Hoch & Raven			•				S1	S1
Epilobium halleanum Hausskn.		•					S1	S1
Epilobium lactiflorum Hausskn.			•	•			S2	S2

Appendix 2

Scientific Name	Canadian Shield	Boreal Forest	Rocky Mtn.	Foot-hills	Park-land	Grass-lands	S Rank 1998	S Rank 2000
Epilobium leptocarpum Hausskn.			•	•			S1	S1
Epilobium luteum Pursh			•				S1	S1
Epilobium mirabile Trel.			•				SR	S?
Epilobium oreganum Greene			•				SR	SRF
Epilobium saximontanum Hausskn.			•	•			S1S2	S1
Erigeron divergens T. & G.			•				S1	S1
Erigeron flagellaris A. Gray			•		•		S1	S1
Erigeron hyssopifolius Michx.		•					S1	S1
Erigeron lackschewitzii Nesom & Weber			•				S1	S1
Erigeron ochroleucus Nutt. var. *scribneri* (Canby) Cronq.			•				S2?	S2
Erigeron pallens Cronq.			•				S2	S2
Erigeron radicatus Hook.			•	•		•	S2	S2
Erigeron trifidus Hook.			•				S2	S1S2
Eriogonum cernuum Nutt.						•	S2	S2
Eriogonum ovalifolium Nutt. var. *ovalifolium*			•				S2	S3
Eriophorum callitrix Cham. ex C.A. Mey			•				S2	S2
Eriophorum scheuchzeri Hoppe			•	•			S3	S3
Erysimum pallasii (Pursh) Fern.			•				S3	S3
Eupatorium maculatum L.		•					S1S2	S1S2
Festuca altaica Trin. ex Ledeb.			•				S2	S2
Festuca minutiflora Rydberg			•				S2	S2
Festuca occidentalis Hook.			•				S1	S1
Festuca subulata Trin.			•				S1	S1
Festuca viviparoidea Krajina ex Pavlick ssp. *krajinae* Pavlick			•				S1	S1
Galium bifolium S. Wats.			•				S1	S1
Gayophytum racemosum Nutt. ex T. & G.			•				S1	S1
Gentiana fremontii Torr.					•		S1	S2S3
Gentiana glauca Pallas			•				S2	S2
Gentianopsis detonsa (Rottb.) Ma spp. *raupii* (Porsild) A. Löve & D. Löve		•					S1	S1
Geranium carolinianum L.		•			•		S1	S1
Geranium erianthum DC.			•				S1	SH
Glyceria elata (Nash) A.S. Hitchc.		•	•	•			S2	S2
Gnaphalium microcephalum Nutt.					•		S1	SH
Gnaphalium viscosum Kunth			•				SH	SH
Gratiola neglecta Torr.					•	•	S3	S2S3

Rare *Vascular* Plants of Alberta 399

Appendix 2

Scientific Name	Canadian Shield	Boreal Forest	Rocky Mtn.	Foot-hills	Park-land	Grass-lands	S Rank 1998	S Rank 2000
Gymnocarpium jessoense (Koidz.) Koidz.	•						S1	S1
Halimolobos virgata (Nutt.) O.E. Schulz					•		S1	S1
Hedyotis longifolia (Gaertn.) Hook.		•			•		S2	S2
Heliotropium curassavicum L.					•		S1	S1
Heuchera glabra Willd. *ex* R. & S.			•				S1	S1
Hieracium cynoglossoides Arv.-Touv. *ex* A. Gray			•		•		S2	S2S3
Hippuris montana Ledeb.			•				S1	S1
Hordeum pusillum Nutt.						•	SH	SH
Huperzia haleakalae (Brack.) Holub			•	•			S2	S2
Huperzia selago (L.) Bernh. *ex* Schrank & Mart.			•	•			S1	S1
Hydrophyllum capitatum Dougl. *ex* Hook.			•				S3	S2S3
Hymenopappus filifolius Hook. var. *polycephalus* (Osterh.) B.L. Turner						•	S2	S2
Hypericum majus (Gray) Britt.	•	•					S2	S2
Hypericum scouleri Hook.			•				S1	S1
Iliamna rivularis (Dougl. *ex* Hook.) Greene					•		S2	S2
Iris missouriensis Nutt.					•	•	S1	S1
Isoetes bolanderi Engelm.			•				S1	S1
Isoetes echinospora Durieu	•	•					S1	S1
Isoetes maritima L. Underwood			•				S1	S1
Isoetes occidentalis L.F. Henderson			•				S1	S1
Isoetes x *truncata* (A.A. Eaton) Clute			•				S1	S1
Juncus biglumis L.			•				S2	S2
Juncus brevicaudatus (Engelm.) Fern.	•	•		•			S2	S2
Juncus confusus Coville			•		•	•	S2	S2S3
Juncus filiformis L.	•	•		•			S2S3	S2S3
Juncus nevadensis S. Wats.			•				S1	S1
Juncus parryi Engelm.			•				S2	S2
Juncus regelii Buch.			•				S1	S1
Juncus stygius L. ssp. *americanus* (Buch.) Hult.			•		•		S2	S2
Koenigia islandica L.			•				S1	S1
Lactuca biennis (Moench) Fern.				•	•		S2S3	S2
Larix occidentalis Nutt.			•				S2	S2

Appendix 2

Scientific Name	Canadian Shield	Boreal Forest	Rocky Mtn.	Foothills	Parkland	Grasslands	S Rank 1998	S Rank 2000
Lesquerella arctica (Wormskj. *ex* Hornem.) S. Wats. var. *purshii* Wats.			•				S2	S2
Lewisia pygmaea (A. Gray) B.L. Robins. ssp. *pygmaea*			•				S2	S2
Lewisia rediviva Pursh			•				S1	S1
Leymus mollis (Trin.) Pilger ssp. *mollis*	•						S1	S1
Lilaea scilloides (Poir.) Hauman			•			•	S1	S1
Linanthus septentrionalis H.L. Mason			•			•	S2	S2
Listera caurina Piper			•				S1S2	S1
Listera convallarioides (Sw.) Nutt.			•				S2	S2
Lithophragma glabrum Nutt.			•				S2	S2
Lithophragma parviflorum (Hook.) Nutt. *ex* T. & G.			•		•		S2	S2
Lobelia dortmanna L.		•					S1	S1
Lobelia spicata Lam.					•		S1	S1
Loiseleuria procumbens (L.) Desv.			•				S1S2	S1S2
Lomatium cous (S. Wats.) Coult. & Rose			•				S1	S1S2
Lomatogonium rotatum (L.) Fries *ex* Nyman		•	•	•	•	•	S2	S2
Lupinus lepidus Dougl. *ex* Lindl.			•				S2	SRF
Lupinus minimus Dougl. *ex* Hook.			•				S2	S1
Lupinus polyphyllus Lindl.			•				S1	S1
Lupinus wyethii S. Wats.			•				S1	S1
Luzula acuminata Raf.		•		•			S1	S1
Luzula groenlandica Bocher	•						S1	S1
Luzula rufescens Fisch. *ex* E. Mey.		•		•			S1	S1
Lycopodiella inundata (L.) Holub	•						S1	S1
Lycopus americanus Muhl. *ex* W.C. Barton					•	•	S2	S2
Lysimachia hybrida Michx.					•	•	S2	S2
Machaeranthera tanacetifolia (Kunth) Nees						•	SR	SX
Malaxis monophylla (L.) Sw. var. *brachypoda* (A. Gray) Morris & Ames		•			•		S2	S2
Malaxis paludosa (L.) Sw.		•			•		S1S2	S1
Marsilea vestita Hook. & Grev.						•	S2	S2
Melica smithii (Porter *ex* A. Gray) Vasey			•		•		S1S2	S1S2
Melica spectabilis Scribn.			•		•		S2	S2

Appendix 2

Scientific Name	Canadian Shield	Boreal Forest	Rocky Mtn.	Foot- hills	Park- land	Grass- lands	S Rank 1998	S Rank 2000
Mertensia lanceolata (Pursh) A. DC.			•		•	•	S2	S2
Mertensia longiflora Greene			•		•		S2	S2
Microseris nutans (Hook.) Sch.-Bip.			•				S3	S2S3
Mimulus breweri (Greene) Coville			•				S1	S1
Mimulus floribundus Dougl. *ex* Lindl.			•				S1	S1
Mimulus glabratus Kunth					•		S1	S1
Mimulus guttatus DC.			•		•		SU	SU
Mimulus tilingii Regel			•		•		SU	S1
Minuartia elegans (Cham. & Schlecht.) Schischkin			•				S1	S1
Monotropa hypopithys L.		•	•				S2	S2
Montia linearis (Dougl. *ex* Hook.) Greene			•		•	•	S1S2	S1
Montia parvifolia (Moc. *ex* DC.) Greene			•				S1	S1
Muhlenbergia asperifolia (Nees & Meyen *ex* Trin.) Parodi						•	S2	S2
Muhlenbergia racemosa (Michx.) BSP.		•			•		S1	S1
Munroa squarrosa (Nutt.) Torr.						•	S1	S1
Najas flexilis (Willd.) Rostk. & Schmidt		•			•		S1S2	S1S2
Nemophila breviflora A. Gray			•		•	•	S1S2	S1S2
Nothocalais cuspidata (Pursh) Greene						•	S1	S1
Nuttallanthus canadensis (L.) D.A. Sutton						•	S1	S1
Nymphaea leibergii Morong		•					S1	S1
Nymphaea tetragona Georgi	•	•					S1	S1
Oenothera flava (A. Nels.) Garrett					•	•	S1S2	S2
Onosmodium molle Michx. var. *occidentale* (Mack.) I.M. Johnston					•	•	S2	S2
Oplopanax horridus (Sm.) T. & G. *ex* Miq. var. *horridus*		•	•	•			S3	S3
Orobanche ludoviciana Nutt.						•	S2	S2
Orobanche uniflora L.			•		•		S2	S2
Oryzopsis canadensis (Poir.) Torr.		•			•		S1	S1
Oryzopsis exigua Thurb.			•				S1	S1
Oryzopsis micrantha (Trin. & Rupr.) Thurb.						•	S2	S2

Appendix 2

Scientific Name	Canadian Shield	Boreal Forest	Rocky Mtn.	Foot-hills	Park-land	Grass-lands	S Rank 1998	S Rank 2000
Osmorhiza longistylis (Torr.) DC.			•		•	•	S2	S2
Osmorhiza purpurea (Coult. & Rose) Suksd.			•				S2	S2
Oxytropis jordalii Porsild var. *jordalii*			•				S1	SRF
Oxytropis lagopus Nutt. var. *conjugens* Barneby						•	S1	S1
Packera subnuda Trock & Barkley		•					S1	S2
Panicum acuminatum Swartz		•					SX	SU
Panicum leibergii (Vasey) Scribn.					•		S1	S1
Panicum wilcoxianum Vasey					•		S1	S1
Papaver pygmaeum Rydb.			•				S1S2	S2
Papaver radicatum Rottb. ssp. *kluanense* (D. Löve) D.F. Murray			•				S2	S2
Parietaria pensylvanica Muhl. *ex* Willd.			•			•	S2	S2
Parnassia parviflora DC.	•	•	•				S2	S2
Pedicularis capitata J.E. Adams			•				S2	S2
Pedicularis flammea L.			•				S3	S2
Pedicularis lanata Cham. & Schlecht.			•				S2	S2
Pedicularis langsdorfii Fisch. *ex* Stev. ssp. *arctica* (R. Br.) Pennell			•				S2	S2
Pedicularis oederi Vahl *ex* Hornem.			•					S1
Pedicularis racemosa Dougl. *ex* Hook.			•				S1	S1
Pedicularis sudetica Willd. ssp. *interioides* Hult.		•					S1	S1
Pellaea gastonyi Windham			•				S1	S1
Pellaea glabella Mett. *ex* Kuhn ssp. *occidentalis* (E.E. Nelson) Windham					•	•		S1
Pellaea glabella Mett. *ex* Kuhn ssp. *simplex* (Butters) A. Löve & D. Löve		•	•					S2
Penstemon fruticosus (Pursh) Greene var. *scouleri* (Lindl.) Cronq.			•				S2	S2
Phacelia linearis (Pursh) Holzinger			•		•	•	S2	S2
Phacelia lyallii (A. Gray) Rydb.			•				S2	S2
Phegopteris connectilis (Michx.) Watt			•	•			S2	S2

Appendix 2

Scientific Name	Canadian Shield	Boreal Forest	Rocky Mtn.	Foothills	Parkland	Grasslands	S Rank 1998	S Rank 2000
Philadelphus lewisii Pursh			•				S1	S1
Phlox gracilis (Hook.) Greene ssp. *gracilis*			•				S1S2	S1
Physocarpus malvaceus (Greene) Kuntze			•				S1	S1
Physostegia ledinghamii (Boivin) Cantino		•			•		S2	S2
Picradeniopsis oppositifolia (Nutt.) Rydb. *ex* Britt.						•	S1	S1
Pinguicula villosa L.	•	•					S1S2	S1
Pinus monticola Dougl. *ex* D. Don.			•				SU	SU
Plantago canescens J.E. Adams		•	•			•	S2	S2
Plantago maritima L.			•				S1	S1
Platanthera stricta Lindl.			•				S2	S2
Poa gracillima Vasey			•				S2	S2
Poa laxa Haenke ssp. *banffiana* Soreng			•				SR	S1
Poa leptocoma Trin.			•	•		•	S3	S3
Poa lettermanii Vasey			•				S1	S1
Poa nervosa (Hook.) Vasey var. *wheeleri* (Vasey) C.L. Hitchc.			•		•	•	S3	S3
Poa stenantha Trin.			•				SU	SU
Polanisia dodecandra (L.) DC. ssp. *trachysperma* (T. & G.) Iltis						•	S1S2	S2
Polygala paucifolia Willd.		•					S1	S1
Polygonum minimum S. Wats.			•		•		S1S2	S2
Polygonum polygaloides Meissn. ssp. *confertiflorum* (Nutt. *ex* Piper) Hickman			•			•	S2	S2
Polypodium hesperium Maxon			•				S1	S1S2
Polypodium sibiricum Siplivinsky	•						S1	S1?
Polypodium virginianum L.	•	•		•			S2	S2?
Populus angustifolia James						•	S2	S3
Potamogeton foliosus Raf.		•	•				S2	S2
Potamogeton natans L.	•	•	•	•	•		S2	S2
Potamogeton obtusifolius Mert. & W.D.J. Koch	•	•					S2	S2
Potamogeton praelongus Wulfen		•		•	•		S2	S2
Potamogeton robbinsii Oakes	•	•					S1	S1
Potamogeton strictifolius A. Bennett		•					S2	S2
Potentilla drummondii Lehm. ssp. *drummondii*			•				S2	S2
Potentilla finitima Kohli & Packer					•		S1	S1

Appendix 2

Scientific Name	Canadian Shield	Boreal Forest	Rocky Mtn.	Foot- hills	Park- land	Grass- lands	S Rank 1998	S Rank 2000
Potentilla hookeriana Lehm.	•		•	•			S2	S2
Potentilla macounii Rydb.		•					S1	S1
Potentilla multifida L.	•	•					S1	S1
Potentilla multisecta (S. Wats.) Rydb.			•				S1	S2
Potentilla paradoxa Nutt.						•	S2	S2
Potentilla plattensis Nutt.					•	•	S1?	S1
Potentilla subjuga Rydb.			•				S1	S1
Potentilla villosa Pallas *ex* Pursh			•				S2	S2
Prenanthes alata (Hook.) D. Dietr.				•			S1	S1
Prenanthes sagittata (A. Gray) A. Nels.			•		•		S2	S2
Primula egaliksensis Wormskj.			•	•			S2	S2
Primula stricta Hornem.			•				S1S2	S1
Psilocarphus elatior A. Gray						•	S2	S2
Psoralea argophylla Pursh			•		•	•	S3	S3
Pterospora andromeda Nutt.			•		•		S2	S3
Pyrola grandiflora Radius		•					S2	S2
Pyrola picta J.E. Smith			•				S1	S1
Ranunculus gelidus Karel. & Kiril. ssp. *grayi* (Britt.) Hult.			•				S3	S3
Ranunculus glaberrimus Hook.			•			•	S2	S2
Ranunculus nivalis L.			•				S1	S1
Ranunculus occidentalis Nutt. var. *brevistylis* Greene							S2	S2
Ranunculus uncinatus D. Don			•	•	•		S2	S2
Rhododendron lapponicum (L.) Wahlenb.			•				S2	S2
Rhynchospora capillacea Torr.		•			•		S1	S1
Ribes laxiflorum Pursh			•				S1S2	S2
Romanzoffia sitchensis Bong.			•				S2	S2
Rorippa curvipes Greene var. *curvipes*			•				SU	SU
Rorippa curvipes Greene var. *truncata* (Jepson) Rollins					•		S1	S1
Rorippa sinuata (Nutt. *ex* T. & G.) A.S. Hitchc.						•	S1	S1
Rorippa tenerrima E. Greene		•				•	S1	S1
Rumex paucifolius Nutt. *ex* Wats.		•					S1	S1
Ruppia cirrhosa (Petagna) Grande		•			•	•	S1	S1S2
Sagina decumbens (Ell.) T. & G.						•	S1	S1
Sagina nivalis (Lindbl.) Fries		•					SU	SU

Appendix 2

Scientific Name	Canadian Shield	Boreal Forest	Rocky Mtn.	Foot-hills	Park-land	Grass-lands	S Rank 1998	S Rank 2000
Sagina nodosa (L.) Fenzl ssp. *borealis* Crow	•						S1	S1
Sagittaria latifolia Willd.		•		•		•	S1	S1
Salix alaxensis (Anderss.) Coville ssp. *alaxensis*			•	•			S2	S2
Salix commutata Bebb			•	•			S2	S2
Salix lanata L. ssp. *calcicola* (Fern. & Wieg.) Hult.			•				S1	S1
Salix raupii Argus			•	•			S1	S1
Salix sitchensis Sanson *ex* Bong.		•	•				S1	S1
Salix stolonifera Coville			•				S1	S1
Salix tyrellii Argus		•						S1
Sarracenia purpurea L. ssp. *purpurea*		•					S1S2	S2
Saussurea americana D.C. Eat.			•				S1	S1
Saxifraga ferruginea Graham			•				S2	S2
Saxifraga flagellaris Willd. ssp. *setigera* (Pursh) Tolm.			•				S2	S2
Saxifraga nelsoniana D. Don ssp. *porsildiana* (Calder & Savile) Hult.			•				S2	S2
Saxifraga nivalis L.			•	•			S2	S2
Saxifraga odontoloma Piper			•				S1?	S1
Saxifraga oregana J.T. Howell var. *montanensis* (Small) C.L. Hitchc.			•				SU	SU
Schizachyrium scoparium (Michx.) Nash ssp. *scoparium*					•	•	S3	S3
Scirpus pallidus (Britt.) Fern.		•				•	S1	S1
Sedum divergens S. Wats.			•				S2	S2
Selaginella wallacei Hieron.			•				S1	S1
Senecio megacephalus Nutt.			•				S3	S3
Shinneroseris rostrata (A. Gray) S. Tomb					•	•	S2	S2
Silene involucrata (Cham. & Schlecht.) Bocquet ssp. *involucrata*			•				S1S2	S1S2
Sisyrinchium septentrionale Bicknell			•	•	•	•	S2S3	S2S3
Sparganium fluctuans (Engelm. *ex* Morong) B.L. Robins.		•					S1	S1
Sparganium glomeratum (Beurling *ex* Laest.) L.M. Neuman		•					S1	S1
Sparganium hyperboreum Beurling *ex* Laest.			•	•			S1	S1

Appendix 2

Scientific Name	Canadian Shield	Boreal Forest	Rocky Mtn.	Foot-hills	Park-land	Grass-lands	S Rank 1998	S Rank 2000
Spartina pectinata Link		•				•	S1	S1
Spergularia salina J. & K. Presl var. *salina*		•	•		•	•	S2	S2
Sphenopholis obtusata (Michx.) Scribn.				•	•	•	S2	S2
Spiraea splendens Baumann ex K. Koch var. *splendens*			•				S1	S1
Spiranthes lacera (Raf.) Raf. var. *lacera*		•					S1	S1
Stellaria americana (Porter *ex* B.L. Robins.) Standl.			•				S1	S1
Stellaria arenicola Raup	•						S1	S1
Stellaria crispa Cham. & Schlecht.		•	•	•			S2	S2
Stellaria obtusa Engelm.			•				S1	S1
Stellaria umbellata Turcz. *ex* Kar. & Kir.			•				S1	S1
Stephanomeria runcinata Nutt.						•	S2	S2
Streptopus roseus Michx.		•		•			S1	S1
Streptopus streptopoides (Ledeb.) Frye & Rigg				•			S1	S1
Suaeda moquinii (Torr.) Greene						•	S2	S2
Suckleya suckleyana (Torr.) Rydb.						•	S1	S1
Suksdorfia ranunculifolia (Hook.) Engl.			•				S1S2	S2
Suksdorfia violacea A. Gray			•				S1	S1
Tanacetum bipinnatum (L.) Schultz-Bip. ssp. *huronense* (Nutt.) Breitung	•						S1	S1
Taxus brevifolia Nutt.			•				S1	S1
Tellima grandiflora (Pursh) Dougl. *ex* Lindl.			•				S1	S1
Thelesperma subnudum A. Gray var. *marginatum* (Rydb.) T.E. Melchert *ex* Cronq.						•	S1	S1
Thuja plicata Donn *ex* D. Don			•				S1S2	S1S2
Torreyochloa pallida (Torr.) Church var. *pauciflora* (J. Presl.) J.I. Davis					•		S1	S1
Townsendia condensata Parry *ex* A. Gray			•				S1	S2
Townsendia exscapa (Richards.) Porter						•	S1S2	S2
Tradescantia occidentalis (Britt.) Smyth var. *occidentalis*						•	S1	S1

Appendix 2

Scientific Name	Canadian Shield	Boreal Forest	Rocky Mtn.	Foot- hills	Park- land	Grass- lands	S Rank 1998	S Rank 2000
Triantha occidentalis (S. Wats.) Gates ssp. *brevistyla* (C.L. Hitchc.) Packer			•				S1	S1
Triantha occidentalis (S. Wats.) Gates ssp. *montana* (C.L. Hitchc.) Packer			•				S1	S1
Trichophorum clintonii (A. Gray) S.G. Smith		•		•			S1	S1
Trichophorum pumilum (Vahl) Schinz & Thellung		•	•	•			S2	S2
Trillium ovatum Pursh			•				S1	S1
Tripterocalyx micranthus (Torr.) Hook.						•	S2	S2
Trisetum cernuum Trin. var. *canescens* (Buckl.) Beal			•				S1	S1
Trisetum cernuum Trin. var. *cernuum*			•		•		S2?	S2
Trisetum montanum Vasey			•				S1	S1
Trisetum wolfii Vasey			•		•		S1	S1
Tsuga heterophylla (Raf.) Sarg.			•				S1	S1
Utricularia cornuta Michx.	•	•					S1	S1
Vaccinium ovalifolium J.E. Smith			•				S2	S2
Vaccinium uliginosum L.	•	•					S2	S2
Veronica catenata Pennell						•	S1S2	S1S2
Veronica serpyllifolia L.			•	•	•		S3	S3
Viola pallens (Banks *ex* DC.) Brainerd		•	•				S1S2	S1
Viola pedatifida G. Don					•	•	S2	S2
Viola praemorsa Dougl. *ex* Lindl. var. *linguifolia* (Nutt.) M.S. Baker & J.C. Klausen *ex* M.E. Peck			•				S2	S2
Wolffia columbiana Karsten		•					S2	S2
Woodsia glabella R. Br. *ex* Richardson			•				S1	S1
Yucca glauca Nutt. *ex* Fraser						•	S1	S1
TOTALS: 488	33	114	312	76	101	124		

Appendix 3

APPENDIX 3
SPECIES IN THIS GUIDE THAT ARE NOT FOUND IN THE *FLORA OF ALBERTA* (SECOND EDITION)
The following list of native vascular plants is new to the flora of Alberta since Moss (1983). The additions are of two types: those representing new discoveries, and others resulting from taxonomic revision to existing groups. This list includes only those taxa that have been documented by specimens and does not include those reported without supporting documentation.

New Discoveries
Agoseris lackschewitzii D. Henderson & R. Moseley
Aruncus dioicus (Walt.) Fern.
Bacopa rotundifolia (Michx.) Wettst.
Botrychium pinnatum H. St. John
Brasenia schreberi J.F. Gmelin
Carex pedunculata Muhl.
Carex supina Willd. *ex* Wahlenb.
Erigeron lackschewitzii Nesom & Weber
Galium bifolium S. Wats.
Isoetes maritima L. Underwood
Isoetes occidentalis L. f. Henderson
Isoetes x *truncata* (A.A. Eaton) Clute
Juncus regelii Buch.
Lewisia rediviva (Gray) B.L. Robins.
Lupinus wyethii S. Wats.
Luzula groenlandica Bocher
Luzula rufescens Fisch. *ex* E. Mey.
Mimulus brewerii (Greene) Coville
Mimulus glabratus Kunth
Nuttallanthus canadensis (L.) D.A. Sutton
Pedicularis oederi Vahl *ex* Hornem.
Poa laxa Haenke ssp. *banffiana* Soreng
Rorippa curvipes Greene var. *truncata* (Jepson) Rollins
Salix raupii Argus
Salix tyrellii Argus
Scirpus rufus (Huds.) Schrad.
Spiranthes lacera (Raf.) Raf. var. *lacera*
Tellima grandiflora (Pursh) Dougl. *ex* Lindl.
Tradescantia occidentalis (Britt.) Smyth var. *occidentalis*
Wolffia borealis (Engelm. *ex* Hegelm.) Landolt *ex* Landolt & Wildi
Wolffia columbiana Karst

Appendix 3

Changes Resulting from Taxonomic Revision

Antennaria aromatica Evert (newly described, Bayer 1989)

Botrychium campestre W.H. Wagner & Farrar *ex* W.J. & F. Wagner (newly described, Wagner and Wagner 1986)

Botrychium crenulatum W.H. Wagner (newly described, Wagner and Wagner 1981)

Botrychium michiganense

Botrychium minganense Victorin (formerly included with description of *B. dusenii*)

Botrychium paradoxum W.H. Wagner (newly described, Wagner and Wagner 1981)

Botrychium pedunculosum W.H. Wagner (newly described, Wagner and Wagner 1986)

Botrychium spathulatum W.H. Wagner (newly described, Wagner and Wagner 1990)

Botrychium x *watertonense* W.H. Wagner (newly described, Wagner et al. 1984)

Carex utriculata Boott (formerly recognized as *C. rostrata* Stokes)

Cerastium brachypodum (Engelm. *ex* A. Gray) B.L. Rob. (a segregate from *C. nutans*)

Festuca hallii (Vasey) Piper (formerly included within *F. scabrella* Trin.)

Festuca minutiflora Rydberg (formerly identified as *F. brachyphylla* Schultes)

Festuca viviparoidea Krajina *ex* Pavlick ssp. *krajinae* Pavlick (formerly *F. vivipara* (L.) Sm. ssp. *glabra* Frederiksen)

Gymnocarpium disjunctum (Ruprecht) Ching (formerly included within *G. dryopteris* (L.) Newman; see *Flora of North America* 1993)

Huperzia haleakalae (Brachenridge) Holub (formerly included within *Lycopodium selago*; see *Flora of North America* 1993)

Mimulus tilingii Regel (formerly included within *M. guttatus*)

Nymphaea tetragona Georgi ssp. *tetragona*

Oxytropis campestris (L.) DC. var *davisii* Welsh

Pellaea gastonyi Windham (formerly included within *P. atropurpurea* (L.) Link; see *Flora of North America* 1993)

Polypodium sibiricum Siplivinsky (formerly included within *P. virginianum* L.; see *Flora of North America* 1993)

Triantha occidentalis (S. Wats.) Gates ssp. *brevistyla* (C.L. Hitchc.) Packer (Packer 1993)

Triantha occidentalis (S. Wats.) Gates ssp. *montana* (C.L. Hitchc.) Packer (Packer 1993)

Appendix 4

APPENDIX 4
TAXA PREVIOUSLY REPORTED AS RARE FOR ALBERTA BUT NOT INCLUDED IN THIS BOOK

This list includes taxa that have been considered rare in other publications on Alberta's flora. Many of the taxa are now known to have more than twenty occurrences in the province and large population sizes; they are no longer considered rare. Several other taxa are included on a watch list. These are taxa that have restricted distributions within Alberta but are common within their range. Information is collected to ascertain trends in populations. A taxon may move from the watch list to the tracking list if information suggests that the taxon is in decline. Taxa that are currently on the watch list are indicated with a w.

SCIENTIFIC NAME	SOURCE	COMMENTS
Agoseris grandiflora (Nutt.) Greene	1	lacks documentation
Agropyron violaceum Hornem. Lange ssp. *andinum* (Scribn. & Melderis)	1	
Alopecurus carolinianus Walt.	3	w
Androsace occidentalis Pursh	1, 2	
Angelica dawsonii Wats.	1, 3	w
Antennaria alpina (L.) Gaetn. var. *media* (Greene) Jeps	1	
Antennaria dimorpha (Nutt.) Torr. & Gray	2, 3	
Arnica alpina (L.) Olin ssp. *angustifolia* (M. Vahl) Maguire	1	
Aruncus sylvester Kostel. *ex* Maxim.	2, 3	introduced
Balsamorhiza sagittata (Pursh) Nutt.	1	
Braya humilis (C.A. Mey.) Robins. var. *maccallae* Harris	3	not a published name
Bupleurum americanum C. & R.	1, 2, 3	w
Calochortus apiculatus Baker	1, 3	w
Carex athabascensis F.J. Herm. (*C. scirpoidea* Michx.)	1, 3	
Carex geyeri Boott	2, 3	w
Castilleja hispida Benth.	2, 3	
Ceanothus velutinus Dougl.	1, 2, 3	w
Chrysothamnus nauseosus (Pall.) Britt.	1	
Crataegus douglasii Lindl.	1, 2, 3	w
Disporum hookeri (Torr.) Nichols.	2, 3	w
Disporum oreganum Wats.	1, 2	included within *D. hookeri* (Torr.) Nichols.
Draba verna L.	1	introduced
Dryopteris fragrans (L.) Schott	1, 2, 3	w
Festuca viridula (Vasey)	1	no Alberta records known
Fragaria vesca L. var. *bracteata* (Heller) Davis	1	
Galium palustre L.	1	no Alberta records known
Gaultheria humifusa (Graham) Rydb.	1	
Gentiana calycosa Griseb. var. *obtusiloba* (Rydb.) C.L. Hitche.	1, 2, 3	w
Habenaria unalaschensis (Spreng.) S. Wats. (*Piperia unalascensis* (Spreng.) Rydb.)	2, 3	

Appendix 4

Scientific Name	Source	Comments
Haplopappus uniflorus (Hook.) T. & G.	2	W
Hedeoma hispidum Pursh	2, 3	
Ledum glandulosum Nutt.	2, 3	
Lesquerella alpina (Nutt.) Wats.	1	
Lomatium sandbergii C. & R.	1, 2, 3	W
Lonicera utahensis S. Wats.	2, 3	
Luetkea pectinata (Pursh) Kuntze	1	W
Lupinus arbustus Dougl. ssp. *pseudoparviflorus* (Rydb.) Dunn	1	no Alberta records known
Lupinus nootkatensis Donn	1	W
Lupinus pusillus Pursh	2	
Luzula arcuata (Wahl.) Sw.	1	
Luzula hitchcockii L. Hamet-Ahti	2, 3	W
Melica subulata (Griseb.) Scribn.	2, 3	
Minuartia nuttallii (Pax) Briq.	1, 2	W
Mirabilis nyctaginea (Michx.) MacM.	2, 3	introduced
Myosurus aristatus Benth.	1, 3	
Osmorhiza occidentalis (Nutt.) Tor.	1, 3	W
Pachistima myrsinites (Pursh) Raf.	1, 2, 3	W
Pedicularis macrodonta Richards	1	included within *P. parviflora* S. ex Rees
Penstemon albertinus Greene	1, 3	W
Penstemon eriantherus Pursh	3	W
Penstemon lyallii Gray	1	W
Phlox alyssifolia Greene	1, 3	
Plantago patagonica Jacq. var. *spinulosa* (Dcne.) Gray	1	
Poa lanata Scribn. & Merr.	1	no Alberta records known
Polemonium viscosum Nutt.	1, 2, 3	
Potamogeton pusillus L. var. *tenuissimus* Mert. & Koch	1	
Potentilla yukonensis Hult. (*P. anserina* L.)	1	
Pteridium aquilinum (L.) Kuhn var. *pubescens* Underw.	1, 2, 3	W
Pyrola bracteata Hook.	1, 2, 3	W
Ribes inerme Rydb.	1, 2, 3	
Ribes viscosissimum Pursh	1, 2, 3	W
Salix boothii Dorn	2, 3	
Saxifraga debilis Engelm. (*S. hyperborea* R. Br.)	1	
Saxifraga mertensiana Bong.	1, 2, 3	W
Schedonnardus paniculatus (Nutt.) Trel.	1, 2, 3	
Scirpus cyperinus (L.) Kunth	2	
Scirpus nevadensis S. Wats.	2, 3	

Appendix 4

Scientific name	Source	Comments
Sedum douglasii Hook. (*S. stenopetalum* Pursh)	2	W
Selaginella rupestris (L.) Spring	2, 3	
Senecio conterminus Greenm.	1, 3	W
Senecio hydrophiloides Rydb.	1, 2	
Silene douglasii Hook.	1	lacks documentation
Sorbus sitchensis M. Roemer		
Sporobolus neglectus Nash	3	introduced
Thalictrum dasycarpum Fisch. & Ave-Lall.	2	
Tiarella trifoliata L.	1	W
Tofieldia coccinea Rich.	1	no Alberta records known
Viola glabella Nutt.	2, 3	
Viola selkirkii Pursh	1, 2, 3	
Vulpia octoflora (Walt.) Rydb.	2	
Xerophyllum tenax (Pursh.) Nutt.	2, 3	W

1 Argus, G W. and D.J. White, 1978
2 Packer, J.G. and C.E. Bradley, 1984
3 Federation of Alberta Naturalists, 1992

Appendix 5
Species reported from or expected to occur in Alberta but not yet verified

Several species have been reported from Alberta but lack supporting documentation (i.e., a specimen). Other species occur in close proximity to the province and should be expected to occur here in the appropriate habitat.

Species reported for Alberta but lacking sufficient documentation

Carex piperi Mack.
Carex sitchensis Prescott *ex* Bong. (*Carex aquatilis* Wahlenb. var. *dives* (Holm) Kukenth.)
Cerastium beeringianum Cham. & Schlecht. ssp. *terrae-novae* (Fern. & Wieg.) Hult.
Cornus unalaschkensis Ledeb.
Cryptantha affinis (A. Gray) Greene
Eleocharis nitida Fern.
Elymus vulpinus Rydb.
Erigeron uncialis Blake
Eriogonum pauciflorum Pursh
Festuca lenensis Drob.
Hackelia ciliata (Dougl. *ex* Lehm.) I.M. Johnson
Melica hitchcockii Boivin
Minuartia yukonensis Hulten
Phippsia algida (Phipps) R. Br.
Saxifraga subapetala E. Nels.
Senecio integerrimus Nutt. var. *ochroleucus* (Gray) Cronq.
Stellaria nitens Nutt.
Torreyochloa pallida Church var. *fernaldiii* (Hitchc.) Dore

Species that may occur in Alberta

Achillea millefolium L. var. *megacephala* (Raup) Boivin
Carex bicolor Bellardi *ex* All.
Erigeron ochroleucus Nutt. var. *ochroleucus* Nutt.
Isoetes howellii Engelm.
Minuartia rossii (R. Br.) Graebn.
Ranunculus eschscholtzii Schlecht. var. *suksdorfii* (A. Gray) L.D. Benson
Rhynchospora alba (L.) Vahl

Appendix 6
Alberta taxa classified as nationally rare by Argus and Pryer (1990)

The following taxa are reported to occur in Alberta by Argus and Pryer and were considered to be nationally rare in 1990. These taxa are all considered provincially rare unless otherwise indicated (x). Many of those that are not considered rare at this time are included on a watch list. These taxa have restricted distributions within Alberta but are common within their range. Information is collected to ascertain trends in populations. A taxon may move from the watch list to the tracking list if information suggests that the taxon is in decline. Taxa that are currently on the watch list are indicated with a w.

Species Name	Synonym(s)	Notes
Achillea millefolium L. var. *megacephala* (Raup) Boivin		x
Allium geyeri S. Wats. var. *geyeri*	*Allium rubrum* Osterh., *Allium rydbergii* Macbr.	
Angelica dawsonii S. Wats.		w
Antennaria corymbosa E. Nels.		
Aquilegia jonesii Parry		
Arnica longifolia D.C. Eat.		
Astragalus kentrophyta A. Gray var. *kentrophyta*		
Atriplex canescens (Pursh) Nutt.	*Atriplex aptera* A. Nels.	
Atriplex powellii S. Wats.		
Bacopa rotundifolia (Michx.) Wettst.		
Boisduvalia glabella (Nutt.) Walp.	*Epilobium pygmaeum* (Speg.) P. Hoch & Raven	
Botrychium ascendens Wagner		
Botrychium campestre W.H. Wagner & D.R. Farrar		
Botrychium hesperium (Maxon & R.T. Clausen) W.H. Wagner & Lellinger	*Botrychium matricariifolium* (Döll) W.D.J. Koch ssp. *hesperium* Maxon & R.T. Clausen	
Botrychium paradoxum W.H. Wagner		
Botrychium pedunculosum W.H. Wagner		
Boykinia heucheriformis (Rydb.) Rosendahl	*Telesonix heucheriformis* Rydb., *Telesonix jamesii* (Torr.) Raf.	
Braya humilis (C.A. Mey.) B.L. Robins. var. *maccallae* Harris in ed.	Not a published name	x
Brickellia grandiflora (Hook.) Nutt.		
Calochortus apiculatus Baker		w
Camassia quamash (Pursh) Greene var. *quamash*		
Camissonia andina Nutt. (Raven)	*Oenothera andina* Nutt.	
Camissonia breviflora (Torr. & Gray) Raven	*Oenothera breviflora* T. & G., *Taraxia breviflora* (T. & G.) Nutt.	

Appendix 6

Species Name	Synonym(s)	Notes
Carex enanderi Hult.	*Carex lenticularis* Michx. var. *dolia* (M.E. Jones) L.A. Standley	
Carex epapillosa F.J. Herm.	*Carex heteroneura* Boott var. *epapillosa* (Mack.) F.J. Hermann	
Carex geyeri Boott		w
Carex nebrascensis Dewey	*Carex nebraskensis* Dewey, *Carex jamesii* T. & G.	
Carex paysonis Clokey		
Castilleja cusickii Greenm.		w
Chenopodium leptophyllum (Nutt. *ex* Moq.) S. Wats.		
Chenopodium subglabrum (S. Wats.) A. Nels.		
Chenopodium watsonii A. Nels.		
Cirsium scariosum Nutt.	*Cirsium hookerianum* Nutt., *Cirsium foliosum* (Hook.) DC., *Cnicus scariosus* (Nutt.) A. Gray, *Carduus scariosus* (Nutt.) Heller	
Conimitella williamsii (D.C. Eaton) Rydb.		
Crepis occidentalis Nutt.		
Cryptantha affinis (Gray) Greene		x
Cryptantha kelseyana Greene		x
Cryptantha minima Rydb.		
Dichanthelium wilcoxianum (Vasey) Freckman	*Panicum wilcoxianum* Vasey, *Panicum oligosanthes* Schultes, *Dichanthelium oligosanthes* (Schultes) Gould var. *wilcoxianum* (Vasey) Gould & Clark	
Douglasia montana A. Gray		
Downingia laeta Greene		
Draba densifolia Nutt *ex* T. & G.		
Draba kananaskis Mulligan		
Draba ventosa A. Gray		
Eleocharis engelmannii Steud.	*Eleocharis ovata* (Roth) R. & S. var. *engelmanii* Britt., *Eleocharis obtusa* (Willd.) Schultes var. *engelmanii* (Steud.) Gilly	
Elodea bifoliata St. John	*Elodea longivaginata* St. John	
Epilobium glaberrimum Barbey ssp. *fastigiatum* (Nutt.) Hoch & Raven	*Epilobium platyphyllum* Rydb., *Epilobium fastigiatum* var. *glaberrimum* (Barbey) Piper, *Epilobium affine* Bong. var. *fastigiatum* Nutt.	
Epilobium halleanum Hausskn.	*Epilobium glandulosum* Lehm. var. *macounii* (Trel.) C.L. Hitchc.	

Appendix 6

Species Name	Synonym(s)	Notes
Epilobium mirabile Trel.	*Epilobium clavatum* Trel. var. *glareosum* (G.N. Jones) Munz	
Erigeron ochroleucus Nutt. var. *scribneri* (Canby) Cronq.		
Erigeron radicatus Hook.		
Erigeron trifidus Hook.		
Eriogonum cernuum Nutt.		
Festuca minutiflora Rydberg	*Festuca brevifolia* var. *endotera* Saint-Yves, *Festuca brevifolia* var. *utahensis* Saint-Yves, *Festuca ovina* var. *minutiflora* (Rydberg) Howell	
Galium bifolium S. Wats.		
Gayophytum racemosum T. & G.	*Gayophytum helleri* Rydb. var. *glabrum* Munz	
Gentiana calycosa Griseb.		w
Gymnocarpium jessoense (Koidz.) Koidz.		
Hackelia ciliata (Dougl.) Johnst.		x
Halimolobus virgata (Nutt.) O.E. Schulz		
Haplopappus uniflorus (Hook.) Torr. & Gray ssp. *uniflorus*		w
Hymenopappus filifolius Hook. var. *polycephalus* (Osterh.) B.L. Turner	*Hymenopappus polycephalus* Osterh.	
Hypericum scouleri Hook.	*Hypericum formosum* HBK. var. *scouleri* (Hook.) C.L. Hitchc.	
Iliamna rivularis (Dougl. *ex* Hook.) Greene	*Sphaeralcea rivularis* (Dougl. *ex* Hook.) Torr. *ex* Gray	
Iris missouriensis Nutt.		
Isoetes bolanderi Engelm.		
Juncus nevadensis S. Wats. var. *nevadensis*		
Juncus regelii Buch.		
Lewisia pygmaea (A. Gray) B.L. Robins. ssp. *pygmaea*		
Lilaea scilloides (Poir.) Hauman	*Lilaea subulata* Humb. & Bonpl.	
Lomatium cous (S. Wats.) Coult. & Rose	*Lomatium montanum* Coult. & Rose	
Lomatium sandbergii (Coult. & Rose) Coult. & Rose		w
Lupinus lepidus Dougl. *ex* Lindl.		
Lupinus minimus Dougl. *ex* Hook.		
Lupinus wyethii S. Wats.		
Luzula hitchcockii Hamey-Ahti		w
Machaeranthera tanacetifolia (HBK.) Nees	*Aster tanacetifolius* Kunth	
Malaxis paludosa (L.) Sw.	*Ophrys paludosa* L., *Hammarbya paludosa* (L.) Kuntze	

Appendix 6

Species Name	Synonym(s)	Notes
Melica spectabilis Scribn.		
Mimulus breweri (Greene) Rydb.		
Mimulus glabratus HBK.		
Minuartia nuttallii (Pax) Briq.		w
Munroa squarrosa (Nutt.) Torr.	*Monroa squarrosa*	
Nemophila breviflora A. Gray		
Nothocalais cuspidata (Pursh) Greene	*Microseris cuspidata* (Pursh) Schultz-Bip., *Agoseris cuspidata* (Pursh) Raf.	
Osmorhiza occidentalis (Nutt. *ex* Torr. & Gray) Torr.		w
Oxytropis lagopus Nutt. var. *conjugens* Barneby		
Papaver freedmanianum Löve	*Papaver radicatum* Rottboll ssp. *kluanensis* (D. Löve) D.F. Murray	
Papaver pygmaeum Rydb.	*Papaver alpinum* L., *Papaver nudicaule* L. ssp. *radicatum* (Rottb.) Fedde var. *pseudocorydalifolium*, *Papaver radicatum* Rottb. var. *pygmaeum* (Rydb.) S.L. Welsh	
Pellaea gastonyi Windham		
Penstemon albertinus Greene		w
Penstemon eriantherus Pursh var. *eriantherus*		w
Penstemon lyallii (Gray) Gray		w
Phacelia lyallii (A. Gray) Rydb.		
Phlox alyssifolia Greene		x
Physocarpus malvaceus (Greene) Kuntze		
Poa nevadensis Vasey *ex* Scribn.		x
Polygonum douglasii Greene ssp. *austinae* (Greene) E. Murray		x
Polygonum engelmannii Greene	*Polygonum douglasii* Greene ssp. *engelmannii* (Greene) Kartesz & Gandhi	x
Polygonum polygaloides Meissn. ssp. *confertiflorum* (Nutt. *ex* Piper) Hickman	*Polygonum watsonii* Small, *Polygonum imbricatum* Nutt., *Polygonum kelloggii* Greene var. *confertiflorum* (Nutt. *ex* Piper) Dorn, *Polygonum confertiflorum* Nutt. *ex* Piper	
Prenanthes sagittata (A. Gray) A. Nels.		
Psilocarphus elatior A. Gray	*Psilocarphus oregonus* Nutt. var. *elatior* A. Gray	

Appendix 6

Species Name	Synonym(s)	Notes
Puccinellia distans (L.) Parl. ssp. *hauptiana* (Krecz.) W.E. Hughes		
Ranunculus verecundus B.L. Robins.		x
Rorippa tenerrima E. Greene		
Rorippa truncata (Jepson) R. Stuckey	*Rorippa curvipes* Greene var. *truncata* (Jepson) Rollins	
Rumex paucifolius Nutt. *ex* Wats.		
Salix raupii Argus		
Saxifraga oregana Howell		
Senecio conterminus Greenm.		w
Senecio cymbalarioides Buek non Nutt.		
Senecio foetidus T.J. Howell var. *hydrophiloides* (Rydb.) T.M. Barkl. *ex* Cronq.		x
Senecio megacephalus Nutt.		
Sparganium glomeratum Laest.		
Stellaria americana (Porter *ex* B.L. Robins.) Standl.		
Stellaria arenicola Raup		
Stellaria obtusa Engelm.		
Stellaria umbellata Turcz. *ex* Kar. & Kir.		
Stephanomeria runcinata Nutt.		
Suaeda moquinii (Torr.) Greene	*Suaeda intermedia* S. Wats.	
Tanacetum bipinnatum (L.) Schultz-Bip. ssp. *huronense* (Nutt.) Breitung	*Tanacetum huronense* Nutt. var. *bifarium* Fern., *Chrysanthemum bipinnatum* L. var. *huronense* (Nutt.) Hult.	
Thelesperma subnudum A. Gray var. *marginatum* (Rydb.) T.E. Melchert *ex* Cronq.	*Thelesperma marginatum* Rydb. (Rydb.) T.E. Melchert *ex* Cronq.	
Townsendia condensata D.C. Eat.		
Tradescantia occidentalis (Britt.) Smyth		
Trichophorum pumilum (Vahl) Schinz & Thellung	*Scirpus rollandii* Fern., *Scirpus pumilus* Vahl ssp. *rollandii* (Fern.) Raymond	
Tripterocalyx micranthus (Torr.) Hook.	*Abronia micrantha* Torr.	
Trisetum montanum Vasey		
Trisetum wolffii Vasey		
Viola praemorsa Dougl. *ex* Lindl. ssp. *linguifolia* (Nutt.) M.S. Baker & J.C. Klausen *ex* M.E. Peck	*Viola nuttallii* Pursh var. *linguaefolia* (Nutt.) Jepson & J.C. Klausen *ex* M.E. Peck	
Xerophyllum tenax (Pursh) Nutt.		w

Appendix 7

APPENDIX 7
RARE NATIVE PLANT REPORT FORM

Please enter all information available to you and attach a detailed sketch or map showing the location of the population.

SCIENTIFIC NAME: _____
COMMON NAME: _____
OBSERVER NAME, ADDRESS AND TELEPHONE NUMBER: _____
_____
_____

OBSERVATION DATE(S): _____
PHOTOGRAPH TAKEN: Y / N
SPECIMEN COLLECTED: Y / N   **COLLECTION NUMBER:** _____
IF YES, NAME HERBARIUM WHERE DEPOSITED: _____

LOCATION INFORMATION:
SITE NAME: _____

TOPOGRAPHIC MAP NUMBER: _____
DIRECTIONS TO POPULATION (*include descriptions of landmarks and distances if possible*):
_____
_____

ELEVATION (*Please do not use elevation from GPS unit*): _____ ft/m (*circle one*)

(*Complete one of the following and accompany with map or sketch*)
UTM EASTING: _____ UTM NORTHING: _____ GRID ZONE: _____
NORTH AMERICAN DATUM: 27 / 83 (*circle one*)

LEGAL: TWP: _____ RGE: _____ W: _____ M   SECTION: _____ LSD: _____

LATITUDE: _____ LONGITUDE: _____
Was the location determined using a GPS? Y / N

POPULATION INFORMATION (*include information on extent in cm^2/m^2 (circle one), number of individuals*):
_____
_____
_____

PHENOLOGY (*based on average development phase of population--see over*):
_____ vegetative; _____ reproductive (*E.G. V6. R7 for a species with leaves fully unfolded and in full bloom*)

SITE/HABITAT DESCRIPTION (*include information on habitat [alpine, aquatic, cliff, forest, grassland, peatland], plant communities / dominant species / associated species / other rare species / substrate / soils / phenology of dominant species*):
_____
_____
_____

ASPECT: _____ SLOPE: _____ MOISTURE: _____
OWNERSHIP (*if known. Include name/address/phone number*): _____
_____

CURRENT LAND USE: _____
_____

HABITAT THREATS/MANAGEMENT CONCERNS: _____
_____
_____
_____

Return to: *Alberta Natural Heritage Information Centre, 2nd Floor, 9820 106 Street, Edmonton, AB T5K 2J6 (780) 427-5209. Thank You.*

Appendix 7

Phenology Codes (after Dierschke, 1972)

VEGETATIVE		REPRODUCTIVE

Deciduous Tree or Shrub

0 Closed Bud
1 Buds with green tips
2 Green leaf out but not unfolded
3 Leaf unfolding up to 25%
4 Leaf unfolding up to 50%
5 Leaf unfolding up to 75%
6 Full leaf unfolding
7 First leaves turned yellow
8 Leaf yellowing up to 50%
9 Leaf yellowing over 50%
10 Bare

Conifer

0 Closed Bud
1 Swollen bud
2 Split bud
3 Shoot capped
4 Shoot elongate
5 Shoot full length, lighter green
6 Shoot mature, equally green

0 Without blossom buds
1 Blossom buds recognizable
2 Blossom buds strongly swollen
3 Shortly before flowering
4 Beginning flowering
5 In bloom up to 25%
6 In bloom up to 50%
7 Full bloom
8 Fading
9 Completely faded
10 Bearing green fruit
11 Bearing ripe fruit
12 Bearing overripe fruit
13 Fruit or seed dispersal

Herbs

0 Without shoots above ground
1 Shoots without unfolded leaves
2 First leaf unfolds
3 2 or 3 leaves unfolded
4 Several leaves unfolded
5 Almost all leaves unfolded
6 Plant fully developed
7 Stem and/or first leaves fading
8 Yellowing up to 50%
9 Yellowing over 50%
10 Dead

0 Without blossom buds
1 Blossom buds recognizable
2 Blossom buds strongly swollen
3 Shortly before flowering
4 Beginning bloom
5 Up to 25% in blossom
6 Up to 50% in blossom
7 Full bloom
8 Fading
9 Completely faded
10 Bearing green fruit
11 Bearing ripe fruit
12 Bearing overripe fruit
13 Fruit or seed dispersal

Grasses

0 Without shoots above ground
1 Shoots without unfolded leaves
2 First leaf unfolded
3 2 or 3 leaves unfolded
4 Beginning development of blades of grass
5 Blades partly formed
6 Plant fully developed
7 Blades and/or first leaves turning yellow
8 Yellowing up to 50%
9 Yellowing over 50%
10 Dead

0 Without recognizable inflorescence
1 Inflorescence recognizable, closed
2 Inflorescence partly visible
3 Inflorescence fully visible, not unfolded
4 Inflorescence unfolded
5 First blooms pollenizing
6 Up to 50% pollenized
7 Full bloom
8 Fading
9 Fully faded
10 Bearing fruit
11 Fruit or seed dispersal

Ferns

0 Without shoots above ground
1 Rolled fronds above ground
2 First frond unfolds
3 2 or 3 fronds unfold
4 Several fronds unfolded
5 Almost all fronds unfolded
6 Plant fully developed
7 First fronds fading
8 Yellowing up to 50%
9 Yellowing over 50%
10 Dead

0 sori absent
1 sori green, forming
2 sori mature, darker, drier
3 sori depressing, strobili forming in lycopodium

Glossary

achene: a small, dry, thin-walled, single-seeded fruit (see Figs. 2, 12, 15, 29)

acrid: bitter and pungent

agamospermous: producing seeds asexually (without fertilization)

apogamous: producing seeds asexually (without fertilization)

alkaloid: a nitrogenous, organic base; many alkaloids have strong physiological effects (e.g., morphine, strychnine)

alternate: attached singly rather than in pairs or whorls (see Figs. 1, 13); cf. opposite, whorled

androgynous: with male flowers at the tip of the spike and female flowers at the base; cf. gynaecandrous (see Fig. 15)

annual ring: the wood laid down in a single year in a tree trunk, visible as a ring when viewed in cross-section because of alternating dense winter wood and less-dense summer wood

annual: completing its life cycle in a single year; cf. biennial, perennial

anther: the pollen-bearing part of a stamen (see Figs. 2, 3, 4, 6, 7, 12, 13, 15, 16)

aquatic: living in water

armed: bearing spines, prickles or thorns

ascending: growing obliquely upwards (see Fig. 26)

auricle: a small, blunt or pointed, ear-like lobe, usually at the base of leaf blade (see Figs. 1, 3, 20)

awn: a slender, bristle-like appendage (see Figs. 16, 23)

axil: the angle between 2 organs, such as between a leaf and stem (see Fig. 1)

axis: the main stem of a plant or the central line of an organ

banner: the upper petal of a 2-lipped flower of the pea family; a standard (see Fig. 6)

beak: a prolonged, more or less slender tip on a thicker organ such as fruit or seed (see Figs. 15, 29)

bearded: conspicuously hairy, usually with long, stiff hairs (see Fig. 8)

berry: a fleshy, single- to many-seeded fruit developed from a single ovary (see Fig. 29)

biennial: living for 2 years, often producing a rosette of leaves the first year and flowering and fruiting the second year; cf. annual, perennial

bilaterally symmetric: divisible into equal halves along 1 line only (see Fig. 8); zygomorphic; cf. radially symmetric

bipinnate: twice pinnately divided (see Fig. 26)

bisexual flower: with both male and female sex organs (see Fig. 15); cf. perfect flower

biternate: twice divided in 3s

blade: the broad, flat part of an organ, usually of a leaf or petal (see Figs. 1, 16, 20, 21)

Glossary

bloom: a whitish, often waxy powder covering the surface, usually on leaves and fruits

bog: an acidic, nutrient-poor wetland dominated by heath shrubs and *Sphagnum* mosses

bract: a small, specialized leaf or scale below a flower or flower cluster (see Figs. 13, 15, 27)

bracteole or bractlet: a small bract

bulb: a short, vertical underground stem covered with layers of leaves (often fleshy) or leaf-bases (e.g., an onion) (see Fig. 30)

bulblet or bulbil: a small bulb-like structure, often located in a leaf axil or replacing a flower

bur: a fruit or compact cluster of fruits with barbs or bristles that enable it to attach to passing animals (see Fig. 29)

calcareous: calcium-rich; rich in lime

calciphile: a calcium-lover, preferring sites rich in calcium

callosity: a small, grain-like projection on the valve of a dock or sorrel (*Rumex*) flower

callus: a small, hard, thickening, usually at the base of a lemma in grasses

callus hair: the fine hair attached to the callus at the base of a lemma, usually tufted (see Fig. 16)

calyx [calyxes]: the outer ring of the petals and sepals; the sepals collectively (see Figs. 3, 9)

capitate: head-like, forming a dense cluster

capsule: a dry fruit that splits open at maturity and is produced by a compound ovary (see Figs. 13, 14, 29)

carpel: a fertile leaf bearing the undeveloped seed(s) (see Fig. 2); 1 or more fused carpels form a pistil

carpophore: a stalk-like elongation of the receptacle (see Fig. 3)

caruncle: a small appendage at or near the point of attachment of a seed

catkin: a scaly, spike-like cluster of small flowers, usually of a single sex and without petals (see Fig. 13)

caudex: the persistent, thickened stem base of a perennial plant

channelled: marked with at least 1 deep, lengthwise groove (see Figs. 15, 21)

chasmogamous: with flowers that open for normal pollination; cf. cleistogamous

ciliate: fringed with cilia (see Fig. 22)

cilium [cilia]: a tiny, hair-like outgrowth (see Fig. 20)

circumscissile: splitting open along a transverse, circular line, so that the top comes off like a lid (see Fig. 29)

clasping: embracing or surrounding, usually in reference to a leaf base around a stem (see Fig. 1)

Glossary

claw: the slender, stalk-like base of a petal or sepal (see Figs. 3, 7)

cleft: a hollow between 2 lobes; cut about halfway, or slightly deeper, to the middle or base, deeply lobed, a sinus (see Fig. 1)

cleistogamous: with self-pollinating flowers that do not open; cf. chasmogamous

closed sheath: a sheath that is fused into a cylinder (see Figs. 15, 21); cf. open sheath

collar: the outer side of a grass leaf at the point where the blade and sheath join (see Fig. 20)

column: the prominent central structure of an orchid flower, comprised of the fused stamens, style and stigma

coma: a tuft of soft hairs at the tip of a seed

comose: with a coma

compound: composed of 2 or more smaller parts, such as leaves divided into leaflets and flower clusters consisting of several smaller groups (see Figs. 26, 27); cf. simple

cone: a dense, fruiting or spore-producing structure with overlapping scales or bracts arranged around a central axis, generally woody when bearing seeds and non-woody when bearing spores or pollen (see Figs. 17, 18, 29)

coniferous: bearing cones

corm: a short, thickened, fleshy underground stem without thickened leaves (see Fig. 30)

corolla: the inner ring of the petals and sepals; the petals collectively (see Fig. 9)

corona: the 5-hooded, crown-like structure at the tip of the stamen tube in a milkweed (*Asclepias*) flower (see Fig. 10)

corymb: a flat- or round-topped flower cluster in which the flowers on the outer (lower) branches bloom first (see Fig. 27)

cotyledon: the first leaf (or 1 of the first pair or whorl of leaves) produced by the embryo of a seed plant, developed within the seed; a seed leaf

creeping: growing along (or beneath) the surface of the ground and rooting at intervals, usually at the nodes (see Fig. 17)

crisped: with irregularly curled and rippled edges, like a potato chip (see Fig. 22)

cross-partition: a horizontal line between 2 parallel veins, arranged like an irregular rung on a ladder, often visible towards the base of sedge (*Carex*) leaves; a septum; nodulose septae are thickened cross-partitions (see Fig. 21)

cross-wall: horizontal cell walls separating individual cells in linear series; appearing as thin, purple bands in the hairs of some fleabanes (*Erigeron* spp.)

cyme: a flat- or round-topped flower cluster in which the flowers on the inner (upper) branches bloom first (see Fig. 27)

deciduous: shed after completion of its normal function, usually at the end of the growing season; cf. persistent

Glossary

decumbent: lying on the ground but with ascending stem tips

decurrent: extending downward; usually describing leaves whose lower edges extend down the stem as 2 ridges or wings (see Fig. 24)

dendritic: branched like a tree (see Fig. 26)

diadelphus: with stamens joined by their filaments into 2, often unequal sets; cf. monadelphous

dichotomous: forked in 2 equal parts, like a Y; cf. trichotomous (see Fig. 26)

dicot: a plant whose seeds have 2 cotyledons or seed leaves; most dicots have net-veined leaves; a dicotyledonous plant; cf. monocot

digitate: finger-like; usually referring to leaflets arranged like fingers on a hand (see Fig. 25)

dimorphic: with 2 forms, as in pteridophytes with sterile and fertile plants or leaves

dioecious: with male and female flowers or cones on separate plants

disc floret: a small tubular flower in a flowerhead of the aster family (Asteraceae) (see Fig. 12)

discoid [in reference to flowerheads in asters (Asteraceae)]: with disc florets only; cf. ligulate, radiate

disjunct: separated; referring to plant or animal populations that are a significant distance from all other populations of the same species

dorsal: on the lower side or outer side (i.e., away from the stem); cf. ventral (see Fig. 1)

drupe: a fleshy or pulpy, single-seeded fruit in which the seed is encased in a hard or stony covering (see Fig. 29)

drupelet: a small drupe, usually borne in clusters (e.g., raspberries) (see Fig. 29)

ecotype: a subspecies or race that is specially adapted to a particular set of environmental conditions

ellipsoid: a 3-dimensional form in which every plane is an ellipse or a circle

entire: smooth-edged, without teeth or lobes (see Figs. 1, 22)

epidermis: the outermost layers of cells

epidermal: of the epidermis

evergreen: always with green leaves

farinose: coated with a mealy powder

fascicle: a compact bundle or cluster (see Fig. 26)

fen: a mineral-rich wetland with slow-moving, often calcareous water with sedge and brown moss (not *Sphagnum*) peat

fertile: capable of producing viable pollen, ovules or spores; cf. sterile

fibrous root: a slender, thread-like root, usually 1 of many in a clump (see Figs. 16, 30)

Glossary

filament: the stalk of a stamen, bearing an anther at its tip (see Figs. 2, 3, 6, 7, 13)

filiform: thread-like

fleshy: succulent, firm and pulpy; plump and juicy

floret: a small flower in a cluster; usually applied to single flowers of the grass, sedge, or aster family (see Figs. 12, 13, 14, 15, 16)

flowerhead: a dense cluster of florets, often appearing as a single flower in the aster family (see Fig. 12)

follicle: a dry, pod-like fruit that splits open along only 1 side when mature (see Fig. 29)

forbs: broad-leaved herbs

fornix [fornices]: a small appendage in the throat of a flower corolla (e.g., in borages)

frond: a fern leaf (see Fig. 19)

fruit: the ripened, seed-bearing organ of a plant, together with any other structures that join with it as a unit (see Fig. 29)

fusiform: spindle-shaped, slender and gradually tapered at both ends (see Fig. 25)

galea: a hooded upper lip of a 2-lipped flower (see Fig. 9)

gametophyte: the part or phase of a plant that produces sexual reproductive structures, often small and inconspicuous in pteridophytes; cf. sporophyte

gemma [gemmae]: a small, often bud-like body used for vegetative reproduction

gemmiferous: with gemmae

generic name: the first part of the scientific species name, denoting the genus to which the species belongs; cf. specific epithet

germinate: to sprout

gland: a bump, appendage or depression that produces secretions such as nectar or oil

glandular: with glands

glandular-hairy: with hairs bearing glands (usually at their tips)

glaucous: covered with a whitish, waxy powder (bloom) that can usually be scraped or rubbed off

glomerule: a dense, head-like cluster

glume: 1 of 2 chaff-like bracts at the base of a grass spikelet (see Fig. 16)

glycoside: a chemical compound derived from a sugar, in which 1 carbon is replaced by another component, alcohol or phenol

grain: a hard, seed-like fruit or kernel, usually referring to the fruit of a grass

graminoids: narrow-leaved herbs

Glossary

gynaecandrous: with female flowers at the tip of the spike and male flowers at the base (see Fig. 15); cf. androgynous

herb: a plant without persistent woody parts (at least none above ground), with stems that die back to the ground each year

herbaceous: herb-like, with a leaf-like texture and colour

heteromorphic: with many different forms and sizes

hybrid: the offspring of parents of 2 different kinds (usually a cross between 2 species)

hybridization: the process of creating a hybrid

hypanthium: a ring or cup around the ovary, usually bearing sepals, petals and stamens at its upper edge (see Figs. 5, 29)

hypha [hyphae]: a tiny fungal thread, part of the main body of a fungus

inclined: leaning to 1 side, diverging from vertical

indusium [indusia]: an outgrowth covering and protecting a spore cluster in ferns (see Fig. 19); a protective membrane

inferior ovary: an ovary with the petals, sepals and stamens attached at its top or otherwise above it (see Fig. 7); cf. superior ovary

inflated: hollowed or puffed out (see Fig. 3)

inflorescence: a flower cluster (see Figs. 15, 16, 27)

insectivorous: feeding on insects

internode: the portion of a stem between 2 joints (nodes) (see Figs. 16, 20)

interrupted: discontinuous

introduced: brought in from another region (e.g., Europe)

involucral: of an involucre

involucre: a set of bracts at the base of a flower cluster (as in the carrot family [Apiaceae]) or flowerhead (as in the aster family [Asteraceae]) (see Figs. 12, 27)

irregular flower: a flower in which the members in a series of organs (e.g., the petals or sepals) have different forms or orientations (see Fig. 8); cf. regular flower

joint: a place where 1 part is attached or joined to another; a node (see Figs. 1, 16, 20)

keel: a conspicuous, lengthwise ridge, like the keel of a boat (see Figs. 16, 21); the 2 lower petals of a flower in the pea family (Fabaceae) (see Fig. 6)

lacuna [lacunae]: a small empty space or gap in a tissue

lateral: on the side of (see Figs. 1, 16)

latex: the milky juice of some plants, often containing rubber-like compounds

leaflet: a single part of a compound leaf (see Figs. 19, 25)

Glossary

lemma: the lower of the 2 bracts immediately enclosing a single grass flower (see Fig. 16); cf. palea

lenticel: a slightly raised, often elongated opening (pore) on the bark of young branches or roots

lenticular: lens-shaped (see Fig. 29)

ligulate [in reference to flowerheads of the aster family (Asteraceae)]: with ligules only; cf. discoid, radiate

ligule: a flat, linear, ray floret in a flowerhead of the aster family (Asteraceae) (see Fig. 12); the thin, membranous projection from the top of a leaf sheath, as in the grass family (Poaceae) (see Fig. 20)

limb: the expanded, often lobed section at the tip of a tubular calyx or corolla (see Figs. 3, 7)

linear: long and narrow with essentially parallel sides (see Fig. 25)

lip: the upper or lower lobe of a 2-lobed (2-lipped) flower, such as an orchid or violet flower (see Figs. 8, 9, 11)

lobe: a rounded division, too large to be called a tooth (see Figs. 1, 3)

locule: a cavity or compartment within an ovary or anther

lyrate: pinnately divided with a large lobe at the tip and gradually smaller lobes towards the base of the leaf (see Fig. 25)

marcescent: with persistent, withered remains

marsh: a nutrient-rich wetland that is periodically inundated by standing or slow-moving water, characterized by emergent vegetation from mineral soil

megaspore: in pteridophytes with 2 types of spores, the larger spores, which develop into female gametophytes; cf. microspore

membranous: in a thin, pliable, usually translucent sheet; like a membrane

mericarp: a single carpel of a schizocarp (see Fig. 29)

mesic: with intermediate moisture levels, neither very wet nor very dry

micron: a thousandth of a millimetre; a millionth of a metre

microspore: in pteridophytes with 2 types of spores, the smaller spores, which develop into male gametophytes; cf. megaspore

midrib: the middle rib of an organ such as a leaf (see Figs. 1, 15, 16)

monadelphous: with stamens joined by their filaments into a single set; cf. diadelphus

monocot: a plant whose seeds have a single cotyledon or seed leaf; most monocots have leaves with parallel veins; a monocotyledonous plant; cf. dicot

monoecious: with male and female parts in separate flowers or cones but on the same plant

mucilaginous: producing sticky or gelatinous secretions

Glossary

mycorrhiza [mycorrhizae]: the fungal partner in a symbiotic association between certain fungi and seed plants, in which the fungi permeate the roots and assist in nutrient absorption

naked: lacking hairs, scales or other appendages

nectary: a nectar-secreting gland, usually in a flower (see Fig. 13)

nerve: a prominent line or vein in a leaf or other organ (see Fig. 20)

net-veined: with a network of branched veins; reticulate (see Fig. 1); cf. parallel-veined

node: the point of attachment of a leaf or branch; a joint (see Figs. 1, 16, 20)

nut: a dry, thick-walled fruit that does not split open at maturity, usually single-seeded (see Fig. 29)

nutlet: a small nut, distinguished from an achene by its thicker outer covering

obovoid: with a 3-dimensional, egg-shaped form that is broadest above the middle

ochrea [ochreae]: a sheath of fused stipules at the base of a leaf; common in the buckwheat family (Polygonaceae) (see Fig. 24)

open sheath: a sheath that is split down 1 side rather than fused into a cylinder (see Figs. 16, 20); cf. closed sheath

opposite: attached across from each other on an axis (e.g., opposite leaves, see Fig. 1), or directly in front of one another (e.g., staminodia opposite petals, see Fig. 4); cf. alternate, whorled

ovary: the organ containing ovules or young, undeveloped seeds, located at the base of the pistil (see Figs. 3, 4, 6, 7)

ovoid: with a 3-dimensional, egg-shaped form that is broadest below the middle

ovule: an organ that develops into a single seed after fertilization

palea: the upper of the 2 bracts immediately enclosing an individual grass flower; cf. lemma (see Fig. 16)

palmate [leaves]: with 3 or more lobes or leaflets arising from a single point, like fingers of a hand (see Fig. 25); cf. pinnate

panicle: a branched flower cluster, with lower blooms developing first (see Fig. 27)

papilla [papillae]: a tiny, wart-like projection

papillose: with papillae

pappus: the hairs or bristles on the tip of an achene (see Fig. 12, 29)

parallel-veined: with veins running parallel to one another, not branching to form a network (see Fig. 1); cf. net-veined

pectinate: divided into narrow, closely set, parallel parts arranged like teeth on a comb (see Fig. 26)

Glossary

pedicel: the stalk of a single flower in a flower cluster (see Figs. 3, 11)

peduncle: a main flower stalk, supporting a single flower or flower cluster (see Figs. 2, 8, 12)

perennial: living for 3 or more years

perfect flower: a flower with functional ovaries and stamens; cf. bisexual flower

perianth: the sepals and petals of a flower collectively

perigynium [perigynia]: the sac-like membrane enclosing the achene of a sedge (*Carex* spp.) (see Fig. 15)

persistent: remaining attached after normal function has been completed; cf. deciduous

petal: a member of the inner whorl of the perianth, usually white or brightly coloured (see Figs. 2, 3, 4, 5, 6, 7, 8, 10, 11, 12)

petiole: the stalk of a single leaf (see Figs. 1, 26)

pinna [pinnae]: the main division of a pinnately divided leaf or frond (see Figs. 19, 26)

pinnate: feather-formed; with parts (usually leaflets) arranged on either side of a common axis (see Fig. 26)

pinnatifid: cut at least halfway to the middle in a pinnate pattern (see Fig. 25)

pinnule: a segment of a pinnately divided pinna; the smallest segment of a leaf that is twice pinnately divided (see Figs. 19, 26)

pistil: the female part of a flower, composed of one or more ovaries, styles and stigmas (see Figs. 3, 6, 7); cf. stamen

pistillate: with pistil(s), usually applied to flowers that are functionally female (see Figs. 13, 15); cf. staminate

pith: the soft, spongy centre of a stem or branch

pleated: with parallel folds, like pleats in an accordion

pod: a dry fruit that opens to release its seeds, usually applied to pea-like fruits (see Fig. 29)

pollen: the tiny, powdery grains contained in an anther; the male reproductive units which form pollen tubes and fertilize the ovule

pollen cone: a cone producing pollen; a male cone; cf. seed cone

pollination: the transfer of pollen from anthers to stigmas, leading to fertilization

pollinium [pollinia]: a waxy mass of pollen grains carried as a unit during pollination, as in milkweed and orchid flowers (see Figs. 10, 11)

pome: a fleshy fruit with a core (e.g., an apple) (see Fig. 29)

potherb: a leafy herb that is cooked as a vegetable or thickener

poultice: a moist mass of leaves, bark or other plant material, applied hot or cold to the body to stop bleeding, speed healing, reduce inflammation or relieve pain

Glossary

prostrate: lying on the ground

prothallus [prothalli]: a small, usually disc-shaped, thallus-like growth resulting from the germination of a spore and producing the sexual organs of the plant

pteridophyte: a member of a division of seedless plants, including the horsetails, club-mosses and ferns

pustule: a small, blister-like swelling

raceme: a flower cluster with an unbranched, elongated central stalk bearing few to many stalked blooms, and with lower flowers developing first (see Fig. 27)

rachilla: the axis of an individual grass or sedge floret (see Fig. 16)

rachis: the main axis of a compound leaf or flower cluster (see Figs. 15, 16, 19)

radially symmetric: with parts arranged like spokes on a wheel (see Fig. 5); a regular flower; cf. bilaterally symmetric

radiate [in reference to flowerheads of the aster family (Asteraceae)]: with strap-like ray florets around a central cluster of disc florets (see Fig. 12); cf. discoid, ligulate

rank: a vertical row, usually referring to leaves on a stem (see Fig. 15)

ray [branch]: 1 of 3 or more branches originating from a common point, as in an umbel or similar flower cluster (see Fig. 27)

ray [floret]: a small, strap-like floret in a flowerhead of the aster family (Asteraceae) (see Fig. 12)

receptacle: the enlarged end of a stem to which the flower parts (in the aster family [Asteraceae] the flowers) are attached (see Figs. 2, 3, 12)

recurved: curved under or downward (see Fig. 26)

reflexed: abruptly bent or turned backward or downward (see Fig. 26)

regular flower: a flower in which the members in a series of organs (e.g., the petals or sepals) have similar forms or orientations; a radially symmetric flower (see Fig. 5); cf. irregular flower

reticula: a mesh-like netting of persistent leaf veins, often covering the bulbs of some members of the lily family (Liliaceae)

reticulate : with a network of branched veins; net-veined

rhizomatous: with rhizomes

rhizome: a somewhat lengthened underground stem; distinguished from a root by the presence of buds or scale-like leaves (see Figs. 15, 18, 19, 30)

rhombic: shaped like a rhombus, a figure with 4 equal, parallel sides and 2 oblique angles; diamond-shaped (see Fig. 25)

rib: a prominent, usually longitudinal vein (see Fig. 15)

Glossary

rosette: a circular cluster of organs (usually leaves), usually at the base of a plant (see Fig. 1)

runner: an elongated, slender, prostrate branch, rooting at the joints and/or at the tip; a stolon

samara: a dry, winged fruit that does not split open at maturity, usually single-seeded, often borne in pairs, as in maples (*Acer* spp.) (see Fig. 29)

saprophyte: a plant that takes its food from dead organic matter

scale: a small, flat structure, usually thin and membranous (see Figs. 13, 15, 19)

scape: a leafless flower stalk arising from the plant base

schizocarp: a fruit that splits into separate carpels when mature (see Fig. 29); cf. mericarp

scorpioid: rolled lengthwise in bud and uncoiling as the structure develops, like a fiddlehead

scree: rock debris at the foot of a rock wall; talus

scurfy: covered with tiny scales

seed cone: a cone producing seeds, a female cone; cf. pollen cone

sepal: a member of the outermost whorl of the perianth, usually small, green and more or less leaf-like (see Figs. 2, 5, 6, 7, 10, 11)

sexual reproduction: producing offspring from the union of male and female cells, as in the fertilization of the egg cell in an ovule (female) by a cell from a pollen grain (male); cf. vegetative reproduction

sheath: a tubular organ that partly or completely surrounds some part of a plant, as the sheath of a grass leaf surrounds the stem (see Figs. 15, 16, 18, 20, 21)

sheathing: forming a sheath

silique: a pod-like fruit of a member of the mustard family (Brassicaceae) (see Fig. 29)

simple: all in a complete piece, not divided; cf. compound (see Fig. 13)

sinus: a hollow between 2 lobes; a cleft (see Fig. 1)

sobole: a shoot or sucker growing from a rhizome or stem base, usually numerous

sorus [sori]: a cluster of sporangia, often small and dot-like (see Fig. 19)

spadix: a dense spike of small flowers with a thick, fleshy axis (see Figs. 27, 28)

spathe: a large, leaf-like bract enclosing a flower cluster (usually a spadix) (see Fig. 27)

specific epithet: the second part of the species scientific name, distinguishing the species from other members of the same genus; cf. generic name

spike: a more or less elongated flower cluster with essentially stalkless flowers or spikelets (see Figs. 15, 27)

spikelet: a small or secondary spike; the characteristic unit of a flower cluster in the grass family (Poaceae) (see Fig. 16)

Glossary

spine: a sharp, stiff outgrowth from the stem (e.g., a thorn)

sporangiaster: a modified spore case

sporangium [sporangia]: a spore sac or spore case (see Fig. 19); spore cluster

spore: the tiny, usually single-celled, reproductive body of a non-seed plant (e.g., a fern or horsetail), producing a gametophyte

sporocarp: the firm spore case of some fern allies (e.g., *Marsilea*)

sporophyll: a spore-bearing leaf (see Fig. 17)

sporophyte: the spore-bearing part or phase of a plant, usually leafy (see Figs. 17, 18, 19); cf. gametophyte

spreading: diverging widely from the vertical, approaching horizontal (see Fig. 26)

spur: a hollow extension, usually at the base of a flower, petal or sepal, often containing nectar (see Figs. 8, 11)

stamen: the pollen-bearing (male) organ of a flower, usually consisting of an anther and a filament (see Figs. 2, 3, 5, 6, 7, 13); cf. pistil

staminate: with stamen(s); usually applied to flowers that are functionally male (see Figs. 13, 15); cf. pistillate

staminode, staminodium [staminodia]: a modified sterile stamen (see Fig. 4)

standard: the large upper petal of a flower in the pea family (Fabaceae); a banner (see Fig. 6)

sterile: without viable pollen, ovules or spores; cf. fertile

stigma: the tip of the female organ (pistil) of a flower, where the pollen lands and adheres (see Figs. 6, 7, 12, 13, 14, 15)

stigmatic disc: the broad, flat, round surface of a stigma

stipule: an appendage at the base of a leaf stalk, ranging from slender and awl-shaped to broad and leaf-like (see Figs. 1, 24)

stolon: a trailing, horizontal stem, rooting at nodes and/or tips; a runner

stoloniferous: with stolons

stoma, stomate [stomata]: a tiny pore in the plant epidermis, bounded by 2 guard cells which open and close the pore by changing shape

strobilus [strobili]: a cone-like cluster of sporophylls (see Figs. 17, 18)

style: the middle part of a pistil, connecting the stigma and ovary (see Figs. 3, 6, 7, 13, 15, 29)

stylopodium: a disc-like expansion at the base of a style, as in the carrot family (Apiaceae) (see Fig. 29)

succulent: fleshy, soft and juicy; a plant that stores water in fleshy stems or leaves

sucker: a vertical vegetative shoot arising from the lower part of a plant, typically from the trunk base or from spreading underground stems of trees and shrubs

Glossary

superior ovary: an ovary with the petals, sepals and stamens attached at its base or otherwise below it (see Fig. 6); cf. inferior ovary

swamp: a wetland that is permanently waterlogged below the surface, and periodically inundated by standing or gently moving water, nutritionally intermediate between a bog and a fen

talus: rocky debris at the foot of a rock wall; scree

taproot: a main root growing vertically downwards (see Fig. 30)

tepal: a sepal or petal, in flowers where these structures are almost identical (see Fig. 14)

terete: round in cross-section, usually referring to cylindrical structures

ternate: divided or arranged in threes (see Fig. 25)

tetragonal: 4-sided

tetraploid: with 4 sets of chromosomes in the cell nucleus

thalloid: like or with a thallus

thallus: the main body of a plant that is not differentiated into stems and leaves (as in *Wolffia* and in pteridophyte gametophytes)

throat: the opening in the middle of a flower, at the centre of a ring of petals or petal-like sepals (see Figs. 8, 9); the opening at the upper edge of a leaf sheath (see Fig. 21)

tomentum: a matted covering of woolly hairs

torulose: cylindrical with alternating swellings and constrictions (see Fig. 29)

trailing: spreading on the ground, but not rooting along the stems

translucent: semi-transparent

trichotomous: branching into 3 equal parts (see Fig. 26); cf. dichotomous

trifoliate: divided into 3 leaflets (see Fig. 25)

tripinnate: 3-times pinnately divided into leaflets (see Fig. 26)

tuber: a short, thick underground branch, usually with several buds or 'eyes' (see Fig. 30)

tubercle: a small, usually rounded, swelling or projection

tuberous: like a tuber

turion: a small bulb-like growth at the base of the stem, often present in some willowherb (*Epilobium*) species

tussock: a compact clump of plants, especially grasses or sedges

umbel: an often flat-topped flower cluster with few to many stalks arising from a common point, like the stays of an umbrella (see Fig. 27)

umbellet: the smaller cluster in a repeatedly branched (compound) umbel (see Fig. 27)

unarmed: without thorns or prickles

Glossary

unisexual: with 1 sex only, either male or female (see Fig. 15)

utricle: a small single-seeded fruit, thin-walled and somewhat inflated (see Fig. 29)

valve: a section of the wall of a pod or capsule, separating from the other section(s) when the fruit splits open (see Fig. 29); 1 of the enlarged inner sepals that enclose the seed of a dock or sorrel (*Rumex*) flower

vascular: containing vascular tissues (xylem and phloem) that transport water and nutrients within the plant

vascular bundle: a strand of vascular fibers (xylem and phloem) and associated tissues

vegetative reproduction: producing new plants from asexual parts (e.g., rhizomes, leaves, bulbils) rather than from fertilized ovules (usually seeds); cf. sexual reproduction

vein: a thread of conducting tubes (a vascular bundle), especially if visible on the surface of a petal or leaf (see Figs. 1, 16, 20, 21)

velum: a thin, membranous covering, over the spore clusters near the leaf base in *Isoetes* species

ventral: on the upper side or inner side (i.e., closest to the stem) (see Fig. 1); cf. dorsal

viviparous: producing young plants in the flower cluster of the parent plant from sprouting bulbils or germinating seeds

whorl: a circle of 3 or more similar structures (e.g., leaves, flowers) attached at the same node (see Fig. 1)

whorled: arranged in whorls; cf. alternate, opposite

wing: a thin, flattened expansion from the side or tip of an organ (see Fig. 29); 1 of the 2 side petals of a flower (see Figs. 6, 8)

wintergreen: with leaves that remain green through winter but die the following summer

zygomorphic: divisible into 2 equal parts along 1 line only, usually with the upper half unlike the lower and each side a mirror image of the other (see Fig. 8); bilaterally symmetric

Glossary

FIGURE 1: VASCULAR PLANT PARTS

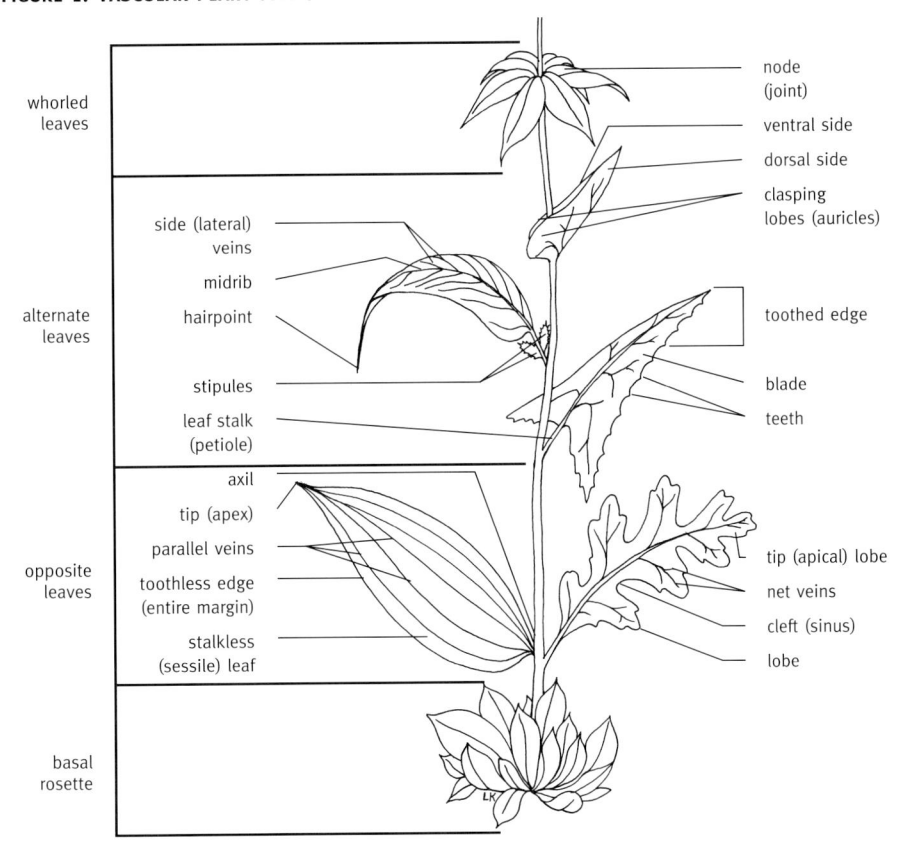

- whorled leaves
- alternate leaves
- opposite leaves
- basal rosette

- side (lateral) veins
- midrib
- hairpoint
- stipules
- leaf stalk (petiole)
- axil
- tip (apex)
- parallel veins
- toothless edge (entire margin)
- stalkless (sessile) leaf

- node (joint)
- ventral side
- dorsal side
- clasping lobes (auricles)
- toothed edge
- blade
- teeth
- tip (apical) lobe
- net veins
- cleft (sinus)
- lobe

FIGURE 2: PARTS OF A BUTTERCUP (*RANUNCULUS*) FLOWER

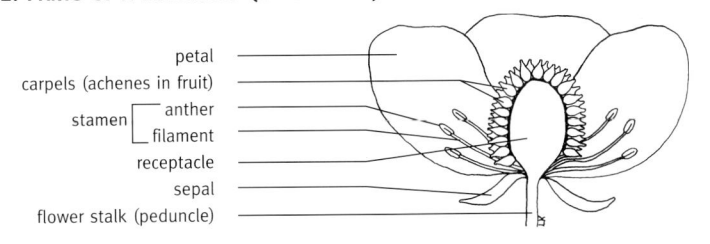

- petal
- carpels (achenes in fruit)
- stamen
  - anther
  - filament
- receptacle
- sepal
- flower stalk (peduncle)

Glossary

FIGURE 3: PARTS OF A BLADDER CAMPION (*SILENE*) FLOWER

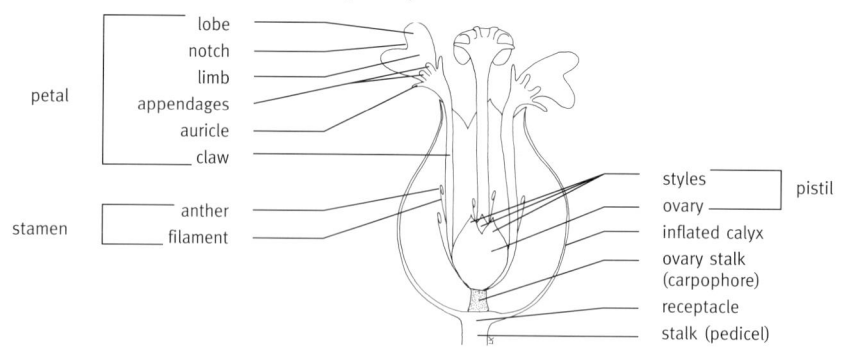

petal
- lobe
- notch
- limb
- appendages
- auricle
- claw

stamen
- anther
- filament

pistil
- styles
- ovary

- inflated calyx
- ovary stalk (carpophore)
- receptacle
- stalk (pedicel)

FIGURE 4: PARTS OF A GRASS-OF-PARNASSUS (*PARNASSIA*) FLOWER

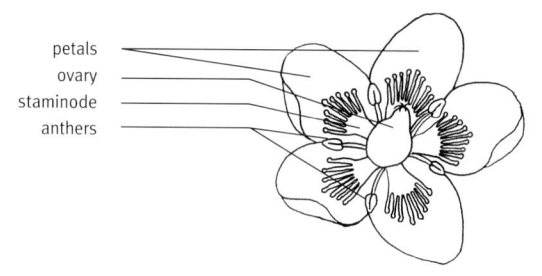

- petals
- ovary
- staminode
- anthers

FIGURE 5: PARTS OF A ROSE (*ROSA*) FLOWER (A RADIALLY SYMMETRIC [REGULAR] FLOWER)

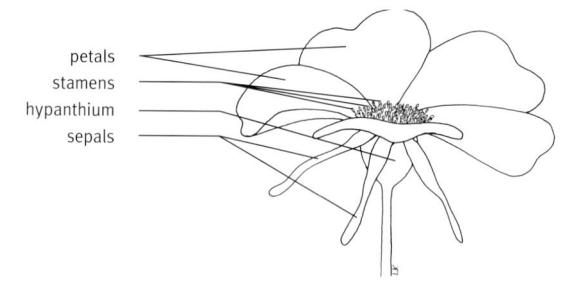

- petals
- stamens
- hypanthium
- sepals

FIGURE 6: PARTS OF A PEA (FABACEAE) FLOWER

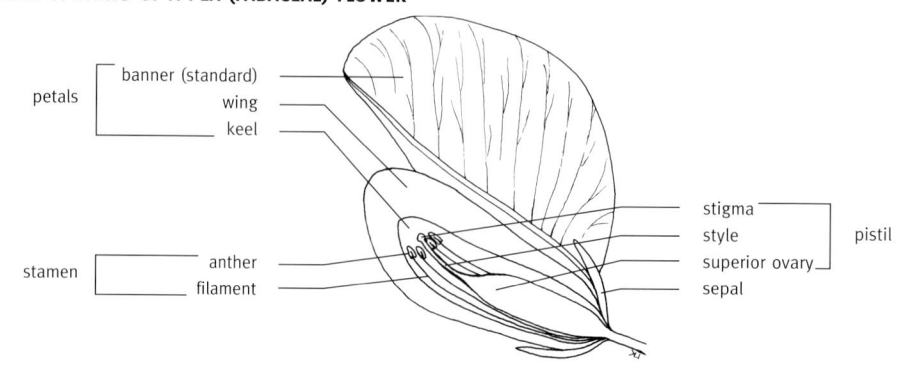

petals
- banner (standard)
- wing
- keel

stamen
- anther
- filament

pistil
- stigma
- style
- superior ovary

- sepal

Glossary

FIGURE 7: PARTS OF A WILLOWHERB (*EPILOBIUM*) FLOWER

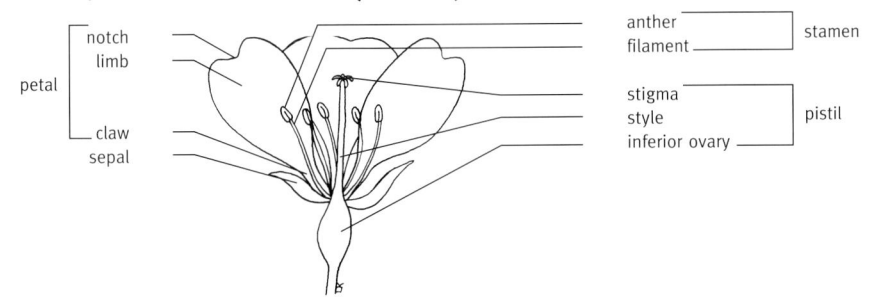

- petal
  - notch
  - limb
  - claw
- sepal
- stamen
  - anther
  - filament
- pistil
  - stigma
  - style
  - inferior ovary

FIGURE 8: PARTS OF A VIOLET (*VIOLA*) FLOWER (A BILTERALLY SYMMETRIC [IRREGULAR] FLOWER)

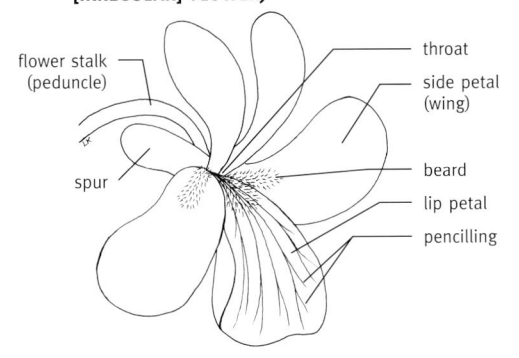

- flower stalk (peduncle)
- spur
- throat
- side petal (wing)
- beard
- lip petal
- pencilling

FIGURE 9: PARTS OF A LOUSEWORT (*PEDICULARIS*) FLOWER

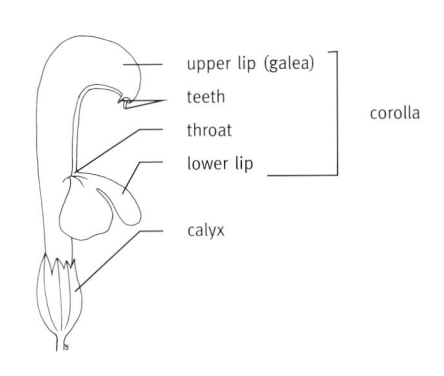

- corolla
  - upper lip (galea)
  - teeth
  - throat
  - lower lip
- calyx

FIGURE 10: PARTS OF A MILKWEED (*ASCLEPIAS*) FLOWER

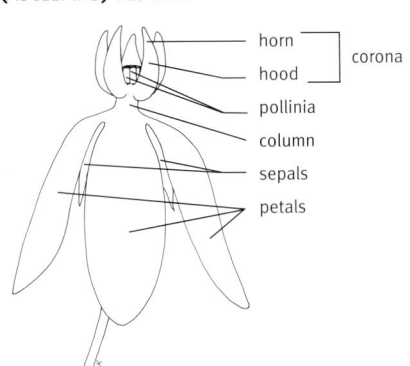

- corona
  - horn
  - hood
- pollinia
- column
- sepals
- petals

FIGURE 11: PARTS OF A BOG ORCHID (*PLATANTHERA*) FLOWER

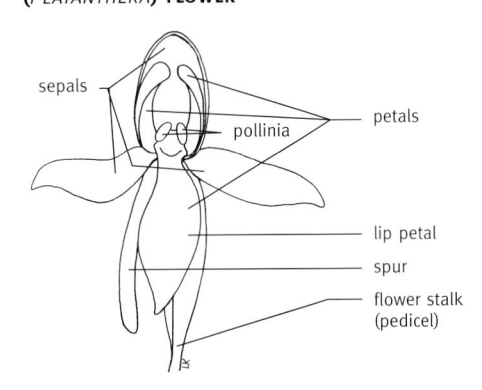

- sepals
- pollinia
- petals
- lip petal
- spur
- flower stalk (pedicel)

Rare *Vascular* Plants of Alberta   441

Glossary

FIGURE 12: PARTS OF AN ASTER (ASTERACEAE) FLOWERHEAD

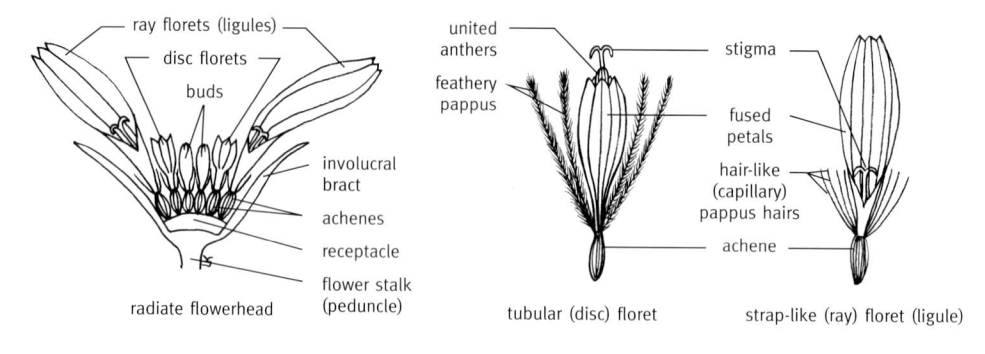

- ray florets (ligules)
- disc florets
- buds
- involucral bract
- achenes
- receptacle
- flower stalk (peduncle)

radiate flowerhead

- united anthers
- feathery pappus
- stigma
- fused petals
- hair-like (capillary) pappus hairs
- achene

tubular (disc) floret

strap-like (ray) floret (ligule)

FIGURE 13: PARTS OF A WILLOW (*SALIX*) SHRUB

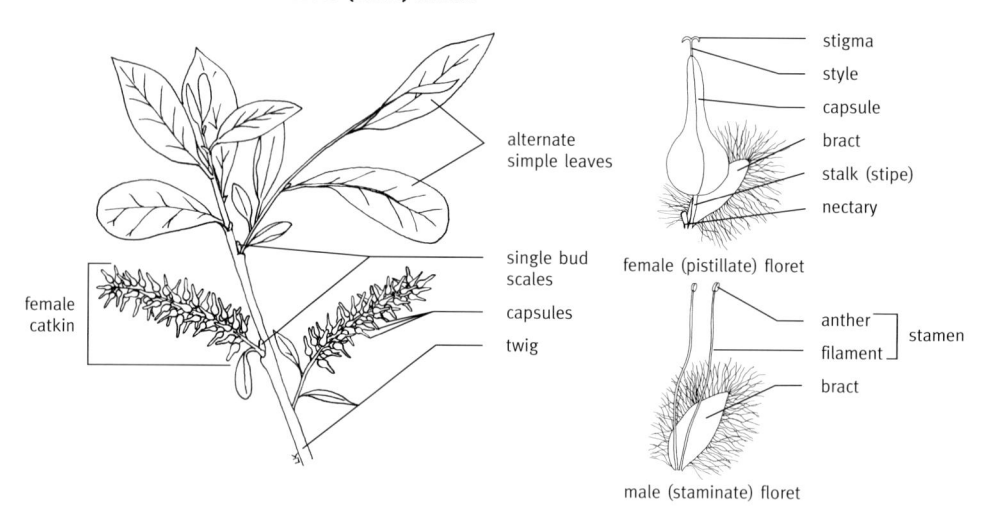

- alternate simple leaves
- single bud scales
- capsules
- twig

female catkin

- stigma
- style
- capsule
- bract
- stalk (stipe)
- nectary

female (pistillate) floret

- anther
- filament
- bract

stamen

male (staminate) floret

FIGURE 14: PARTS OF A RUSH (*JUNCUS*) FLORET

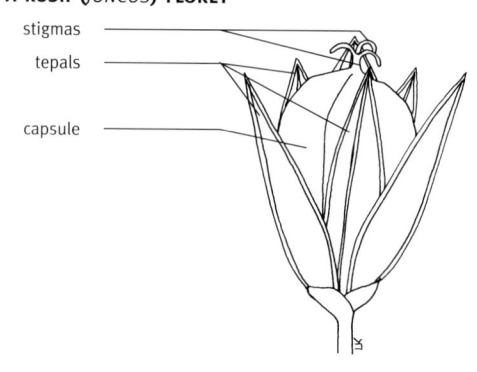

- stigmas
- tepals
- capsule

Glossary

FIGURE 15: PARTS OF A SEDGE (*CAREX*) PLANT

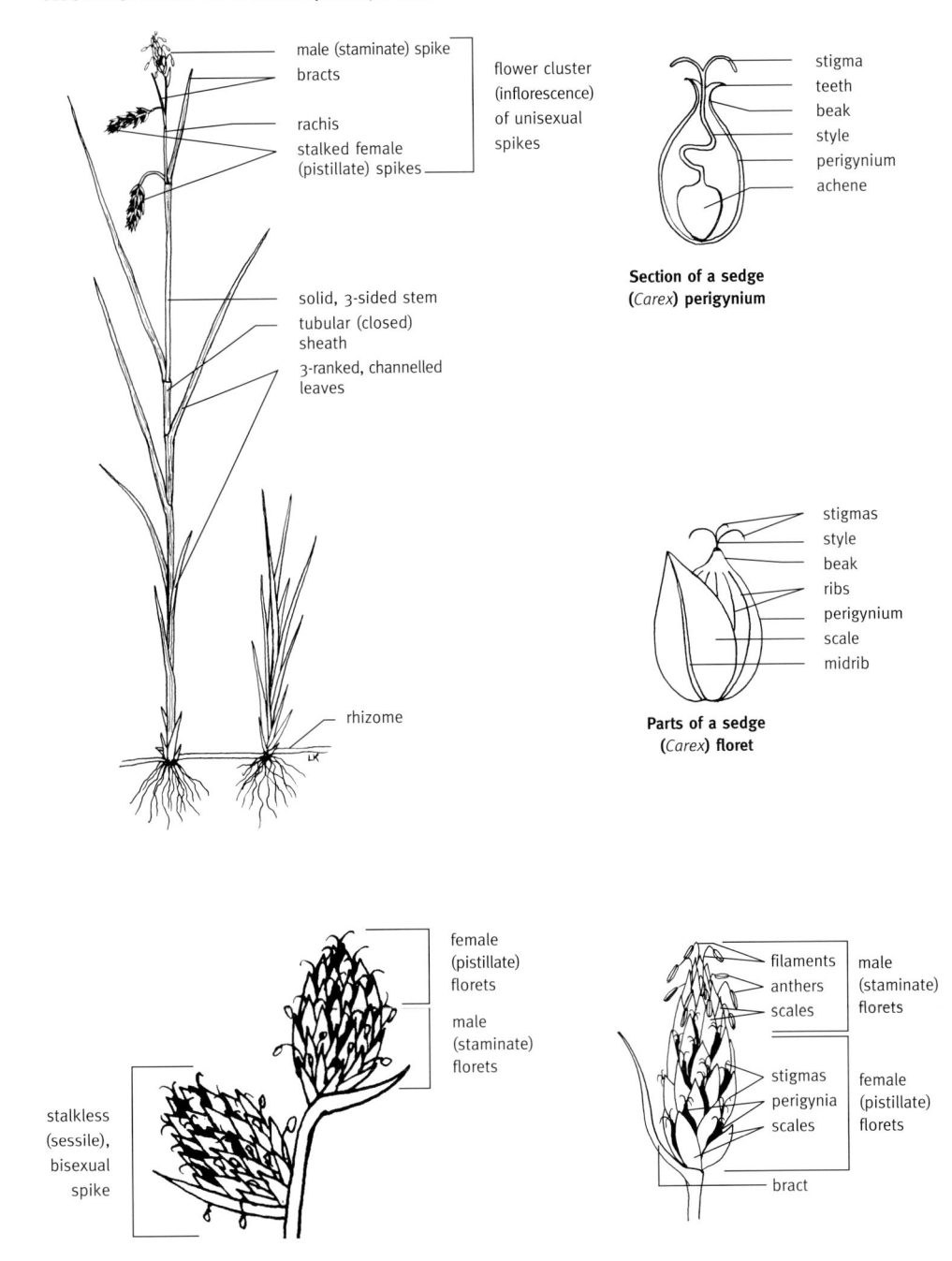

Labels on main plant illustration:
- male (staminate) spike
- bracts
- rachis
- stalked female (pistillate) spikes
- flower cluster (inflorescence) of unisexual spikes
- solid, 3-sided stem
- tubular (closed) sheath
- 3-ranked, channelled leaves
- rhizome

Section of a sedge (*Carex*) perigynium
- stigma
- teeth
- beak
- style
- perigynium
- achene

Parts of a sedge (*Carex*) floret
- stigmas
- style
- beak
- ribs
- perigynium
- scale
- midrib

Inflorescence of gynaecandrous spikes
- female (pistillate) florets
- male (staminate) florets
- stalkless (sessile), bisexual spike

Inflorescence of a single androgynous spike
- filaments
- anthers
- scales
- male (staminate) florets
- stigmas
- perigynia
- scales
- female (pistillate) florets
- bract

Rare *Vascular* Plants of Alberta

Glossary

FIGURE 16: PARTS OF GRASS (POACEAE) PLANTS

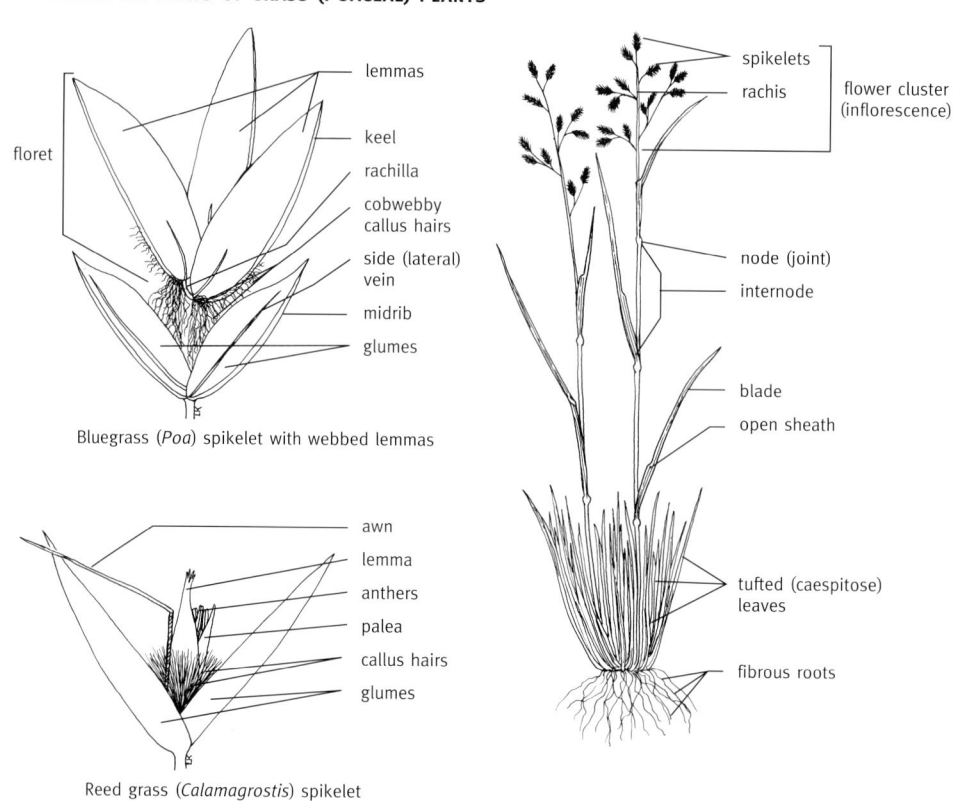

Bluegrass (*Poa*) spikelet with webbed lemmas

- floret
- lemmas
- keel
- rachilla
- cobwebby callus hairs
- side (lateral) vein
- midrib
- glumes

Reed grass (*Calamagrostis*) spikelet

- awn
- lemma
- anthers
- palea
- callus hairs
- glumes

- spikelets
- rachis
- flower cluster (inflorescence)
- node (joint)
- internode
- blade
- open sheath
- tufted (caespitose) leaves
- fibrous roots

FIGURE 17: PARTS OF A CLUB-MOSS (*LYCOPODIELLA*) PLANT (SPOROPHYTE)

- sporophyll
- spore-bearing cone (strobilus)
- leaves
- erect branch
- creeping stem (rhizome)

Glossary

FIGURE 18: PARTS OF A HORSETAIL (*EQUISETUM*) PLANT (SPOROPHYTE)

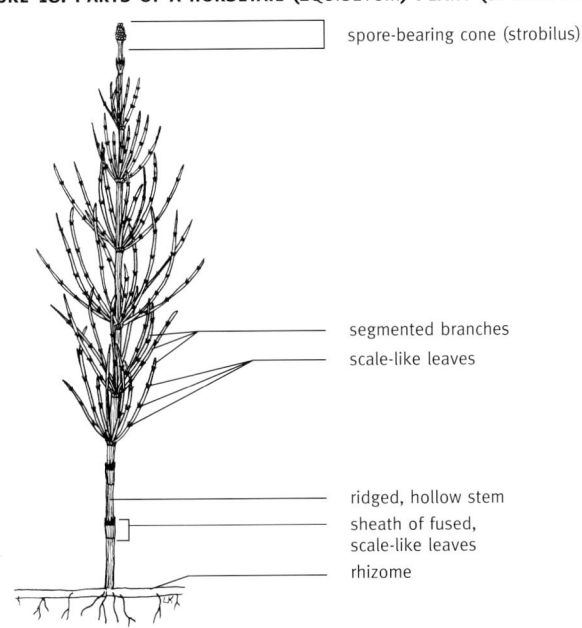

- spore-bearing cone (strobilus)
- segmented branches
- scale-like leaves
- ridged, hollow stem
- sheath of fused, scale-like leaves
- rhizome

FIGURE 19: PARTS OF A WOOD FERN (*DRYOPTERIS*) PLANT (SPOROPHYTE)

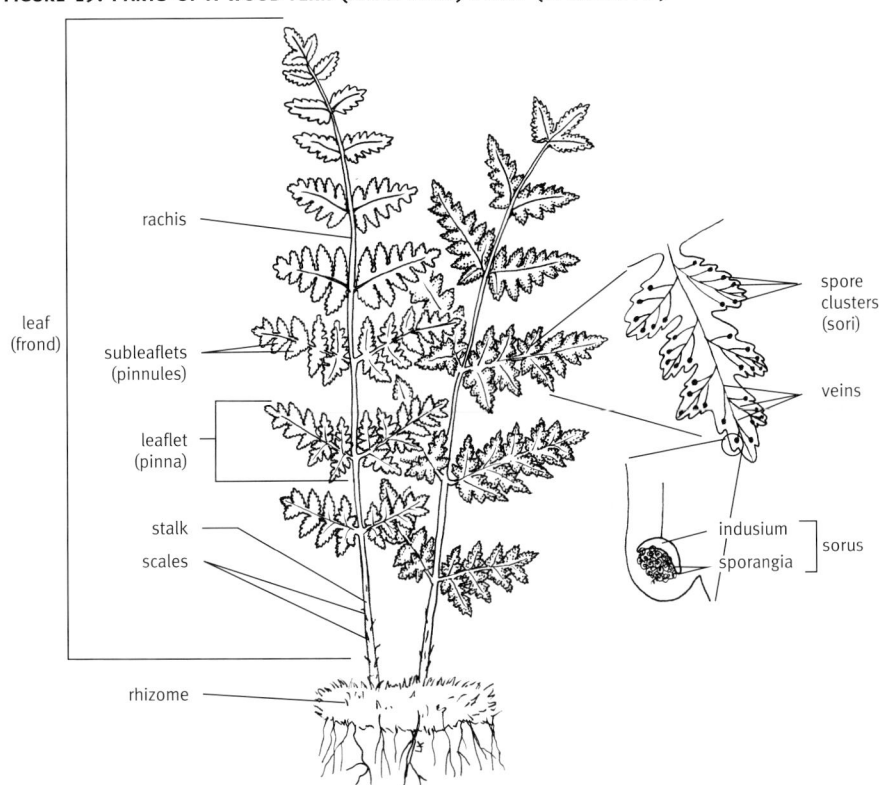

- leaf (frond)
- rachis
- subleaflets (pinnules)
- leaflet (pinna)
- stalk
- scales
- rhizome
- spore clusters (sori)
- veins
- indusium
- sporangia
- sorus

Rare *Vascular* Plants of Alberta 445

Glossary

FIGURE 20: Parts of a grass leaf and stem

- node (joint)
- flat blade
- parallel veins (nerves)
- stem (culm)
- cilia
- ligule
- collar
- auricles
- split (open) sheath
- internode
- node (joint)

FIGURE 21: Parts of a sedge leaf and stem

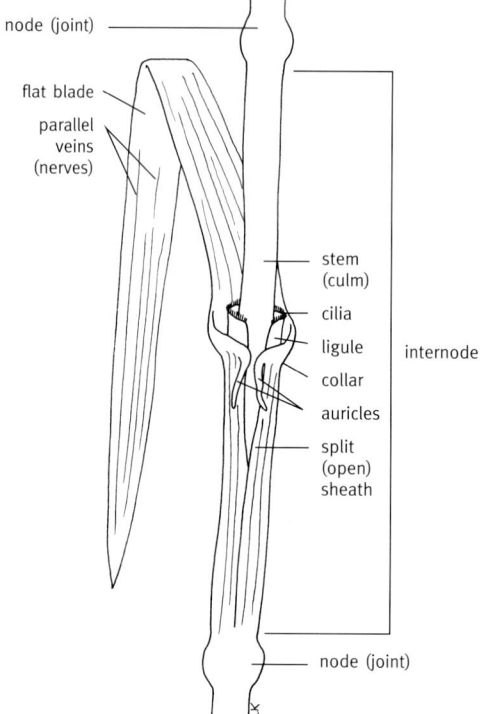

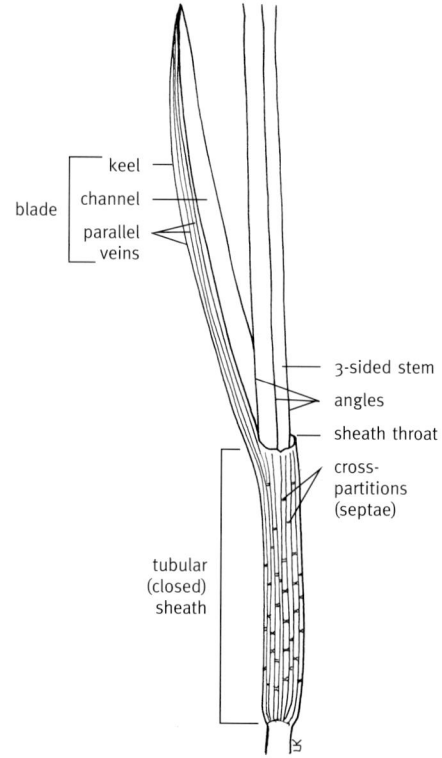

- keel
- channel
- parallel veins
- blade
- 3-sided stem
- angles
- sheath throat
- cross-partitions (septae)
- tubular (closed) sheath

FIGURE 22: Leaf edges

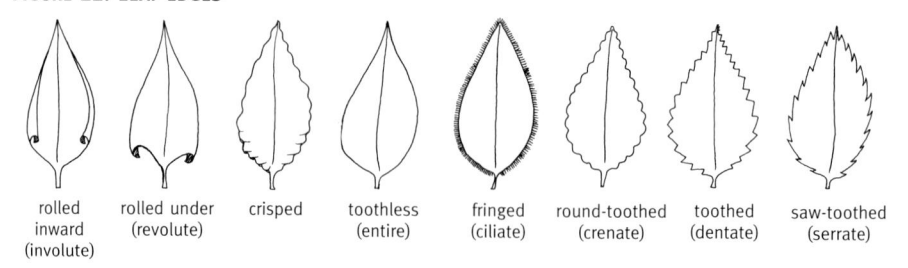

- rolled inward (involute)
- rolled under (revolute)
- crisped
- toothless (entire)
- fringed (ciliate)
- round-toothed (crenate)
- toothed (dentate)
- saw-toothed (serrate)

FIGURE 23: Leaf tips

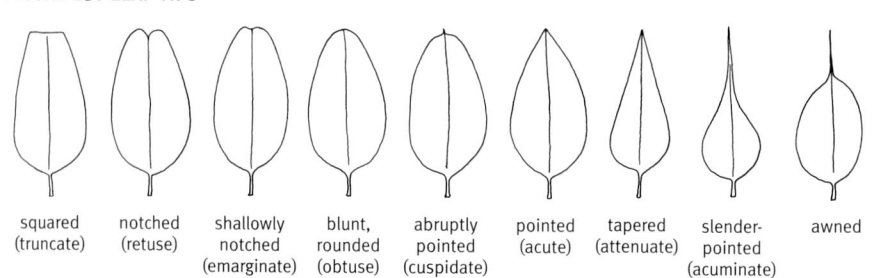

- squared (truncate)
- notched (retuse)
- shallowly notched (emarginate)
- blunt, rounded (obtuse)
- abruptly pointed (cuspidate)
- pointed (acute)
- tapered (attenuate)
- slender-pointed (acuminate)
- awned

Glossary

FIGURE 24: LEAF BASES

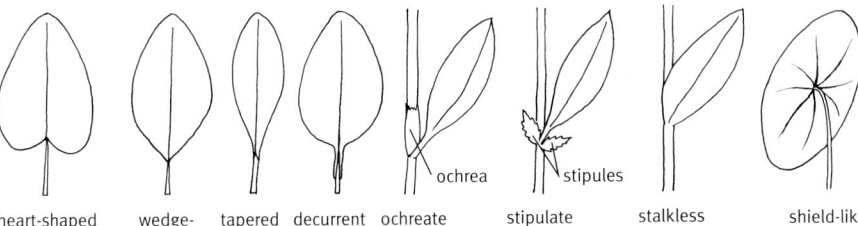

heart-shaped (cordate)   wedge-shaped (cuneate)   tapered (attenuate)   decurrent   ochreate   stipulate   stalkless (sessile)   shield-like (peltate)

FIGURE 25: LEAF SHAPES

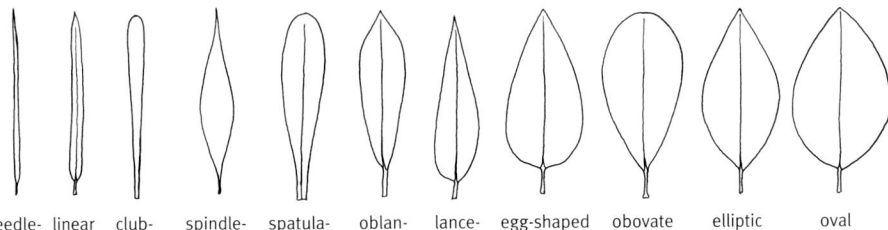

needle-like (acicular)   linear   club-shaped (clavate)   spindle-shaped (fusiform)   spatula-shaped (spatulate)   oblanceolate   lance-shaped (lanceolate)   egg-shaped (ovate)   obovate   elliptic   oval

round (orbicular)   kidney-shaped (reniform)   deltate   triangular   fan-shaped (flabellate)   arrowhead-shaped (sagittate)

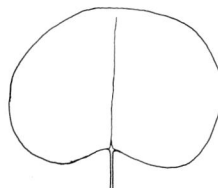

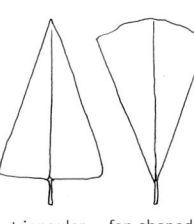

hastate   diamond-shaped (rhombic)   oblong   lyrate   pinnatifid   trifoliate (ternate)   hand-shaped (palmate) with finger-like (digitate) leaflets

Rare *Vascular* Plants of Alberta   447

Glossary

FIGURE 26: BRANCHING PATTERNS

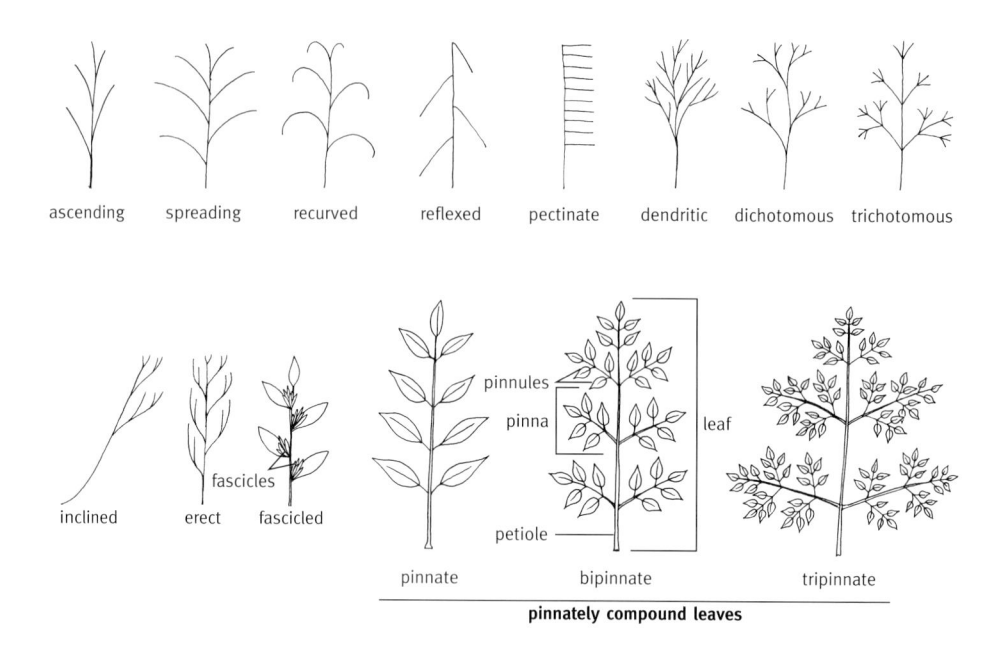

ascending spreading recurved reflexed pectinate dendritic dichotomous trichotomous

inclined erect fascicled

pinnules pinna leaf petiole

pinnate bipinnate tripinnate

pinnately compound leaves

FIGURE 27: TYPES OF FLOWER CLUSTERS (INFLORESCENCES)

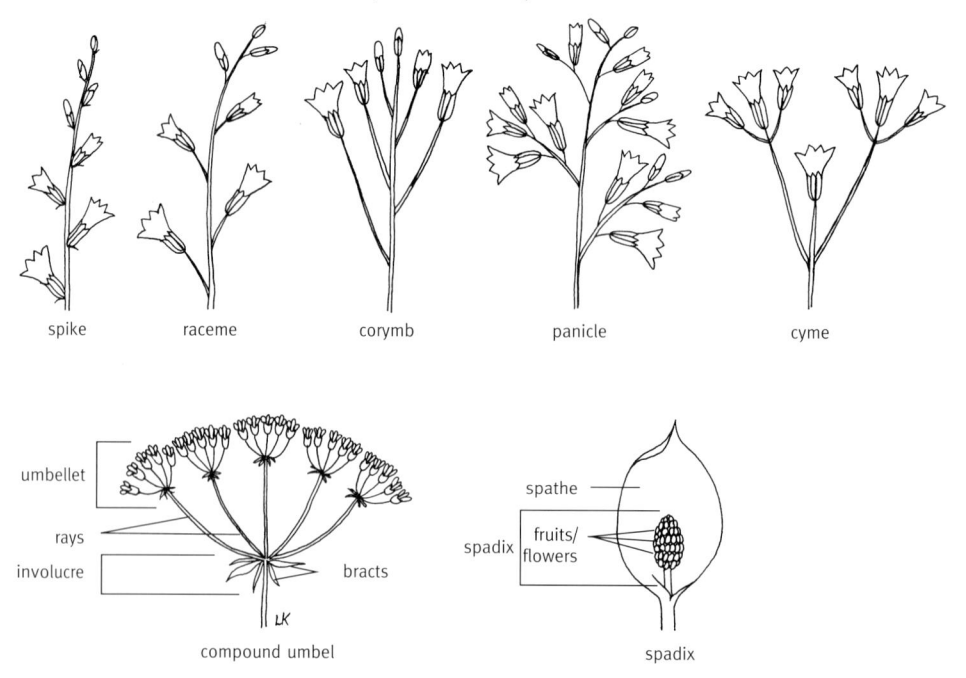

spike raceme corymb panicle cyme

umbellet rays involucre bracts

compound umbel

spathe spadix fruits/flowers

spadix

448 Rare *Vascular* Plants of Alberta

Glossary

FIGURE 28: FLOWER SHAPES

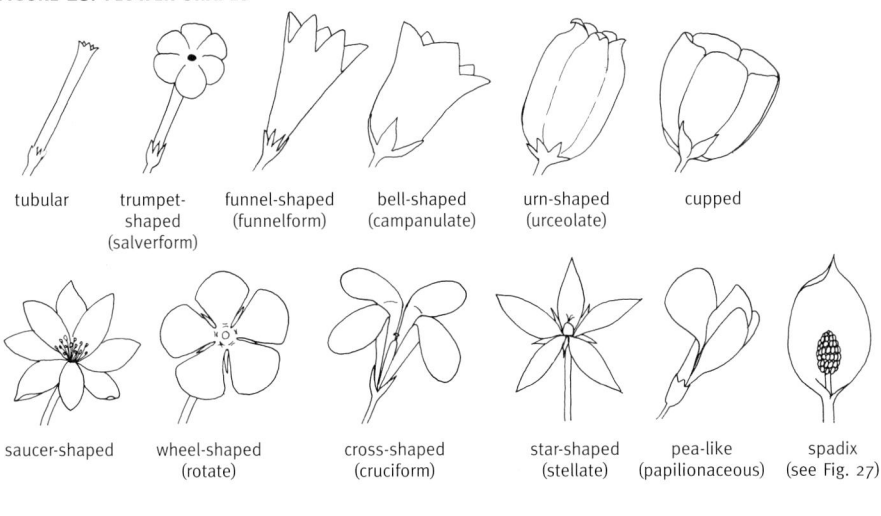

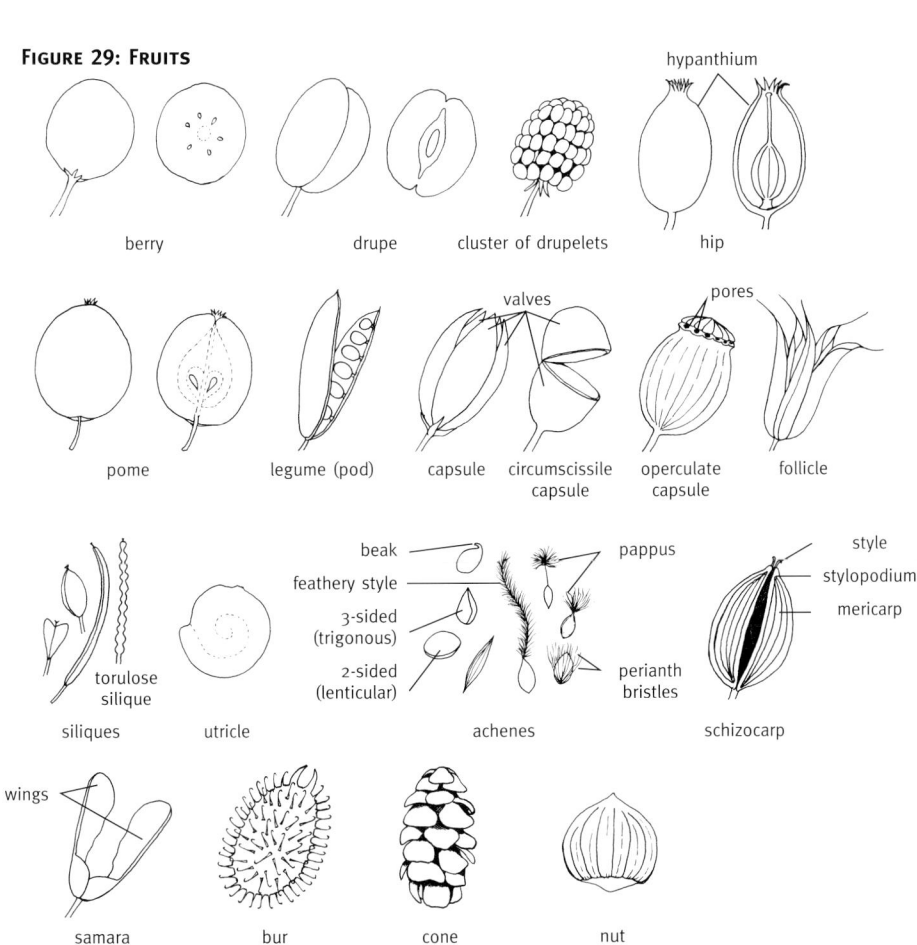

tubular — trumpet-shaped (salverform) — funnel-shaped (funnelform) — bell-shaped (campanulate) — urn-shaped (urceolate) — cupped

saucer-shaped — wheel-shaped (rotate) — cross-shaped (cruciform) — star-shaped (stellate) — pea-like (papilionaceous) — spadix (see Fig. 27)

FIGURE 29: FRUITS

berry — drupe — cluster of drupelets — hip (hypanthium)

pome — legume (pod) — capsule (valves) — circumscissile capsule — operculate capsule (pores) — follicle

torulose siliques — utricle — achenes (beak, feathery style, 3-sided (trigonous), 2-sided (lenticular), pappus, perianth bristles) — schizocarp (style, stylopodium, mericarp)

samara (wings) — bur — cone — nut

Rare *Vascular* Plants of Alberta

Glossary

FIGURE 30: UNDERGROUND PARTS

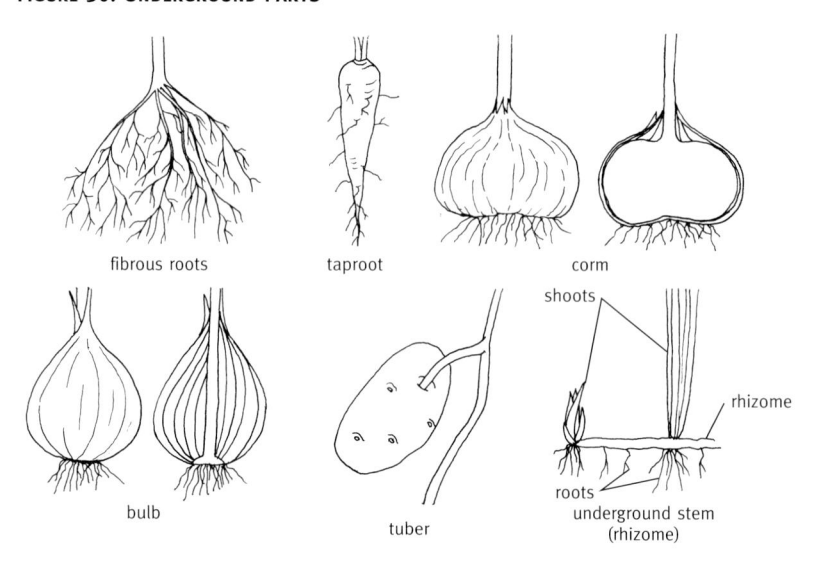

fibrous roots

taproot

corm

bulb

tuber

shoots

rhizome

roots

underground stem (rhizome)

References

Abrams, L. and R.S. Ferris. 1960. An illustrated flora of the Pacific states: Washington, Oregon and California. 4 volumes. Stanford University Press, Stanford, California.

Achuff, P.L. 1987. Rare vascular plants in the Rocky Mountains of Alberta: summary. Pages 109–10 in G.L. Holroyd et al., eds. Proceedings of the workshop on endangered species in the prairie provinces. Provincial Museum of Alberta, Edmonton, Alberta. Natural History Occasional Paper No. 9.

Achuff, P.L. 1992. Natural regions, subregions and natural history themes of Alberta: a classification for protected areas management. Parks Services, Alberta Environmental Protection, Edmonton, Alberta.

Achuff, P.L. 1996. Status report on the large-flowered *Brickellia* (*Brickellia grandiflora*) in Canada. Committee on the Status of Endangered Wildlife in Canada, Ottawa, Ontario.

Achuff, P.L. and I.G.W. Corns. 1985. Plants new to Alberta from Banff and Jasper national parks. Canadian Field-Naturalist 99:94–98.

Aiken, S.G. and S.J. Darbyshire. 1990. Fescue grasses of Canada. Agriculture Canada, Ottawa, Ontario. Publication no. 1844/E.

Aiken, S.G., L.P. Lefkovitch, S.E. Gardiner and W.W. Mitchell. 1994. Evidence against the existence of varieties in *Arctagrostis latifolia* ssp. *aundinacea* (Poaceae). Canadian Journal of Botany 72:1039–50.

Alberta Forestry, Lands and Wildlife. 1991. Western blue flag. Alberta Forestry, Lands and Wildlife, Alberta Environmental Protection, Edmonton, Alberta. Alberta's threatened wildlife.

Alberta Environmental Protection. 1993. Alberta plants and fungi: master species list and species group checklists. Alberta Environmental Protection, Edmonton, Alberta.

Alberta Forestry, Lands and Wildlife. 1993. Big sagebrush (*Artemisia tridentata* Nutt.). Alberta Environmental Protection, Edmonton, Alberta. Alberta's Rare Plants Fact Sheet No. 6.

Alberta Forestry, Lands and Wildlife. 1993. Dwarf fleabane (*Erigeron radicatus* Hook.). Alberta Environmental Protection, Edmonton, Alberta. Alberta's Rare Plants Fact Sheet No. 8.

Alberta Forestry, Lands and Wildlife. 1993. Geyer's wild onion (*Allium geyeri* S. Wats.). Alberta Environmental Protection, Edmonton, Alberta. Alberta's Rare Plants Fact Sheet No. 4.

Alberta Forestry, Lands and Wildlife. 1993. Hare-footed locoweed (*Oxytropis lagopus* Nutt.). Alberta Environmental Protection, Edmonton, Alberta. Alberta's Rare Plants Fact Sheet No. 9.

Alberta Forestry, Lands and Wildlife. 1993. Mountain ladies'-slipper (*Cypripedium montanum* Dougl. ex Lindl.). Alberta Environmental Protection, Edmonton, Alberta. Alberta's Rare Plants Fact Sheet No. 2.

Alberta Forestry, Lands and Wildlife. 1993. Western blue flag (*Iris missouriensis* Nutt.). Alberta Environmental Protection, Edmonton, Alberta. Alberta's Rare Plants Fact Sheet No. 3.

Alberta Forestry, Lands and Wildlife. 1993. Western spiderwort (*Tradescantia occidentalis* (Britt.) Smyth). Alberta Environmental Protection, Edmonton, Alberta. Alberta's Rare Plants Fact Sheet No. 7.

Alberta Forestry, Lands and Wildlife. 1993. What is a rare plant? Alberta Environmental Protection, Edmonton, Alberta. Alberta's Rare Plants Fact Sheet No. 1.

Alberta Forestry, Lands and Wildlife. 1993. Yellow paintbrush (*Castilleja cusickii* Greenm.). Alberta Environmental Protection, Edmonton, Alberta. Alberta's Rare Plants Fact Sheet No. 5.

Alberta Forestry, Lands and Wildlife. 1994. Alpine poppy (*Papaver pygmaeum* Rydb.). Alberta Environmental Protection, Edmonton, Alberta. Alberta's Rare Plants Fact Sheet No. 15.

Alberta Forestry, Lands and Wildlife. 1994. Bog adder's-mouth (*Malaxis paludosa* (L.) Sw.). Alberta Environmental Protection, Edmonton, Alberta. Alberta's Rare Plants Fact Sheet No. 10.

Alberta Forestry, Lands and Wildlife. 1994. Engelmann's spike-rush (*Eleocharis ovata* (Roth) R.& S.). Alberta Environmental Protection, Edmonton, Alberta. Alberta's Rare Plants Fact Sheet No. 13.

Alberta Forestry, Lands and Wildlife. 1994. Jones' columbine (*Aquilegia jonesii* Parry). Alberta Environmental Protection, Edmonton, Alberta. Alberta's Rare Plants Fact Sheet No. 11.

Alberta Forestry, Lands and Wildlife. 1994. Nebraska sedge (*Carex nebraskensis* Dewey). Alberta Environmental Protection, Edmonton, Alberta. Alberta's Rare Plants Fact Sheet No. 12.

Alberta Forestry, Lands and Wildlife. 1994. Smooth boisduvalia (*Boisduvalia glabella* (Nutt.) Walp.). Alberta Environmental Protection, Edmonton, Alberta. Alberta's Rare Plants Fact Sheet No. 14.

References

Alberta Forestry, Lands and Wildlife. 1994. Upland evening-primrose (*Oenothera andina* Nutt.). Alberta Environmental Protection, Edmonton, Alberta. Alberta's Rare Plants Fact Sheet No. 16.

Alberta Native Plant Council. 1988. Alberta's first rare plant monitoring project: western blue flag (*Iris missouriensis*). Iris 2:3–4.

Alberta Native Plant Council. 1989. Native plant collecting: conservation guidelines. Iris 5:19–20.

Alberta Natural and Protected Areas Program 1991. Threatened plant management plan—A first! Natural Areas Newsletter 15:1.

Alberta Natural Heritage Information Centre. 1998. Rare plant data files. Recreation and Protected Areas, Alberta Environmental Protection, Edmonton, Alberta.

Allen, L.J. 1987. Alberta rare plants. Canadian Plant Conservation Programme Newsletter 2(1):11–12.

Allison, J.L. 1945. *Selenophoma bromigena* leaf spot on *Bromus inermis*. Phytopathology 35: 233–40.

Argus, G.W. 1986. *Salix raupii*, Raup's willow, new to the flora of Alberta and the Northwest Territories. Canadian Field-Naturalist 100:386–388.

Argus, G.W. and C.J. Keddy. 1984. Atlas of the rare vascular plants of Ontario. Botany division, National Museum of Natural Sciences, National Museums of Canada, Ottawa, Ontario.

Argus, G.W. and D.J. White. 1977. The rare vascular plants of Ontario. Botany division, National Museum of Natural Sciences, National Museums of Canada, Ottawa, Ontario.

Argus, G.W. and D.J. White. 1978. The rare vascular plants of Alberta. National Museum of Natural Sciences, Ottawa, Ontario. Syllogeus No. 17.

Argus, G.W. and K.M. Pryer. 1990. Rare vascular plants in Canada: our natural heritage. Canadian Museum of Nature. Ottawa, Ontario.

Babcock, E.B. 1947. The genus *Crepis*. Part 2. Systematic treatment. University of California Publications in Botany 22:199–1030.

Barkley, T.M., ed. 1977. Atlas of the flora of the Great Plains. Iowa State University Press, Ames, Iowa.

Barkworth, M.E. and D.R. Dewey. 1985. Genomically based genera in the perennial Triticeae of North America: identification and membership. American Journal of Botany 72: 767–76.

Barneby, R.C. 1964. Atlas of North American *Astragalus*. Part 2. The Ceridotherix, Hypoglottis, Piptoloboid, Trimeniaeus and Orophaca Astragali. Memoirs of the New York Botanical Garden 13:597–1188.

Bassett, I.J. 1973. The Plantagos of Canada. Canada Department of Agriculture, Ottawa, Ontario.

Bassett, I.J. and C.W. Crompton. 1982. The genus *Chenopodium* in Canada. Canadian Journal of Botany 60:586–610.

Bayer, R.J. 1989. A systematic and phytogeographic study of *Antennaria aromatica* and *A. densifolia* (Asteraceae: Inuleae) in the western North American Cordillera. Madrono 36(4):248–59.

Bayer, R.J. 1993. A synopsis with keys for the genus *Antennaria* (Asteraceae: Inuleae: Gnaphaliinae) of North America. Canadian Journal of Botany 71:1589–1604.

Beaman, J.H. 1957. The systematics and evolution of *Townsendia* (Compositae). Contributions from the Gray Herbarium 183:1–151.

Benson, L. 1962. Plant taxonomy: methods and principles. Ronald Press Co., New York, New York. A Chronica Botanica Publication.

Berch, S.M., S. Gameit and E. Doem. 1987. Mycorrhizal status of some plants in southwestern British Columbia. Canadian Journal of Botany 66: 1924–28.

Bliss, L.C. 1971. Arctic and alpine plant life cycles. Annual Reviews Inc., Palo Alto, California. Annual Review of Ecology and Systematics Vol. 2.

Boivin, B. 1968. Flora of the Prairie Provinces. Phytologia 15:121–59, 329–446; 16:1–47, 219–61, 265–339; 17:58–112; 18:281–93.

Booth, W.E. and J.C. Wright. 1959. Flora of Montana. Part 1. Conifers and monocots. Montana State University, Bozeman, Montana.

References

Booth, W.E. and J.C. Wright. 1959. Flora of Montana. Part 2. Dicotyledons. Montana State University, Bozeman, Montana.

Borror, D.J. 1960. Dictionary of word roots and combining forms. Mayfield Publishing Co., Mountain View, California.

Bouchard, A., D. Barabe, M. Dumais and S. Hay. 1983. The rare vascular plants of Quebec. National Museum of Natural Sciences, Ottawa, Ontario. Syllogeus No. 48.

Bradley, C. 1987. Disappearing cottonwoods: the social challenge. Pages 119–21 in G.L. Holroyd et al., eds. Proceedings of the Workshop on Endangered Species in the Prairie Provinces. Provincial Museum of Alberta, Edmonton, Alberta. Natural History Occasional Paper No. 9.

Braidwood, B. 1987. A framework for the identification, evaluation, and protection of sites of special interest in Alberta. M.Sc. thesis, Department of Forest Science, University of Alberta, Edmonton, Alberta.

Brayshaw, T.C. 1985. Pondweeds and bur-reeds, and their relatives: aquatic families of monocotyledons in British Columbia. British Columbia Provincial Museum, Victoria, British Columbia. Occasional Paper No. 26.

Brayshaw, T.C. 1989. Buttercups, waterlilies and their relatives in British Columbia. Royal British Columbia Museum, Victoria, British Columbia. Memoir No. 1.

Brayshaw, T.C. 1996. Trees and shrubs of British Columbia. University of British Columbia Press, Vancouver, British Columbia. Royal British Columbia Museum Handbook.

Breitung, A.J. 1954. A botanical survey of the Cypress Hills. Canadian Field-Naturalist 68:55–92.

Breitung, A.J. 1957. Plants of Waterton Lakes National Park, Alberta. Canadian Field-Naturalist 71:39–71.

Britton, D.M. and D.F. Brunton. 1993. *Isoetes* x *truncata*: a newly considered pentaploid hybrid from western North America. Canadian Journal of Botany 71: 1016–25.

Brooke, R.C. and S. Kojima. 1985. An annotated vascular flora of areas adjacent to the Dempster Highway, central Yukon Territory. Volume 2. Dicotyledonae. British Columbia Provincial Museum, Victoria, British Columbia. Contributions to Natural Science No. 4.

Brunton, D.F. 1984. Status of western larch, *Larix occidentalis*, in Alberta. Canadian Field-Naturalist 98:167–70.

Brunton, D.F. 1994. Status report on Bolander's quillwort (*Isoetes bolanderi* Engelm.) in Canada. Committee on the Status of Endangered Wildlife in Canada, Ottawa, Ontario.

Budd, A.C. 1957. Wild plants of the Canadian prairies. Canada Department of Agriculture, Ottawa, Ontario. Publication No. 983.

Burnfield, S.J. and F.D. Johnson. 1990. Cytological, morphological, ecological and phenological support for specific status of *Crataegus suksdorfii*. Madrono 37(4):274–82.

Burt, P. 1991. Barrenland beauties: showy plants of the Arctic coast. Outcrop Ltd., Yellowknife, Northwest Territories.

Calder, J.A. and D.B.O. Savile. 1960. Studies in Saxifragaceae. III. *Saxifraga odontoloma* and *S. lyallii*, and North American subspecies of *S. punctata*. Canadian Journal of Botany 38:409–35.

Case, F.W. Jr. 1987. Orchids of the western Great Lakes region. Cranbrook Institute of Science. Bloomfield Hills, Michigan. Bulletin No. 48.

Catling, P.M. and W. Wojtas. 1986. The waterweeds (*Elodea* and *Egeria*, Hydrocharitaceae) in Canada. Canadian Journal of Botany 64:1525–41.

Catling, P.M. 1991. Morphometrics and phytogeography of the *Eleocharis elliptica* group in Ontario. Canadian Botanical Association, Edmonton, Alberta, and Canadian Society of Plant Physiology, Edmonton, Alberta, Joint Meeting, June 23–27, 1991, Edmonton, Alberta.

Chambers, J.C. and R.C. Sidle. 1991. Fate of heavy metals in an abandoned lead-zinc tailings pond. Part 1. Vegetation. Journal of Environmental Quality 20: 745–51.

Chase, A. 1918. Axillary cleistogenes in some American grasses. American Journal of Botany 5: 254–58.

References

Chmielewski, J.G. 1993. *Antennaria pulvinata* Greene: the legitimate name for *A. aromatica* Evert (Asteraceae: Inuleae). Rhodora 95:261–76.

Chmielewski, J.G., C.C. Chinnappa and J.C. Semple. 1990. The genus *Antennaria* (Asteraceae: Inuleae) in western North America: morphometric analysis of *Antennaria alborosea*, *A. corymbosa*, *A. marginata*, *A. microphylla*, *A. parvifolia*, *A. rosea*, and *A. umbrinella*. Plant Systematics and Evolution 169:141–75.

Church, G.L. 1949. A cytotaxonomic study of *Glyceria* and *Puccinellia*. American Journal of Botany 36: 155–65.

Clark, L.J. 1975. Lewis Clark's field guide to wild flowers of the mountains in the Pacific Northwest. Gray's Publishing Ltd., Sidney, British Columbia.

Clark, L.J. 1976. Wild Flowers of the Pacific Northwest from Alaska to northern California. Evergreen Press Ltd., Vancouver, British Columbia.

Clausen, R.T. 1975. *Sedum* of North America north of the Mexican plateau. Cornell University Press, Ithaca, New York.

Cody, W.J. 1956. New plant records for northern Alberta and the southern Mackenzie District. Canadian Field-Naturalist 70:101–30.

Cody, W.J. 1971. A phytogeographical study of the floras of the Continental Northwest Territories and Yukon. Naturaliste Canadien 98: 145–58.

Cody, W.J. 1996. Flora of the Yukon Territory. National Research Press, Ottawa, Ontario.

Cody, W.J. and D.M. Britton. 1989. Ferns and fern allies of Canada. Research Branch, Agriculture Canada, Ottawa, Ontario. Publication No. 1829/E.

Cody, W.J. and S.S. Talbot. 1973. The pitcher plant, *Sarracenia purpurea* L., in the northwestern part of its range. Canadian Field-Naturalist 88:229–30.

Cody, W.J., B. Boivin and G.W. Scotter. 1974. *Loisleuria procumbens* (L.) Desv., alpine azalea, in Alberta. Canadian Field-Naturalist 88:229–30.

Coffey, T. 1993. The history and folklore of North American wildflowers. Houghton Mifflin, Boston, Massachusetts.

Colorado Native Plant Society. 1989. Rare plants of Colorado. Colorado Native Plant Society and Rocky Mountain Nature Association, Rocky Mountain National Park, Estes Park, Colorado.

Committee on the Status of Endangered Wildlife in Canada. 1997. Canadian species at risk. Canadian Wildlife Service, Ottawa, Ontario.

Constance, L. and R.H. Shan. 1948. The genus *Osmorhiza* (Umbelliferae): a study in geographic affinities. University of California Publications in Botany 23:111–156.

Constance, L. 1941. The genus *Nemophila* Nutt. University of California Publications in Botany 19:341–345.

Constance, L. 1942. The genus *Hydrophyllum*. American Midland Naturalist 27:710–731.

Coombes, A.J. 1987. Dictionary of plant names. Timber Press, Portland, Oregon.

Coombes, A.J. 1992. The Hamlyn guide to plant names. Hamlyn, London, England.

Cormack, R.G.H. 1948. Orchids of the Cypress Hills. Canadian Field-Naturalist 88:229–30.

Cormack, R.G.H. 1977. Wildflowers of Alberta. Hurtig Publishers, Edmonton, Alberta.

Coupé, R., C.A. Ray, A. Comeau, M.V. Ketcheson and R.M. Annas. 1982. A guide to some common plants of the Skeena area of British Columbia. Province of British Columbia, Ministry of Forestry, Victoria, British Columbia. Land Management Handbook No. 4.

Craighead, J.J., F.C. Craighead, Jr. and R.J. Davis. 1963. Rocky Mountain wildflowers. Houghton Mifflin Company, Boston, Massachusetts.

Crawford, D.J. 1975. Systematic relationships in the narrow-leaved species of *Chenopodium* of the western United States. Brittonia 27:279–88.

Crawford, D.J. and J.F. Reynolds. 1974. A numerical study of the common narrow-leaved taxa of *Chenopodium* occurring in the western United States. Brittonia 26:398–410.

Crow, G. 1978. A taxonomic revision of *Sagina* (Caryophyllaceae) in North America. Rhodora 80:1–91.

References

Crow, G.E. and C.B. Hellquist. 1985. Aquatic vascular plants of New England. Part 8. Lentibulariaceae. New Hampshire Agricultural Experiment Station, University of New Hampshire, Durham, New Hampshire. Station Bulletin No. 528.

Crum, H. 1988. A focus on peatlands and peat mosses. The University of Michigan Press, Ann Arbor, Michigan.

Daubenmire, R.F. 1970. Steppe vegetation of Washington. Washington Agricultural Experimental Station, Pullman, Washington. Technical Bulletin No. G2.

Davis, J.I. 1991. A note on North American *Torreyochloa* (Poaceae), including a new combination. Phytologia 70: 361–65.

Davis, R.J. 1966. The North American perennial species of *Claytonia*. Brittonia 18:285–303.

de Vries, B. 1966. *Iris missouriensis* Nutt. in southwestern Alberta and in central and northern British Columbia. Canadian Field-Naturalist 80:158–60.

Dewey, D.R. 1963. Natural hybrids of *Agropyron trachycaulum* and *Agropyron scribneri*. Bulletin of the Torrey Botanical Club 90: 111–22.

Dewey, D.R. 1967. Genome relations between *Agropyron scribneri* and *Sitanion hystrix*. Bulletin of the Torrey Botanical Club 94: 395–404.

Dewey, D.R. 1976. Cytogenetics of *Agropyron pringlei* and its hybrids with *A. spicatum*, *A. scribneri*, *A. violaceum*, and *A. dasystachyum*. Botanical Gazette 137: 179–85.

Douglas, G.W. 1982. The sunflower family (Asteraceae) of British Columbia. Volume 1. Senecioneae. Royal British Columbia Museum, Victoria, British Columbia. Occasional Paper No. 23.

Douglas, G.W. 1995. The sunflower family (Asteraceae) of British Columbia. Volume 2. Astereae, Anthemideae, Eupatorieae and Inuleae. Royal British Columbia Museum, Victoria, British Columbia.

Douglas, G.W., G.B. Straley and D. Meidinger. 1989. The vascular plants of British Columbia. Part 1. Gymnosperms and Dicotyledons (Aceraceae through Cucurbitaceae). British Columbia Ministry of Forests, Victoria, British Columbia.

Douglas, G.W., G.B. Straley and D. Meidinger. 1990. The vascular plants of British Columbia. Part 2. Dicotyledons (Diapensiaceae through Portulacaceae). British Columbia Ministry of Forests, Victoria, British Columbia.

Douglas, G.W., G.B. Straley and D. Meidinger. 1991. The vascular plants of British Columbia. Part 3. Dicotyledons (Primulaceae through Zygophyllaceae) and Pteridophytes. British Columbia Ministry of Forests, Victoria, British Columbia.

Douglas, G.W., G.B. Straley and D. Meidinger. 1994. The vascular plants of British Columbia. Part 4. Monocotyledons. British Columbia Ministry of Forests, Victoria, British Columbia.

Douglas, G.W., G.W. Argus, H.L. Dickson and D.F. Brunton. 1981. The rare vascular plants of the Yukon. National Museum of Natural Sciences, Ottawa, Ontario. Syllogeus No. 28.

Dunn, D.B. 1965. The inter-relationships of the Alaskan lupines. Madrono 18:1–17.

Dunn, D.B. and J.M. Gillett. 1966. The lupines of Canada and Alaska. Canada Department of Agriculture, Ottawa, Ontario. Monograph No. 2.

Duran, R. and G.W. Fisher. 1961. The genus *Tilletia*. Washington State University Press, Pullman, Washington.

Eastman, J. 1995. The book of swamp and bog. Stackpole Books, Mechanicsburg, Pennsylvania.

Elisens, W.J. and J.G. Packer. 1980. A contribution to the taxonomy of the *Oxytropis campestris* complex in northwestern North America. Canadian Journal of Botany 58:1820–31.

Ellison, W.L. 1964. A systematic study of the genus *Bahia* (Compositae). Rhodora 66:67–86, 177–215, 281–311.

Elvander, P.E. 1984. The taxonomy of *Saxifraga* (Saxifragaceae) Section *Boraphila* subsection *Integrifolia* in western North America. Systematic Botany Monographs 3:1–44.

Erichsen-Brown, C. 1979. Use of plants for the past 500 years. Breezy Creek Press, Aurora, Ontario.

References

Evert, E.F. 1984. A new species of *Antennaria* (Asteraceae) from Montana and Wyoming. Madrono 13(2):109–12.

Fairbarns, M.D. 1984. Status report on the soapweed (*Yucca glauca*) in Canada. Committee on the Status of Endangered Wildlife in Canada, Ottawa, Ontario.

Fairbarns, M.D. 1986. Conservation values and management concerns in the Candidate South Castle natural area. Public Lands Division, Alberta Forestry, Lands and Wildlife, Edmonton, Alberta. Technical Report No. T/127.

Fairbarns, M.D. 1989. Bog adder's-mouth orchid (*Malaxis paludosa*) in the Wagner Bog natural area vicinity. Alberta Forestry, Lands and Wildlife, Edmonton, Alberta.

Fairbarns, M., D. Loewen and C. Bradley. 1987. The rare vascular flora of Alberta. Volume 1. A summary of the taxa occurring in the Rocky Mountain natural region. Alberta Forestry, Lands and Wildlife, Edmonton, Alberta. Natural Areas Technical Report No. 29.

Farrar, D.R. and C.L. Johnson-Groh. 1990. Subterranean sporophytic gemmae on moonwort ferns, *Botrychium* subgenus *Botrychium*. American Journal of Botany 77: 1168–75.

Farrar, J.L. 1995. Trees in Canada. Fitzhenry & Whiteside Ltd., Markham, Ontario, and Canadian Forest Service, Ottawa, Ontario.

Federation of Alberta Naturalists. 1992. Potential species for inclusion in the rare vascular flora of Alberta. Alberta Naturalist 22(4), Supplement No. 3:1A–11A.

Fernald, M.L. 1905. The North American species of *Eriophorum*. Rhodora 7:81–92, 129–36.

Fernald, M.L. 1935. Critical plants of the upper Great Lakes region of Ontario and Michigan. Rhodora 37:197–262.

Fernald, M.L. 1950. Gray's manual of botany. Eighth edition. American Book Company, New York, New York.

Fisher, G.W. and M.N. Levine. 1941. Summary of the recorded data on the reaction of wild and cultivated grasses to stem rust (*Puccinia graminis*), leaf rust (*P. rubigo-vera*), stripe rust (*P. glumarum*), and crown rust (*P. coronata*) in the United States and Canada. The Plant Disease Reporter. Supplement No. 130.

Flora of North America Editorial Committee, eds. 1993. Flora of North America, north of Mexico. Volume 2. Pteridophytes and Gymnosperms. Oxford University Press, New York, New York.

Flora of North America Editorial Committee, eds. 1997. Flora of North America, north of Mexico. Volume 3. Magnoliidae and Hamamelidae. Oxford University Press, New York, New York.

Foster, S. and J.A. Duke. 1990. A field guide to medicinal plants of eastern and central North America. Houghton Mifflin Co., Boston, Massachusetts.

Frankton, C. and I.J. Bassett. 1970. The genus *Atriplex* (Chenopodiaceae) in Canada. Part 2. Four native western annuals: *A. argentea*, *A. truncata*, *A. powellii*, and *A. dioica*. Canadian Journal of Botany 48:981–89.

Gale, S. 1944. *Rhynchospora*, section *Eurhynchospora*, in Canada, the United States and the West Indies. Rhodora 46:89–143, 159–97, 207–49, 255–78.

Galloway, L.A. 1975. Systematics of the North American desert species of *Abronia* and *Tripterocalyx* (Nyctaginaceae). Brittonia 27:328–47.

Gerrand, M. and D. Sheppard. 1995. Rare and endangered species of the Castle Wilderness. Castle-Crown Wilderness Coalition, Pincher Creek, Alberta. Special Publication No. 4.

Gill, J.D. and F.L. Pogge. 1974. *Physocarpus maxim*. Ninebark. Seeds of woody plants in the United States. Forest Service, United States Department of Agriculture, Washington, D.C. Handbook No. 450.

Gillet, J. 1979. New combinations in *Hypericum*, *Triandenum*, and *Gentianopsis*. Canadian Journal of Botany 57:185–86.

Gillet, J. and N. Robson. 1981. The St. John's-worts of Canada (Guttiferae). National Museum of Natural Sciences, Ottawa, Ontario. Publications in Botany No. 11.

Gillett, G.W. 1960. A systematic treatment of the *Phacelia franklinii* group. Rhodora 62:205–22.

References

Gillett, G.W. 1962. Evolutionary relationships of *Phacelia linearis*. Brittonia 14:231–36.

Gillett, J.M. 1957. A revision of the North American species of *Gentianella* Moench. Annals of the Missouri Botanical Garden 44(3):195–269.

Gillett, J.M. 1963. The gentians of Canada, Alaska and Greenland. Canada Department of Agriculture Research Branch, Ottawa, Ontario. Publication No. 1180.

Gillett, J.M. 1968. The milkworts of Canada. Canada Department of Agriculture Research Branch, Ottawa, Ontario. Monograph No. 5.

Gleason, H.A. 1952. The new Britton and Brown illustrated flora of the northeastern United States and adjacent Canada. Hafner Press, New York, New York.

Gleason, H.A. and A. Cronquist. 1963. Manual of vascular plants of northeastern United States and adjacent Canada. Van Nostrand Reinhold Co., Toronto, Ontario.

Gornall, R.J. and B.A. Bohm. 1985. A monograph of *Boykinia*, *Peltoboykinia*, *Bolandra* and *Suksdorfia* (Saxifragaceae). Botanical Journal of the Linnaean Society 90:1–71.

Gould, F.W. and C.A. Clark. 1978. *Dichanthelium* (Poaceae) in the United States and Canada. Annals of the Missouri Botanical Garden 65:1088–1132.

Gould, F.W. and R.B. Shaw. 1983. Grass systematics. Second edition. Texas A&M University Press, College Station, Texas.

Gould, J. 1998. Rare plant conservation in Alberta. J. Thorpe, T.A. Steeves, and M. Gollop, eds. Proceedings of the Fifth Prairie Conservation and Endangered Species Conference, February 1998, Saskaton, Saskatchewan. Provincial Museum of Alberta, Edmonton, Alberta. Natural History Occasional Paper No. 24.

Gould, J. 1996. The status of rare plant conservation in Alberta. W.D. Willms and J.F. Dormaar, eds. Pages 244–46 in Proceedings of the Fourth Prairie Conservation and Endangered Species workshop. Provincial Museum of Alberta, Edmonton, Alberta. Natural History Occasional Paper No. 23.

Gould, J. 1998. Alberta Natural Heritage Information Centre plant species of special concern. Alberta Environment, Edmonton, Alberta.

Gould, J. 2000. Alberta Natural Heritage Information Centre plant species of special concern. Alberta Environment, Edmonton, Alberta.

Great Plains Flora Association. 1977. Flora of the Great Plains. University of Kansas Press, Lawrence, Kansas.

Grieve, M. 1931. A modern herbal. Penguin Books Ltd., Harmondsworth, Middlesex, England.

Griffiths, D.E. and G.C.D. Griffiths. 1990. Further notes on *Wolffia* (Lemnaceae), with the addition of *W. borealis* to the Alberta flora. Alberta Naturalist 20:59–64.

Griffiths, G.C.D. 1988. *Wolffia arrhiza* and *W. columbiana* (Lemnaceae) in Alberta. Alberta Naturalist 18:18–20.

Griffiths, G.C.D. 1989. The true *Carex rostrata* (Cyperaceae) in Alberta. Alberta Naturalist 19(3):105–08.

Guard, B.J. 1995. Wetland plants of Oregon and Washington. Lone Pine Publishing, Edmonton, Alberta.

Gunther, E. 1973. Ethnobotany of western Washington: the knowledge and use of indigenous plants by native Americans. Revised edition. University of Washington Press, Seattle, Washington.

Hafliger, E. and H. Scholz. 1981. Grass weeds. Volume 2. Documenta. Ciba-Geigy Ltd., Basle, Switzerland.

Hallworth, B. and C.C. Chinnappa. 1997. Plants of Kananaskis Country in the Rocky Mountains of Alberta. University of Alberta Press, Edmonton, Alberta, and University of Calgary Press, Calgary, Alberta.

Harms, V.L. 1983. Chenopodiaceae (goosefoot family). W.P. Fraser Herbarium, University of Saskatchewan, Saskatoon, Saskatchewan. Preliminary Saskatchewan Flora Contribution.

Harms, V.L., J.H. Hudson and G.F. Ledingham. 1986. *Rorippa truncata*, the blunt-fruited yellow cress, new for Canada and *R. tenerrima*, the slender yellow cress, in southern Saskatchewan and Alberta. Canadian Field-Naturalist 100(1):45–51.

Harms, V.L., P.A. Ryan and J.A. Haraldson. 1992. The rare and endangered native vascular plants of Saskatchewan. Saskatchewan Natural History Society, Saskatoon, Saskatchewan.

References

Harrington, H.D. 1967. Edible native plants of the Rocky Mountains. University of New Mexico Press, Albuquerque, New Mexico.

Haufler, C.H. and M.D. Windham. 1991. New species of North American *Cystopteris* and *Polypodium*, with comments on their reticulate relationships. American Fern Journal 81:7–23.

Hermann, F.J. 1970. Manual of the Carices of the Rocky Mountains and Colorado Basin. United States Department of Agriculture Forest Service, Washington, D.C. Agriculture Handbook No. 374.

Hermann, F.J. 1975. Manual of the rushes (*Juncus* spp.) of the Rocky Mountains and Colorado Basin. United States Department of Agriculture Forest Service, Rocky Mountain Forest and Range Experiment Station, Fort Collins, Colorado. General Technical Report No. RM–18.

Hickman, J.C., ed. 1993. The Jepson manual: higher plants of California. University of California Press, Berkeley, California.

Hinds, H.R. 1983. The rare vascular plants of New Brunswick. National Museum of Natural Sciences, Ottawa, Ontario. Syllogeus No. 50.

Hiratsuka, Y. 1987. Forest tree diseases of the prairie provinces. Canadian Forestry Service, Northern Forest Research Centre, Edmonton, Alberta. Information Report No. NOR–X–286.

Hitchcock, A.S. 1950. Manual of the grasses of the United States. Second edition. Government Printing Office, Washington, D.C. United States Department of Agriculture Publication No. 200.

Hitchcock, C.L. and A. Cronquist. 1973. Flora of the Pacific Northwest: an illustrated manual. University of Washington Press, Seattle, Washington.

Hitchcock, C.L., A. Cronquist, M. Ownbey and J.W. Thompson. 1955–69. Vascular plants of the Pacific Northwest. 5 volumes. University of Washington Press, Seattle, Washington.

Holroyd, G.L., W.B. McGillivray, P.H.R. Stepney, D.M. Ealey, G.C. Trottier and K.E. Eberhart, eds. 1987. Proceedings of the Workshop on Endangered Species in the Prairie Provinces. Natural History Section, Provincial Museum of Alberta, Edmonton, Alberta.

Hosie, R.C. 1979. Native trees of Canada. Fitzhenry & Whiteside Ltd., Markham, Ontario.

Hrapko, J.O. 1989. Prairie spiderwort (*Tradescantia occidentalis*) in Alberta. Iris 4:4.

Hrapko, J.O. 1991. Rare and endangered vascular plants. Provincial Museum of Alberta, Edmonton, Alberta. Natural History Reference List No. 209.

Hudson, J.H. 1977. *Carex* in Saskatchewan. Bison Publishing House, Saskatoon, Saskatchewan.

Hudson, J.H. 1978. *Atriplex powellii* at Cabri Lake. Blue Jay 36(3):137–138.

Hulten, E. 1958. The amphi-Atlantic plants and their phytogeographical connections. Kungliga Svenska Vetenskapsakademiens Handlingar 7:1–340.

Hulten, E. 1968. Flora of Alaska and neighboring territories. Stanford University Press, Stanford, California.

Hulten, E. 1971. The circumpolar plants. Volume 2. Dicotyledons. Kungliga Svenska Vetenskapsakademiens Handlingar 13:1–463.

Hunter, A.A. 1992. Utilization of *Hordeum pusillum* (little barley) in the midwest United States: applying Rindos' co-evolutionary model of domestication. Ph.D. thesis, University of Missouri, Columbia, Missouri.

Hurd, E.G., N.L. Shaw, J. Mastrogiuseppe, L.C. Smithman and S. Goodrich. 1998. Field guide to intermountain sedges. United States Department of Agriculture Forest Service, Rocky Mountain Research Station, Ogden, Utah. General Technical Report No. RMRS–GTR–10.

Jackson, L.E. and L.C. Bliss. 1984. Phenology and water relations of three plant life forms (*Penstemon heterodoxus*, *Polygonum minimum*, *Saxifraga aprica*) in a dry tree-line meadow, Washington. Ecology 65:1302–14.

Johnson, D.M. 1986. Systematics of the New World species of *Marsilea* (Marsileaceae). American Society of Plant Taxonomists, Ann Arbor, Michigan. Systematic Botany Monographs Vol. 11.

Johnson, D., L. Kershaw, A. MacKinnon and J. Pojar. 1995. Plants of the western boreal forest and aspen parkland. Lone Pine Publishing, Edmonton, Alberta.

Johnson, H. and B. Hallworth. 1975. Further discoveries of sand verbena in Alberta. Blue Jay 33(1):13–15.

References

Johnson, J.D. 1989. Uncommon plants from Ram Mountain, Alberta. Alberta Naturalist 19:31–34.

Johnston, A. 1987. Plants and the Blackfoot. Lethbridge Historical Society, Lethbridge, Alberta. Occasional Paper No. 15.

Jones, V.P., D.W. Davis, S.L. Smith and D.B. Allred. 1989. Phenology of apple maggot (Diptera: Tephritidae) associated with cherry and hawthorn in Utah. Journal of Entomology 83(3):788–92.

Kartesz, J.T. 1994. A synonomized checklist of the vascular flora of the United States, Canada and Greenland. Timber Press, Portland, Oregon.

Kartesz, J.T. and K.N. Gandhi. 1990. Nomenclatural notes for the North American flora. Part 2. Phytologia 68:421–27.

Kelly, I.T. 1932. Ethnography of the Surprise Valley Paiutes. University of California Publications in American Archaeology and Ethnography 31: 67–210.

Kerik, J. and S. Fisher. 1982. Living with the land: use of plants by the native people of Alberta. Provincial Museum of Alberta, Edmonton, Alberta.

Kershaw, L.J. 1976. A phytogeographical survey of rare, endangered, and extinct vascular plants in the Canadian Flora. M.Sc. thesis, University of Waterloo, Waterloo, Ontario.

Kershaw, L.J. and J.K. Morton. 1976. Rare and potentially endangered species in the Canadian flora: a preliminary list of vascular plants. Canadian Botanical Association Bulletin 9(2):26–30.

Kershaw, L.J., A. MacKinnon and J. Pojar. 1998. Plants of the Rocky Mountains. Lone Pine Publishing, Edmonton, Alberta.

Kerstetter, T.A. 1994. Taxonomic Investigation of *Erigeron lackschewitzii*. M.Sc. thesis. Montana State University, Bozeman, Montana.

Klinka, K., V.J. Krajina, A. Ceska and A.M. Scagel. 1989. Indicator plants of coastal British Columbia. University of British Columbia Press, Vancouver, British Columbia.

Kott, L. and D.M. Britton. 1983. Spore morphology and taxonomy of *Isoetes* in northeastern North America. Canadian Journal of Botany 61:3140–63.

Krause, D.L. and K.I. Beamish. 1973. Notes on *Saxifraga occidentalis* and closely related species in British Columbia. Syesis 6:105–13.

Krenzer, E.G. Jr., D.N. Moss and R.K. Crookston. 1975. Carbon dioxide compensation points of flowering plants. Plant Physiology 56: 194–206.

Kuijt, J. 1973. New plant records in Waterton Lakes National Park, Alberta. Canadian Field-Naturalist 87:67–69.

Kuijt, J. 1982. A flora of Waterton Lakes National Park. University of Alberta Press, Edmonton, Alberta.

Kuijt, J. and G.R. Michner. 1985. First record of the bitterroot, *Lewisia rediviva*, in Alberta. Canadian Field Naturalist 99:264–66.

Kuijt, J. and J.A. Trofymow. 1975. Range extensions of two rare Alberta shrubs—Rocky Mountain juniper (*Juniperus scopulorum*) and mock orange (*Philadelphus lewisii*). Blue Jay 33:96–98.

Lackschewitz, K. 1991. Vascular plants of west-central Montana: identification guidebook. United States Department of Agriculture Forest Service, Intermountain Research Station, Ogden, Utah. General Technical Report No. INT–277.

Lamanova, T.G. 1991. *Puccinellia* spp. in halophyte phytocenoses in the Karasuk River valley and their nutritive value. Rastitel'nye Resursy 26: 400–09. (in Russian, translated abstract in Biological Abstracts 91 (11), ref. 114185).

Lancaster, J., ed. 1997. Guidelines for rare plant surveys. Alberta Native Plant Council, Edmonton, Alberta.

Larson, G.E. 1993. Aquatic and wetland vascular plants of the northern great plains. United States Department of Agriculture Forest Service, Rocky Mountain Forest and Range Experiment Station, Fort Collins, Colorado. General Technical Report No. RM–238.

Lauriault, J. 1989. Identification guide to the trees of Canada. Fitzhenry & Whiteside Ltd., Markham, Ontario.

References

Lee, P.G. 1980. *Boschniakia rossica*, northern ground cone, a vascular plant new to Alberta. Canadian Field-Naturalist 94:341.

Legasy, K., S. LaBelle-Beadman and B. Chambers. 1995. Forest plants of northeastern Ontario. Lone Pine Publishing, Edmonton, Alberta.

Lellinger, D.B. 1981. Notes on North American ferns. American Fern Journal 71:90–94.

Lesica, P. 1985. Checklist of the vascular plants of Glacier National Park, Montana, U.S.A. Monograph No. 4. Montana Academy of Sciences. Supplement to the Proceedings Vol. 44.

Lesica, P. and J.S. Shelly. 1991. Sensitive, threatened and endangered vascular plants of Montana. Montana State Library, Helena, Montana. Occasional Publications of the Montana Natural Heritage Program No. 1.

Lesica, P. and K. Ahlenslager. 1994. Demographic monitoring of three species of *Botrychium* (Ophioglossaceae) in Waterton Lakes National Park, Alberta: 1993 progress report. Division of Biological Sciences, University of Montana, Missoula, Montana.

Lewis, H. and J. Szweykowski. 1964. The genus *Gayophytum* (Onagraceae). Brittonia 16:343–91.

Looman, J. 1982. Prairie grasses identified and described by vegetative characters. Agriculture Canada, Ottawa, Ontario. Publication No. 1413.

Looman, J. and K.F. Best. 1979. Budd's flora of the Canadian prairie provinces. Agriculture Canada Research Branch, Ottawa, Ontario. Publication No. 1662.

Löve, D. 1969. *Papaver* at high altitudes in the Rocky Mountains. Brittonia 21:1–10.

Löve, D. and N.J. Freedman. 1956. A plant collection from the southwestern Yukon. Botaniska Notiser 109:153–211.

Luer, C.A. 1975. The native orchids of the United States and Canada excluding Florida. New York Botanical Garden, Ipswich, New York.

Lyons, C.P. and B. Merilees. 1995. Trees, shrubs and flowers to know in British Columbia and Washington. Lone Pine Publishing, Edmonton, Alberta.

Mabey, R. 1977. Plantcraft: a guide to the everyday use of wild plants. Universe Books, New York, New York.

Macdonald, I.D. and B. Smith. 1995. Status report on the Nebraska sedge (*Carex nebrascensis*) in Canada. Committee on the Status of Endangered Wildlife in Canada, Ottawa, Ontario.

Mackenzie, K.K. 1940. North American Cariceae. 2 volumes. New York Botanical Garden, New York, New York.

MacKinnon, A., J. Pojar and J.R. Coup. 1992. Plants of northern British Columbia. Lone Pine Publishing, Edmonton, Alberta.

Maher, R.V., D.L. White, G.-W. Argus and P.A. Keddy. 1978. The rare vascular plants of Nova Scotia. National Museum of Natural Sciences, Ottawa, Ontario. Syllogeus No. 18.

Maher, R.V., G.W. Argus, V.L. Harms and J.H. Hudson. 1979. The rare vascular plants of Saskatchewan. National Museum of Natural Sciences, Ottawa, Ontario. Syllogeus No. 20.

Maquire, B. 1946. Studies in the Caryophyllaceae: II. *Arenaria nuttallii* and *Arenaria filiorum*, section *Alsini*. Madrono 8:258–63.

Mason, H.L. 1941. The taxonomic status of *Microsteris* Greene. Madrono 6:122–27.

McCone, M.J. 1985. Reproductive biology of several bromegrasses (*Bromus*): breeding system, pattern of fruit maturation, and seed set. American Journal of Botany 72:1334–39.

McCone, M.J. 1989. Intraspecific variation on pollen yield in bromegrass (Poaceae: *Bromus*). American Journal of Botany 76:231–37.

McJannet, C.L., G.W. Argus and W.J. Cody. 1995. Rare vascular plants in the Northwest Territories. Canadian Museum of Nature, Ottawa, Ontario. Syllogeus No. 73.

McJannet, C.L., G.W. Argus, S. Edlund and J. Cayouette. 1993. Rare vascular plants in the Canadian Arctic. Canadian Museum of Natural Sciences, Ottawa, Ontario. Syllogeus No. 72.

References

McNeill, J. 1980. The delimitation of *Arenaria* (Caryophyllaceae) and related genera in North America, with 11 new combinations in *Minuartia*. Rhodora 82:495–502.

Mitchell, W.W. 1962. Variation and relationships in some rhizomatous species of *Muhlenbergia*. Ph.D. thesis, Iowa State University, Ames, Iowa.

Mohlenbrock, R.H., ed. 1976. The illustrated flora of Illinois. Southern Illinois University Press, Carbondale and Edwardsville, Illinois.

Moore, M. 1979. Medicinal plants of the mountain west. Museum of New Mexico Press, Santa Fe, New Mexico.

Moore, R.J. and C. Frankton. 1967. Cytotaxonomy of foliose thistles (*Cirsium* spp. aff. *C. foliosum*) of western North America. Canadian Journal of Botany 45:1733–49.

Morris, M.S., J.E. Schmautz and P.F. Stickney. 1986. Winter field key to the native shrubs of Montana. Montana Forest and Conservation Experiment Station, Montana State University and Intermountain Forest and Range Experiment Station, United States Department of Agriculture Forest Service, Montana. Bulletin No. 23.

Mosquin, T. and C. Suchal, eds. 1977. Canada's threatened species and habitats: proceedings of the symposium on Canada's threatened species and habitats, May 20–24, 1976, Ottawa, Ontario. Canadian Nature Federation, Ottawa, Ontario. Publication No. 6.

Moss, E. 1944. *Lilaea scilloides* in southeastern Alberta. Rhodora 46:205–06.

Moss, E.H. 1959. Flora of Alberta. University of Toronto Press, Toronto, Ontario.

Moss, E.H. 1983. Flora of Alberta. Second edition. University of Toronto Press, Toronto, Ontario.

Moss, E.H. and G. Pegg. 1963. Noteworthy plant species and communities in west central Alberta. Canadian Journal of Botany 48:1431–37.

Mueggler, W. and W. Stewart. 1980. Grassland and shrubland habitat types of western Montana. United States Department of Agriculture Forest Service, Intermountain Forest and Range Experiment Station, Ogden, Utah. General Technical Report No. INT–66.

Muenscher, W.C. 1944. Aquatic plants of the United States. Cornell University Press, Ithaca, New York.

Mulligan, G. A. 1970. A new species of *Draba* in the Kananaskis range of southwestern Alberta. Canadian Journal of Botany 48:1897–98.

Mulligan, G.A. 1970. Cytotaxonomic studies of *Draba glabella* and its close allies in Canada and Alaska. Canadian Journal of Botany 48(8):1431–37.

Mulligan, G.A. 1971. Cytotaxonomic studies of *Draba* species of Canada and Alaska: *D. ventosa*, *D. ruaxes*, and *D. paysonii*. Canadian Journal of Botany 49:1455–60.

Mulligan, G.A. 1974. Cytotaxonomic studies of *Draba nivalis* and its close allies in Canada and Alaska. Canadian Journal of Botany 52:1793–1801.

Mulligan, G.A. 1976. The genus *Draba* in Canada and Alaska: key and summary. Canadian Journal of Botany 54:1386–93.

Mulligan, G.A. and D.B. Munro. 1990. Poisonous plants of Canada. Biosystematics Research Centre, Agriculture Canada, Ottawa, Ontario. Publication No. 11842/E.

Murray, D.F. 1970. *Carex podocarpa* and its allies in North America. Canadian Journal of Botany 48:313–24.

Nelson, J.R. 1984. Rare plant field survey guidelines. J.P. Smith and R. York, eds. Inventory of rare and endangered vascular plants of California. Third edition. California Native Plant Society, Berkeley, California.

Nelson, J.R. 1986. Rare plant surveys: techniques for impact assessment. Natural Areas Journal 5(3):18–30.

Nelson, J.R. 1987. Rare plant surveys: techniques for impact assessment. Pages 159–66 in T.S. Elias, ed. Conservation and management of rare and endangered plants. California Native Plant Society, Sacramento, California.

Newmaster, S.G., A.G. Harris and L.J. Kershaw. 1997. Wetland plants of Ontario. Lone Pine Publishing, Edmonton, Alberta.

References

Niehaus, T.F. and C.L. Ripper. 1976. A field guide to the Pacific states wildflowers. Houghton Mifflin Co., New York, New York.

Nilsson, O. 1971. Studies in *Montia* L. and *Claytonia* L. and allied genera. Part 5. The genus *Montiastrum* (Gray) Rydb. Botaniska Notiser 124:119–214.

Ogilvie, R.T. 1962. Notes on plant distribution in the Rocky Mountains. Canadian Journal of Botany. 40:1091–94.

Packer, J.G. 1972. A taxonomic and phytogeographic review of some arctic and alpine *Senecio* species. Canadian Journal of Botany 50:507–18.

Packer, J.G. 1993. Two new combinations in *Triantha* (Liliaceae). Novon 3:278–79.

Packer, J.G. and C.E. Bradley. 1984. A checklist of the rare vascular plants of Alberta. Provincial Museum of Alberta, Edmonton, Alberta. Natural History Occasional Paper No. 5.

Packer, J.G. and M.G. Dumais. 1972. Additions to the flora of Alberta. Canadian Field-Naturalist 86:269–74.

Packer, J.G. and D.H. Vitt. 1974. Mountain Park: a plant refugium in the Canadian Rocky Mountains. Canadian Journal of Botany 52:1393–1409.

Parish, R., R. Coupe and D. Lloyd. 1996. Plants of southern interior British Columbia. Lone Pine Publishing, Edmonton, Alberta.

Pavlik, L. 1995. *Bromus* L. of North America. Royal British Columbia Museum, Victoria, British Columbia.

Peterson, L.A. 1977. A field guide to edible wild plants of eastern and central North America. Houghton Mifflin Co., New York, New York.

Peterson, R.T. and M. McKenny. 1968. Northeastern wildflowers: a fieldguide to wildflowers. Houghton Mifflin Co., New York, New York.

Petrides, G.A. 1992. A fieldguide to western trees, western United States and Canada. Houghton Mifflin Co., New York, New York.

Petrie, W. 1981. Guide to orchids of North America. Hancock House, Vancouver, British Columbia.

Pohl, R.W. 1969. *Muhlenbergia*, subgenus *Muhlenbergia* (Gramineae) in North America. American Midland Naturalist 82:512–42.

Polunin, N. 1959. Circumpolar arctic flora. Clarendon Press, Oxford, England.

Porsild, A.E. 1959. Botanical excursion to Jasper and Banff national parks, Alberta. National Museum of Canada, Ottawa, Ontario.

Porsild, A.E. 1964. Illustrated flora of the Canadian Arctic Archipelago. Second edition. National Museums of Canada, Ottawa, Ontario. Bulletin No. 146.

Porsild, A.E. and W.J. Cody. 1980. Vascular plants of continental Northwest Territories, Canada. National Museum of Natural Sciences, National Museums of Canada, Ottawa, Ontario.

Porsild, A.E. and D.T. Lid. 1979. Rocky Mountain wildflowers. National Museum of Natural Sciences, National Museums of Canada, Ottawa, Ontario.

Potter, M. 1996. Central Rockies wildflowers. Luminous Press, Banff, Alberta.

Primack, R.B. 1980. Phenotypic variation of rare and widespread species of *Plantago*. Rhodora 82:87–95.

Purdy, B.G. and S.E. Macdonald. 1992. Status report on the sand stitchwort (*Stellaria arenicola* Raup.) in Canada. Committee on the Status of Endangered Wildlife in Canada, Ottawa, Ontario.

Purdy, B.G., R.J. Bayer and S.E. MacDonald. 1994. Genetic variation, breeding system evolution, and conservation of the narrow sand dune endemic *Stellaria arenicola* and the widespread *S. longipes* (Caryophyllaceae). American Journal of Botany 81(7):904–11.

Rabinowitz, D. 1981. Seven forms of rarity. The biological aspects of rare plant conservation. H. Synge, ed. John Wiley and Sons Ltd., New York, New York.

Ramaley, F. 1939. Sand hill vegetation of northeastern Colorado. Ecological Monographs 9:1–51.

Raven, P. 1969. A revision of the genus *Camissonia* (Onagraceae). Smithsonian Institute Press, Washington, D.C.

References

Raven, P. and D. Moore. 1965. A revision of *Boisduvalia* (Onagraceae). Brittonia 17:238–54.

Reddoch, J.M. and A.H. Reddoch. The orchids in the Ottawa district. Canadian Field-Naturalist 111(1):165–67.

Reynolds, D.N. 1984. Alpine annual plants: phenology, germination, photosynthesis, and growth of three Rocky Mountain species (*Koenigia islandica, Polygonum confertiflorum, Polygonum douglasii*). Ecology 65:759–66.

Reynolds, D.N. 1984. Population dynamics of three annual species of alpine plants in the Rocky Mountains (*Koenigia islandica, Polygonum confertiflorum, Polygonum douglasii*). Oecologia 62:250–55.

Ritchie, J.C. 1956. The vegetation of northern Manitoba. Part 1. Studies in the southern spruce forest zone. Canadian Journal of Botany 34:523–61.

Robertson, A. 1984. *Carex* of Newfoundland. Canadian Forestry Service, Newfoundland Forest Research Centre, St. John's, Newfoundland. Information Report No. N–X–219.

Rollins, R.C. 1993. The Cruciferae of continental North America. Stanford University Press, Stanford, California.

Romuld, M.A. 1991. Rare plants of Dinosaur Provincial Park, Alberta. Alberta Naturalist 21:16–22.

Rossbach, R.P. 1940. *Spergularia* in North and South America. Rhodora 42:57–83, 105–43, 158–93, 203–13.

Russell, W.B. 1979. Survey of native trees and shrubs on disturbed lands in the eastern slopes. Alberta Forest Service, Edmonton, Alberta.

Rydberg, P.A. 1932. Flora of the prairies and plains of central North America. Hafner Publishing Co., New York, New York.

Rydberg, P.A. 1969. Flora of the Rocky Mountains and adjacent plains. Second edition. Hafner Publishing Co., New York, New York.

Savile, D.B.O. 1969. Interrelationships of *Ledum* species and their rust parasites in western Canada and Alaska. Canadian Journal of Botany 47:1085–1100.

Savile, D.B.O. 1972. Arctic adaptations in plants. Canada Department of Agriculture Research Branch, Ottawa, Ontario. Monograph No. 6.

Savile, D.B.O. 1975. Evolution and biogeography of Saxifragaceae with guidance from their rust parasites. Annals of the Missouri Botanical Garden 62:354–61.

Schemske, D.W., B.C. Husband, M.H. Ruckelshaus, C. Goodwillie, I.M. Parker and J.G. Bishop. 1994. Evaluation approaches to the conservation of rare and endangered plants. Ecology 75(3):584–606.

Schofield, J.J. 1989. Discovering wild plants: Alaska, western Canada, the Northwest. Alaska Northwest Books, Anchorage, Alaska.

Schofield, W.B. 1959. The salt marsh vegetation of Churchill, Manitoba, and its phytogeographic implications. National Museum of Canada Bulletin 160:107–32.

Scoggan, H.J. 1979. Flora of Canada. 4 volumes. National Museum of Natural Sciences, Ottawa, Ontario. Publications in Botany No. 7.

Scotter, G.W. and H. Flygare. 1986. Wildflowers of the Canadian Rockies. Hurtig Publishers, Edmonton, Alberta.

Scotter, G.W. and J.H. Hudson. 1975. *Carex illota* L.H. Bailey in Alberta. Canadian Field-Naturalist 89:74–75.

Sczawinski, A.F. and N.J. Turner. 1980. Wild green vegetables of Canada. National Museum of Natural Sciences, Ottawa, Ontario.

Shaw, K. 1966. A new species of *Oxytropis* in Alberta. Blue Jay 24:40.

Shaw, P.J. and D. On. 1979. Plants of Waterton–Glacier national parks and the northern Rockies. Mountain Press Publishing Company, Missoula, Montana.

Shay, J.M. 1974. Preliminary list of endangered and rare species in Manitoba. University of Manitoba Botany Department, Winnipeg, Manitoba.

References

Slack, A. 1979. Carnivorous plants. The MIT Press, Cambridge, Massachusetts.

Smith, B.M. 1992. Status report on the Kananaskis whitlow-cress (*Draba kananaskis*) in Canada. Committee on the Status of Endangered Wildlife in Canada, Ottawa, Ontario.

Smith, B.M. 1992. Status report on the slender mouse-ear-cress (*Halimolobos virgata* (Nutt.) O.E. Schulz) in Canada. Committee on the Status of Endangered Wildlife in Canada, Ottawa, Ontario.

Smith, B.M. 1993. Status report on the little barley (*Hordeum pusillum*) in Canada. Committee on the Status of Endangered Wildlife in Canada, Ottawa, Ontario.

Smith, B.M. 1994. Status report on blue phlox (*Phlox alyssifolia* Greene) in Canada. Committee on the Status of Endangered Wildlife in Canada, Ottawa, Ontario.

Smith, B.M. 1994. Status report on dwarf fleabane *(Erigeron radicatus* Hook.) in Canada. Committee on the Status of Endangered Wildlife in Canada, Ottawa, Ontario.

Smith, B.M. 1995. Status report on the hare-footed locoweed (*Oxytropis lagopus* Nutt.) in Canada. Committee on the Status of Endangered Wildlife in Canada, Ottawa, Ontario.

Smith, B.M. 1996. Status report on the rush pink (*Stephanomeria runcinata*) in Canada. Committee on the Status of Endangered Wildlife in Canada, Ottawa, Ontario.

Smith, B.M. 1997. Status report on tiny cryptanth (*Cryptantha minima*) in Canada. Committee on the Status of Endangered Wildlife in Canada, Ottawa, Ontario.

Smith, B.M. and C. Bradley. 1992. Status report on sand verbena (*Abronia micrantha* Torr.) in Canada. Committee on the Status of Endangered Wildlife in Canada, Ottawa, Ontario.

Smith, B.M. and C. Bradley. 1992. Status report of smooth goosefoot (*Chenopodium subglabrum* (S. Wats.) A. Nels.) in Canada. Committee on the Status of Endangered Wildlife in Canada, Ottawa, Ontario.

Smith, B.M. and C. Bradley. 1992. Status report on the western spiderwort (*Tradescantia occidentalis* (Britt) Smyth) in Canada. Committee on the Status of Endangered Wildlife in Canada, Ottawa, Ontario.

Smith, B.W. 1968. Cytogeography and cytotaxonomic relationships of *Rumex paucifolius*. American Journal of Botany 55:673–83.

Smithsonian Institution. 1975. Report on endangered and threatened plant species of the United States. United States Government Printing Office. Washington, D.C. Serial No. 94–A.

Smreciu, E.A. and R.S. Currah. 1989. A guide to the native orchids of Alberta. University of Alberta Devonian Botanic Garden, Edmonton, Alberta.

Smreciu, E.A., J. Hobden and R. Hermesh. 1992. A checklist of the vascular flora in the vicinity of the Oldman River Dam. Alberta Environmental Centre, Vegreville, Alberta.

Soper, J.H. and M.L. Heimburger. 1982. Shrubs of Ontario. Royal Ontario Museum, Toronto, Ontario.

Soreng, R.J., J.I. Davis and J.J. Doyle. 1990. A phylogenetic analysis of chloroplast DNA restriction site variation in Poaceae subfam. Pooideae. Plant Systematics and Evolution 172:83–97.

Stearn, W.T. 1966. Botanical Latin. Thomas Nelson and Sons (Canada) Ltd., Don Mills, Ontario.

Stearn, W.T. 1983. Botanical Latin: history, grammar, syntax, terminology and vocabulary. Third revised edition. David and Charles, Vermont.

Steyermark, J.A. 1963. Flora of Missouri. Iowa State University Press, Ames, Iowa.

Stoddart, L.A. 1941. The Palouse grassland association in northern Utah. Ecology 22:158–63.

Straley, G.B., R.L. Taylor and G.W. Douglas. 1985. The rare vascular plants of British Columbia. National Museum of Natural Sciences, Ottawa, Ontario. Syllogeus No. 59.

Stubbendieck, J., S.L. Hatch and K.J. Kjar. 1982. North American range plants. Second edition. University of Nebraska Press, Lincoln, Nebraska.

Stuckey, R. 1972. Taxonomy and distribution of the genus *Rorippa* (Cruciferae) in North America. Sida 4:279–428.

Stutz, H.C. and L. Pope. 1973. The origin of spreading wheatgrass (*Agropyron scribneri* Vasey). Proceedings of the Western Grass Breeders Work Planning Conference, Kansas State University, Manhattan, Kansas.

Svenson, H.K. 1957. *Eleocharis*. North American Flora 18(9):509–40.

References

Szczawinski, A.F. 1975. Orchids of British Columbia. British Columbia Provincial Museum, Victoria, British Columbia.

Tannas, K. 1997. Common plants of the western rangelands. 2 volumes. Lethbridge Community College, Lethbridge, Alberta.

Taylor, R. 1965. The genus *Lithophragma* (Saxifragaceae). University of California Publications in Botany 37:1–122.

Taylor, R.L. and B. MacBride. 1977. Vascular plants of British Columbia: a descriptive resource inventory. University of British Columbia Press, Vancouver, British Columbia.

Taylor, T.M.C. 1970. Pacific Northwest ferns and their allies. University of Toronto Press, Toronto, Ontario, and University of British Columbia Press, Vancouver, British Columbia.

Taylor, T.M.C. 1983. The sedge family of British Columbia. British Columbia Provincial Museum, Victoria, British Columbia. Handbook No. 43.

Tingley, D. 1987. Endangered plants in Alberta: alternatives for legal protection. Pages 115–18 in G.L. Holroyd et al., eds. Proceedings of the Workshop on Endangered Species in the Prairie Provinces. Provincial Museum of Alberta, Edmonton, Alberta. Natural History Occasional Paper No. 9.

Tisdale, E.W. and A.C. Budd. 1948. Range extensions for three grasses in western Canada. Canadian Field-Naturalist 62:173–75.

Turner, B.L. 1956. A cytotaxonomic study of the genus *Hymenopappus* (Compositae). Rhodora 58:208–42.

Turner, N.J. 1978. Food plants of British Columbia Indians. Part 2. Interior Peoples. Royal British Columbia Museum, Victoria, British Columbia.

Uhl, C.H. 1977. Cytogeography of *Sedum lanceolatum* and its relatives. Rhodora 79:95–114.

Uphof, J.C. 1968. Dictionary of economic plants. Verlag Von J. Cramer, New York, New York.

Viereck, L.A. and E.L. Little, Jr. 1972. Alaska trees and shrubs. United States Department of Agriculture, Washington, DC. Agriculture Handbook No. 410.

Vitt, D.H. and R.J. Belland. 1996. Attributes of rarity among Alberta mosses: patterns and prediction of species diversity. The Bryologist 100(1):1–12.

Voss, E.G. 1972. Michigan flora. Part 1. Gymnosperms and monocots. Cranbrook Institute of Science, Bloomfield Hills, Michigan. Bulletin No. 55.

Voss, E.G. 1985. Michigan Flora. Part 2. Dicots (Saururaceae–Cornaceae). Cranbrook Institute of Science, Bloomfield Hills, Michigan. Bulletin No. 59.

Wagner, W.H., Jr. and F.S. Wagner. 1981. Three new species of moonworts, *Botrychium* subg. *Botrychium* (Ophioglossaceae), from North America. American Fern Journal 71:20–30.

Wagner, W.H., Jr. and F.S. Wagner. 1983. Genus communities as a systematic tool in the study of New World *Botrychium* (Ophioglossaceae). Taxon 32:51–63.

Wagner, W.H., Jr. and F.S. Wagner. 1983. Two moonworts of the Rocky Mountains: *Botrychium hesperium* and a new species formerly confused with it. American Fern Journal 73:53–62.

Wagner, W.H., Jr. and F.S. Wagner. 1986. Three new species of moonworts (*Botrychium* subg. *Botrychium*) endemic in western North America. American Fern Journal 76:33–47.

Wagner, W.H., Jr. and F.S. Wagner. 1990. Moonworts (*Botrychium* subg. *Botrychium*) of the upper Great Lakes region, United States and Canada, with descriptions of two new species. Contributions of the University of Michigan Herbarium 17:313–25.

Wagner, W.H., Jr., F.S. Wagner, C.H. Haufler and J.L. Emerson. 1984. A new nothospecies of moonwort (Ophioglossaceae, *Botrychium*). Canadian Journal of Botany 62:629–34.

Wallis, C. 1977. Preliminary lists of the rare flora and fauna of Alberta. Alberta Provincial Parks, Edmonton, Alberta.

Wallis, C. 1987. Critical, threatened and endangered habitats in Alberta. Pages 49–63 in G.L. Holroyd et al., eds. Proceedings of the Workshop on Endangered Species in the Prairie Provinces. Provincial Museum of Alberta, Edmonton, Alberta. Natural History Occasional Paper No. 9.

References

Wallis, C. 1987. The rare vascular flora of Alberta. Volume 2. A summary of the taxa occurring in the grasslands, parkland and boreal forest. Cottonwood Consultants Ltd., Calgary, Alberta. Alberta Forestry, Lands and Wildlife, Edmonton, Alberta. Publication No. T/164.

Wallis, C. and C. Bradley. 1989. Status report on the western blue flag (*Iris missouriensis* Nutt.). Committee on the Status of Endangered Wildlife in Canada, Ottawa, Ontario.

Wallis, C. and C. Wershler. 1988. Rare wildlife and plant conservation studies in sandhill and sand plain habitats of southern Alberta. Alberta Forestry, Lands and Wildlife, Edmonton, Alberta. Publication No. T/176.

Wallis, C., C. Bradley, M. Fairbarns and V. Loewen. 1987. The rare flora of Alberta. Volume 3. Species summary sheets. Alberta Energy, Forestry, Lands and Wildlife, Edmonton, Alberta. Publication No. T/155.

Wallis, C., M. Fairbarns, V. Loewen and C. Bradley. 1987. The rare flora of Alberta. Volume 4. Bibliography. Alberta Energy, Forestry, Lands and Wildlife, Edmonton, Alberta. Publication No. T/155.

Wallis, C., C. Bradley, M. Fairbarns, J. Packer and C. Wershler. 1986. Pilot rare plant monitoring program in the Oldman regional plan area of southwestern Alberta. Alberta Forestry, Lands and Wildlife, Edmonton, Alberta. Publication No. T/148.

Weber, W.A. 1990. Colorado flora: eastern slope. University Press of Colorado, Niwot, Colorado.

Welsh, S.L. 1974. Anderson's flora of Alaska and adjacent parts of Canada. Brigham Young University Press, Provo, Utah.

White, D.J. and K.L. Johnson. 1980. The rare vascular plants of Manitoba. National Museums of Canada, Ottawa, Ontario. Syllogeus No. 27.

White, H.R. 1979. Endangered—*Cypripedium montanum*. Alberta Naturalist 9, Supplement 1:1.

Whiting, R.E. and P.M. Catling. 1986. Orchids of Ontario. The Canadian Collections Foundation, Ottawa, Ontario.

Whitney, S. 1989. The Audubon Society nature guides. Western forests. Random House of Canada Ltd., Toronto, Ontario.

Wilkinson, K. 1987. The mountain lady's-slipper in Alberta: concerns and recommendations for an endangered plant species. Alberta Forestry, Lands and Wildlife, Edmonton, Alberta. Publication No. T/157.

Wilkinson, K. 1990. Trees and shrubs of Alberta. Lone Pine Publishing, Edmonton, Alberta.

Willard, T. 1992. Edible and medicinal plants of the Rocky Mountains and neighbouring territories. Wild Rose College of Natural Healing, Calgary, Alberta.

Williams, J.G. and A.E. Williams. 1983. Field guide to the orchids of North America. Universe Books, New York, New York.

Willms, W.D. and J.F. Dormaar, eds. 1996. Proceedings of the fourth prairie conservation and endangered species workshop, February 1995 at the University of Lethbridge and Lethbridge Community College, Lethbridge, Alberta. Curatorial Section, Provincial Museum of Alberta, Edmonton, Alberta.

Windham, M.D. 1993. New taxa and nomenclatural changes in the North American fern flora. Contributions of the University of Michigan Herbarium 19:31–61.

Wiseman, E.F. and S.J. Hurst. 1980. Pictures and descriptions of selected seeds not illustrated in Agriculture Handbook 30. Journal of Seed Technology 5(1):1–16.

Wolf, S.J., J.G. Packer and K.E. Denford. 1979. The taxonomy of *Minuartia rossii* (Caryophyllaceae). Canadian Journal of Botany 57:1673–86.

Wong, S.Y., Y. Oshima, J. Pezzuto, H. Fong and N. Farnsworth. 1986. Plant anticancer agents—Triterpenes from *Iris missouriensis*: Iridaceae. Journal of Pharmaceutical Sciences 75(3): 317–20.

Wood, C.E., Jr. 1972. Morphology and phytogeography: the classical approach to the study of disjunctions. Annals of the Missouri Botanical Garden 59:107–24.

Index

Abronia micrantha, 72
Acorus
  americanus, 49
  calamus, 49
adder's-mouth, white, 54
adder's-tongue family, 354–58, 376–77
Adenocaulon bicolor, 387
Adiantum
  aleuticum, 359
  pedatum, 359
Agavaceae, 48
agave family, 48
Agoseris
  aurantiaca, 254
  cuspidata, 248
  glauca, 248
  glauca, 254
  lackschewitzii, 254
Agropyron
  albicans, 342
  scribneri, 342
  spicatum, 342
Agrostis
  borealis, 321
  exarata, 321
  humilis, 320
  mertensii, 321
  thurberiana, 320
alder, green, 204
Alismataceae, 39
Allium spp., 192
  cernuum, 43
  geyeri, 43
  rubrum, 43
  rydbergii, 43
  textile, 43
Alopecurus
  alpinus, 318
  borealis, 318
  occidentalis, 318
alumroot, 122, 124
  alpine, 121
  smooth, 121
amaranth family, 71
amaranth, Californian, 71
Amaranthaceae, 71
Amaranthus californicus, 71
Ambrosia
  acanthicarpa, 215
  artemisiifolia, 215
  psilostchya, 215
  trifida, 215
Anagallis minima, 168
Anaphalis margaritacea, 233
Andropogon scoparius, 310

Anemone
  nemorosa var. *bifolia*, 84
  quinquefolia var. *bifolia*, 84
  quinquefolia var. *interior*, 84
anemone, wood, 84
Antennaria
  aromatica, 230
  corymbosa, 231
  luzuloides, 232
  monocephala, 231
  pulvinata, 230
Anthoxanthum monticola, 314
Apiaceae, 161–63
Aquilegia
  canadensis var. *formosa*, 90
  columbiana, 90
  flavescens, 90
  formosa var. *formosa*, 90
  jonesii, 89
Arabidopsis salsuginea, 103
Arabis
  lemmonii, 104
  lyallii, 104
Araliaceae, 19
Arctagrostis
  arundinacea, 319
  latifolia
    ssp. *arundinacea*, 319
Arenaria
  cylindricarpa, 75
  humifusa, 75
  longipedunculata, 75
  pedunculata, 75
  rossii ssp. *elegans*, 76
Aristida
  longiseta, 313
  purpurea var. *longiseta*, 313
Arnica
  amplexicaulis, 242
  longifolia, 242
  parryi, 243
arnica
  long-leaved, 242
  nodding, 243
  stem-clasping, 242
arrow-grass family, 38
arrowhead
  arum-leaved, 39
  broad-leaved, 39
Artemisia spp., 205
  borealis, 375
  campestris, 375
  campestris ssp. *borealis* var. *borealis*, 375
  furcata var. *furcata*, 240
  herriotii, 241
  hyperborea, 240

  tilesii ssp. *elatior*, 241
  tridentata, 25
  trifurcata, 240
Asclepiadaceae, 175–76
Asclepias
  ovalifolia, 176
  viridiflora, 175
Aster
  campestris, 219
  eatonii, 220
  pauciflorus, 219
  subspicatus, 220
  tanacetifolius, 220
  umbellatus, 221
aster family, 25, 215–55, 387–88
aster
  Eaton's, 220
  few-flowered, 219
  flat-topped white, 221
  leafy-bracted, 220
  meadow, 219
  tansy-leaved, 220
Asteraceae, 25, 215–55, 387–88
Astragalus spp., 135
  alpinus, 132
  bodinii, 132
  kentrophyta
    var. *kentrophyta*, 134
  lotiflorus, 133
  purshii var. *purshii*, 133
  yukonis, 132
Athyrium
  alpestre
    var. *americanum*, 363
  americanum, 363
  distentifolium
    ssp. *americanum*, 363
  filix-femina, 363
Atriplex
  aptera, 16
  argentea, 65
  canescens var. *aptera*, 16
  canescens, 65
  nuttallii, 65
  powellii, 65
  truncata, 65
Avena spp., 326
azalea, alpine, 24
baby-blue-eyes, small, 182
Bacopa rotundifolia, 194
Bahia oppositifolia, 237
Balsamorhiza sagittata, 192
balsamroot, arrow-leaved, 192
Barbarea orthoceras, 102
barley
  foxtail, 344
  little, 344

Index

beak-rush
  slender, 269
  white, 269
beardtongue
  elliptic-leaved, 192
  shrubby, 192
bedstraw
  Labrador, 210
  small, 210
  thin-leaved, 210
beech fern, northern, 368
beggarticks
  common, 234
  tall, 234
begonia, wild, 72
bellflower, arctic, 212
bent grass
  low, 320
  northern, 321
  Thurber's, 320
Bidens frondosa, 234
bilberry
  blue, 23
  bog, 23
  low, 23
  tall, 23
biscuit-root, 161
bishop's-cap, 122, 124
bitter cress
  alpine, 108
  meadow, 110
  Pennsylvanian, 109
  small, 109
bitter-root, 73
  dwarf, 73
bladder catchfly, alpine, 78
bladder fern,
  mountain, 364, 365
bladderpod, northern, 99
bladderwort family, 206–07
bladderwort, horned, 207
blue flag, western, 49
blue-eyed grass
  common, 50
  pale, 50
bluebell family, 212
bluebells,
  long-flowered, 185–86
blueberry, oval-leaved, 23
bluegrass
  bog, 333
  Canby, 379
  early, 332
  Letterman's, 333
  narrow-leaved, 334
  Pacific, 379
  Sandberg, 334, 379
  wavy, 334
  Wheeler's, 332
bluejoint, 322
bluestem, little, 310
bluets, long-leaved, 211
Blysmus rufus, 273
bog adder's-mouth, 54
bog orchid, slender, 55
Boisduvalia glabella, 150
boisduvalia, smooth, 150
Bolboschoenus
  fluviatilis, 274
  maritimus
    ssp. *paludosus*, 274
borage family, 185–89
Boraginaceae, 185–89
Boschniakia rossica, 204
Botrychium
  ascendens, 355
  boreale
    ssp. *obtusilobum*, 377
  boreale, 376
  campestre, 354–55
  crenulatum, 355
  dusenii, 355
  hesperium, 356–57
  lanceolatum ssp.
    angustisegmentum, 356
  lanceolatum
    ssp. *lanceolatum*, 356
  lunaria, 355
  lunaria var. *minganense*, 354
  matricariifolium
    ssp. *hesperium*, 356
  minganense, 354–55
  multifidum, 358
  paradoxum, 356, 357
  pedunculosum, 357
  pinnatum, 377
  simplex, 358
  spathulatum, 354
  tenebrosum, 358
  x *watertonense*, 357
Boykinia heucheriformis, 119
boykinia, heuchera-like, 119
Brasenia schreberi, 83
Brassicaceae, 382, 94–110
Braya
  americana, 107
  glabella, 107
  humilis, 107
  purpurascens, 107, 108
braya
  alpine, 107, 108
  leafy, 107
  purple-leaved, 107
breadroot, Indian, 137
Brickellia grandiflora, 217, 252
brickellia,
  large-flowered, 217, 252
brome
  Canada, 341
  Colombian, 378
  woodland, 378
Bromus
  altissimus, 341
  ciliatus var. *latiglumis*, 341
  incanus, 341
  latiglumis, 341
  purgans var. *incanus*, 341
  vulgaris, 378
broomrape family, 204–05
broomrape, Louisiana, 205
Buchloe dactyloides, 329
buckwheat family, 59–64
buffalo grass, 329
  false, 329
bulrush
  Clinton's, 272
  dwarf, 271–72
  pale, 275
  prairie, 274
  red, 273
  river, 274
  small-fruited, 275
  tufted, 272
bur-reed family, 29–30
bur-reed
  floating-leaved, 29
  globe, 29
  northern, 30
butte-primrose, common, 153
buttercup family, 84–90
buttercup
  alpine, 87
  dwarf, 87
  early, 86
  hairy, 87
  mountain, 87, 88
  snow, 88
  western, 86
butterweed
  alpine meadow, 244–45
  arctic, 244
butterwort
  common, 206
  hairy, 206
  small, 206
Cabombaceae, 83
Calamagrostis
  canadensis, 322
  inexpansa, 322
  lapponica, 322

Index

stricta, 322
stricta ssp. *inexpansa*, 322
Calylophus serrulatus, 151
camas, blue, 45
Camassia quamash
    var. *quamash*, 45
Camissonia
    andina, 152
    breviflora, 153
    hilgardii, 152
Campanula
    lasiocarpa, 212
    uniflora, 212
Campanulaceae, 212
cancer-root,
    one-flowered, 205
caper family, 93
Capparidaceae, 93
Cardamine
    bellidifolia, 108
    oligosperma
        var. *kamtschatica*, 109
    parviflora, 109
    pensylvanica, 109
    pratensis, 110
    umbellata, 109
Carduus scariosus, 246
Carex
    adusta, 279
    aperta, 298
    aquatilis
        var. *aquatilis*, 298, 308
    arcta, 286
    atherodes, 304
    atrata, 299
    atrosquama, 300
    aurea, 293
    backii, 292
    backii var. *saximontana*, 292
    bebbii, 279, 283
    bipartita, 286
    bonplandii var. *minor*, 281
    capitata, 276
    columbiana, 299
    concinna, 290
    crawei, 293
    crawfordii, 283
    deflexa, 291
    depreauxii, 306
    deweyana ssp. *leptopoda*, 285
    deweyana, 285
    dieckii, 281
    disperma, 288
    enanderi, 301
    epapillosa, 299
    eurystachya, 301
    fuliginosa var. *misandra*, 294
    glacialis, 289
    hagiana, 299
    haydeniana, 281
    heleonastes, 286
    heteroneura
        var. *epapillosa*, 299–300
    heterostachya, 293
    hoodii, 277
    hookerana, 277
    houghtoniana, 303
    houghtonii, 303
    hystericina, 305–06
    hystricina, 305
    illota, 281–82
    incurviformis
        var. *incurviformis*, 277
    inflata, 309
    interior, 282
    kelloggii, 301
    lachenalii, 286–87
    lacustris, 304
    lagopina, 286
    lanuginosa, 303
    lenticularis var. *dolia*, 301
    leptopoda, 285
    loliacea, 287–88
    macloviana
        ssp. *haydeniana*, 281
    maritima
        var. *incurviformis*, 277
    mertensii ssp. *mertensii*, 299
    microglochin, 307
    microptera, 284
    mirabilis var. *tincta*, 283
    misandra, 294
    monile, 309
    montanensis, 297
    multiflora, 278
    nebrascensis, 302
    nebraskensis, 302
    norvegica, 296
    oligosperma, 306
    pachystachya, 284
    parryana
        var. *parryana*, 295–96
    pauciflora, 307
    paysonis, 296–97
    pedunculata, 290
    pellita, 303
    petasata, 280
    petricosa, 294
    phaeocephala, 280
    pinguis, 279
    piperi, 284
    platylepis, 284
    podocarpa, 297
    praticola, 280, 284
    preslii, 281
    pseudo-cyperus, 305
    pseudocyperus, 305
    raeana, 309
    raynoldsii, 295
    retrorsa, 309
    riparia var. *lacustris*, 304
    rossii, 291
    rostrata, 308
    rugosperma var. *tonsa*, 291
    saximontana, 292
    scoparia, 282–83
    scopulorum, 298
    setacea, 278
    spaniocarpa, 289
    spectabilis, 297
    stipata, 278
    supina ssp. *spaniocarpa*, 289
    tenera, 283
    tenuiflora, 288
    tincta, 283
    tonsa var. *tonsa*, 291–92
    trisperma, 287
    turgidula, 298
    umbellata, 291
    umbellata var. *tonsa*, 291
    utriculata, 308
    vesicaria var. *vesicaria*, 309
    vulpinoidea, 278
    xerantica, 279
carrot family, 161–63
Caryophyllaceae, 75–82, 380
Castilleja
    cusickii, 384
    lutea, 384
    lutescens, 202, 203
    pallida ssp. *caudata*, 203
    septentrionalis, 202, 203, 384
    sessiliflora, 202
cat's-breeches, 383
cattails, 49
cedar family, 5
cedar, western red, 5
Centunculus minimus, 168
Cerastium
    brachypodum, 380
    nutans
        var. *brachypodum*, 380
chaffweed, 168
Cheilanthes
    feei, 361
    gracillima, 361
Chenopodiaceae, 16, 65–70
Chenopodium
    atrovirens, 66
    dessicatum, 67
    fremontii var. *atrovirens*, 66
    fremontii var. *incanum*, 68
    incanum, 68
    leptophyllum, 66
    pratericola, 66
    subglabrum, 67
    watsonii, 68

Index

chickweed
 American, 80
 long-stalked, 82
 meadow, 81
 sand-dune, 82
 shining, 80
 short-stalk mouse-ear, 380
 umbellate, 82
 wavy-leaved, 81
Chrysanthemum bipinnatum
 var. *huronense*, 239
cinquefoil
 branched, 127
 bushy, 129–30
 Colorado, 126
 feather-leaved, 128
 hairy, 131–32
 Hooker's, 131–32
 low, 130
 Macoun's, 126
 mountain, 128
 one-flowered, 131
 prairie, 127, 128
 sandhills, 130
 sheep, 126, 128
 smooth-leaved, 128
 snow, 131
 staghorn, 127
Cirsium
 foliosum, 246
 hookerianum, 246
 scariosum, 246
clammyweed, 93
Claytonia parvifolia, 74
cliff brake
 Gaston's, 362
 purple, 362
 smooth, 362
club-moss family, 349–50
club-moss
 bog, 349
 mountain, 350
club-rush
 Clinton's, 272
 dwarf, 271–72
 tufted, 272
Clusiaceae, 145
Cnicus scariosus, 246
cockle, nodding, 78
columbine
 crimson, 90
 Sitka, 90
 western, 90
 yellow, 90
comfrey, wild, 188–89
Commelinaceae, 42

Compositae, 25, 215–55, 387–88
Conimitella williamsii, 122
conimitella, Williams', 122
Convolvulaceae, 177
Coptis trifolia, 85
cord grass
 alkali, 328
 prairie, 327–28
Coreopsis tinctoria, 235
cotton-grass
 beautiful, 270–71
 close-sheathed, 270
 dense, 271
 one-spike, 271
 sheathed, 271
cottonwood
 black, 10
 narrow-leaved, 10
 plains, 10
Crassulaceae, 114
Crepis
 atrabarba, 250
 atribarba, 250–51
 exilis, 250
 intermedia, 251
 occidentalis var. *gracilis*, 250
 occidentalis, 251
cress
 mountain, 109
 mouse-ear 103
Cruciferae, 94–110, 382
Cryptantha
 affinis, 187
 celosioides, 375
 fendleri, 187
 interrupta, 375
 macounii, 375
 minima, 187
cryptanthe
 Fendler's, 187
 Macoun's, 375
 slender, 187
 small, 187
 tiny, 187
Cryptogramma stelleri, 360
cuckoo flower, 110
cudweed
 clammy, 233
 common, 233
 tall, 233
Cupressaceae, 5
currant family, 18
currant
 mountain, 18
 trailing black, 18

Cuscuta gronovii, 177
Cynoglossum
 boreale, 188
 officinale, 188
 virginianum
  var. *boreale*, 188–89
Cyperaceae, 266–309
Cyperus
 aristatus, 266
 inflexus, 266
 schweinitzii, 266
 squarrosus, 266
Cypripedium
 acaule, 51
 calceolus, 52
 montanum, 52
 passerinum, 52
Cystopteris
 montana, 364, 365
Danthonia
 californica, 326
 intermedia, 326
 spicata, 326–27
 unispicata, 326
Deschampsia elongata, 323
devil's club, 19
Dichanthelium
 acuminatum
  var. *acuminatum*, 311
 leibergii, 312
 oligosanthes
  var. *wilcoxianum*, 312
 wilcoxianum, 312
 sitchense, 349
ditch-grass family, 37
dodder, common 177
Douglasia montana, 169
douglasia, mountain, 169
Downingia laeta, 213
downingia, 213
Draba
 borealis, 98
 densifolia, 95
 fladnizensis, 95
 glabella, 98
 kananaskis, 96
 longipes, 96
 macounii, 96
 porsildii, 382
 reptans, 94
 ventosa, 97
Dracocephalum
 nuttallii, 191
 parviflorum, 191
dragonhead
 American, 191
 false, 191

Index

Drosera
  anglica, 111
  linearis, 112
  longifolia, 111
Droseraceae, 111–12
Drummond's cinquefoil, 129
dryad, white, 169
Dryas octopetala, 169
Dryopteridaceae, 363–68
Dryopteris
  carthusiana, 367
  cristata, 367–68
  filix-mas, 367
  phegopteris, 368
duckmeal
  Columbian, 41
  northern, 41
duckweed family, 41
duckweed
  common, 41
  larger, 41
dune grass, American 345
dwarf primrose, mountain, 169
Elatinaceae, 146
Elatine
  brachysperma, 146
  triandra, 146
Eleocharis spp., 353
  compressa var. *borealis*, 268
  elliptica, 268
  engelmannii, 267
  nitida, 268
  obtusa var. *engelmannii*, 267
  ovata var. *engelmannii*, 267
  tenuis var. *borealis*, 268
Ellisia nyctelea, 181
Elodea
  bifoliata, 40
  canadensis,
  canadensis, 40
  longivaginata, 40
Elymus
  arenarius ssp. *mollis*, 345
  elymoides, 343
  mollis, 345
  scribneri, 342
  spicatus, 342
  virginicus, 345
Epilobium spp., 213
  affine var. *fastigiatum*, 156
  ciliatum
    var. *glandulosum*, 155
  clavatum, 157
  clavatum var. *glareosum*, 156
  exaltatum, 155
  fastigiatum var.
    glaberrimum, 156
  glaberrimum
    ssp. *fastigiatum*, 156
  glandulosum
    var. *macounii*, 154, 156
  halleanum, 154–55
  lactiflorum, 157
  leptocarpum, 155–56
  luteum, 158
  mirabile, 156
  oreganum, 155
  platyphyllum, 156
  pygmaeum, 150
  saximontanum, 154
Equisetum spp., 160
Ericaceae, 22–24
Erigeron
  compositus, 223
  divaricatus, 227
  divergens, 227
  flagellaris, 228
  grandiflorus, 224
  hyssopifolius, 222
  lackschewitzii, 224
  lanatus, 223
  ochroleucus, 224
  ochroleucus, 226
  pallens, 225
  purpuratus ssp. *pallens*, 225
  radicatus, 226
  trifidus, 223
Eriogonum
  cernuum, 59
  multiceps, 60
  ovalifolium
    var. *ovalifolium*, 60
  pauciflorum, 60
  brachyantherum, 270
  callitrix, 270–71
  scheuchzeri, 271
  vaginatum ssp. *spissum*, 271
Erysimum pallasii, 106
Eupatorium
  bruneri, 216
  maculatum, 216
  purpureum, 216
evening-primrose
  family, 150–59
evening-primrose
  low yellow, 153
  short-flowered, 153
  shrubby, 151
  upland, 152
everlasting
  aromatic, 230
  corymbose, 231
  pearly, 233
  silvery, 232

Fabaceae, 132–40
false dandelion
  orange, 254
  pink, 254
  prairie, 248
  yellow, 248, 254
false gromwell, western, 186
false-asphodel
  sticky, 44
  western, 44
felwort, 174
  marsh, 174
fern family, marsh, 368
fern
  lace, 361
  lady, 363
  long beech, 368
  male, 367
  water, 371
fescue
  alpine, 337, 338
  arctic, 338
  bearded, 340
  bluebunch, 339
  northern rough, 340
  tiny-flowered, 337
  viviparous, 338
  western, 339
Festuca
  altaica, 340
  baffinensis, 338
  brachyphylla, 337
  brachyphylla, 338
  brevifolia var. *endotera*, 337
  brevifolia var. *utahensis*, 337
  idahoensis, 339
  jonesii, 340
  minutiflora, 337
  occidentalis, 339
  ovina var. *minutiflora*, 337
  scabrella, 340
  subulata, 340
  vivipara ssp. *glabra*, 338
  viviparoidea
    ssp. *krajinae*, 338
figwort family, 192–203, 384–86
fir-moss
  alpine, 350
  northern, 350
fleabane
  creeping, 228
  dwarf, 226
  front-range, 224
  large-flowered, 224
  pale alpine, 225
  spreading, 227
  three-forked, 223

Index

trifid-leaved, 223
wild daisy, 222
yellow alpine, 224, 226
flowering-quillwort family, 38
flowering-quillwort, 38
four o'clock family, 72
foxtail, alpine, 318
Franseria acanthicarpa, 215
fringe-cups, 124
fringed gentian
   common, 171
   northern, 171
Galium
   bifolium, 210
   labradoricum, 210
   trifidum, 210
Gayophytum
   helleri var. *glabrum*, 159
   racemosum, 159
gentian family, 171–74
gentian
   alpine, 173
   glaucous, 173
   lowly, 172
   marsh, 172
   pale, 173
   Raup's fringed, 171
Gentiana
   aquatica, 172
   detonsa, 171
   fremontii, 172
   glauca, 173
   lutea, 172
   prostrata, 172
Gentianaceae, 171–74
Gentianella
   amarella, 174
   detonsa ssp. *raupii*, 171
Gentianopsis
   crinita, 171
   detonsa spp. *raupii*, 171
Geraniaceae, 141–42
Geranium
   bicknellii, 141
   carolinianum var.
      sphaerospermum, 141
   carolinianum, 141
   erianthum, 142
   pratense var. *erianthum*, 142
   sphaerospermum, 141
   viscosissimum, 142
geranium family, 141–42
geranium
   Bicknell's, 141
   Carolina wild, 141
   sticky purple, 142
   woolly, 142

ginseng family, 19
Glyceria spp., 336
   elata, 335
   grandis, 335
   pauciflora, 336
   striata, 335
Gnaphalium
   microcephalum, 233
   thermale, 233
   viscosum, 233
goldthread, 85
Goodyera spp., 164
goosefoot family, 16, 65–70
goosefoot
   dark-green, 66
   dried, 67
   hoary, 68
   meadow, 66
   narrow-leaved, 66
   smooth narrow-leaved, 67
   Watson's, 68
Gramineae, 310–45, 378–79
grape fern
   ascending, 355
   dwarf, 358
   field, 354–55
   lance-leaved, 356
   leather, 358
   Mingan, 354
   northern, 376
   northwestern 377
   paradoxical, 356
   scalloped, 355
   spatulate, 354
   stalked, 357
   western, 356–57
grass family, 310–45, 378–79
grass, polar, 319
grass-of-Parnassus family, 125
grass-of-Parnassus
   northern, 125
   small northern, 125
Gratiola neglecta, 385
green bog orchid,
   northern, 55
greenthread, 236
Grossulariaceae, 18
ground-cone, Russian, 204
ground-fir, 349
groundsmoke, 159
Gymnocarpium
   disjunctum, 364
   dryopteris, 364, 365
   jessoense, 364

Habenaria
   hyperborea, 55
   saccata, 55
Hackelia ciliata, 188
hair grass, slender, 323
Halimolobos virgata, 105
halimolobos, 105
Hammarbya paludosa, 54
harebell
   Alaska, 212
   alpine, 212
   arctic, 212
hawk's-beard
   intermediate, 251
   slender, 250–51
   small-flowered, 251
hawkweed, woolly, 253
heath family, 22–24
hedge-hyssop, clammy, 385
Hedyotis longifolia, 211
heliotrope,
   spatulate-leaved, 189
Heliotropium
   curassavicum, 189
hemlock, western, 8
Heuchera spp., 122, 124
   glabra, 121
heuchera
   alpine, 121
   smooth, 121
Hieracium
   albertinum, 253
   cynoglossoides, 253
Hierochloe
   alpina, 314
   odorata, 314
Hippuridaceae, 160
Hippuris
   montana, 160
   vulgaris, 160
hollyhock, mountain, 144
holygrass, 314
Hordeum
   jubatum, 344
   pusillum, 344
horehound, 190
horsetails, 160
hound's-tongue, 188–89
   common, 188
Houstonia longifolia, 211
huckleberry
   oval-leaved, 23
   tall, 23

Index

Huperzia
   haleakalae, 350
   selago, 350
hydrangea family, 17
Hydrangeaceae, 17
Hydrocharitaceae, 40
Hydrophyllaceae, 180–84, 383
Hydrophyllum capitatum, 383
Hymenopappus
   filifolius
      var. *polycephalus*, 238
   polycephalus, 238
hymenopappus, tufted, 238
Hypericaceae, 145
Hypericum
   formosum var. *scouleri*, 145
   majus, 145
   scouleri, 145
Hypopitys monotropa, 165
hyssop, 222
   water, 194
Hyssopus officinalis, 222
Iliamna rivularis, 144
Indian pipe family, 165–66
Indian pipe, 165
Indian-paintbrush, Cusick's, 384
Iridaceae, 49–50
iris family, 49–50
Iris missouriensis, 49
Isoetaceae, 352–53
Isoetes spp., 38, 214, 352–53
   bolanderi, 352
   echinospora, 352
   howellii, 352–53
   maritima, 352, 353
   occidentalis, 352, 353
   x *truncata*, 353
Joe-pye weed, spotted, 216
Jones' columbine, 89
Juncaceae, 259–65
Juncaginaceae, 38
Juncus
   biglumis, 262
   brevicaudatus, 263
   confusus, 260
   filiformis, 259
   longistylis, 260
   nevadensis, 263
   parryi, 259
   regelii, 260
   stygius ssp. *americanus*, 262
June grass, 325

knotweed
   least, 62
   white-margined, 62
Koeleria macrantha, 325
Koenigia islandica, 61
koenigia, 61
Labiatae, 190–91
Lachenal's sedge, 286–87
Lactuca
   biennis, 255
   sativa, 255
   spicata, 255
ladies'-tresses
   hooded, 56
   northern slender, 56
lady fern, American alpine, 363
lady's-slipper
   mountain, 52
   pink, 51
   sparrow's-egg, 52
   stemless, 51
   yellow, 52
Lamiaceae, 190–91
larch, western, 6
Larix occidentalis, 6
Leguminosae, 132–40
Lemna
   minor, 41
   turionifera, 41
Lemnaceae, 41
Lentibulariaceae, 206–07
Lesquerella arctica, 99
lettuce, 255
   tall blue, 255
   western white, 251–52
Lewisia
   pygmaea ssp. *pygmaea*, 73
   rediviva, 73
Leymus mollis ssp. *mollis*, 345
Lilaea
   scilloides, 38
   subulata, 38
Lilaeaceae, 38
Liliaceae, 43–47
lily family, 43–47
Linanthus
   harknessii var.
      septentrionalis, 178
   septentrionalis, 178
linanthus, northern, 178
Linaria canadensis, 195
lip fern, slender, 361

Listera
   caurina, 53
   convallarioides, 53
Lithophragma
   bulbifera, 123
   glabrum, 123
   parviflorum, 123
little club-moss, Wallace's, 351
Lobelia dortmanna, 214
Lobelia spicata, 214
lobelia family, 213–14
lobelia
   spiked, 214
   water, 214
Lobeliaceae, 213–14
locoweed
   hare-footed, 135
   late yellow, 136
   purple mountain, 136
Loiseleuria procumbens, 24
Lomatium
   cous, 161
   montanum, 161
Lomatogonium rotatum, 174
loosestrife
   fringed, 170
   lance-leaved, 170
lousewort
   arctic, 200, 201
   coil-beaked, 199
   flame-coloured, 386
   large-flowered, 199
   leafy, 199
   sickletop, 199
   woolly, 200, 201
lungwort
   lance-leaved, 185
   large-flowered, 185–86
   prairie, 185
   tall, 185
lupine
   alpine, 139, 140
   arctic, 138
   large-leaved, 139
   least, 140
   Wyeth's, 139
Lupinus
   arcticus ssp. *subalpinus*, 138
   lepidus, 140
   minimus, 139
   minimus, 140
   polyphyllus, 139
   wyethii, 139
Luzula
   acuminata, 264
   groenlandica, 265
   multiflora, 265

Index

rufescens, 264
saltuensis, 264
Lychnis
  affinis, 78
  apetala, 78
  furcata, 78
Lycopodiaceae, 349–50
Lycopodiella inundata, 349
Lycopodium
  inundatum, 349
  sabinaefolium, 349
  sabinifolium
    var. *sitchense*, 349
  selago, 350
  sitchense, 349
Lycopus americanus, 190
Lygodesmia
  juncea, 249
  rostrata, 249
  spp., 247
lyme grass, sea, 345
Lysimachia
  ciliata, 170
  hybrida, 170
  lanceolata ssp. *hybrida*, 170
Machaeranthera
  tanacetifolia, 220
madder family, 210–11
maidenhair fern family, 359–62
maidenhair fern, western, 359
Malaxis
  brachypoda, 54
  monophylla
    var. *brachypoda*, 54
  paludosa, 54
mallow family, 144
Malvaceae, 144
manna grass, 336
  common tall, 335
  fowl, 335
  tufted tall, 335
mare's-tail family, 160
mare's-tail
  common, 160
  mountain, 160
Marrubium vulgare, 190
marsilea family, 371
Marsilea
  mucronata, 371
  vestita, 371
Marsileaceae, 371
meadowsweet
  pink, 21
  white, 21

Melandrium
  affine, 78
  apetalum, 78
melic grass, 330
melic, Smith's, 330
Melica
  smithii, 330
  spectabilis, 330
Mentha spp., 190
Mertensia
  lanceolata, 185
  longiflora, 185–86
  paniculata, 185
Microseris
  cuspidata, 248
  nutans, 388
  gracilis, 179
milk vetch, 135
  alpine, 132
  Bodin's, 132
  low, 133
  prickly, 134
  Pursh's, 133
milkweed family, 175–76
milkweed
  green, 175
  low, 176
milkwort family, 143
milkwort, fringed, 143
millet
  hot-springs, 311
  Leiberg's, 312
  thermal, 311
Mimulus
  breweri, 198
  floribundus, 196
  glabratus, 196
  guttatus, 196, 197
  lewisii, 198
  tilingii, 197
mint family, 190–91
Minuartia, 75
  austromontana, 76
  elegans, 76
  rossii, 76
  rossii var. *elegans*, 76
mistmaiden, 180
  Sitka, 180
Mitella spp., 122, 124
moccasin-flower, 51
mock-orange, Lewis', 17
Moldavica parviflora, 191
monkeyflower
  Brewer's, 198
  large mountain, 197
  red, 198

  small yellow, 196
  smooth, 196
  yellow, 196, 197
Monotropa
  hypopitys, 165
  uniflora, 165
Monotropaceae, 165–66
Montia
  linearis, 74
  parvifolia, 74
montia
  linear-leaved, 74
  small-leaved, 74
moonwort
  boreal, 376
  common, 355
  least, 358
  Mingan, 354
  northern, 376
  northwestern 377
  paradox, 357
  prairie, 354–55
  triangle, 356
  western, 356–57
Moquin's sea-blite, 69
morning-glory family, 177
moss gentian, 172
moss phlox, 169
mud crud, 146
Muhlenbergia
  asperifolia, 317
  glomerata, 317
  racemosa, 317–18
muhly
  bog, 317
  marsh, 317–18
Munroa squarrosa, 329
mustard family, 94–110, 382
naiad, slender, 31
Najadaceae, 31
Najas flexilis, 31
Nemophila breviflora, 182
nettle family, 92
ninebark, mallow-leaved, 20
Nothocalais cuspidata, 248
nut-grass
  awned, 266
  sand, 266
Nuttallanthus canadensis, 195
Nyctaginaceae, 72
Nymphaea
  leibergii, 83
  tetragona, 83
  tetragona ssp. *leibergii*, 83
Nymphaeaceae, 83

Index

oak fern
  California, 326
  common, 364, 365
  Nahanni 364
  northern, 364
  one-spike, 326
  poverty, 326–27
  timber, 326
  western, 364
oats, 326
Oenothera
  andina, 152
  breviflora, 153
  caespitosa, 153
  flava, 153
  serrulata, 151
Onagraceae, 150–59
one-headed everlasting, 231
onion grass
  purple, 330
  showy, 330
onion
  nodding, 43
  prairie, 43
Onosmodium
  molle var. *occidentale*, 186
  occidentale, 186
Ophioglossaceae, 354–58, 376–77
Ophrys
  caurina, 53
  convallarioides, 53
  paludosa, 54
Oplopanax horridus
  var. *horridus*, 19
orchid family, 51–56
Orchidaceae, 51–56
Orobanchaceae, 204–05
Orobanche
  ludoviciana, 205
  uniflora, 205
Oryzopsis
  canadensis, 315
  exigua, 315
  micrantha, 316
  pungens, 315
Osmorhiza
  berteroi, 162
  chilensis var. *purpurea*, 162
  depauperata, 162
  longistylis, 162
  purpurea, 162–63
Oxytropis
  campestris var. *davisii*, 136
  jordalii var. *jordalii*, 136
  lagopus var. *conjugens*, 135
  monticola, 136

Packera
  buekii, 244
  cymbalaria, 244
  cymbalarioides, 244
  subnuda, 244–45
paintbrush
  downy, 202
  pale greenish, 203
  stiff yellow, 202, 203
  yellow, 384
painted-cup, downy, 202
Panicum
  acuminatum, 311
  leibergii, 312
  oligosanthes, 312
  thermale, 311
  wilcoxianum, 312
Papaver
  alpinum, 91
  freedmanianum, 381
  kluanensis, 381
  nudicaule ssp. *radicatum* var. *pseudocorydalifolium*, 91
  pygmaeum, 91
  radicatum ssp. *kluanensis*, 381
  radicatum var. *pygmaeum*, 91
Papaveraceae, 91, 381
Parietaria
  officinalis, 92
  pensylvanica, 92
Parnassia
  palustris, 125
  palustris var. *montanensis*, 125
  parviflora, 125
Parnassiaceae, 125
pathfinder, 387
pea family, 132–40
pearlwort
  knotted, 77
  snow, 77
  spreading, 77
Pedicularis
  albertae, 386
  arctica, 200
  arctica, 201
  capitata, 199
  contorta, 199
  flammea, 386
  lanata, 200, 201
  langsdorfii
    ssp. *arctica*, 200, 201
  oederi, 386
  oederi var. *albertae*, 386
  racemosa, 199

sudetica, 201
sudetica ssp. *interioides*, 200
Pellaea
  atropurpurea, 362
  gastonyi, 362
  glabella, 362
pellitory, American, 92
pellitory-of-the-wall, 92
Penstemon
  ellipticus, 192
  fruticosus var. *scouleri*, 192
penstemon, shrubby, 192
pepperwort, hairy, 371
Phacelia
  linearis, 183
  lyallii, 184
  sericea, 184
Phegopteris connectilis, 368
Philadelphus lewisii, 17
Phippsia algida, 320
phippsia, frigid, 320
phlox family, 178–79
Phlox
  gracilis ssp. *gracilis*, 179
  hoodii, 169
phlox, slender, 179
Physocarpus
  malvaceus, 20
  ledinghamii, 191
Physostegia
  parviflora, 191
  virginiana
    var. *ledinghamii*, 191
Picradeniopsis oppositifolia, 237
picradeniopsis, 237
pimpernel, false, 168
Pinaceae, 6–8
pine family, 6–8
pine
  silver, 7
  western white, 7
pine-drops, 166
pine-sap, 165
Pinguicula
  villosa, 206
  vulgaris, 206
pink family, 75–82, 380
Pinus monticola, 7
pitcher-plant family, 113
pitcher-plant, 113
Plantaginaceae, 208–09
Plantago
  canescens, 208
  elongata, 209

Index

eriopoda, 208
maritima, 209
septata, 208
plantain family, 208–09
plantain
  linear-leaved, 209
  saline, 208
  seaside, 209
Platanthera
  hyperborea, 55
  stricta, 55
Poa
  canbyi, 379
  cusickii, 332
  gracillima, 379
  laxa ssp. *banffiana*, 334
  leptocoma, 333
  lettermanii, 333
  nervosa, 332
  sandbergii, 334, 379
  secunda, 334
  stenantha, 334
Poaceae, 310–45, 378–79
Podagrostis humilis, 320
Polanisia
  dodecandra
    ssp. *trachysperma*, 93
  trachysperma, 93
Polemoniaceae, 178–79
Polygala paucifolia, 143
Polygalaceae, 143
Polygonaceae, 59–64
Polygonum
  confertiflorum, 62
  imbricatum, 62
  kelloggii
    var. *confertiflorum*, 62
  minimum, 62
  polygaloides.
    ssp. *confertiflorum*, 62
  watsonii, 62
Polypodiaceae, 359–62,
  363–68, 369–70
Polypodium
  hesperium, 369
  hesperium, 370
  sibiricum, 370
  virginianum, 369, 370
  vulgare
    var. *columbianum*, 369
  vulgare
    var. *virginianum*, 370
polypody family, 369–70
polypody
  rock, 369, 370
  Siberian, 370
  western, 369, 370

pondweed family, 32–36
pondweed
  blunt-leaved, 33
  floating-leaved, 32
  leafy, 33
  linear-leaved, 34
  Robbins', 36
  sago, 37
  white-stemmed, 35
poplar, balsam, 10
poppy family, 91, 381
poppy
  alpine, 381
  dwarf alpine, 91
Populus
  angustifolia, 10
  balsamifera L.
    ssp. *balsamifera*, 10
  balsamifera L.
    ssp. *trichocarpa*, 10
  deltoides, 10
Portulacaceae, 73–74
Potamogeton
  foliosus, 33
  natans, 32
  obtusifolius, 33
  pectinatus, 37
  praelongus, 35
  robbinsii, 36
  strictifolius, 34
Potamogetonaceae, 32–36
Potentilla
  bimundorum, 127
  concinna var. *divisa*, 126
  concinna var. *macounii*, 126
  concinna var. *rubripes*, 126
  diversifolia, 128
  drummondii
    ssp. *drummondii*, 129
  finitima, 130
  hookeriana, 131–32
  lasiodonta, 130
  macounii, 126
  multifida, 127
  multisecta, 128
  nivea, 131
  nivea var. *villosa*, 131
  ovina, 126, 128
  paradoxa, 129–30
  pensylvanica, 127, 128
  pensylvanica var. *arida*, 130
  plattensis, 130
  rubripes, 126
  subjuga, 126
  uniflora, 131
  villosa, 131–32
  villosula, 131
  virgulata, 127

Powell's saltbush, 65
Prenanthes
  alata, 251–52
  hastata, 252
  sagittata, 217, 252
primrose family, 167–70
primrose
  erect, 167
  Greenland, 167
Primula
  egaliksensis, 167
  stricta, 167
Primulaceae, 167–70
Psilocarphus
  elatior, 229
  oregonus var. *elatior*, 229
Psoralea
  argophylla, 137
  esculenta, 137
psoralea, silverleaf, 137
Pteridiaceae, 359–62
Pterospora andromedea, 166
Puccinellia pauciflora
  var. *holmii*, 336
purple rattle, 200, 201
purslane family, 73–74
Pyrola
  asarifolia, 163
  grandiflora, 163
  picta, 164
Pyrolaceae, 163–64
quillwort family, 352–53
quillwort, 38, 214, 352–53
  Bolander's, 352
  coastal, 352
  Howell's, 352–53
  northern, 352
  western, 352
ragweed
  bur, 215
  brook, 245
  large-flowered, 245
Ranunculaceae, 84–90
Ranunculus
  eschscholtzii, 87, 88
  gelidus ssp. *grayi*, 87
  glaberrimus, 86
  grayi, 87
  karelinii, 87
  nivalis, 88
  occidentalis var. *brevistylis*, 86
  pygmaeus, 87
  uncinatus, 87
  verecundus, 87
rattlesnake-plantain, 164

Index

rattlesnakeroot
 purple, 217, 252
 wing-leaved, 251–52
redrattle, 386
redtop, spike, 321
reed grass
 Lapland, 322
 narrow 322
 northern, 322
Rhododendron
 albiflorum, 22
 lapponicum, 22
rhododendron, white-flowered, 22
Rhynchospora
 alba, 269
 capillacea, 269
Ribes
 glandulosum, 18
 laxiflorum, 18
ribgrass, western, 208
rice grass
 Canadian, 315
 little, 315
 little-seed, 316
 northern, 315
rock brake, Steller's, 360
rock cress
 Lemmon's, 104
 Lyall's, 104
rocket
 purple alpine, 106
 purple, 106
rockstar
 small-flowered, 123
 smooth, 123
Romanzoffia sitchensis, 180
romanzoffia
 cliff, 180
 Sitka, 180
Rorippa
 curvipes, 100
 sinuata, 101
 sylvestris, 101
 tenerrima, 100
 truncata, 100
Rosaceae, 20–21, 126–32
rose family, 20–21, 126–32
rose mandarin, 46
rose-bay, Lapland, 22
Rubiaceae, 210–11
Rumex
 acetosa, 64
 paucifolius, 64
 venosus, 72

Ruppia
 cirrhosa, 37
 maritima, 37
 occidentalis, 37
Ruppiaceae, 37
rush family, 259–65
rush
 few-flowered, 260
 long-styled, 260
 marsh, 262
 Nevada, 263
 Parry's, 259
 Regel's, 260
 short-tailed, 263
 thread, 259
 two-glumed, 262
rush-pink, 247
sagebrush, big, 25
sagewort, 205
 Herriot's, 241
 mountain, 241
Sagina
 decumbens, 77
 intermedia, 77
 nivalis, 77
 nodosa ssp. *borealis*, 77
Sagittaria
 cuneata, 39
 latifolia, 39
 sagittifolia var. *latifolia*, 39
Salicaceae, 10–15
Salix
 alaxensis ssp. *alaxensis*, 13
 arctica, 15
 barclayi, 11, 15
 calcicola, 12
 candida, 13
 commutata, 13
 farriae, 11
 glauca, 15
 lanata ssp. *calcicola*, 12
 pedicellaris, 11
 raupii, 11
 reticulata, 15
 scouleriana, 14
 sitchensis, 14
 stolonifera, 15
Salsola kali, 134
salt sage, 65
salt-meadow grass, small-flowered, 336
saltbush, 16, 65
 four-wing, 65
 silver, 65
sand millet, 312

sand spurry
 salt-marsh, 79
 two-stamened, 79
sandwort
 green alpine, 76
 low, 75
 purple alpine, 76
 Ross', 76
Sarracenia purpurea
 ssp. *purpurea*, 113
Sarraceniaceae, 113
Saussurea americana, 245
saw-wort, American, 245
Saxifraga spp., 180
 aizoides, 115
 ferruginea, 117
 flagellaris ssp. *setigera*, 115
 lyallii, 116, 118
 nelsoniana
  ssp. *porsildiana*, 117
 nivalis, 116–17
 occidentalis, 116
 odontoloma, 117–18
 oregana, 116
 oregana var. *subapetala*, 116
 setigera, 115
 subapetala, 116
Saxifragaceae, 115–24
saxifrage family, 115–24
saxifrage, 180
 Alaska, 117
 alpine, 116–17
 cordate-leaved, 117
 mountain, 121
 Oregon, 116
 red-stemmed, 116, 118
 rhomboid-leaved, 116
 stream, 117–18
 yellow mountain, 115
 yellowstone, 116
Schizachyrium scoparium
 ssp. *scoparium*, 310
 var. *scoparium*, 310
Schoenoplectus fluviatilis, 274
Scilla, 38
Scirpus
 cespitosus, 272
 clintonii, 272
 fluviatilis, 274
 microcarpus, 275
 pallidus, 275
 paludosus, 274
 pumilus ssp. *rollandii*, 272
 rollandii, 272
 rubrotinctus, 275
 rufus, 273

Index

scorpionweed
 linear-leaved, 183
 Lyall's, 184
 silky, 184
scorzonella, nodding, 388
scratch grass, 317
Scrophulariaceae, 192–203, 384–86
sea-blite, western, 69
sedge family, 266–309
sedge
 alpine, 297
 awl-fruited, 278
 awned, 304
 Back's, 292
 bald, 291–92
 beaked, 307
 beautiful, 290
 Bebb's, 279, 283
 bent, 291
 blackened, 299–300
 blister, 309
 bottle, 308
 broad-fruited, 283
 broad-scaled, 284
 broom, 282–83
 browned, 279
 capitate, 276
 Crawe's, 293
 Crawford's, 283
 cyperus-like, 305
 dark-scaled, 300
 Dewey's, 285
 few-flowered, 307
 few-fruited, 306
 fox, 278
 glacier 289
 golden, 293
 Hayden's, 281
 head-like, 280
 Holm's Rocky
  Mountain, 298
 Hood's, 277
 Hooker's, 277
 Hudson Bay, 286
 inland, 282
 lakeshore, 304
 lens-fruited, 301
 meadow, 280, 284
 narrow 286
 Nebraska, 302
 nodding, 294
 Norway 296
 open, 298
 Parry's, 295–96
 pasture, 280
 Payson's, 296–97
 Piper's, 284
 porcupine, 305–06
 Presl's, 281

 purple, 299
 Raynolds', 295
 Rocky Mountain, 292
 Ross', 291
 rye-grass, 287–88
 sand, 303
 seaside, 277
 shaved, 291–92
 short-awned, 307
 showy, 297
 small-headed, 281–82
 small-winged, 284
 stalked, 290
 stone, 294
 taper-fruit short-scale, 285
 thick-headed, 284
 thin-flowered, 288
 three-seeded, 287
 tinged, 283
 turned, 309
 two-seeded, 288
 umbellate, 291
 water, 298, 308
 weak, 289
 white-scaled, 279
 woolly, 303
Sedum
 divergens, 114
 lanceolatum, 114
 stenopetalum, 114
Selaginella
 densa, 351
 wallacei, 351
selaginella, prairie, 351
Selaginellaceae, 351
Senecio
 conterminus, 244
 cymbalaria, 244
 cymbalarioides, 244
 megacephalus, 245
 resedifolius, 244
 streptanthifolius, 244
 triangularis, 245
shadscale, 16
sheep sorrel
 alpine, 64
 mountain, 64
shield fern
 crested, 367–68
 narrow spinulose, 367
Shinneroseris rostrata, 249
Silene
 furcata, 78
 involucrata
  ssp. *involucrata*, 78
 uralensis ssp. *attenuata*, 78
silver-plant, 60
silverpuffs, nodding, 388

Sisyrinchium
 angustifolium, 50
 montanum, 50
 sarmentosum, 50
 septentrionale, 50
Sitanion hystrix, 343
skeletonweed, 247
 annual, 249
 common, 249
skunk currant, 18
soapweed, 48
sorrel, green, 64
Sparganiaceae, 29–30
Sparganium
 fluctuans, 29
 glomeratum, 29
 hyperboreum, 30
Spartina
 gracilis, 328
 pectinata, 327–28
speedwell
 marsh, 193
 thyme-leaved, 193
 water, 193–94
Spergularia
 diandra, 79
 marina var. *leiosperma*, 79
 salina var. *salina*, 79
Sphaeralcea rivularis, 144
Sphenopholis
 intermedia, 331
 obtusata, 331
spiderplant, 115
spiderwort family, 42
spiderwort, western, 42
spike-moss family, 351
spike-moss
 prairie, 351
 Wallace's, 351
spike-rush, 353
 Engelmann's, 267
 quill, 268
 slender, 268
Spiraea
 betulifolia, 21
 densiflora, 21
 splendens var. *splendens*, 21
Spiranthes
 gracilis, 56
 lacera var. *lacera*, 56
 romanzoffiana, 56
 x *simpsonii*, 56
Spirodela polyrhiza, 41
spleenwort, alpine, 363

Index

spring beauty
   slender-leaved, 74
   small-leaved, 74
squill, 38
squirreltail, 343
St. John's-wort family, 145
St. John's-wort
   large, 145
   western, 145
Steironema
   hybridum, 170
   lanceolatum
      var. *hybridum*, 170
Stellaria
   americana, 80
   arenicola, 82
   crispa, 81
   longipes, 82
   nitens, 80
   obtusa, 81
   umbellata, 82
Stephanomeria runcinata, 247
stickseed, fringed, 188
stonecrop family, 114
stonecrop
   lance-leaved, 114
   narrow-petalled, 114
   spreading, 114
Streptopus
   amplexifolius, 46
   roseus, 46
   streptopoides, 46
Suaeda
   calceoliformis, 69
   depressa, 69
   intermedia, 69
   moquinii, 69
Suckleya suckleyana, 70
suckleya, poison, 70
sugar-scoop, 124
Suksdorfia
   ranunculifolia, 120
   violacea, 120
suksdorfia
   blue, 120
   white, 120
sundew family, 111–12
sundew
   English, 111
   great, 111
   linear-leaved, 112
   oblong-leaved, 111
   slender-leaved, 112
swamp potato, 39

sweet cicely
   blunt-fruited, 162
   purple, 162–63
   smooth, 162
   spreading, 162
sweetflag, 49
sweetgrass
   alpine, 314
   common, 314
Tanacetum
   bipinnatum
      ssp. *huronense*, 239
   huronense bifarium, 239
   vulgare, 239
tansy
   common, 239
   Indian, 239
Taraxia breviflora, 153
taraxia, 153
Taxaceae, 9
Taxus brevifolia, 9
Telesonix
   heucheriformis, 119
   jamesii, 119
telesonix, 119
Tellima grandiflora, 124
Thelesperma
   marginatum, 236
   subnudum
      var. *marginatum*, 236
Thellungiella salsuginea, 103
Thelypteridaceae, 368
Thelypteris phegopteris, 368
thistle
   elk, 246
   Russian, 134
three-awn, red, 313
Thuja plicata, 5
Tiarella unifoliata, 124
tickseed, 236
   common, 235
toadflax
   blue, 195
   field, 195
Tofieldia
   glutinosa, 44
   occidentalis, 44
Torreyochloa pallida
   pallida var. *fernaldii*, 336
   pallida var. *pauciflora*, 336
   pauciflora, 336
Townsendia
   condensata, 218
   exscapa, 218

townsendia
   alpine, 218
   low, 218
Tradescantia occidentalis
   var. *occidentalis*, 42
trailplant, American, 387
Triantha
   glutinosa, 44
   occidentalis, 44
Trichophorum
   cespitosum, 272
   clintonii, 272
   pumilum, 271–72
Trillium ovatum, 47
Tripterocalyx micranthus, 72
Trisetum
   canescens, 324
   cernuum, 324
   montanum, 324
   wolfii, 325
trisetum
   mountain 324
   nodding, 324
   Wolf's, 325
Tsuga heterophylla, 8
twayblade
   broad-lipped, 53
   western, 53
twisted-stalk
   clasping-leaved, 46
   small, 46
Typha spp., 49
Umbelliferae, 161–63
umbrella-plant
   few-flowered, 60
   nodding, 59
Urticaceae, 92
Utricularia cornuta, 207
Vaccinium
   membranaceum, 23
   myrtillus, 23
   ovalifolium, 23
   uliginosum, 23
verbena, sand, 72
Veronica
   anagallis-aquatica, 193
   catenata, 193–94
   comosa var. *glaberrima*, 193
   salina, 193
   scutellata, 193
   serpyllifolia var. *humifusa*, 193
Viola
   macloskeyi, 147
   nuttallii, 148
   nuttallii var. *linguifolia*, 147
   pallens, 147–48

Index

palustris, 147
pedatifida, 149
praemorsa
    ssp. *linguifolia*, 147
renifolia, 147
Violaceae, 147–49
violet family, 147–49
violet
    broad-leaved yellow
        prairie, 147
    common yellow prairie, 148
    crowfoot, 149
    kidney-leaved, 147
    Macloskey's, 147–48
    marsh, 147
    northern white, 147–48
wakerobin, western, 47
wapato, 39
water-clover, hairy, 371
water-horehound, American, 190
water-lily family, 83
water-lily
    pygmy, 83
    small white, 83
water-nymph family, 31
water-nymph, slender, 31
water-plantain family, 39
water-shield family, 83
water-shield, 83
waterleaf family, 180–84, 383
watermeal, 41
waterpod, 181
waterweed family, 40
waterweed
    Canada, 40
    long-sheathed, 40
    two-leaved, 40
waterwort family, 146
waterwort, three-stamened, 146
wedge grass
    prairie, 331
    slender, 331
wedgescale, 65
wheat grass
    awned northern, 342
    blue-bunch, 342
    Scribner's, 342
    spreading, 342
whitlow-grass
    Austrian, 95
    creeping, 94
    dense-leaved, 95
    Kananaskis, 96

    long-stalked, 96
    Macoun's, 96
    northern, 98
    Porsild's, 382
    smooth, 98
    windy, 97
widgeon-grass, 37
wild onion, 192
    Geyer's, 43
wild rye, Virginia, 345
willow family, 10–15
willow
    Alaska, 13
    arctic, 15
    Barclay's, 11, 15
    bog, 11
    changeable, 13
    Farr's, 11
    felt-leaved, 13
    hoary, 13
    lime, 12
    Raup's, 11
    Scouler's, 14
    Sitka, 14
    smooth, 15
    snow, 15
    stoloniferous, 15
    woolly, 12
willowherb, 213
    club, 157
    Hall's, 154–55
    low, 159
    Oregon, 155
    pale, 156
    Rocky Mountain, 154
    slender-fruited, 155–56
    white, 157
    wonderful, 156
    yellow, 158
winter cress, American, 102
wintergreen family, 163–64
wintergreen
    arctic, 163
    common pink, 163
    white-veined, 164
Wolffia
    arhiza, 41
    borealis, 41
    columbiana, 41
wood-rush
    field, 265
    Greenland, 265
    reddish, 264
    sharp-pointed, 264
woodbrome, hairy 341
woodfern family, 363–68

Woodsia
    glabella, 366
    ilvensis, 366
woodsia
    rusty, 366
    smooth, 366
woollen-breeches, 383
woollyheads, 229
wormwood
    forked, 240
    northern, 375
yellow cress
    blunt-leaved, 100
    creeping, 101
    slender, 100
    spreading, 101
yew family, 9
yew, western, 9
Yucca glauca

Illustration Credits

All line drawings reproduced in this volume are used by permission of the rights holders.

Pages 337, 338: Aiken, S.G. and S.J. Darbyshire. Fescue grasses of Canada. Biosystematics Research Centre. Research Branch, Agriculture Canada, Ottawa, Ontario. Publication No. 1844/E.

Page 65 (bottom): Bassett, I.J., C.W. Crompton, J. McNeill and P.M. Taschereau. 1983. The genus *Atriplex* (Chenopodiaceae) in Canada. Agriculture Canada, Ottawa, Ontario.

Pages 29 (top), 34: Brayshaw, C.T. 1985. Pondweeds and bur-reeds, and their relatives. British Columbia Provincial Museum, Victoria, BC. Occasional Papers of the British Columbia Museum No. 26.

Pages 11, 15: Brayshaw, C.T. 1996. Catkin-bearing plants of British Columbia. Second edition. British Columbia Provincial Museum, Victoria, BC. Occasional Papers of the British Columbia Museum No. 18.

Pages 77 (bottom), 96 (middle), 118 (middle), 138, 265, 308, 320 (bottom), 354: Cody, W.J. 1996. Flora of the Yukon Territory. NRC Research Press, Ottawa, Ontario.

Page 370: Cody, W.J. and D.M. Britton. 1989. Ferns and fern allies of Canada. Research Branch, Agriculture Canada. Ottawa, Ontario. Publication No. 1829/E.

Pages 231 (bottom), 239: Douglas, G.W. 1995. The sunflower family (Asteraceae) of British Columbia. Volume 2. British Columbia Provincial Museum, Victoria, BC.

Pages 291 (top), 292 (right), 334 (bottom): Douglas, G.W., G.B. Straley and D.V. Meidinger. 1998. Rare Native Vascular Plants of British Columbia. BC Environment, Victoria, BC.

Pages 60 (bottom), 75, 78 (bottom), 356: Douglas, G.W., D. Meidinger and J. Pojar. 1998–2000. Illustrated flora of British Columbia. 5 volumes. BC Ministry of Environment, Lands and Parks and BC Ministry of Forestry, Victoria, BC.

Page 230: Evert, E.F. 1984. A new species of *Antennaria* (Asteraceae) from Montana and Wyoming. Madrono 13(2):109-112.

Pages 33 (top), 54 (top, bottom), 67 (bottom), 77 (middle), 175, 176, 211, 235, 263 (bottom), 272 (bottom), 290: Gleason, H.A. 1952. The Britton & Brown illustrated flora of the northeastern United States and adjacent Canada. Hafner Press, New York, New York.

Page 285: Hermann, F.J. 1970. Manual of the Carices of the Rocky Mountains and Colorado Basin. United States Department of Agriculture Forest Service, Washington, D.C. Agriculture Handbook No. 374.

Pages 5, 6, 7, 8, 9, 24, 31, 32, 33 (bottom), 35, 36, 37, 38, 39, 40 (top, bottom), 41 (top, bottom), 43, 44, 45, 46 (top, bottom), 47, 50, 53 (top, bottom), 55, 259 (top, bottom), 260, 261, 263 (top), 266 (top, bottom), 267, 268 (top), 269, 270 (bottom), 274, 275, 276, 278, 282, 283 (top), 286 (top, bottom), 292 (left), 293, 294 (top), 295, 296, 297 (top), 298, 299, 300, 301, 302, 305 (top), 307, 309 (top, bottom), 310, 311, 312 (top), 313, 314, 315 (top), 316, 317, 318 (top, bottom), 320 (top, middle), 321 (top, bottom), 323, 324 (top), 325, 326 (top, bottom), 327, 328, 329, 330 (left, right), 331, 332, 333 (top, bottom), 334 (top), 335, 336 (top, bottom), 339, 340 (top, bottom), 342, 343, 344, 345 (top, bottom), 349 (bottom), 350 (bottom), 351, 352 (top, bottom), 353 (top right, middle), 359, 360, 361, 363, 365, 367 (top, bottom), 369, 371, 378, 379, 384: Hitchcock, C.L. et al. 1955–69. Vascular plants of the Pacific Northwest. University of Washington Press, Seattle, Washington. Volume 1.

Pages 10, 13 (bottom), 14, 16, 59, 60 (top), 62, 63, 64, 65 (top), 66 (bottom), 67 (top), 68, 69, 71, 73 (top), 74 (top, bottom), 79, 80 (top, bottom), 81 (top, bottom), 82 (top), 83 (top, bottom), 85, 86 (top), 87 (top, bottom), 90, 92, 93, 94, 95 (top, bottom), 97, 101, 102, 103, 104, 105, 108, 109 (bottom), 111, 114: Hitchcock, C.L. et al. 1955–69. Vascular plants of the Pacific Northwest. University of Washington Press, Seattle, Washington. Volume 2.

Pages 17, 18, 20, 21, 116 (bottom), 117, 118 (top, bottom), 120 (top, bottom), 121, 122, 123 (top, bottom), 124, 125, 126 (top, bottom), 128, 129 (top, bottom), 130 (bottom), 131, 132 (top), 133 (top, bottom), 134, 135, 137, 139 (top, bottom), 140, 141, 144, 145 (top, bottom), 146, 147, 148, 151, 152, 153, 154 (top), 154 (bottom), 156 (top, bottom), 157 (top, bottom), 158, 159, 160, 161, 162: Hitchcock, C.L. et al. 1955–69. Vascular plants of the Pacific Northwest. University of Washington Press, Seattle, Washington. Volume 3.

Pages 23 (top, bottom), 164, 168, 169 (bottom), 170, 171, 172 (top), 174, 177 (right), 179, 180, 181, 182, 183, 184, 185 (top), 186, 187 (bottom), 188, 189, 190, 191, 192, 193 (top, bottom), 194, 195, 196 (top, bottom), 197 (top, bottom), 198, 199, 200 (bottom), 202, 205 (top, bottom), 209, 210, 212, 213, 214 (top), 383, 385: Hitchcock, C.L. et al. 1955–69. Vascular plants of the Pacific Northwest. University of Washington Press, Seattle, Washington. Volume 4.

Photo Credits

Pages 25, 215, 216, 217 (top), 218 (bottom), 219 (bottom), 220 (top, bottom), 223, 224 (bottom), 227, 228, 233 (top, bottom), 234, 236, 237, 238, 240, 241, 242 (bottom), 244, 245 (top, bottom), 246, 247, 248, 250 (left, right), 251, 252 (top, bottom), 253, 255, 388: Hitchcock, C.L. et al. 1955–69. Vascular plants of the Pacific Northwest. University of Washington Press, Seattle, Washington. Volume 5.

Pages 86 (bottom), 88 (bottom), 107, 155, 268 (bottom), 272 (top): Hultén, E. 1968. Flora of Alaska and neighboring territories. Stanford University Press. Stanford, California.

Page 178: Kujit, J. 1982. The flora of Waterton Lakes National Park. University of Alberta Press, Edmonton, Alberta.

Page 70: Looman, J. and K.F. Best. Budd's Flora of the Canadian prairie provinces. Research Branch, Agriculture Canada, Ottawa, Ontario. Publication No. 1662.

Pages 30, 61, 78 (top), 98, 106, 113, 132 (bottom), 136, 163 (bottom), 167 (bottom), 173, 203, 206, 208, 225 (right), 262, 264: Porsild, A.E. and W.J. Cody. Vascular plants of Northwest Territories, Canada. National Museum of Natural Sciences, National Museums of Canada, Ottawa, Ontario.

Pages 277 (top, bottom), 281 (bottom), 283 (bottom), 284 (top, bottom), 287, 291 (bottom), 294 (bottom), 297 (bottom), 303, 305 (bottom): MacKenzie, Kenneth K. The Cariceae of North America. New York Botanical Garden, Bronx, New York.

Pages 29 (bottom), 130 (top), 312 (bottom), 315 (bottom), 349 (top), 353 (top left): Holmgren, N.H. 1998. Illustrated companion to Gleason and Cronquist's manual of northeastern United States and adjacent Canada. New York Botanical Garden. Bronx, New York.

Pages 376, 377, 380, 381, 382, 386, 387, 439–50: courtesy of Linda Kershaw.

Pages 12, 13 (top), 19, 22, 42, 48, 49, 52, 56, 66 (top), 68 (top), 72, 73 (bottom), 76, 77 (top), 82 (bottom), 84, 89, 91, 96 (bottom), 99, 100 (top, bottom), 109 (top), 110, 112, 115, 116, 119, 127, 142, 143, 150, 165, 166, 167 (top), 169 (top), 172 (bottom), 177 (left), 185 (bottom), 187 (top), 200 (top), 201, 204, 207, 214 (bottom), 217 (bottom), 218 (bottom), 219 (top), 221, 222, 225 (left), 226, 229, 231 (top), 232, 242 (top), 243, 249, 264 (top), 270 (top), 273, 279, 280, 281 (top), 288, 289 (top, bottom), 304, 306, 319, 324 (bottom), 341, 350 (top), 353 (bottom), 362 (top, bottom), 364, 366 (top, bottom), 368: John R. Maywood, courtesy of ANHIC.

Pages 355 (top, middle, bottom), 357 (top, middle, bottom), 358 (top, bottom): courtesy of D. Wagner.

Pages 51, 96 (top), 149, 163 (top), 224 (top), 254, 269 (top), 322: courtesy of Joan Williams.

Photo Credits

All photographs reproduced in this volume are used by permission of the rights holders. Photograph sources are identified by initials running alongside the photograph.

AL	A. Landals	JP	Jim Pojar	SS	Steve Shelly
BH	Bonnie Heidel	JR	John Rintoul	SW	Steve Wirt
BM	Bob Moseley	JS	J. Smith	SZ	S.C. Zoltai
CWa	Cliff Wallis	JV	Jim Vanderhorst	TC	Terry Clayton
DG	Dierdrie Griffiths	LA	Lorna Allen	TD	Teresa Dolman
DJ	Derek Johnson	LK	Linda Kershaw	TK	Tulli Kerststetter
ES	Kathy Tannas	PA	Peter Achuff	TT	T.W. Thormin
GCT	G.C. Trottier	PL	Peter Lesica	WJC	W.J. Crins
JD	Joseph Duft	PLe	Peter Lee	WLNP	Waterton Lakes National Park
JG	Joyce Gould	RB	R. Bird		
JH	Julie Hrapko	SM	S. Myers	WM	William Merilees
JJ	John Joy	SMB	Montana Natural Heritage Program	WW	Warren Wagner
JL	Jane Lancaster				

Contributors

Peter Achuff
Lorna Allen 
Elisabeth Beaubien
Cheryl Bradley 
Dana Bush 
Patsy Cotterill 
Dave Downing 
Dave Ealey 
Gina Fryer 
Graham Griffiths 
Coral Grove 
Julie Hrapko 
Duke Hunter 
Ruth Johnson 
Kim Krause 
Peter Lee 
Dan MacIssac 
Karen Mann 
Heather Mansell 
Marge Meijer 
Christie Sarafinchin 
Arthur (Art) G. Schwarz 
Cindy Verbeek 
Dragomir Vujnovic 
Ksenija Vujnovic 
Cliff Wallis 
Anne Weerstra 
Kathleen Wilkinson 
Joan Williams 
Bill Richards 

Rare *Vascular* Plants of Alberta   483

About the Editors

LINDA KERSHAW

Linda's interest in rare species began in 1976 with her MSc thesis on rare and endangered Canadian plants. Over the past 25 years, her work has focussed on biophysical inventories in the Yukon, NWT and Alberta. She has authored and contributed to many field guides and papers. Linda now works as a writer and editor when not pursuing her two favourite pastimes: photography and illustration.

JOYCE GOULD

Joyce Gould is currently employed as a biologist with Alberta Environment where she works on protected areas issues and as the botanist with the Alberta Natural Heritage Information Centre (ANHIC). She has conducted plant surveys in Alberta, Ontario, Yukon and Nunavut. Joyce has a BSc in Botany from the University of Alberta and her MSc from the University of Toronto.

DEREK JOHNSON

Derek is a lifetime resident of Alberta. He received his MSc in botany from the University of Calgary. He is currently employed by the Canadian Forest Service in Edmonton where he is a plant taxonomist and ecologist and curates the herbarium at the Northern Forestry Centre. His work has taken him to the arctic, as well as many years studying peatlands in the boreal forest of the prairie provinces.

JANE LANCASTER

An independent consulting botanist, Jane holds a Bsc in plant biology from the University of Calgary. She lives with her husband and family near Cochrane, Alberta.

Ian MacDonald